THE WEST CO.

Rec'd DEC 1 1987

GLASS:
SCIENCE AND TECHNOLOGY

VOLUME 1
Glass-Forming Systems

Contributors

C. A. Angell
George H. Beall
A. Bondi
David A. Duke
Richard A. Eppler
R. L. Freed
N. J. Kreidl
G. W. Scherer
P. C. Schultz
D. R. Uhlmann
J. B. Vander Sande
H. Yinnon

GLASS: SCIENCE AND TECHNOLOGY

Edited by D. R. UHLMANN

DEPARTMENT OF MATERIALS SCIENCE
  AND ENGINEERING
MASSACHUSETTS INSTITUTE OF TECHNOLOGY
CAMBRIDGE, MASSACHUSETTS

N. J. KREIDL

DEPARTMENT OF CHEMICAL ENGINEERING
UNIVERSITY OF NEW MEXICO
ALBUQUERQUE, NEW MEXICO

VOLUME 1
Glass-Forming Systems

 1983

ACADEMIC PRESS, INC.
Harcourt Brace Jovanovich, Publishers
Orlando San Diego New York
Austin Boston London Sydney
Tokyo Toronto

COPYRIGHT © 1983, BY ACADEMIC PRESS, INC.
ALL RIGHTS RESERVED.
NO PART OF THIS PUBLICATION MAY BE REPRODUCED OR
TRANSMITTED IN ANY FORM OR BY ANY MEANS, ELECTRONIC
OR MECHANICAL, INCLUDING PHOTOCOPY, RECORDING, OR ANY
INFORMATION STORAGE AND RETRIEVAL SYSTEM, WITHOUT
PERMISSION IN WRITING FROM THE PUBLISHER.

ACADEMIC PRESS, INC.
Orlando, Florida 32887

United Kingdom Edition published by
ACADEMIC PRESS, INC. (LONDON) LTD.
24/28 Oval Road, London NW1 7DX

Library of Congress Cataloging in Publication Data
Main entry under title:

Glass-forming systems.

(Glass–science and technology ; v. 1)
Includes index.
1. Glass. 2. Ceramic materials. I. Uhlmann, D. R.
(Donald Robert) II. Kreidl, N. J. III. Series: Glass --
science and technology (Academic Press) ; v. 1.
TP848.G56 vol. 1 [TP858] 661'.ls [661'.1042] 83-11914
ISBN 0-12-706701-9

PRINTED IN THE UNITED STATES OF AMERICA

87 88 89 90   9 8 7 6 5 4 3 2

Contents

LIST OF CONTRIBUTORS ix
PREFACE xi

Chapter 1 The Formation of Glasses
D. R. Uhlmann and H. Yinnon

I.	Introduction	1
II.	Nucleation and Crystal Growth in Glass-Forming Systems	8
III.	Kinetic Models of Glass Formation: Growth Rate and Nucleation Rate	14
IV.	Kinetic Models for Glass Formation: The Perspective of Transformation Kinetics	20
V.	Glass-Forming Ability and Material Properties	41
	References	44

Chapter 2 Unusual Methods of Producing Glasses
G. W. Scherer and P. C. Schultz

I.	Introduction	49
II.	Unconventional Melting	50
III.	Deposition Methods	60
IV.	Solution Methods	86
V.	Solid State Transformations	89
	References	97

Chapter 3 Inorganic Glass-Forming Systems*
N. J. Kreidl

I.	Vitreous Silica	107
II.	Alkali Silicate Glasses	122
III.	Soda-Lime Glasses	147
IV.	Other Cations in Silicate Glasses	153
V.	Borate Glasses	160
VI.	Borosilicate Glasses	171
VII.	Aluminosilicate Glasses	181
VIII.	Phosphate Glasses	191
IX.	Other Oxide Glasses	196
X.	Ionic Salt and Solution Glasses	209

* Section X, Ionic Salt and Solution Glasses, is by C. A. Angell.

XI.	Fluoride and Oxyhalide Glasses	226
XII.	Chalcogenide Glasses	231
	References	260

Chapter 4 Glazes and Enamels
Richard A. Eppler

I.	Introduction	301
II.	Ceramic Glazes	303
III.	Porcelain Enamels	328
IV.	Summary	336
	References	336

Chapter 5 Organic Glasses (Molecular Glasses)
A. Bondi

I.	Introduction	339
II.	Effects of Cooling Rate	342
III.	The Glass Transition Temperature (T_g)	342
IV.	The Glass Transition Temperature of Mixtures	348
V.	Heterogeneous Mixtures	350
VI.	Detailed Effects of Molecular Structure	351
VII.	Equilibrium Properties	356
VIII.	Effect of Admixtures	358
IX.	Transport Phenomena	359
	References	362

Chapter 6 Metallic Glasses
J. B. Vander Sande and R. L. Freed

I.	Introduction	365
II.	Metallic Glass Alloy Types	366
III.	Structure of Metallic Glasses	367
IV.	Theories for the Formation of Metallic Glasses	369
V.	Investigations of Properties and Behavior of Noncrystalline Phases	373
VI.	Theoretical Mechanisms of Deformation and Fracture	381
VII.	The Effect of Crystallization on Mechanical Properties	385
VIII.	Metallic Glasses at High Temperature	393
IX.	Summary and Conclusions	398
	References	399

Chapter 7 Glass-Ceramic Technology
George H. Beall and David A. Duke

I.	Introduction	404
II.	Nucleation of Crystals in Glass	405

CONTENTS

III.	Crystallization Phenomena in Glass-Ceramics	413
IV.	Properties of Glass-Ceramics	422
V.	Glass-Ceramic Processing	436
VI.	Applications	441
	References	444

MATERIALS INDEX 447
SUBJECT INDEX 453

List of Contributors

Numbers in parentheses indicate the pages on which the authors' contributions begin.

C. A. ANGELL (209), *Department of Chemistry, Purdue University, West Lafayette, Indiana 47907*

GEORGE H. BEALL (403), *Technical Staffs Division, Corning Glass Works, Corning, New York 14830*

A. BONDI[†] (339), *Shell Development Company, Houston, Texas 87501*

DAVID A. DUKE (403), *Technical Products Division, Corning Glass Works, Corning, New York 14830*

RICHARD A. EPPLER (301), *Pemco Products, Mobay Chemical Corporation, Baltimore, Maryland 21224*

R. L. FREED (365), *Engineering Technology Laboratory, E. I. du Pont de Nemours and Company, Inc., Wilmington, Delaware 19898*

N. J. KREIDL (105), *Department of Chemical Engineering, University of New Mexico, Albuquerque, New Mexico 87106*

G. W. SCHERER (49), *Research and Development Division, Corning Glass Works, Corning, New York 14831*

P. C. SCHULTZ (49), *Research and Development Division, Corning Glass Works, Corning, New York 14831*

D. R. UHLMANN (1), *Department of Materials Science and Engineering, Massachusetts Institute of Technology, Cambridge, Massachusetts 02139*

J. B. VANDER SANDE (365), *Department of Materials Science and Engineering, Massachusetts Institute of Technology, Cambridge, Massachusetts 02139*

H. YINNON (1), *Department of Materials Science and Engineering, Massachusetts Institute of Technology, Cambridge, Massachusetts 02139*

---

† Deceased.

Preface

This volume of the treatise *Glass: Science and Technology* is focused on glass-forming systems and glass-ceramic materials. In addition to the topics of principal concern, specific attention is directed to glass formation, techniques of forming glasses, and glazes and enamels. Detailed information is given on the glass-forming regions in oxide, fused salt, aqueous, and organic systems, as well as on the newly important classes of metal alloy, fluoride, and chalcogenide glasses.

The present volume, designated Volume 1 of the treatise, is the second to appear. The previous one, designated Volume 5, *Elasticity and Strength in Glasses,* was published in 1980. The next two volumes—*Viscous Flow and Relaxation* and *Glass Processing*—are scheduled for publication within the next year.

The editors are saddened to announce the premature death of one of the contributors to the present volume, Dr. Arnold Bondi. Dr. Bondi was an outstanding scientist in the field of organic glasses; his incisive intellect, breadth of knowledge, and warm personality will be sorely missed. To preserve the flavor of his contribution, the editors have effected only minor changes in the manuscript as originally submitted by Dr. Bondi.

CHAPTER 1

The Formation of Glasses

D. R. Uhlmann
H. Yinnon

DEPARTMENT OF MATERIALS SCIENCE AND ENGINEERING
MASSACHUSETTS INSTITUTE OF TECHNOLOGY
CAMBRIDGE, MASSACHUSETTS

I.	Introduction	1
	A. Structural Models	2
	B. Thermodynamic Models	5
II.	Nucleation and Crystal Growth in Glass-Forming Systems	8
	A. Crystal Formation	8
	B. Crystal Growth	10
	C. Crystallization with Change of Composition	12
III.	Kinetic Models of Glass Formation: Growth Rate and Nucleation Rate	14
	A. Growth Rate	15
	B. Nucleation Rate	16
IV.	Kinetic Models for Glass Formation: The Perspective of Transformation Kinetics	20
	A. General Formulation	20
	B. Dependence of Critical Cooling Rate on Material Parameters	26
	C. Comparison of Theory with Experimental Data	28
	D. Experimental Determinations of Nucleation Frequencies and Effects of Nucleating Heterogeneities	32
	E. Crystallization on Reheating a Glass	37
V.	Glass-Forming Ability and Material Properties	41
	References	44

I. Introduction

At least some glasses have been formed of materials with all types of bonding. These include covalent (SiO_2), ionic [$0.4Ca(NO_3)_2-0.6KNO_3$], metallic ($0.4Fe-0.4Ni-0.14P-0.06B$), van der Waals (toluene), and hydrogen (H_2O). As discussed in Chapter 2, glasses can be formed using a variety of techniques, including cooling from the liquid state, condensation from the vapor, pressure quenching, solution hydrolysis, anodization, gel formation, and bombardment of crystals by high-energy particles or by shock waves. Of these the technique of cooling from the liquid state is by far the most important and most widely used. As such, it will be the exclusive focus of the present chapter. For treatments of glass formation by other techniques, see Gutzow and Avramov (1977).

The formation of glasses requires cooling to a sufficiently low temperature—below the glass transition—without the occurrence of detectable crystallization. In treating this phenomenon it has been suggested by some authors that specific structural features or properties of the materials will result in glasses being formed. This has led to classifications of materials as glass formers or non–glass formers, where reference is implicitly made to cooling bulk samples at reasonable rates. By others it has been suggested that the critical factor in glass formation is the rate of cooling relative to the kinetics of crystallization. This directs attention to kinetic characteristics of the materials and suggests that nearly all liquids will form glasses if cooled rapidly and will crystallize if cooled slowly.

On this basis the various models that have been advanced to describe glass formation can be grouped into three categories, depending on the factors that are viewed as decisive in the formation of glasses. These categories are (1) structural, (2) thermodynamic, and (3) kinetic. It will be seen, however, that the distinctions among the groups are often rather nebulous, since the structural and thermodynamic models often have related kinetic considerations, and the kinetic models often utilize structural and thermodynamic concepts.

In discussing the principal models for glass formation, kinetic treatments will be considered in greatest detail. Such treatments are preferred because of their potential for providing quantitative predictions of glass-forming behavior and because glass formation is essentially a kinetic phenomenon.

A. Structural Models

Perhaps the best known model of glass formation is that due to Zachariasen (1932) and Warren (1937, 1941). Speaking of oxides, Zachariasen expressed the "ultimate condition" for the information of glasses as follows: "the substance can form extended three-dimensional networks lacking periodicity with an energy content comparable with that of the corresponding crystal network." This led to the formulation of his celebrated rules for glass formation, according to which an oxide glass will be formed

if the sample contains a high percentage of cations which are ounded by oxygen tetrahedra or by oxygen triangles; (2) if these edra or triangles share only corners with each other, and (3) if some oxygen atoms are linked to only two such cations and do not form further bonds with any other cations.

The requirement that the network be three-dimensional led to an additional rule:

(4) at least three corners in each oxygen polyhedron must be shared.

The structural implications of this *random network model* led to the classification of cations as network formers, network modifiers, and intermediates; and a sizable literature developed around descriptions of the network-forming or network-modifying character of various cations. This approach is summarized well by Stevels (1957); and the classification of various cations is given in Table I.

The structural model of a random network received support from x-ray diffraction studies of a variety of glasses, although these studies did not establish the model as a unique representation of structure. More recent

TABLE I

CLASSIFICATION OF CATIONS AS NETWORK FORMERS, NETWORK MODIFIERS, AND INTERMEDIATES[a]

Glass formers:	B	Modifiers:	Sc
	Si		La
	Ge		Y
	Al		Sn
	B		Ga
	P		In
	V		Th
	As		Pb
	Sb		Mg
	Zr		Li
Intermediates:	Ti		Pb
	Zn		Zn
	Pb		Ba
	Al		Ca
	Th		Sr
	Be		Cd
	Zr		Na
	Cd		Cd
			K
			Rb
			Hg
			Cs

[a] After Kingery *et al.* (1976).

diffraction work on a variety of oxide glasses has provided effective support for this model [see Mozzi and Warren (1969), Porai-Koshits (1977), Milner and Wright (1980)]. As noted by Uhlmann (1980), however, the random network may consist of randomly organized structural units larger than the basic oxygen polyhedra (e.g., a random network of boroxyl units rather than BO_3 triangles in the case of glassy B_2O_3). Further constraints on the use of the random network model to describe glass structure and glass formation have been provided by the widespread occurrence of liquid–liquid phase separation [see reviews by Vogel (1977) and by Uhlmann and Kolbeck (1976)]. Such phase separation results in the glasses consisting of two or more phases and imposes constraints on the use of the random network model that go beyond those recognized by Zachariasen (1932). It seems likely that the structures of the individual amorphous phases in multiphase glasses may be describable by a random network model, although this point remains to be established in detail.

The random network model has also been generalized as a *random array* picture in which the structural elements are randomly arranged and in which no unit of the structure is repeated at regular intervals in three dimensions. In this form it has been used with considerable success in describing the structures of a wide variety of glasses, including metal alloys, chalcogenides, and polymers [see Chaudhari and Turnbull (1978); Uhlmann (1980)].

The utility of the random network or random array models for representing the structures of various glasses will be discussed at length in the volume of this treatise on glass formation. Our present concern, however, is directed not at the structural implications of these models, but at their descriptions of glass formation. In this regard, the original paper of Zachariasen (1932) presented a brief rationale: "Glasses which do not devitrify very rapidly will have an energy only slightly greater than that of the (corresponding) crystal." The justification for the model as an approach to glass formation is then related to a kinetic parameter, the driving force for crystallization. Materials with small differences in energy between liquid and crystal will, at a given undercooling, have smaller driving forces for crystallization than those with large differences in energy. Although Zachariasen also mentioned the mobility ("In melts where there are highly associated groups . . . the viscosity will be high"), no use was made of this factor.

As a general approach to glass formation, the criterion of a small energy difference between liquid and crystal is inadequate. As noted by Morey the assumption of a small energy difference "is, as far as the writer , without experimental foundation." Examples of good oxide ers with large energy differences include B_2O_3 and $K_2O \cdot 3SiO_2$. is no simple correlation between energy difference (or heat of fusion) and glass-forming ability, the model cannot be used to describe the relative ease of forming different materials as glasses.

Other structural models have directed attention to structural units (coordination polyhedra) that are nearly regular but are not space filling. Notable among these are the pentagonal dodecahedral models of Tilton (1957) and Robinson (1965) and the models for simple liquids of Frank (1952), Bernal (1960), and Bagley (1965). All of these models suggest the importance in liquids of structural elements that exhibit fivefold symmetry. Such structural elements are expected to correlate with resistance of the liquids to crystallization, both because of the large number of configurations possible by combining different types of polyhedra and because of the change in topology required for the formation of crystalline arrays. Although such models thus direct attention to important characteristics of liquids and glasses, they provide little insight into the relative ease of forming different materials as glasses.

B. THERMODYNAMIC MODELS

The importance of thermodynamic factors in viscous flow and glass formation has been most strongly stated by Gibbs and DiMarzio (1959) and by Adam and Gibbs (1965), who associated the decrease in molecular mobility with falling temperature of liquids with a decrease in their configurational entropy. The configurational entropy was suggested to vanish at a finite temperature, designated T_2; and the occurrence of the glass transition at a somewhat higher temperature (about 50°C above T_2 for most organic liquids) reflects an approach of the system toward the state of zero configurational entropy, with a limited number of available configurations and large changes in topology required for transitions from one configuration to another.

The relatively small configurational entropy at the glass transition had previously been noted by Kauzmann (1948), who pointed out the paradoxical situation of an amorphous phase with a smaller entropy than its corresponding crystal, which would result if the heat capacity of the equilibrium liquid continues unchanged over a range of temperature below the observed glass transition. A striking example of this paradox is provided by data on lithium acetate (Fig. 1). As shown in the figure, some 90% of the entropy of fusion has been lost from the liquid when the glass transition temperature T_g is reached. If the glass transition did not occur, extrapolation of the equilibrium liquid heat capacity over only 15–20°C would produce an amorphous phase with the same entropy as the crystal. Even if the equilibrium heat capacity begins to decrease at temperatures only a few degrees below the glass transition, the decrease would have to be quite sharp to avoid the paradoxical situation.

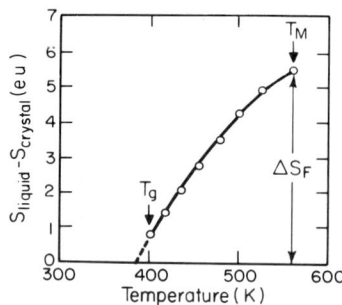

FIG. 1. Entropy difference between liquid and crystal for lithium acetate. (After Wong and Angell, 1977.)

Although the residual entropy difference at the glass transition for most glasses is comparable to the entropy of fusion of metals or SiO_2, and hence is by no means insignificant, the problem posed by Kauzmann remains significant. Kauzmann attempted to avoid the problem by suggesting that the amorphous state of zero configurational entropy would never be reached (even with infinitely slow cooling) because transition to a fine-grained crystalline state would become more likely than continued relaxation to amorphous states of progressively lower energy. Cohen and Turnbull (1964) alternatively suggested that the glassy state is metastable rather than unstable, and hence should have zero entropy at 0 K. For this to be true

> each microscopic structural unit of the glass must lie at a position of static equilibrium, the totality of which is randomly distributed. If one such structure exists there must be a large number of similar random structures of equal energy. Nevertheless, the entropy of each is zero, because all these structures are mutually inaccessible.

As a model of glass formation, the views of Gibbs and his co-workers lead to the expectation of the glass transition as a universal feature of liquid behavior, provided crystallization does not occur. It has been used with success to predict the glass transition temperatures of copolymers of varying chain stiffness and leads to predicted viscosity–temperature relations of the Williams–Landel–Ferry (1955) form in the region around the glass transition. The approach provides little insight, however, into the relative glass-forming abilities of different materials; and data on the flow behavior of a variety of liquids indicate a variety of temperature ⟨ences as the glass transition is approached (Cukierman *et al.*,

extension of the free volume model by Grest and Cohen (1980) ⟩uids to be composed of solidlike cells and liquidlike cells, with the kinetic behavior being dominated by the latter. The model reduces to the familiar free volume form at high temperatures, where it closely describes the flow behavior of a wide variety of liquids; but unlike the simple free volume model, it also describes the observed behavior in the low-temperature region as the glass transition is approached. This model suggests the occurrence of a thermodynamic first-order transition below T_g, in contrast to the second-order transition suggested by Gibbs and his co-workers. This aspect of the model remains, however, subject to some question; and the model does not predict the relative ease of forming different materials as glasses.

Energetic bases for the distinction between network-forming and network-modifying cations have been suggested by several authors. Dietzel (1948) proposed that the field strength z/a^2 is the critical parameter. Here z is the charge on the cation in electron units and a is the distance in angstroms between the ion centers. On this basis, cations with field strengths greater than 1.3 are regarded as network formers, whereas those with field strengths smaller than 0.5 are considered network modifiers. Exceptions to this classification scheme were suggested for cases where symmetric configurations are formed.

An alternative classification, based on single bond strength (the dissociation energy of the oxide divided by the coordination number), was advanced by Sun and Huggins (1947). On this basis, cations with single bond strengths greater than 80 kcal/mole were suggested to be network formers, whereas those with bond strengths less than 35 kcal/mole were classified as network modifiers. A third approach to classifying cations based on electronegativity was suggested by Stanworth (1946). On this basis, network formers are suggested to have electronegativities between 1.8 eV and 2.2 eV, whereas network modifiers have electronegativities less than 1 eV.

All of these approaches group the important network-forming and network-modifying cations in the appropriate categories. As noted by Stevels (1957), however, the relative rankings within each category differ with the classification scheme employed. Further, all three approaches rank certain ions (e.g., As^{5+} and Sb^{5+}) as good glass formers, whose oxides are in fact very difficult to prepare as glasses. The utility of the methods in describing glass formation is restricted to the rather limited class of glasses to which they refer (oxides) and by their lack of reliability in predicting the relative glass-forming abilities of different materials. For detailed criticisms of the three approaches, see Weyl and Marboe (1962).

Other approaches, based primarily on kinetic considerations, have directed attention to thermodynamic quantities such as the melting point and boiling point. These are discussed in Sections III and IV. Since the kinetic models discuss glass formation in terms of the avoidance of crystallization, and hence direct attention to the processes of nucleation and crystal growth, it seems appropriate to discuss briefly these processes before undertaking the formal treatment of glass formation from a kinetic perspective.

II. Nucleation and Crystal Growth in Glass-Forming Systems

A. Crystal Nucleation

Recognizing that the formation of glasses requires the avoidance of detectable crystallization, the basic concepts of crystallization of supercooled liquids will be discussed in this section. It is well established that the overall process of crystallization involves two individual kinetic processes, namely, nucleation and crystal growth. The steady-state rate of homogeneous nucleation can be expressed (Turnbull, 1956)

$$I_v(\text{ho}) = N_v \nu \exp\left(-\frac{16\pi}{3} \frac{\sigma^3}{\Delta G_v^2}\right). \tag{1}$$

Here I_v is the frequency of nucleation per unit volume, N_v is the number of molecules per unit volume in the liquid, ν is the frequency of atom transport at the nucleus–matrix interface, σ is the specific crystal–liquid surface free energy, and ΔG_v is the difference in Gibbs free energy per unit volume between the liquid and the crystal phases.

The steady-state frequency of homogeneous nucleation can alternatively be expressed

$$I_v(\text{ho}) = N_v \nu \exp(-0.0205 B/\Delta T_r^2 T_r^3), \tag{2}$$

where the nucleation barrier in the exponent of Eq. (1) has been replaced by a single parameter B. Here the nucleation barrier is BkT^*, where $T^* = 0.8\, T_m$. The expression in Eq. (2) takes into account a constant difference in heat capacity between liquid and crystal in evaluating the driving force for nucleation. In Eq. (2), $T_r = T/T_m$, $\Delta T_r = \Delta T/T_m$, $\Delta T = T_m - T$ is the undercooling, and T_m is the melting point.

For materials where crystallization involves molecular reorientation or the breaking of directional bonds at the interface, ν can be related to the bulk viscosity η as

$$\nu = b/\eta. \tag{3}$$

...ple organic liquids, the constant b in Eq. (3) can usefully be represented by the Stokes–Einstein relation

$$b_{\text{org}} \approx kT/3\pi a_0^3. \tag{3a}$$

For complex oxide liquids, where molecular transport is a more complicated atomistic process, crystal growth data (Scherer and Uhlmann, 1975a) have suggested

$$b_{ox} \approx 10kT/3\pi a_0^3. \tag{3b}$$

Here a_0 is a molecular diameter in Eq. (3a) and the diameter of the rate-limiting species in Eq. (3b).

Combining Eqs. (2) and (3), a convenient expression can be obtained for the steady state rate of homogeneous nucleation:

$$I_v = K/\eta \, \exp(-0.0205B/\Delta T_r^2 T_r^3), \tag{4}$$

where $K \sim 10^{30}$ P/cm³ sec.

The steady-state rate of heterogeneous nucleation may be expressed as follows:

$$I_v(\text{het}) = n_v N_v^{2/3} \nu \, A \, \exp(-0.0205B\phi/\Delta T_r^2 T_r^3), \tag{5}$$

where n_v is the number of active nucleating heterogeneities per unit volume; A is the surface area per heterogeneity; and ϕ, the efficiency of the nucleating phase, is expressed in terms of the contact angle θ between the heterogeneity and the nucleated crystal in the presence of the liquid (Turnbull, 1956):

$$\phi = (2 + \cos \theta)(1 - \cos \theta)^2/4. \tag{6}$$

Here θ is the contact angle, given by

$$\cos \theta = (\gamma_{HC} - \gamma_{HL})/\gamma_{CL}, \tag{6a}$$

where γ_{HC}, γ_{HL}, and γ_{CL} are, respectively, the heterogeneity–crystal, heterogeneity–liquid, and crystal–liquid surface energies.

The number of active nucleating heterogeneities per unit volume, n_v, does not remain constant during the course of nucleation. Rather, the available heterogeneities are depleted as nucleation proceeds. This depletion should be most marked for small values of θ, and its effect on n_v can be approximated by

$$n_v(t) = n_v^0 \, (1 - I_v(\text{het})Vt), \tag{7}$$

where V is the volume of the sample, $n_v(t)$ is the concentration of nucleating heterogeneities at time t, and n_v^0 is the initial concentration of such heterogeneities. This expression assumes that each heterogeneity provides a single site for nucleation.

When the crystallization time is short, a steady state concentration of subcritical embryos is not achieved, producing a nucleation rate that is time dependent (Becker and Döring, 1935). The time dependence of the nucleation frequency has been expressed in several ways (see Gutzow and Toschev, 1972). For most purposes, however, the expression of Zeldovich is used as an approximation (Zeldovich, 1943; Turnbull, 1948):

$$I_v(t) = I_v(\text{ss}) \exp(-\tau/t), \tag{8}$$

where $I_v(\text{ss})$ is the steady state nucleation frequency considered in Eqs. (1) and (2) and τ is the transient time. The latter quantity can be expressed, to order-of-magnitude accuracy, as

$$\tau(\text{hom}) = (n^*)^2/n_s\nu, \tag{9}$$

where n^* is the number of molecules in the critical nucleus; n_s is the number of molecules on the surface of the critical nucleus; and ν is the frequency of molecular transport at the nucleus–matrix interface, defined in Eq. (3).

It should be noted that the transient time can be quite long for conditions where ν is small—i.e., when the viscosity is high. As pointed out by Uhlmann (1969), however, large values of τ do not imply that steady-state nucleation will not be attained. Rather, the transient times should be compared with the overall times of crystallization, which can be much larger than the transient times when ν is small. Indeed, both the transient time and the overall time for crystallization are expected to scale with the viscosity.

B. Crystal Growth

Once a stable nucleus has formed, it grows until it encounters other crystals and their associated diffusion fields, or until the molecular mobility in the melt is reduced to a sufficiently low level that further growth is effectively cut off. Models used to describe the process of crystal growth as applied to glass-forming systems have been reviewed by Uhlmann (1972b). As noted there, for materials with interfaces that are rough on an atomic scale, the growth rate is usually described by the normal growth relation

$$u = a_0\nu[1 - \exp(-\Delta H_{fM}\Delta T/RTT_E)], \tag{10}$$

where u is the crystal growth rate and ΔH_{fM} is the molar heat of fusion.

For materials with interfaces that are smooth on an atomic scale, the two standard models describe (1) growth by the formation and growth of two-dimensional nuclei on the interface and (2) growth at step sites provided by screw dislocations intersecting the interface. For the latter case (screw dislocation growth), the growth rate is described by the relation

$$a = f\nu a_0[1 - \exp(-\Delta H_{fM}\Delta T/RTT_E)]. \tag{11}$$

Here f is the fraction of preferred growth sites on the crystal–liquid interface, which for screw dislocation growth may be expressed as

$$f \approx (1/2\pi)(\Delta T/T). \tag{12}$$

The temperature dependence of the growth rate for screw dislocation growth is thus similar to that for normal growth, but with an additional factor f, which increases linearly with the undercooling.

For growth by a surface nucleation mechanism, with a finite rate of lateral propagation of the surface across the interface, the growth rate may be expressed (see Calvert and Uhlmann, 1972) as

$$u = C(I_s u_1 u_2)^{1/3}, \tag{13}$$

where C is a constant, u_1 and u_2 are the lateral growth rates in the two directions on the interface plane, and I_s is the rate of formation of the two-dimensional nuclei on the interface. Since I_s represents the dominant contribution to the temperature dependence of the crystal growth rate, the growth rate for this model is expressed as

$$u = D \exp[-c/T \, \Delta T], \tag{14}$$

where D and c are constants.

According to Jackson (1958, 1967), the entropy of fusion is an important parameter in characterizing the crystal growth process. In particular, materials with small entropies of fusion ($\Delta S_{fM} < 2R$) are expected to have crystal–liquid interfaces that are rough on an atomic scale, whereas materials with large entropies of fusion ($\Delta S_{fM} > 4R$) are expected to have smooth interfaces. The correlations between the entropy of fusion and observed crystallization behavior—including interface morphology, growth rate anisotropy, kinetics of growth, and melting rate versus crystallization rate at equal small departures from equilibrium—are illustrated by the results shown in Table II for GeO_2 ($\Delta S_{fM} \sim 1.3R$) and $Na_2O \cdot 2SiO_2$ ($\Delta S_{fM} \sim 4R$). The results are seen to be in good accord with predictions based on the model of Jackson; and the results of more recent computer simulations of crystal growth likewise indicate the importance of the entropy of fusion to the crystallization process, and the importance of defects (screw dislocations) to the growth of high entropy of fusion materials (see Gilmer, 1976, 1980).

The discussion of crystal growth in this section has assumed that the rate of advance of the crystal–liquid interface is limited by attachment kinetics at the interface. Although this is a valid assumption for many congruently melting glass-forming materials, it is not applicable to most materials that melt to liquids of low viscosity. For such materials, including the pure metals and many simple organics, the growth rate is limited

TABLE II

COMPARISON OF CRYSTALLIZATION AND MELTING CHARACTERISTICS BETWEEN GeO_2 AND $Na_2O \cdot 2SiO_2$[a]

	GeO_2	$Na_2O \cdot 2SiO_2$
Interface morphology	Nonfaceted in both crystallization and melting	Faceted in crystallization; nonfaceted in melting
Anisotropy	Largely isotropic; no strongly preferred growth direction	Anisotropic; growth preferentially in b direction
Interface site factor	Independent of undercooling and superheat; described in form by normal growth model	Increases with increasing undercooling; not well described by any standard kinetic model
Continuity at melting point	Crystallization and melting data continuous with similar slope through T_E; melting and crystallization kinetics, corrected for viscosity, equal at equal small departures from equilibrium	Change in slope of kinetic data at melting point; melting more rapid than crystallization at given small departure from equilibrium even after correction for viscosity

[a] After Uhlmann (1972b).

by heat flow rather than by attachment kinetics. In the case of materials whose crystallization involves sizable changes of composition, even with highly viscous melts, the growth can be limited by diffusion in the melt; and the descriptions of the individual kinetic processes involved in crystallization require modification. Such modifications are discussed in the following section.

C. CRYSTALLIZATION WITH CHANGE OF COMPOSITION

In the case of nucleation, the principal change lies in the form of the driving force for crystallization, ΔG_v. The form assumed in Eqs. (2), (4), and (5) is

$$\Delta G_v \approx \Delta H_v \, \Delta T_r T_r, \qquad (15)$$

where ΔH_v is the heat of fusion per unit volume. For oxide systems this expression provides a useful approximation for the driving force; whereas for metallic systems, an expression without the T_r factor seems to provide a better approximation (Thompson and Spaepen, 1979). When the nucleus differs appreciably in composition from the liquid, the driving force for nucleation at a given temperature is usually obtained from the construction shown in Fig. 2. Note that the driving force for nucleation differs from that for bulk crystallization. This results from the small size of the nucleus compared with the volume of the melt (see Christian, 1975).

For materials that undergo sizable changes of composition on crystallization, the crystal growth rate is under most conditions limited by diffusive transport in the melt. Crystallization often takes the form of arrays of dendrites. In many cases, even under isothermal conditions, the tips of neighboring dendrites form a planar crystallization front. As noted by Swift (1947) in his classic study of the crystallization of soda–lime–silica glasses, the growth rate of such parallel arrays of dendrites is often constant with time.

Uhlmann (1972b) suggested that the observation of such growth rates independent of time reflected the scale of the diffusion field in crystallization being independent of time, in accord with previous calculations of Ivantsov (1947) and of Horvay and Cahn (1961) for isolated dendrites growing into a melt. Subsequent treatments by Christensen *et al.* (1973) and by Scherer and Uhlmann (1975b) have supported this suggestion for parallel arrays of dendrites as well. According to the latter treatment, the growth rate is determined by the interdiffusion coefficient in the melt, the size and spacing of the dendrites, and differences between the interface concentration and that in bulk liquid and in the crystal. As shown in Fig. 3, the model is capable of providing a useful description of experimental data on materials crystallizing under conditions of coupled diffusion-controlled growth.

The central remaining problem in treating such growth is the relation between the radius of curvature and the undercooling or growth rate. This is likewise a key issue in treating the growth of isolated dendrites from the melt, where several approaches have been suggested (often based on the

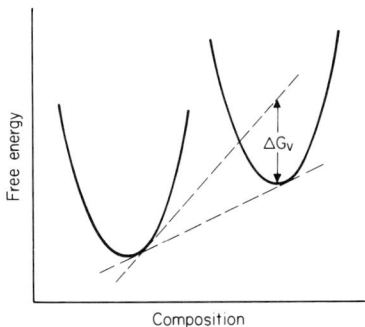

FIG. 2. Free energy versus composition diagram (schematic) for crystallization with a change of composition showing the driving force for nucleation, ΔG_v.

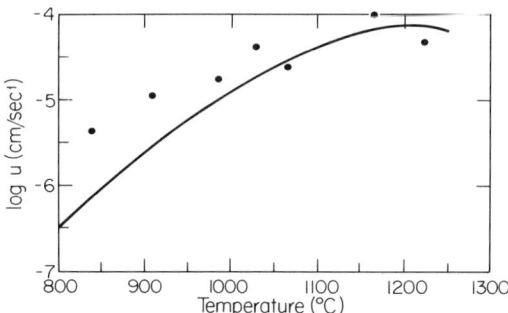

FIG. 3. Variation of crystal growth rate with temperature for $0.10K_2O$–$0.90SiO_2$ showing experimental data (curve) and calculated values. (After Scherer and Uhlmann, 1977.)

principles of irreversible thermodynamics). No one, however, has proved satisfactory; and lacking such insight, it is not possible to predict the growth rate versus undercooling relation without some independent determination of a parameter such as the radius of curvature of the growth front.

A second problem that arises in predicting the growth rate of molecularly complex materials, such as silicates or borates, is the relation between the interdiffusion coefficient (the central transport parameter in treating diffusion-controlled growth) and other, more easily measured quantities. The comparison between theory and experiment shown in Fig. 3 was effected for a composition ($0.10K_2O$–$0.90SiO_2$) for which interdiffusion data were already available. As noted in Section II.A, crystal growth data on congruently melting silicates have indicated a simple inverse relation between the bulk viscosity and the frequency of transport at the crystal–liquid interface [see Eq. (3b)]. Data on interdiffusion coefficients in alkali silicate systems do not always, however, exhibit such a simple inverse relation with the viscosity (Uhlmann *et al.*, 1983); and further work in this area is clearly indicated.

III. Kinetic Models of Glass Formation: Growth Rate and Nucleation Rate

The importance of kinetics to the process of glass formation was recognized long ago by Tammann (1933), who stated: "the melt frequently solidifies as a glass. The rate of cooling is most important for this phenomenon. Many melts which crystallize completely on slow cooling do not crystallize at a faster cooling rate." A similar recognition of glass formation as a kinetic phenomenon was provided by the work of Tammann and

Elbrachter (1932), who considered the dependence of the ability of molten salts to be formed as glasses on sample (droplet) size.

From this perspective the critical question in discussing glass formation is not *whether* a material will form an amorphous solid when cooled from the liquid state, but *how fast* a given liquid must be cooled for detectable crystallization to be avoided. The ultimate concern in discussing glass formation must then be with the rates at which bodies of different materials are cooled and with the kinetic processes involved in crystallization.

A. GROWTH RATE

It has long been recognized that the overall rate of crystallization depends on the rate of formation of crystal nuclei in the melt as well as on the crystal growth rate. Various kinetic views of glass formation have been based on each of these kinetic processes. Dietzel and Wickert (1956), for example, defined "glassiness" as the reciprocal of the growth rate, and carried out an extensive study of growth rates for various compositions in the Na_2O–SiO_2 system. From their results, "glassiness" was associated with (1) high viscosity, (2) large compositional difference from compounds noted in the phase diagram, and (3) low liquidus temperature. Although no quantitative relation was established between "glassiness" and these factors, or between "glassiness" and the cooling rates required to form glasses of different compositions, the work did explicitly identify factors that are critical in glass formation.

The study by Dietzel and Wickert covered the range from 0 to 47% Na_2O and included compositions whose crystallization products were cristobalite alone, cristobalite + $Na_2O \cdot 2SiO_2$, $Na_2O \cdot 2SiO_2$ alone, $Na_2O \cdot 2SiO_2$ + $Na_2O \cdot SiO_2$, and $Na_2O \cdot SiO_2$ alone. Subsequent work by Scherer and Uhlmann (1977) has indicated that, depending on the composition and the temperature of crystallization, the crystal morphology may be spherulitic, dendritic, or fibrillar (needlelike). The crystal growth rate is independent of time, at least in the early stages of crystallization. Except for compositions at or very near congruently melting compounds, the growth rate is diffusion-controlled; and the relevant kinetic constant is the interdiffusion coefficient ($\tilde{D}$) rather than the viscosity (η). See the discussion in Section II above. More generally, although "glassiness" does direct attention to some important kinetic parameters, it is quite limited in its utility for predicting the ease of glass formation.

Turnbull (1969) noted that the formation of glasses would be largely determined by the crystal growth rate for liquids that already contain crystal nuclei at the melting point. A highly simplified model was used to differentiate between regimes of interface-controlled and heat-flow-controlled growth. No specific relations were obtained between the growth rate and the cooling rate required to form glasses; but estimates of cooling rates were obtained for liquids with a spacing of nuclei of 0.1 cm (selected arbitrarily) that have viscosities at the melting point of 10^3 and 10^7 P.

B. Nucleation Rate

Turnbull's quantitative approach to glass formation was directed principally to liquids of low viscosity in which a single nucleation event can lead to rapid and complete crystallization of a sample. The condition for glass formation was taken as the absence of even a single nucleation event. This was expressed (Turnbull, 1969) as

$$n = V \int_0^t I_v \, dt, \quad n < 1, \tag{16}$$

where n is the number of nuclei that form in a sample of volume V cooled over a time t, and I_v is the nucleation frequency per unit volume.

The steady-state rate of homogeneous nucleation was expressed as

$$I_v = \frac{K}{\eta} \exp\left[\frac{-b\alpha^3\beta}{T_r \Delta T_r^2}\right]. \tag{17}$$

Here K is a constant (typically about 10^{30} dyne cm), b is a constant ($b = 16\pi/3$ for a spherical nucleus), and α and β are dimensionless parameters given by

$$\alpha = (N_0 \overline{V}^2)^{1/3} \, \sigma/\Delta H_{fM}, \tag{18}$$

$$\beta = \Delta H_{fM}/RT_M = \Delta S_{fM}/R, \tag{19}$$

where N_0 is Avogadro's number, $\overline{V}$ is the molar volume of the crystal, σ is the crystal–liquid interfacial energy, R is the gas constant, and ΔS_{fM} is the molar entropy of fusion. The expression of Eq. (17) differs in form from that of Eq. (2) through the absence of a factor T_r^2 in the denominator of the exponential function. As noted in the discussion of Eq. (15), the expression without the T_r factor—which assumes that the heat capacities of liquid and crystal are the same—seems to describe thermodynamic data on metals; and the expression of Eq. (15) with the T_r factor—which assumes a constant difference in heat capacity between crystal and liquid—is preferable for nonmetals.

Calculations were carried out for various values of $\alpha\beta^{1/3}$, with the viscosity taken as 10^{-2} P, independent of temperature. For typical sample volumes and cooling rates, it was suggested that for nucleation rates smaller than $10^{-6}/\text{cm}^3$ sec, glasses rather than crystalline bodies would be formed. The calculations indicate (Fig. 4) that liquids with $\alpha\beta^{1/3} > 0.9$

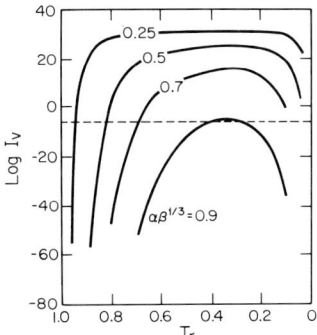

FIG. 4. Calculated variation of nucleation frequency versus reduced temperature for various values of $\alpha\beta^{1/3}$. $I_v = 10^{32} \exp[-16\pi\alpha^3\beta/3T_r(\Delta T_r)^2]$. (After Turnbull, 1969.)

should readily form glasses, whereas those with $\alpha\beta^{1/3} < 0.25$ should almost always crystallize on cooling. Measurements of ΔH_f indicate β values near 1 for metals, SiO_2, GeO_2, organic plastic crystals, and many strong acids and bases, whereas for typical organic and inorganic compounds, β is in the range of 4 to 10, with values about 6 being typical. Results of droplet nucleation experiments on a variety of materials (Jackson, 1965) indicate α values between $\tfrac{1}{4}$ and $\tfrac{1}{2}$, with $\alpha \approx 0.4$–0.5 being typical for metals and $\alpha \approx \tfrac{1}{3}$ being typical for most nonmetals.

These values indicate that $\alpha\beta^{1/3} \approx 0.4$–$0.5$ for metals; and Fig. 4 indicates that copious nucleation of crystals is expected. Hence a glass would not be formed even at very fast cooling rates, save for very small samples or materials with exceptionally high glass transition temperatures (as $T_g/T_M \sim 0.8$), for which the curves shown in Fig. 4 require modification (see below). For typical nonmetals, $\alpha\beta^{1/3} \approx 0.5$–$0.7$, and glass formation should be much easier, albeit not certain. To predict whether a given material will form a glass, it is necessary to specify the cooling rate and sample size, and to modify the treatment to include the temperature dependence of the viscosity.

Insight into the effect of the viscosity changing with temperature was provided by Turnbull. A Vogel–Fulcher form was assumed for the temperature dependence of the viscosity:

$$\eta = 10^{-3.3} \exp[3.34/(T_r - T_{r_g})], \qquad (20)$$

where $T_{rg} = T_g/T_M$. The results of the calculations are shown in Fig. 5 for three different values of T_{r_g}, where $\alpha\beta^{1/3}$ has been taken as 0.5. It is seen that materials with $T_{rg} = \tfrac{2}{3}$, if free of heterogeneous nuclei, should readily form glasses. In contrast, materials with $T_{rg} = \tfrac{1}{2}$ could only be formed as glasses in small volumes with high cooling rates. For higher values of

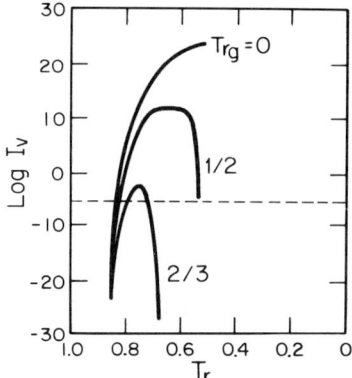

FIG. 5. Calculated variation of nucleation frequency with reduced temperature for $\alpha\beta^{1/3} = \frac{1}{2}$ showing effect of T_{r_g}. (After Turnbull, 1969.)

$\alpha\beta^{1/3}$, glasses should be formed (at a cooling rate) for materials with lower values of T_{rg}.

The assumption that a single stable crystal nucleus must be avoided to form a glass was also adopted by Vreeswijk *et al.* (1974). Transients in the nucleation rate were included by using an expression for the time dependence of the nucleation frequency [see Eq. (8)]:

$$I_v(t) = I_v(ss) \exp(-\tau/t) = (K/\eta) \exp[-b\alpha^3\beta/T_r^3\Delta T_r^2]\exp(-\tau/t). \quad (21)$$

Here τ is the transient time, and the expression for the steady state nucleation rate has the same form as Eq. (2).

The cooling rate R was taken as a constant and was related to the time t as $R = \Delta T/t$. Equation (21) then becomes

$$I_v(t) = I_v(ss) \exp(-\tau R/\Delta T). \quad (22)$$

Using Eqs. (16) and (22) together with the relation between cooling rate and ΔT, the cooling rates at which a single nucleus would form in a 1-cm³ sample of various liquids were calculated. The results for several oxides, obtained using measured values of ΔH_f and an assumed value of $\alpha = 0.32$ are shown in Table III.

The results for these and other materials (including fused salts, molecular liquids, and metals) provide a classification of materials as good and poor glass formers. Small values of the cooling rate are calculated for materials that easily form glasses, and high cooling rates for materials that form glasses only under extreme quench conditions. Limitations to the approach are apparent, however. Considering Table III, for example, the range of indicated cooling rates (a factor of 10^{30}) is excessively wide; the estimated cooling rate for As_2O_3 was noted by the authors to be too large;

the estimate for GeO_2 is much too low; and the relative ranking of SiO_2 and GeO_2 is at variance with experience.

These difficulties may have several origins. Among them the most important are

(1) The use of transient nucleation rates. It is important to view the transient times, even if long, in the perspective of the total crystallization time. For example, with SiO_2 at 1415°C, the nucleation transient time is estimated as about 10^4 sec (Uhlmann, 1969), whereas the time to reach a volume fraction crystallized of 10^{-6} is about 10^6–10^7 sec (Uhlmann, 1972a). Hence the nucleation transient time in this case is insignificant compared with the total time to develop detectable crystallinity.

(2) Consideration of nucleation behavior only neglects the role of the crystal growth rate in affecting the ability of materials to be formed as glasses. As noted in Section II, the rates of nucleation and crystal growth have quite different temperature dependences. The maximum growth rate occurs at a higher temperature than the maximum nucleation rate; and by the time the nucleation rate becomes significant on cooling, the growth rate may be sufficiently small that a glass will still be formed—even for a material that is quite fluid at its melting point.

More generally, treatments of glass formation based on consideration of nucleation kinetics *or* of growth kinetics constitute an advance over structural or thermodynamic models and represent limiting approaches to the problem. The models based on growth rate are, however, either largely qualitative or useful only when the density of nuclei can be specified or when rather restrictive kinetic assumptions are employed. The criterion of the nucleation models of only a single nucleus forming in a sample is appropriate only when the nuclei form while the melts are quite fluid and

TABLE III

CALCULATED CRITICAL COOLING RATES FOR FORMING VARIOUS OXIDES (1 cm³) AS GLASSES[a]

Material	Cooling rate (K/sec)
P_2O_5	10^{-23}
B_2O_3	10^{-16}
GeO_2	10^{-11}
SiO_2	10^{-1}
As_2O_3	10^7

[a] After Vreeswijk *et al.* (1974).

where a single nucleation event can produce rapid crystallization of the entire sample. Neither type of model can be used as a general treatment of glass formation.

To provide a more general and predictive model of glass formation, it is necessary to treat the crystallization process by considering *both* nucleation and crystal growth and to estimate the cooling rates required to form glasses using formalisms developed for describing other phase transformations. Such treatments are discussed in the following section.

IV. Kinetic Models for Glass Formation: The Perspective of Transformation Kinetics

A. GENERAL FORMULATION

Kinetic treatments of glass formation are based on identifying a certain value of the volume fraction crystallized, V_c/V, as borderline between an amorphous and a crystalline body. Thus an operational definition of glass is adopted whereby a body is considered glassy if it contains a degree of crystallinity that cannot be detected by available techniques. A just-detectable degree of crystallinity of 10^{-6} is often used in evaluating critical cooling rates for glass formation, although the estimated cooling rates are not very sensitive to the precise value assumed for the just-detectable degree of crystallinity.

In treating the development of crystallinity within an amorphous body, the formal theory of transformation kinetics due to Johnson and Mehl (1939) and Avrami (1939, 1940, 1941) can be employed. In its general, nonisothermal form, the theory relates the degree of transformation V_c/V to the nucleation frequency I_v, the crystal growth rate u, and time t, as follows (Christian, 1975):

$$\frac{V_c}{V} = 1 - \exp\left(-\int_0^t I_v \left[\int_{t'}^t u\, d\tau\right]^3 dt'\right). \tag{23}$$

Here both the nucleation frequency and the crystal growth rate are assumed to be time-dependent through their dependence on temperature [Eqs. (4), (5), (10), (11), and (14), e.g.].

Although a rigorous treatment of glass formation by cooling a melt requires the use of Eq. (23), useful estimates of critical cooling rates can be obtained by using the simple solution to Eq. (23) for constant nucleation rate and crystal growth rate:

$$\frac{V_c}{V} = 1 - \exp\left(-\frac{\pi}{3} u^3 I_v t^4\right). \tag{24}$$

Knowing the nucleation frequency and the growth rate at a given temperature, the degree of crystallinity can be calculated as a function of time. The results are represented graphically as time–temperature–transformation (TTT) curves that represent two-dimensional projections on the t–T plane of the three-dimensional t–T–V_c/V relations. The TTT curve as shown in Fig. 6 for anorthite $(CaO \cdot Al_2O_3 \cdot 2SiO_2)$ pertains to a degree of crystallinity of 10^{-6}. This value of V_c/V has been taken as a just-detectable degree of crystallinity (Uhlmann, 1972a). Equation (24) indicates that the time t depends only on $(V_c/V)^{1/4}$. Hence TTT curves pertaining to widely different degrees of crystallinity would be narrowly spaced, indicating that the actual choice of the just-detectable degree of crystallinity has little effect on the calculated critical cooling rate.

As seen in Fig. 6, each TTT curve has an extremum or nose. This extremum arises because of the competition between the driving force for crystallization, which increases with increasing undercooling, and the molecular mobility, which decreases with increasing undercooling. The nose represents the least time required to form a given degree of crystallinity; and in the region of the nose, the overall crystallization process—including both nucleation and crystal growth—is much faster than at other temperatures. Recognition of this fact leads to a first-approximation approach to estimating the critical cooling rate R_c required to form a glass (to avoid the just-detectable degree of crystallinity). That is,

$$R_c \approx \Delta T_n/\tau_n, \tag{25}$$

where $\Delta T_n = T_m - T_n$, and T_n and τ_n are, respectively, the temperature and time at the nose of the TTT curve.

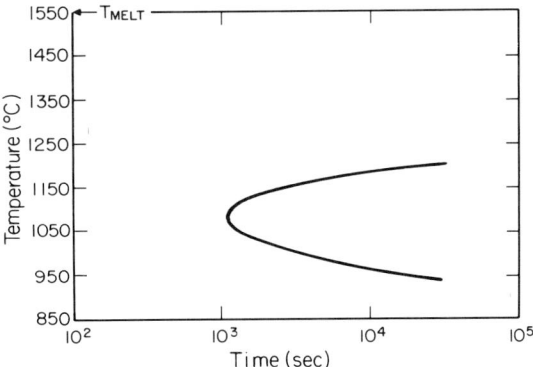

FIG. 6. Isothermal time–temperature transformation (TTT) curve for anorthite $V_c/V = 10^{-3}$; $B = 82kT^*$.

The use of Eq. (25) to estimate the critical cooling rate implicitly assumes that the crystallization kinetics over the full range of temperature between T_m and T_n are as rapid as at the temperature of the nose. Hence it should overestimate the cooling rate required to form a glass.

More realistic estimates of the critical cooling rate can be obtained by taking account of the extent of crystallization that takes place in each incremental region of temperature or cooling. The most widely used approach of this type is the averaging method of Grange and Kiefer (1941), which employs the isothermal information contained in the TTT curves to construct continuous cooling (CT) curves. This method provides highly useful, albeit approximate, estimates of the critical cooling rates. As shown by the representative CT curves in Fig. 7, for lunar composition 79155, a given degree of crystallinity is developed at lower temperatures and longer times than indicated by the TTT curve. (The weight percentages of the various constituents in this and other lunar compositions discussed in this chapter are given in Table IV.)

The critical cooling rates are determined directly from the CT curves by plotting a set of cooling curves on the same T–t axes as shown in Fig. 7. The cooling curve that barely misses the nose of the CT curve represents the critical cooling rate for the particular degree of crystallinity considered. For the lunar composition represented in Fig. 7 the cooling rate required to form a glass ($V_c/V < 10^{-6}$) is estimated as about 1.3×10^2 K/min. As expected, the values of R_c calculated from CT curves are lower than those calculated from TTT curves using Eq. (25). In nearly all cases, however, both estimates are found to be of the same order of magnitude.

In addition to describing crystallization under conditions of constant-rate cooling, CT curves can also provide a description appropriate for

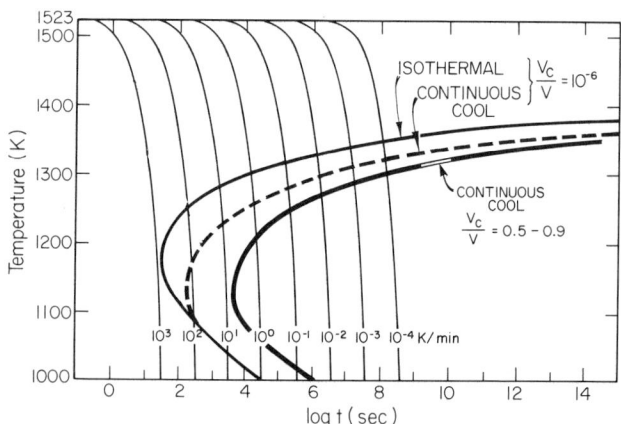

FIG. 7. Isothermal TTT and constant-rate cooling CT curves corresponding to $V_c/V = 10^{-6}$ for the lunar composition 79155.

TABLE IV

Lunar Composition Mentioned in the Text (wt %)

	SiO$_2$	TiO$_2$	Al$_2$O$_3$	FeO	MgO	CaO	Na$_2$O	K$_2$O
10060	42.8	9.4	10.8	16.6	7.4	9.8	0.5	0.1
Apollo 15 green glass	45.2	0.5	7.4	20.3	18.1	8.2	0.2	0.2
15028	48.8	1.4	12.9	14.1	7.4	9.5	0.6	0.4
15086	48.3	1.7	9.6	16.9	10.0	10.0	0.4	0.1
15101	49.2	1.5	15.9	12.1	10.4	9.4	0.5	0.1
15286 Matrix	47.6	1.2	13.3	13.6	13.6	9.8	0.7	0.3
15286 Intrusion	46.1	1.6	14.3	14.2	12.5	10.4	0.8	0.1
15301	48.3	1.2	13.3	14.4	11.5	8.4	0.5	0.1
15498	48.2	2.0	19.8	8.4	7.5	13.0	0.9	—
60095	46.4	—	23.5	6.9	10.5	12.1	—	0.8
60255	48.6	0.6	23.3	6.0	6.0	12.6	0.6	0.1
65016	44.2	0.6	26.5	5.5	7.3	15.3	0.4	0.1
70019	41.3	8.4	12.4	16.1	10.2	11.2	0.3	—
77017	46.0	0.6	23.9	6.4	5.5	14.7	0.4	0.1
79155	35.5	14.3	3.7	23.5	11.6	11.1	0.2	—
Luna 24 highland basalt	47.0	0.3	26.7	4.9	6.9	15.6	0.2	—

other cooling conditions. Examples of the different curves—TTT, CT (constant rate cooling), and CT (logarithmic cooling)—are shown in Fig. 8 for lunar composition 15286 (Matrix). As noted by Onorato and Uhlmann (1976), curves describing crystallization under other conditions of con-

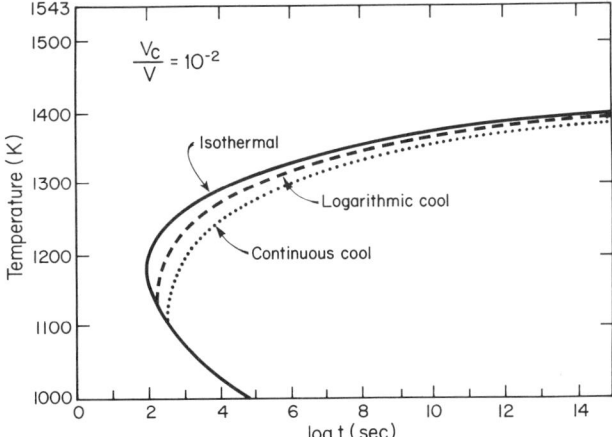

FIG. 8. Isothermal TTT, logarithmic cooling CT, and constant-rate cooling CT curves for the matrix composition of lunar sample 15286, calculated for volume fraction crystallized of 10^{-2}.

tinuous cooling can likewise be constructed using variations on the method of Grange and Kiefer.

Although the analysis of crystallization in terms of CT curves represents an improvement over the simple TTT analysis, it still retains substantial approximations. The averaging method of Grange and Kiefer does not take into account the contribution of crystallization at temperatures and times above the TTT curve to the overall crystallinity; it also neglects any crystallization that takes place at temperatures below the nose of the CT curve.

A notably improved description of crystallization on cooling a body was advanced by Hopper et al. (1974a). These workers introduced a crystal distribution function $\psi(r,t,R)$, which was defined such that the number dn of crystals in a volume dV at r having radii between R and $R + dR$ at time t is given by

$$dn = \psi(r,t,R)\,dV\,dR. \tag{26}$$

This function contains detailed information about the crystal sizes as well as the numbers of crystals and volume fractions crystallized in any portion of a body.

For the particular case where one is concerned with the volume fraction crystallized, this approach reduces to a form of Eq. (23), which can readily be handled by a computer:

$$\frac{V_c}{V}(t_j) = 1 - \exp\left\{-\sum_{i=1}^{j} \frac{4\pi}{3} R_i^3(t_j,t_i) I_{vi}(t_i)\,\Delta t\right\}, \tag{27}$$

where $V_c/V(t_j)$ is the volume fraction crystallized up to time t_j, I_{vi} is the steady state nucleation frequency at time t_i, and $R_i(t_i,t_j)$ is the radius at time t_j of nuclei nucleated at time t_i. The latter quantity is given as

$$R_i = R_i^* + \sum_{k=i}^{j} u_k(t_k)\,\Delta t. \tag{28}$$

Here R_i^* is the size of the critical nucleus at time t_i, u_k is the crystal growth rate at time t_k, and Δt is the duration of the time interval. The time–temperature conversion is evaluated using whatever cooling (or heating) schedule is appropriate—e.g., for constant-rate cooling,

$$T = T_0 - aT, \tag{29}$$

where T_0 is the initial temperature from which cooling begins and a is the cooling rate.

Using this approach, the detailed statistics of the number and size distribution of small crystallites within a supercooled liquid can be calculated as a function of temperature. Hence this analysis is known as the crystallization statistics technique. Knowing the nucleation frequency and the crystal growth rate, the calculations yield for each assumed cooling rate a time and a temperature where the degree of crystallinity V_c/V reaches the just-detectable value of 10^{-6}. The set of these values comprises a continuous cooling curve obtained without the simplifying assumptions used to obtain the CT curve as described earlier.

Although the technique of crystallization statistics is capable of calculating critical cooling rates with a high degree of accuracy, it requires rather detailed information about the materials, which in most cases is not available. Direct use of the TTT or CT curves, which provide less accurate estimates of the critical cooling rates, require similarly detailed kinetic information (on crystal growth rates, liquid viscosities, and nucleation frequencies). To circumvent these difficulties, Uhlmann *et al.* (1979) and Onorato *et al.* (1981) have proposed a simplified model for glass formation that can provide order-of-magnitude estimates of the critical cooling rates from only limited kinetic data on the materials.

This simplified model is based on the following considerations: use of TTT curves to calculate the critical cooling rates relies primarily on the information at the noses of the curves. If a method could be found to predict the time and temperature of the nose, one would not need to calculate the entire TTT curve but could use a form of Eq. (25) to calculate the critical cooling rates. In this regard it was noted that application of the kinetic treatment of glass formation to a wide range of materials had produced a large number of TTT and CT curves. The temperatures of the nose of the TTT curves were observed to occur at approximately 0.77 of the melting point. Thus the temperature of the nose can be approximated as

$$T_n \approx 0.77 T_m. \qquad (30)$$

With the temperature of the nose known, Eq. (24)—modified for small values of V_c/V—can be used to calculate the time required to form a volume fraction crystallized of 10^{-6}. This is done by combining Eqs. (4), (10), (11) or (14), and (25), together with a relation between the barrier to crystal nucleation and the entropy of fusion ΔS_m:

$$\Delta G^* \approx 12.6 \frac{\Delta S_m}{R} kT^*. \qquad (31)$$

This relation was obtained by collating available data on nucleation behavior for a variety of materials, calculating the crystal–liquid surface energies from the range of undercoolings where homogeneous nucleation was observed, and relating these surface energies to the respective molar heats of fusion. The relation of Eq. (31) should be applicable for materials characterized by high entropies of fusion in Jackson's (1958) sense. A similar relation, but with a different coefficient, should be used for materials with small entropies of fusion.

With these assumptions and approximations, one obtains an expression for the critical cooling rate for glass formation:

$$R_c \approx \frac{AT_m^2}{\eta(0.77\ T_m)} \exp(-0.212B) \left[1 - \exp\left(-\frac{0.3\Delta H_{fM}}{RT_m}\right)\right]^{3/4}, \quad (32)$$

where $A = 4 \times 10^5$ erg/cm^3 K.

On this basis, the prediction of the critical cooling rate requires only the viscosity at a single temperature (0.77 T_m) and the heat of fusion. To calculate the viscosity at 0.77 T_m, the glass transition temperature ($\eta = 10^{13}$ P) has to be estimated or determined experimentally, as well as the viscosity at the melting point or over a range of temperature above the melting point. In the case of silicate liquids, use can be made of the parameters suggested by Bottinga and Weill (1972) or Shaw (1972) to estimate the high-temperature viscosities. In any case, the high-temperature viscosity data together with the glass transition point can then be fitted to a Vogel–Fulcher relation (Uhlmann et al., 1979; Davies, 1976) or to a different form of the free volume (Doolitle) relation (Ramachandrarao et al., 1977) to obtain an estimate of the viscosity at 0.77 T_m.

Three or four measurements are needed, then, to estimate the critical cooling rate—the molar heat of fusion, the melting temperature, the glass transition temperature—and when the viscosity over a range of temperature above the melting point cannot be estimated, this quantity should also be measured. The critical cooling rates thus estimated for a variety of oxide liquids, including anorthite, were found to be within an order of magnitude of the values estimated using continuous cooling (CT) curves (Uhlmann et al., 1979).

B. Dependence of Critical Cooling Rate on Material Parameters

The discussion of the previous pages has treated glass formation by considering only homogeneous nucleation. In this way it has led to evaluations of the *minimum* cooling rates required to form glasses. That is, if a material will not form a glass at a given cooling rate when nucleating heterogeneities are absent, it will certainly not form a glass when such heterogeneities are present. As will be seen in Sections IV.C and IV.D below, such minimum cooling rates for glass formation are often remarkably accurate. Apparently nucleation on adventitious heterogeneities at modest undercoolings leads to only small volume fractions crystallized as

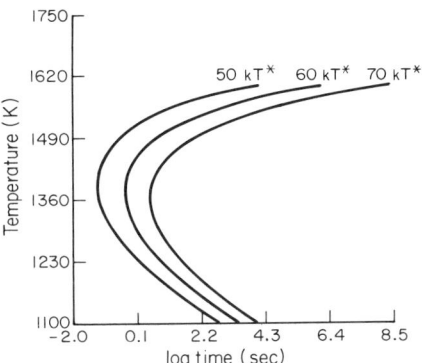

FIG. 9. Time–temperature transformation curves for anorthite corresponding to nucleation barriers of $50kT^*$, $60kT^*$, and $70kT^*$.

the samples are cooled; and in the critical region of large undercoolings ($\Delta T_r > 0.2$), homogeneous nucleation represents the dominant contribution to the formation of crystals and hence to the development of detectable crystallinity in the materials.

The magnitude of the crystal–liquid surface energy, and hence of the barrier to crystal nucleation, can have an important effect on the critical cooling rates for glass formation. This is shown in Fig. 9, where TTT curves are shown for anorthite for $\Delta G^* = 50$, 60, and 70 kT^*. Here ΔG^* is the free energy of forming the critical nucleus. The critical cooling rates corresponding to these values of the nucleation barrier are, respectively, 3×10^3, 3×10^2, and 40 K/sec.

For classes of materials with similar viscosity–temperature relations, glass formation is favored by a low melting point. This is shown in Fig. 10 for materials having the same viscosity–temperature relation, entropy of fusion, and other properties as $Na_2O \cdot 2SiO_2$, but with assumed melting points of 1046, 1146 (the actual melting point), and 1246 K. The minimum

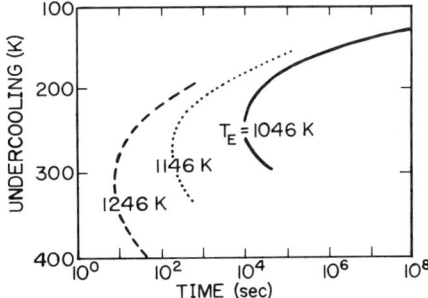

FIG. 10. Time–temperature transformation curves for $Na_2O \cdot 2SiO_2$-like materials having melting points of 1046, 1146, and 1246 K, $V_c/V = 10^{-6}$.

cooling rate for $T_m = 1246$ K exceeds that for $T_m = 1046$ K by about three orders of magnitude. For complex systems glass formation is also favored by a large redistribution of material being required for crystallization (as is the case with many near-eutectic compositions).

Application of the kinetic analysis to a variety of materials—including oxides, simple organics, and metals—indicates that glass formation is favored by a large viscosity at the melting point or liquidus temperature and a viscosity that increases strongly with falling temperature below the melting point. Good oxide glass formers are characterized by the former, whereas good organic glass formers have viscosities that may be small at their melting points but that increase appreciably with falling temperature below the melting point.

C. Comparison of Theory with Experimental Data

The kinetic treatment outlined previously has been applied to a variety of liquidus, including oxides, organics, and metals. As noted earlier, the critical cooling rates estimated using CT curves are generally lower than the values obtained using TTT curves. In turn, the analysis of crystallization statistics often, but not always, yields even lower estimated critical cooling rates. In several cases, the predictions of the theory have been compared with experimental results or with predictions based on independent calculations; these will now be discussed.

As one approach, it has been possible to estimate critical cooling rates from the sizes of bodies that can be obtained as glasses. This approach uses the fact that the maximum thickness Y_c of a body obtainable as a glass can be estimated as

$$Y_c \approx (D_{th}\tau_n)^{1/2}, \tag{33}$$

where D_{th} is the thermal diffusivity of the material and τ_n is the time at the nose of the TTT curve. By observing Y_c, the time τ_n and thus the critical cooling rate can be estimated.

More rigorous calculations of cooling rates for bodies of various shapes cooled under different conditions have yielded improved estimates of the size of bodies that can be formed as glasses. As an example, Hopper *et al.* (1974b) considered the radiational cooling of a spherical body to provide insight into the cooling history of a glassy sample returned from the lunar surface, lunar composition 60095 (see Table IV). It was concluded that a body of the indicated size and optical properties would cool at a rate of ~5 K/sec. By comparison, the critical cooling rate to form a glass of this material, obtained using the analysis of crystallization statistics with the value for the nucleation barrier estimated by Uhlmann *et al.* (1981), is ~2 K/sec.

Critical cooling rates required to form glasses of a variety of lunar compositions have been calculated using the detailed analysis of crystallization statistics (in cases where data on growth rate and viscosity are available) and using the simplified model for glass formation discussed earlier. These predictions have been compared with direct measurements of the cooling rates required to form these materials as glasses, obtained from constant-rate cooling experiments.

The results (Uhlmann et al., 1981) are presented in Table V. As seen from the table, and as indicated in the discussion of the simplified model previously, in cases where the kinetic data are available to permit a comparison between predictions of the simplified model and the analysis of crystallization statistics, the predictions of the two models agree within an order of magnitude. It is also seen that the critical cooling rates calculated from the analysis of crystallization statistics agree remarkably well with the experimentally determined cooling rate. It is also seen that the predic-

TABLE V

A Comparison of Calculated and Experimental Critical Cooling Rates, R_c (K/sec)[a]

	Nucleation barrier (kT^*)	R_c from crystallization statistics analysis	R_c from simplified model	R_c measured
10060	68	—	2	1.3
Apollo 15 green glass	76	8.3	0.8	>4
15028	68	—	1.2	0.9
15086	68	—	4.5	3
15101	68	—	2	0.9
15286 Matrix	72	1.1	7.4	0.3
15286 Intrusion	63	1.3	6.2	2
15301	68	—	5.4	2.7
15498	60	2.8	17	3
60095	68	2	0.9	—
60255	78	—	4.8	1
65016	70	3.7	2.1	—
77017	78	—	11	0.3
Luna 24 highland basalt	67	15	18	—

[a] $V_c/V = 10^{-6}$; nucleation barriers B used in the calculations were either measured from DTA experiments or estimated based on composition.

tions of the simplified model provide reasonable to good agreement with measured critical cooling rates.

The agreement between the experimentally determined critical cooling rates and those estimated from the analysis of crystallization statistics is gratifying. The calculations employed experimental data on crystal growth rates, viscosity, and nucleation barrier; and the agreement indicates that with a full complement of data, highly accurate predictions of critical cooling rates can be effected.

As should be clear from the discussion in Section IV.A, the simplified model of glass formation should be strictly applicable only to materials whose growth rates are limited by interface attachment kinetics. It seems, however, to provide useful estimates of critical cooling rates for compositionally complex materials such as the lunar compositions, whose crystal growth kinetics are dominated by diffusional processes in the melt. Apparently, the growth rates predicted by the interface kinetics model are reasonably close to the actual growth rates—at least at the temperature $(0.77 T_E)$ that is used in the calculations.

Cooling rates required to form glasses of various alkali silicate compositions were reported by Havermans *et al.* (1970). Their results are shown in Fig. 11. As noted there, the critical cooling rates in the Li_2O–SiO_2, Na_2O–SiO_2, K_2O–SiO_2, and $(\frac{1}{2}Na_2O + \frac{1}{2}K_2O)$–$SiO_2$ systems vary in the order $R_c^{Li} > R_c^{Na} > R_c^{K} > R_c^{\frac{1}{2}Na+\frac{1}{2}K}$. The general forms of the critical cooling rate vs. composition relations, both within a given system and among the three single alkali silicate systems, correlate with the liquidus temperatures of the systems—with low liquidus temperatures being associated with small critical cooling rates. The greater ease of glass formation for the

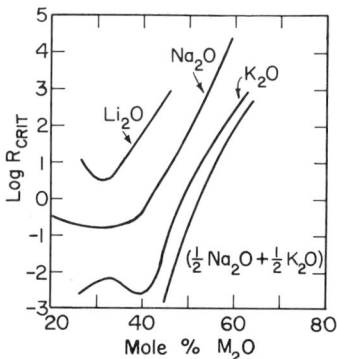

FIG. 11. Critical cooling rates for glass formation in alkali silicate systems. (After Havermans, 1970.)

($\frac{1}{2}$Na$_2$O + $\frac{1}{2}$K$_2$O)–SiO$_2$ compositions relative to the single alkali silicates undoubtedly reflects the effects of solute redistribution favoring glass formation in the mixed alkali materials.

The theoretical evaluation of critical cooling rates has been of particular importance in research on metal alloy systems, where one looks for metal alloys that can be formed as glasses with reasonably low cooling rates and that will remain glassy (stable) at usable temperatures. Considerable effort has been devoted to calculating the critical cooling rates for various metals and metal alloys. Such calculations are, however, quite difficult because of the lack of data on crystal growth kinetics, nucleation kinetics, and liquid viscosity. Although the heat of fusion and nucleation barrier can usually be estimated, the prediction of the viscosity–temperature relationship is fraught with difficulties. An approximate approach was therefore adopted, noting that the viscosities of most metals and metal alloys at the melting temperature can be taken as 0.01–0.1 P; and a viscosity of 10^{13} P can also be associated with the glass transition temperature T_g. Hence, by knowing T_m and T_g and making some assumptions as to the variation of viscosity with temperature, a reasonable $\eta(T)$ relation can be generated.

Davies and various co-workers (Davies, 1976) have used such an approach to estimate the critical cooling rates of several pure liquid metals, such as Te, Ge, and Ni, and a number of metal alloys (e.g., 0.82Pd–0.18Si and 0.775Pd–0.06Ca–0.165Si). The agreement between calculated and observed critical cooling rates was usually within an order of magnitude, which is acceptable in light of the approximations used in the calculations. More important, such calculations seem to agree with the observed relative ease of glass formation of the various compositions.

One metal alloy composition, 0.778Au–0.138Ge–0.084Si, has been amenable to more accurate examination, since its viscosity above T_m and just above T_g has been determined experimentally (Polk and Turnbull, 1972; Chen and Turnbull, 1968). Using these viscosity data, Davies (1975) has calculated the critical cooling rate to be 3×10^7 K/sec, in reasonable agreement with the critical cooling rate of 2×10^6 to 1×10^7 K/sec estimated from the technique used to produce a glassy film of this alloy. Ramachandrarao *et al.* (1977) considered the same alloy and used the free volume theory to fit the experimental viscosity data to a smooth curve in order to estimate the viscosity at the temperature of the nose. Their estimate of the critical cooling rate was 4×10^7 K/sec, in good agreement with the previous estimates.

Tanner and Ray (1979) recently calculated the critical cooling rates for glass formation for the alloy compositions 0.65Zr–0.35Be and 0.63Ti–

0.37Be from consideration of heat transfer conditions and by using the kinetic treatment of glass formation. Because of the lack of information on the nucleation barrier in these systems, they could only show that the TTT curves constructed for the systems led to critical cooling rates within an order of magnitude of the values estimated experimentally. In order to fit the relative glass-forming tendencies observed for the alloys, Tanner and Ray assumed different nucleation barriers for them. Their results indicate also that allowing the crystal–liquid interfacial energy to change with temperature as suggested by Spaepen (1975) and Spaepen and Meyer (1976) leads to a poorer fit between the critical cooling rates determined from the experimental conditions and those calculated using the kinetic treatment of glass formation.

D. EXPERIMENTAL DETERMINATIONS OF NUCLEATION FREQUENCIES AND EFFECTS OF NUCLEATING HETEROGENEITIES

In calculating the cooling rates required to form glasses of various materials using the detailed analysis of crystallization statistics, it is necessary to characterize the system with respect to crystal growth rates and nucleation frequencies. For many materials it is relatively simple to determine the crystal growth rates over wide ranges of temperature; but the determination of the nucleation frequencies is much more tedious and has only been effected in a few cases. In most instances, the nucleation frequencies are calculated from classical nucleation theory, using relations such as Eqs. (2) and (4).

Since these relations are used so often in the detailed kinetic treatments of glass formation, it seems appropriate to review their applicability. In the first place, there is a considerable body of data on a variety of materials (including metals, salts and simple organic compounds), which is in general accord with classical theory (see summary in Jackson, 1965). Further, the celebrated measurements of Turnbull (1952) on the kinetics of crystal nucleation in mercury gave results in substantial accord with the theory. The principal cause of concern lies in the results of James (1974) on crystal nucleation in $Li_2O \cdot 2SiO_2$. As noted by Rowlands and James (1979a,b) and by Neilson and Weinberg (1979), the data on this material are inconsistent, both in the magnitude of the nucleation rate and in its temperature dependence, with the predictions of classical nucleation theory.

In contrast, the results of Gonzalez-Oliver and James (1980) on crystal nucleation in $Na_2O \cdot 2CaO \cdot 3SiO_2$ yielded results in substantial agreement with the predictions of classical nucleation theory. Other results on silicate systems in agreement with theory were provided by Klein and

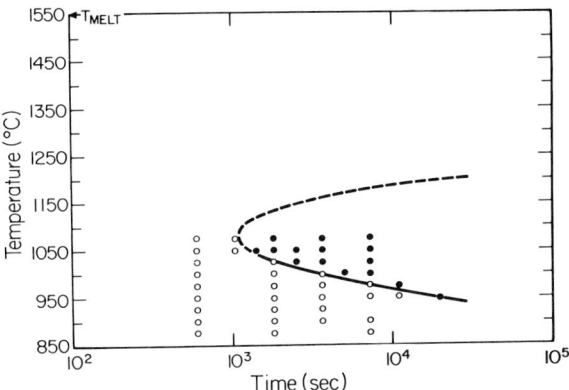

FIG. 12. Experimental isothermal time–temperature transformation (TTT) curve for anorthite: ●, crystallized samples, ○, glassy samples. The dashed part of the curve is extrapolated.

Uhlmann (1976) on lunar composition 70019 (see Table IV) by Klein et al. (1977) on $Na_2O \cdot 2SiO_2$, by Yinnon and Uhlmann (1983a,b) on a number of silicate compositions, and by Cranmer et al. (1981) on anorthite.

The results of James and his co-workers were obtained by directly measuring the number of crystals that formed in various times at a given nucleation temperature, while those of the other investigators were obtained from experimentally determined TTT curves. An example of the latter approach is presented in Fig. 12, where the time to develop a just-detectable degree of crystallinity in anorthite (determined from x-ray diffraction examination of heat-treated samples) was evaluated for a series of temperatures (Cranmer et al., 1981). Knowing the crystal growth rate at each temperature, the nucleation frequency was determined using Eq. (24). When $\log I_v \eta$ is plotted as a function of $T_r^{-3} \Delta T_r^{-2}$, a straight line of negative slope is obtained (Fig. 13), as predicted by classical nucleation theory. The slope of this line indicates a nucleation barrier of $82kT^*$, in excellent agreement with the value of $80kT^*$ obtained from the heating-rate dependence of the crystallization temperature on reheating the glass (see Section IV.E). The intercept of the plot in Fig. 13 indicates a preexponential constant [K in Eq. (4)] of about 10^{27} P/cm³ sec, in reasonable agreement with the value of 10^{30} P/cm³ sec expected from classical nucleation theory.

Besides this agreement with the predictions of theory, the results indicate that homogeneous nucleation represents the dominant contribution to the formation of crystals over the range of undercooling covered by the

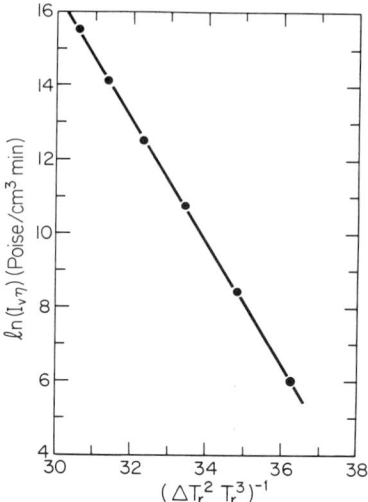

FIG. 13. log $I_v \eta$ versus $(\Delta T_r^2 T_r^3)^{-1}$ for anorthite determined from Fig. 12. The slope gives the nucleation barrier B; the intercept gives the value of K in Eq. (4).

data ($\Delta T_r = 0.26$ to 0.33). Similar conclusions were obtained by Klein and Uhlmann (1976) for lunar composition 70019 and by Klein et al. (1977) for $Na_2O \cdot 2SiO_2$.

The character of the agreement with nucleation theory found in these studies, as well as that of Gonzalez-Oliver and James (1980) on $Na_2O \cdot 2CaO \cdot 3SiO_2$, raises a question about the discrepancy found for $Li_2O \cdot 2SiO_2$. The range of reduced undercooling covered by the data on the last material ($\Delta T_r = 0.39$–0.47) represents a region of exceptionally large undercooling—well beyond that at which homogeneous nucleation has been observed in other materials. It is also noteworthy that the dependence of crystallization temperature on heating rate on reheating $Li_2O \cdot 2SiO_2$ glass indicates "normal" behavior with a nucleation barrier of about $60kT^*$. These results are, in fact, inconsistent with the reported nucleation kinetics of this material and leave unexplained its nucleation behavior. Pending the resolution of this issue for $Li_2O \cdot 2SiO_2$, it seems sensible to continue using classical nucleation theory in kinetic treatments of glass formation.

Turning now to the question of heterogeneous nucleation, it should be recognized that crystal nucleation in many (most) systems takes place heterogeneously—generally on second-phase particles that may be distributed through the bulk of the material or may be associated primarily with the external surfaces. The effects of such nucleation on the process of glass formation have been examined by Onorato and Uhlmann (1976) and by Yinnon and Uhlmann (1981).

Using the results of droplet nucleation experiments on a variety of liquids, Onorato and Uhlmann estimated a typical concentration of nucleating heterogeneities and assumed an average size of heterogeneity of 500 Å. The rates of heterogeneous nucleation were calculated using Eq. (5) for various values of the contact angle θ. These workers constructed TTT and CT curves for 10 systems using measured (where available) or calculated values for the crystal growth rate, viscosity, and nucleation barrier. The range of contact angles was varied between 40 and 160°.

Yinnon and Uhlmann (1981a) used the analysis of crystallization statistics to investigate the effects of various contact angles, concentrations of heterogeneities and distributions of heterogeneities on the crystallization that takes place during continuous cooling, and thus on the critical cooling rates required to form glasses of various materials. Particular attention was directed to anorthite ($CaO \cdot Al_2O_3 \cdot 2SiO_2$) and o-terphenyl, as representative of inorganic and organic glass-forming materials. The effects of depletion of the nucleating heterogeneities were included using Eq. (7).

Both studies have shown that heterogeneities characterized by a contact angle higher than about 100° have no effect on the critical cooling rates for glass formation, at least for concentrations of heterogeneities as large as $10^7/cm^3$. This is shown by the results in Fig. 14, where continuous cooling curves are illustrated for o-terphenyl with homogeneous nucle-

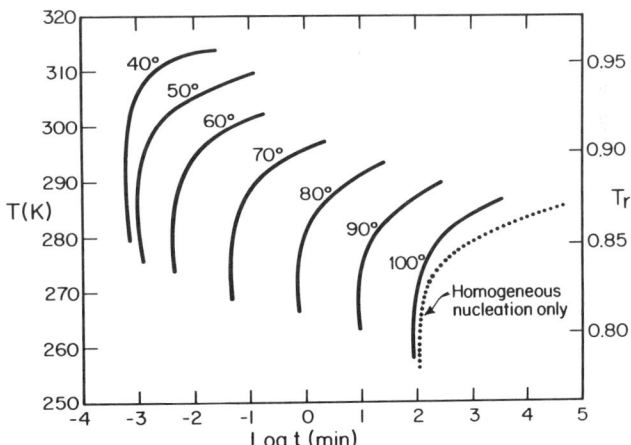

FIG. 14. Continuous cooling curves for homogeneous + bulk heterogeneous nucleation in o-terphenyl with contact angles as indicated, as well as for homogeneous nucleation only. Curves calculated using the analysis of crystallization statistics.

ation only and with homogeneous + heterogeneous nucleation with contact angles between 40 and 100°. The curve for homogeneous nucleation + heterogeneous nucleation with $\theta = 100°$ is very similar to that for homogeneous nucleation only, leading to very similar critical cooling rates. The effect of depletion of the nucleating heterogeneities is clearly shown in the crowding of the cooling curves for small contact angles.

Yinnon and Uhlmann (1981) calculated the nucleation frequency in a system containing heterogeneities distributed with regard to their contact angles. The continuous cooling curves for such a case were found to be dominated by a very small fraction of the heterogeneities having the smallest contact angles. This result confirmed that a small change in the contact angle leads to a larger change in the nucleation frequency than a large change in the concentration of heterogeneities.

In a system containing heterogeneities distributed in contact angle, the nucleation process on cooling is dominated by different types of heterogeneities at different temperatures. As the heterogeneities with small contact angle are depleted, heterogeneities with larger and larger contact angle become effective. Such effects are manifest by a continuous change in the slope of a plot of $\log I_v \eta$ versus $(\Delta T_r^2 T_r^3)^{-1}$, as shown in Fig. 15, reflecting the progressive increase in the value of ϕ in Eq. (5). A change in the dominant nucleation mechanism from heterogeneous to homogeneous will also be reflected in curvature in plots such as those in Fig. 15. Thus a pronounced curvature in experimentally determined $\log I_v \eta$ versus $(\Delta T_r^2 T_r^3)^{-1}$ relations can be taken as indicating the presence of heterogeneities, possibly with a distribution of contact angles.

It has been shown for a wide variety of liquids that steady state nucleation is achieved only after some transient time and that an expression such as Eq. (8) should be used to describe such nucleation [see Gutzow (1980) and Gonzalez-Oliver and James (1980)]. Accordingly, the analysis of crystallization statistics was modified to include transient nucleation, as represented by Eqs. (8) and (9) (Yinnon and Uhlmann, 1982).

In the case of homogeneous nucleation, the critical cooling rates for most materials, as represented by anorthite and o-terphenyl, are sufficiently low that the transient time τ is considerably smaller than typical time interval used in the kinetic calculations of crystallization on cooling. Hence steady state conditions effectively prevail during all the cooling time, and Eq. (4) can be used with confidence. When heterogeneous nucleation is introduced, and faster cooling rates are required, the value of τ can become close to the typical time intervals in the kinetic calculations, and the nucleation rate at high undercoolings can become significantly time-dependent. In such cases, however, the crystallization process is dominated by heterogeneous nucleation that takes place at tempera-

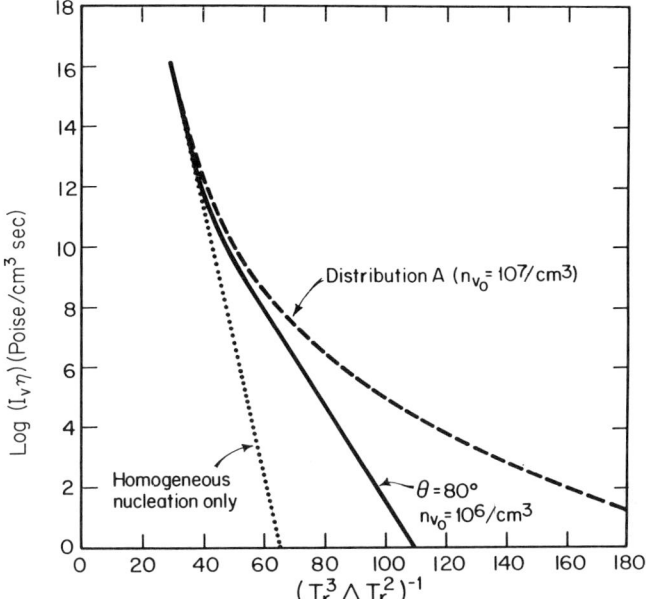

Fig. 15. $\log_{10} I_V \eta$ versus $(\Delta T_r^2 T_r^3)^{-1}$ for homogeneous nucleation only (dotted line), homogeneous + heterogeneous nucleation with a single type of heterogeneity ($\theta = 80°$) (solid line), and homogeneous + heterogeneous nucleation with a Gaussian distribution of heterogeneities ($60° < \theta < 120°$). Calculated using the analysis of crystallization statistics.

tures closer to the melting point, where τ is small (because of the low viscosity); and thus the critical cooling rates even with potent heterogeneous nucleation in both representative materials were found to be affected only to a small extent by transients.

It remains possible that for some materials transients may be of some significance, causing a decrease in the calculated critical cooling rates for glass formation. Such cases are expected to be the exceptions rather than the rule, however, since both the transient time for nucleation and the overall crystallization time increase linearly with increasing viscosity. In nearly all cases, the transient times will be large only when the time to reach a given degree of crystallization is also large, and hence the steady state expression for the nucleation rate will remain a useful representation.

E. Crystallization on Reheating a Glass

In addition to describing crystallization that takes place on cooling a liquid and providing estimates of the cooling rates required to form glasses, the analysis of crystallization statistics can also be used to describe crystallization on reheating a glass. This application of the analysis was originally motivated by the work of Hruby (1972), who carried out studies of the glass-forming abilities of materials by quenching samples to temperatures well below the glass transition, subsequently reheating them—usually at a slower rate—and measuring the temperature of the crystallization exotherm with differential thermal analysis (DTA). In terms of the temperature of crystallization (T_{cr}), Hruby suggested a parameter K_{gl} for use as a measure of the glass-forming tendencies of materials:

$$K_{gl} = \frac{T_{cr} - T_g}{T_m - T_{cr}}. \qquad (34)$$

K_{gl} values were determined for a series of As_2Te_3 compositions with admixtures of Si, Ge, I, and Tl and for a series of $CdAs_2$ compositions with admixtures of Ge and P. Good agreement was found between the K_{gl} values and subjective experimental experience on the relative ease of forming the different compositions as glasses.

Use of the K_{gl} parameter is based on the concept that the thermal stability of a glass on subsequent reheating is directly proportional to the ease of its formation. Implicit in this approach is the assumption that all glasses are in comparable states at T_g. As we shall see, however, this is not in general the case. The detailed state of crystallinity in a nominally glassy sample—i.e., the distribution of stable nuclei and crystallites—depends on the kinetic characteristics of the material and on the rate at which it was cooled.

More generally, differential thermal analysis (DTA) has been used extensively as a tool for studying the devitrification of oxide and chalcogenide glasses as well as the stability of metallic glasses (see references 3–14 in Henderson, 1979b). The development of nuclei and small crystallites within an amorphous body during cooling and subsequent reheating in DTA can be simulated using the analysis of crystallization statistics. The distribution of nuclei and small crystallites are calculated using previously measured crystal growth rates and viscosity–temperature relations and a nucleation barrier estimated from the entropy of fusion [see Eq. (3)]. The volume fraction crystallized is then calculated as a function of temperature by summing over all the crystallites [see Eq. (27)].

It is found from such treatments that glasses at temperatures below the glass transition contain some finite number of crystal nuclei and that the distribution of these nuclei depends on the material characteristics and the thermal history of the sample. On reheating, three processes take place: first, the nuclei already present grow at rates determined by the increasing temperature; second, new nuclei are formed; and third, since the size of the critical nucleus (the smallest crystalline embryo stable at a given temperature) increases with increasing temperature, some nuclei that were stable (supercritical in size) at low temperature become unstable at higher temperature and melt out. Those nuclei are assumed to melt completely and are no longer included in the calculations. Although important conceptually, the contribution of melting nuclei to the overall crystallization behavior of the glass is very small except for conditions of very rapid heating.

When such treatments are carried out, a maximum in the calculated overall crystallization rate is clearly seen, which can be compared with the maximum of the DTA crystallization peak. When the crystal growth rate and viscosity are known as functions of temperature, the variation with heating rate of the temperature of maximum crystallization rate T_{cr} can be used to evaluate the barrier to crystal nucleation B. This is done simply by calculating the crystallization temperature for the experimental heating rates for various values of B. The value that provides the best fit to the measured dependence of T_{cr} on heating rate is taken as the nucleation barrier.

This method has been applied to a variety of materials, including o-terphenyl (Onorato et al., 1980), a number of lunar compositions (Yinnon et al., 1980), alkali disilicate compositions (Yinnon and Uhlmann, 1983a), as well as a number of compositions in the albite–anorthite system (Cranmer et al., 1981; Yinnon and Uhlmann, 1983b). Figure 16 shows the experimental values of T_{cr} as a function of heating rate for o-terphenyl together with the values calculated for a nucleation barrier of $50kT^*$. The analysis is seen to predict closely both the values of T_{cr} and the dependence of T_{cr} on heating rate.

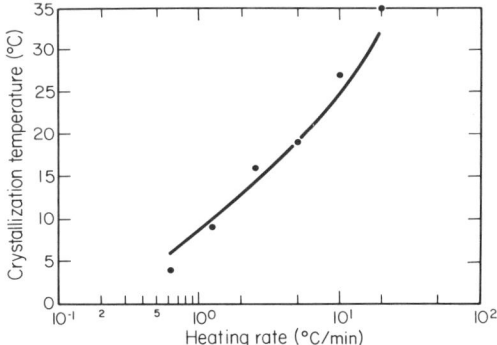

FIG. 16. Measured (points) and calculated (curve) crystallization temperatures in DTA versus heating rate relations for o-terphenyl. The curve was calculated using the analysis of crystallization statistics with $B = 50kT^*$.

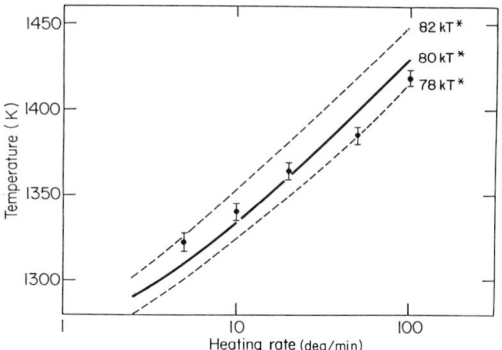

FIG. 17. Measured (points) and calculated (curves) crystallization temperatures in DTA versus heating rate relations for anorthite. The curves were calculated using the analysis of crystallization statistics with nucleation barrier B, as indicated. Best agreement is obtained for $B = 80kT^*$.

Figure 17 shows the calculated and measured variation of crystallization temperature with heating rate for anorthite. In addition to the "best fit" nucleation barrier of $80kT^*$, curves are also shown for barriers of $78kT^*$ and $82kT^*$ to illustrate the sensitivity of the predictions to the magnitude of the nucleation barrier. The value of $80kT^*$ determined in this way is in good agreement with the value of $82kT^*$ obtained from experimental TTT curves (see Section IV.D).

This technique for the determination of the nucleation barrier uses the kinetic information obtained from DTA without undue approximations. Many workers have used the Johnson–Mehl–Avrami equation

$$V_c/V = 1 - \exp(-kt^n) \tag{35}$$

to determine the activation energy of the rate coefficient k [$k = k_0 \exp(-Q/RT)$], and have attributed the activation energy thus obtained to the crystallization process. It should be recognized, however, that Eq. (35) applies only to isothermal phase transformations where the nucleation rate or growth rate may change with time only and cannot be used for the nonisothermal crystallization observed in DTA [see the discussion in Christian (1975) and Turnbull (1956)]. If the appropriate expressions for the nucleation barrier [Eq. (4)] and the crystal growth rate [Eqs. (10), (11), or (14)] are substituted into Eq. (23), it is apparent that only with unrealistic assumptions can Eq. (35) be obtained. For further discussion of this point, see Henderson (1979a,b) and Yinnon and Uhlmann (1983c). The approach of crystallization statistics uses a numerical solution to Eq. (23) and thus avoids these approximations.

The results of applying the analysis of crystallization statistics to a variety of materials have indicated that the crystallization temperature depends on the cooling rate originally used to form the glass, as well as on the heating rate. Such dependences had been noted experimentally by Thornburg (1974) and Lasocka (1976). The dependence on cooling rate is less pronounced than that on heating rate, and is most apparent when poor glass formers are heated at relatively fast rates. Onorato *et al.* (1980) have shown by calculations that the crystallization temperature becomes independent of cooling rate for rates that produce volume fractions crystallized smaller than 10^{-10} at T_g, and have indicated that such a fraction crystallized could be taken as an effective definition of a glass. That is, the thermal stability of the glass would then not depend on the conditions of its initial formation.

Finally, the results of the calculations indicate a general correlation between values of K_{gl} for a given heating rate and the calculated critical cooling rates for forming different materials as glasses. This correlation is only approximate, however; and K_{gl} does not provide a reliable ranking of the glass-forming ability of various materials. It reasonably differentiates between materials that form glasses readily and those that require extreme quenching, and it does serve as a rough, general guide to behavior.

V. Glass-Forming Ability and Material Properties

It has been shown in previous sections that the glass-forming ability of materials can be characterized by a single parameter, the critical cooling rate R_c. The slower the critical cooling rate, the greater is the glass-forming tendency of the material. By analyzing the factors that affect the critical cooling rate, one obtains insight into the material properties that favor the formation of the materials as glasses. These characteristics include

(a) a high viscosity at the nose of the TTT or CT curves requires either a high viscosity at the melting point, as is common for good oxide glass formers such as SiO_2, or a viscosity that increases rapidly with decreasing temperature, as is often found for organic glass formers;

(b) the absence of nucleating heterogeneities, or at least the absence of potent nucleating heterogeneities characterized by a small contact angles;

(c) a large barrier B to crystal nucleation, which implies a large crystal–liquid interfacial energy; and

(d) the requirement of a sizable redistribution of solute for crystallization, such as that found for compositions that differ appreciably from those of the relevant crystalline phases.

The importance of a high viscosity at the melting point or liquidus temperature was recognized by Rawson (1956). It was suggested that

> A melt which forms a glass must have a relatively high viscosity at its liquidus temperature.... Factors which tend to depress the liquidus temperature without at the same time lowering the viscosity, or which increase the viscosity at the liquidus temperature will favor glass formation.

The requirement that the viscosity be large at a temperature between the melting point and the glass transition temperature shifts attention to the ratio of T_g/T_m. The viscosity at T_g being constant, the higher the ratio T_g/T_m, the higher will be the viscosity at the nose of the TTT or CT curves. The importance of a low melting point or liquidus temperature for glass formation within a given class of materials has been noted by several investigators. By comparing experimental data, it was also noted that a good glass-forming ability is correlated with a T_g/T_m ratio of $\frac{2}{3}$ or more for many simple organics (Kauzmann, 1948), as well as for chalcogenide and oxide (Sakka and Mackenzie, 1971) liquids. This so-called $\frac{2}{3}$ rule should not, however, be considered more than a crude indication of glass-forming ability: Recall, for example, the relatively low value of T_g/T_m for the excellent glass former GeO_2. The low values of the T_g/T_m ratio for glassy metals are consistent with their relatively poor glass-forming ability. Davies (1976) has shown that the critical cooling rates of glassy metals scale with the ratio T_g/T_m.

Although the viscosity is the most important kinetic parameter controlling the rate of crystallization, solute redistribution in the liquid may also have a large effect on crystallization kinetics. Thus a liquid having a high T_g/T_m ratio that requires substantial solute redistribution should be an excellent glass former. Compositions near very low-lying eutectics would, then, be good glass formers. This has been observed for metallic glasses, where it was noted (Turnbull, 1974) that the metal alloys that can most easily be cooled to the glassy state are composed of (a) noble or transition metals with filled or nearly filled d bands, alloyed with (b) metalloid or early transition metals. The alloying is usually accompanied by relatively large negative heats of mixing and often by large contractions of volume. It is thus expected that short-range compositional ordering plays an important role in liquid transport. Indeed, as noted by Turnbull (1974):

> The systems which seem most prone to glass formation in melt quenching are those having compositions lying at and near to abnormally low-lying eutectics. These abnormal eutectics would seem to reflect that the alloys are exceptionally stable in their amorphous states.

It is also worth noting that few alloys containing only two elements can be formed as glasses with small cooling rates. The addition of a third element reduces the liquidus temperature and enhances the tendency toward glass formation (Chen, 1980). This is similar to the common practice of oxide glass technologists in formulating glass compositions containing a diversity of constituents. As they have long recognized, such formulations promote glass formation in addition to providing the desired combination of material properties.

Turnbull and Cohen (1958) suggested that T_g should scale roughly as the cohesive energy and hence roughly as the boiling point T_B for materials of a given type. This suggestion is supported by data on organic liquids, which indicate that bulk glasses are usually formed at reasonable cooling rates when $T_B/T_M \geq 2$. For some materials with $T_B/T_M \approx 1.8$ (e.g., cyclopentane and m-xylene), glasses can be obtained of droplet-sized samples. For materials with $T_B/T_M \leq 1.6$, glasses are usually not obtainable, even in droplet form. Data on boiling points are not generally available for inorganic compounds; but data on $ZnCl_2$ (Wong and Angell, 1977), which readily forms a glass in bulk but for which $T_B/T_M = 1.7$, suggest that the criteria may require modification for such liquids.

If the ease of glass formation correlates with T_B/T_M, and T_B is taken as a measure of the cohesive energy, then at a given level of cohesive energy, glass formation should increase with decreasing T_M and hence with decreasing $\Delta H_f/\Delta S_f$. In other words, "the glass-forming tendency of a substance in a particular class is greater the less is the energy, at constant cohesive energy, necessary to produce a given amount of disorder" (Turnbull and Cohen, 1958). Hence asymmetry of organic molecules is viewed as favorable to glass formation, a view in general accord with experience.

A high nucleation barrier is associated with large crystal–liquid interfacial energies. While these energies cannot yet be estimated from first principles, they do seem to scale with the entropy of fusion, at least for materials within a given class [see Eq. (31)]. The absence of nucleating heterogeneities can be associated with compositions that effectively dissolve most second-phase material. Examples of such compositions are provided by silicate melts containing sizable concentrations of PbO. The condition of few nucleating heterogeneities can often be achieved by superheating the melt well above the melting point before commencing cooling.

Before we close, it should be observed again that kinetic treatments of glass formation seem to offer particular promise for describing the formation of glasses. By directing attention to the kinetic processes involved in crystallization, their essential concern deals with the question, "How fast must a liquid be cooled in order to form a glass?" In answering this question the treatments assume that the atom motions involved in crystallization are similar to those involved in viscous flow or interdiffusion. This is expected for most liquids, for which reorientation of the molecules or breaking of directional bonds between atoms must precede incorporation of the molecule or atom in the crystal. For monatomic liquids, such as pure metals and rare gas liquids, the interface process can be quite different from flow in bulk liquid; and crystallization in these materials could take place at temperatures near or below T_g. For this reason as well as their small estimated T_g/T_m values, it may be impossible to form these materials as glasses by cooling from the liquid state.

Despite the success to date of the kinetic treatments, the ultimate treatments of glass formation will undoubtedly expand on present kinetic treatments and describe the processes of transport and crystallization in terms of structure and interatomic interactions. Such treatments remain objects of future work and may well be forthcoming first in the case of molecularly simple liquids such as metals and simple organics.

Acknowledgments

Financial support for our development of kinetic treatments of glass formation was provided by the National Aeronautics and Space Administration and by the National Science Foundation. This support is gratefully acknowledged, as are the contributions of Dr. R. W. Hopper, Dr. G. W. Scherer, Dr. L. C. Klein, Dr. P. I. K. Onorato, and Dr. D. Turnbull.

References

Adam, G., and Gibbs, J. H. (1965). *J. Chem. Phys.* **43,** 139.
Avrami, M. (1939). *J. Chem. Phys.* **7,** 1103.
Avrami, M. (1940). *J. Chem. Phys.* **8,** 212.
Avrami, M. (1941). *J. Chem. Phys.* **9,** 177.
Bagley, B. G. (1965). *Nature (London)* **208,** 674.
Becker, E., and Döring, W. (1935). *Ann. Phys.* **24,** 719.
Bernal, J. D. (1960). *Sci. Am.* **203** (August).
Bottinga, Y., and Weill, D. F. (1972). *Am. J. Sci.* **272,** 438.
Calvert, P. D., and Uhlmann, D. R. (1972). *J.Crystal Growth* **12,** 291.
Chaudhari, P., and Turnbull, D. (1978). *Science* **199,** 11.
Chen, H. S. (1980). *Rep. Prog. Phys.* **43,** 353.
Chen, H. S., and Turnbull, D. (1968). *J. Chem. Phys.* **48,** 2560.
Christensen, N. H., Cooper, A. R., and Rawal, B. S. (1973). *J. Am. Ceram. Soc.* **56,** 557.
Christian, J. W. (1975). "The Theory of Transformations in Metals and Alloys." 2nd ed. Pergamon, Oxford.
Cohen, M. H., and Turnbull, D. (1964). *Nature (London)* **203,** 964.
Cranmer, D., Salomaa, R., Yinnon, H., and Uhlmann, D. R. (1981). *J. Non-Cryst. Solids* **45,** 127.
Cukierman, M., Lane, J. W., and Uhlmann, D. R. (1973). *J. Chem. Phys.* **59,** 3639.

Davies, H. A. (1975). *J. Non-Cryst. Solids* **17**, 266.
Davies, H. A. (1976). *Phys. Chem. Glasses* **17**, 159.
Dietzel, A. (1948). *Glastech. Ber.* **22**, 41.
Dietzel, A., and Wickert, H. (1956). *Glastech. Ber.* **29**, 1.
Frank, F. C. (1952). *Proc. R. Soc. London Ser. A* **215**, 43.
Gibbs, J. H., and DiMarzio, E. A. (1958). *J. Chem. Phys.* **28**, 373.
Gilmer, G. H. (1976). *J. Crystal. Growth* **35**, 15.
Gilmer, G. H. (1980). *J. Crystal. Growth* **49**, 465.
Gonzalez-Oliver, C. J. R., and James, P. F. (1980). *J. Non-Cryst. Solids* **38–39**, 699.
Grange, R. A., and Kiefer, J. M. (1941). *Trans. ASM* **29**, 85.
Grest, G. S., and Cohen, M. H. (1980). *Phys. Rev. B* **21**, 4113.
Gutzow, I. (1980). *Contemp. Phys.* **21**, 121.
Gutzow, I., and Avramov, I. (1977). *Proc. Int. Congr. Glass, 11th, Prague* pp. 299–308.
Gutzow, I., and Toschev, S. (1972). In "Advances in Nucleation and Crystallization in Glasses" (L. L. Hench and S. W. Freiman, eds.), pp. 10–15. American Ceramic Society, Columbus, Ohio.
Havermans, A. C. J., Stein, H. N., and Stevels, J. M. (1970). *J. Non-Cryst. Solids* **5**, 66.
Henderson, D. W. (1979a). *J. Thermal Analysis* **15**, 325.
Henderson, D. W. (1979b). *J. Non-Cryst. Solids* **30**, 301.
Hopper, R. W., Scherer, G. W., and Uhlmann, D. R. (1974a). *J. Non-Cryst. Solids* **15**, 45.
Hopper, R. W., Onorato, P., and Uhlmann, D. R. (1974b). *Proc. Lunar Sci. Conf., 5th* pp. 2257–2273. Pergamon, Oxford.
Horvay, G., and Cahn, J. W. (1961). *Acta Metall.* **9**, 695.
Hruby, A. (1972). *Czech. J. Phys.* **B22**, 1187.
Ivantsov, G. P. (1947). *Dokl. Akad. Nauk SSSR* **58**, 567.
Jackson, K. A. (1958). In "Growth and Perfection of Crystals" (R. H. Doremus, B. W. Roberts, and D. Turnbull, eds.). Wiley, New York.
Jackson, K. A. (1965). In "Nucleation Phenomena" (A. S. Michaels, ed.), pp. 37–40. American Chemical Society, Washington, D.C.
Jackson, K. A. (1967). In "Progress in Solid State Chemistry" (H. Reiss, ed.), pp. 53–80. Pergamon, Oxford.
James, P. F. (1974). *Phys. Chem. Glasses* **15**, 95.
Johnson, W. A., and Mehl, K. F. (1939). *Trans. Am. Inst. Min. Metall. Eng.* **135**, 416.
Kauzmann, H. (1948). *Chem. Rev.* **43**, 219.
Kingery, W. D., Bowen, H. K., and Uhlmann, D. R. (1976). "Introduction to Ceramics," 2nd ed., p. 99. Wiley, New York.
Klein, L., and Uhlmann, D. R. (1976). *Proc. Lunar Sci. Conf., 7th* pp. 1113–1121. Pergamon, Oxford.
Klein, L. C., Handwerker, C. A., and Uhlmann, D. R. (1977). *J. Crystal. Growth* **42**, 47.
Lasocka, M. (1976). *J. Mat. Sci.* **11**, 1771.
Milner, D. F. R., and Wright, A. C. (1980). *J. Non-Cryst. Solids* **42**, 97.
Morey, G. W. (1934). *J. Am. Ceram. Soc.* **17**, 315.
Mozzi, R. L., and Warren, B. E. (1969). *J. Appl. Cryst.* **2**, 164.
Neilson, G. F., and Weinberg, M. C. (1979). *J. Non-Cryst. Solids* **34**, 137.
Onorato, P. I. K., and Uhlmann, D. R. (1976). *J. Non-Cryst. Solids* **22**, 367.
Onorato, P. I. K., Uhlmann, D. R., and Hopper, R. W. (1980). *J. Non-Cryst. Solids* **41**, 189.
Onorato, P. I. K., Scherer, G. W., and Uhlmann, D. R. (1981). A kinetic treatment of glass formation: VI. Simplified model (to be published).
Polk, D. E., and Turnbull, D. (1972). *Acta Metall.* **20**, 493.
Porai-Koshits, E. A. (1977). *J. Non-Cryst. Solids* **25**, 86.

Ramachanrarao, P., Cantor, B., and Cahn, R. W. (1977). *J. Non-Cryst. Solids* **24**, 109.
Rawson, H. (1956). *Trav. Cong. Int. Verre, 4th* p. 62. ICG, Paris.
Robinson, H. A. (1965). *J. Phys. Chem. Solids* **26**, 209.
Rowlands, E. G., and James, P. F. (1979a). *Phys. Chem. Glasses* **20**, 1.
Rowlands, E. G., and James, P. F. (1979b). *Phys. Chem. Glasses* **20**, 9.
Sakka, S., and Mackenzie, J. D. (1971). *J. Non-Cryst. Solids* **6**, 145.
Scherer, G. W., and Uhlmann, D. R. (1975a). *J. Crystal. Growth* **29**, 12.
Scherer, G. W., and Uhlmann, D. R. (1975b). *J. Crystal. Growth* **30**, 304.
Scherer, G. W., and Uhlmann, D. R. (1977). *J. Non-Cryst. Solids* **23**, 59.
Shaw, H. R. (1972). *Am. J. Sci.* **272**, 870.
Spaepen, F. (1975). *Acta Metall.* **23**, 729.
Spaepen, F., and Meyer, R. B. (1976). *Scripta Metall.* **10**, 257.
Stanworth, J. E. (1946). *J. Soc. Glass Technol.* **30**, 56.
Stevels, J. M. (1957). *Handbook Phys.* **20**, 350.
Sun, K. H., and Huggins, M. L. (1947). *J. Phys. Chem.* **51**, 438.
Swift, H. R. (1947). *J. Am. Ceram. Soc.* **30**, 165.
Tammann, G. (1933). "Der Glaszustand Leopold Voss." Leipzig [English translation of section quoted by Weyl and Marboe (1962)].
Tammann, G., and Elbrachter, A. (1932). *Z. Anorg. Allgem. Chem.* **207**, 268.
Tanner, L. E., and Ray, R. (1979). *Acta Metall.* **27**, 1727.
Thompson, C. V., and Spaepen, F. (1979). *Acta Metall.* **27**, 1855.
Thornberg, D. R. (1974). *Mat. Res. Bull.* **9**, 1481.
Tilton, L. W. (1957). *J. Res. Nat. Bur. Std.* **59**, 139.
Turnbull, D. (1948). *Trans. Am. Inst. Min. Metall. Eng.* **175**, 774.
Turnbull, D. (1952). *J. Chem. Phys.* **20**, 411.
Turnbull, D. (1956). *Solid State Phys.* **3**, 225–306.
Turnbull, D. (1969). *Contemp. Phys.* **10**, 473.
Turnbull, D. (1974). *J. Phys.* **35**, Colloq. 4, 1–10.
Turnbull, D., and Cohen, M. H. (1958). *J. Chem. Phys.* **29**, 1049.
Uhlmann, D. R. (1969). *Mat. Sci. Res.* **4**, 172–197.
Uhlmann, D. R. (1972a). *J. Non-Cryst. Solids* **7**, 337.
Uhlmann, D. R. (1972b). *In* "Advances in Nucleation and Crystallization in Glasses" (L. L. Hench and S. W. Freiman, eds.), pp. 91–115. American Ceramic Society, Columbus, Ohio.
Uhlmann, D. R. (1980). *J. Non-Cryst. Solids* **42**, 119.
Uhlmann, D. R., and Kolbeck, A. G. (1976). *Phys. Chem. Glasses* **17**, 146.
Uhlmann, D. R., Onorato, P. I. K., and Scherer, G. W. (1979). *Proc. Lunar Planetary Sci. Conf., 10th* pp. 375–381. Pergamon, Oxford.
Uhlmann, D. R., Yinnon, H., and Fang, C. Y. (1981). *Proc. Lunar Planetary Sci. Conf. 12th* pp. 281–288.
Uhlmann, D. R., Scherer, G. W., and Klein, L. C. (1983). On the relation between interdiffusion and viscosity in silicate melts, *J. Non-Cryst. Solids* (to be published).
Vreeswijk, J. C. A., Gossink, R. G., and Stevels, J. M. (1974). *J. Non-Cryst. Solids* **16**, 15.
Vogel, W. (1977). *J. Non-Cryst. Solids* **25**, 170.
Warren, B. E. (1937). *J. Appl. Phys.* **8**, 645.
Warren, B. E. (1941). *J. Amer. Ceram. Soc.* **24**, 256.
Weyl, W. A., and Marboe, E. C. (1962). "The Constitution of Glasses," Vol. 1, pp. 222–230. Wiley, New York.
Williams, M. C., Landel, R. F., and Ferry, J. D. (1955). *J. Am. Chem. Soc.* **77**, 3701.

Wong, J., and Angell, C. A. (1977). "Vitreous State Spectroscopy," Chapter 1. Dekker, New York.
Yinnon, H., Roshko, A., and Uhlmann, D. R. (1980). *Proc. Lunar Planetary Sci. Conf., 11th* pp. 197–211. Pergamon, Oxford.
Yinnon, H., and Uhlmann, D. R. (1981). *J. Non-Cryst. Solids* **44,** 37.
Yinnon, H., and Uhlmann, D. R. (1982). *J. Non-Cryst. Solids* **50,** 189.
Yinnon, H., and Uhlmann, D. R. (1983a). The nucleation barrier of alkali disilicate (to be published).
Yinnon, H., and Uhlmann, D. R. (1983b). The nucleation barrier of silicates of the albite–anorthite compositions (to be published).
Yinnon, H., and Uhlmann, D. R. (1983c). Applications of thermoanalytical techniques to the study of crystallization kinetics in glass-forming liquids, *J. Non-Cryst. Solids* (to be published).
Zachariasen, W. H. (1932). *J. Am. Chem. Soc.* **54,** 3841.
Zeldovich, J. B. (1943). *Acta Phys.-Chim. URSS* **18,** 1.

CHAPTER 2

Unusual Methods of Producing Glasses

G. W. Scherer
P. C. Schultz

RESEARCH AND DEVELOPMENT DIVISION
CORNING GLASS WORKS
CORNING, NEW YORK

I. Introduction	49
II. Unconventional Melting	50
A. Quenching	50
B. Melting under High Pressure	58
C. Metal Evaporation Technique	59
III. Deposition Methods	60
A. Nonreactive Deposition	60
B. Reactive Deposition	66
IV. Solution Methods	86
V. Solid State Transformations	89
A. Glass Formation by Radiation	89
B. Shock-Induced Glass Formation	90
References	97

I. Introduction

This chapter is concerned with the many unconventional ways that have been devised to produce both usual and unusual inorganic glasses. We shall describe methods that produce glasses from materials that ordinarily crystallize as their melts are cooled, as well as methods that impart special properties to more familiar glass compositions. We regard as a glass any solid that lacks order in its atomic arrangement beyond the two or three nearest neighbors of any atom.

Some of the methods to be described are difficult to categorize, so that any assignment to a class of methods leads to some ambiguity. Nevertheless, we have chosen a scheme of organization for this chapter that, although imperfect, provides a rational framework for discussion. The principal classes of novel glass-forming methods may be divided as follows:

Section II. *Unconventional melting* refers to techniques for rapid quenching from the melt (such as splat cooling) and unusual ways of producing the melt (such as melting under pressure).

Section III. *Deposition methods* fall into two categories—reactive and nonreactive. The latter (e.g., sputtering) are used to produce films, whereas some of the former have been used to make films (e.g., anodic oxidation) and bulk glasses (e.g., flame oxidation). Of special interest are the reactive vapor deposition processes for making optical waveguides.

Section IV. *Solution methods* have been used to produce gels that may be converted to powdered glass with excellent purity and homogeneity; some approaches produce objects with complex shapes from molded gels.

Section V. *Solid state transformations* from the crystalline to the glassy state have been shown to result from heavy doses of radiation and from exposure to shock waves.

There is an enormous literature associated with these glass-forming methods, and we cannot claim to have uncovered all of it. Since the review by Secrist and Mackenzie (1964), the level of effort in the production of unusual glasses by novel methods has risen dramatically. The long list of references we have compiled reflects the magnitude of that effort, and yet we have only collected some of the more interesting examples and reviews from among many others.

II. Unconventional Melting

A. Quenching

The object of quenching a melt is to cool it to the glass transition temperature so quickly that no significant crystallization will occur. For excellent glass formers, such as SiO_2 or B_2O_3, an air quench is sufficient; that is, if the melt is removed from the furnace and allowed to cool by radiation and convection in the air, no crystallization will occur. At the other extreme, very fluid melts, such as metal alloys, will form glasses only when subjected to cooling rates on the order of 10^7 degrees per second. In this section we shall briefly describe the quenching techniques that have been used to produce amorphous materials in a broad range of systems.

1. Liquid Quenching

The most obvious approach to achieving higher cooling rates than in an air quench is to pour the melt into a liquid bath, such as water, liquid nitrogen, or mercury. This technique was used, for instance, by Rao (1963) in a study of the K_2O–SiO_2–TiO_2 system. Sarjeant and Roy (1969) delivered a stream of molten droplets from a Verneuil crystal-growing apparatus into liquid N_2 or water. These authors also used a 10 kVA rf plasma torch with powdered oxide suspended in the argon feed gas to produce molten droplets to be quenched in the liquid baths. With the Verneuil system they produced glasses in the following binary systems: TiO_2 plus BaO, CaO, SrO; Al_2O_3 plus Y_2O_3, Gd_2O_3; MgO–SiO_2.

2. Splat Cooling

A very popular technique for rapid quenching from the melt is "splat cooling," which was invented by Duwez et al. (1960). Using this technique these authors were able to produce noncrystalline metal alloys, a feat that had previously been achieved only by using vapor deposition (Buckel, 1954). The splat cooling apparatus developed by Duwez and Willens (1963) is shown in Fig. 1. The sample is melted in a graphite crucible that has a hole in the bottom; surface tension prevents the melt from dripping through. A chamber above the sample is partitioned with a

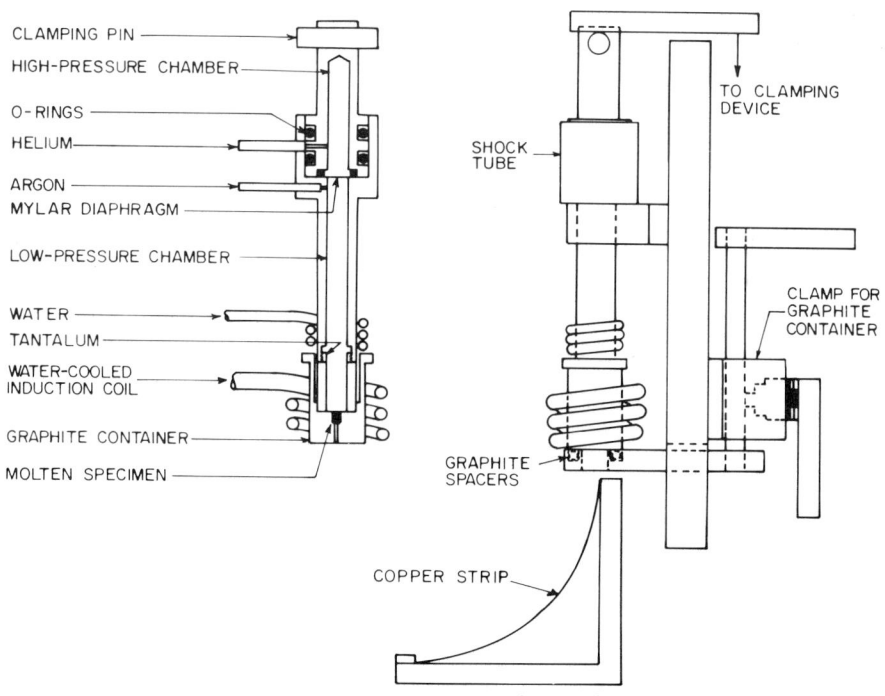

FIG. 1. Schematic drawing of rapid cooling apparatus. (From Duwez and Willens, 63.)

Mylar diaphragm. The upper chamber is pressurized with a gas until the diaphragm bursts and the resulting shock wave propels the molten sample onto a curved copper plate arranged below the crucible. The sample spreads into a thin sheet and cools very rapidly by conduction.

The extraordinary cooling rates obtainable by splat cooling (10^5-10^9 °C/sec) depend on the use of conductive rather than convective heat transfer. This technique produces thin flakes, or "splats," micrometers thick, in good thermal contact with a cold substrate. The quality of the contact between splat and substrate has been shown to be crucial in achieving the highest cooling rates, both in theoretical (Ruhl 1967) and experimental (Sarjeant and Roy, 1967; Scott, 1974) studies. Ruhl (1967) considered a number of other variables, such as initial splat temperature and substrate material and temperature, and showed them to have a much weaker effect on cooling rate.

A different apparatus for splat cooling was invented by Pietrokowsky (1963). As shown in Fig. 2 a molten droplet falls through the beam of a photocell that triggers a piston. The droplet is caught between the piston and a stationary anvil and is smashed into a thin plate. Recent versions (Krepski *et al.*, 1975; Veltri *et al.*, 1979) have used a laser as a heat source. The advantage of this technique is that it produces plates of uniform thickness that are useful for physical property measurements. In contrast,

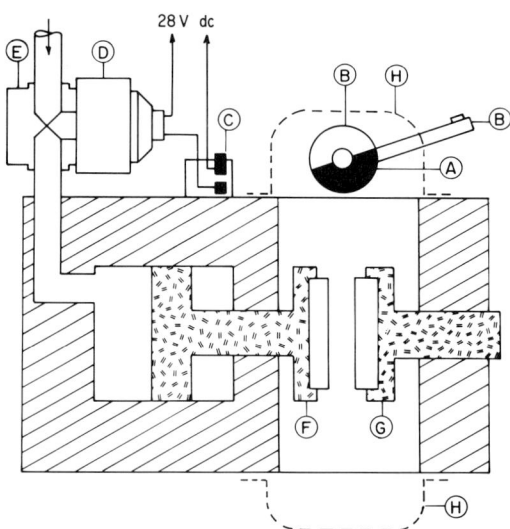

FIG. 2. Rapid cooling equipment, main assembly: A, graphite crucible, B crucible holder assembly and integral trip arm, C, electrical switch, D, energizing solenoid, E, high-speed valve, F, driving anvil, G, reference anvil, and H, cover-plate location. (From Pietrokowsky, 1963.)

the "gun" technique of Duwez *et al.* produces splats whose thicknesses vary drastically and that contain holes. The disadvantage of the technique is that lower cooling rates ($\sim 10^5$ °C/sec) result from the increased thickness of the splats. An analysis of the coolings rates obtainable by both techniques was performed by Bletry (1973). He showed that the solidification of the melt would stop the piston, thereby setting the minimum thickness obtainable. His calculated values were in good agreement with experimental observations. He also demonstrated that the splat thickness produced by the gun technique should be an order of magnitude less than by the piston-and-anvil device.

Chen and Miller (1976) report a method for making amorphous metallic ribbons with approximate dimensions of 0.5 mm × 20 μm × 100 m. The technique consists of ejecting a liquid stream onto the inside surface of a Cu–Be wheel spinning at 300–1800 rpm. The inside surface of the wheel is convex (like a torus), which causes the quenched ribbon to slip out of the wheel under centrifugal force. This method provided enough material of uniform thickness for measurement of thermal, mechanical, magnetic, and superconducting properties. The technique has already been applied to a number of metal alloys. In a related method, called melt extraction, a cold roller in contact with a molten bath, extracts a thin ribbon of quenched material. The dimensional stability of metal ribbons produced by these processes has been analyzed by Anthony and Cline (1979a). They also (1979b) used a fluid and heat flow model to predict the maximum thickness of amorphous ribbon that can be made by melt extraction.

Since Klement *et al.* (1960) succeeded in producing an amorphous metallic gold–silicon alloy, many workers have employed splat cooling to make amorphous alloys. Most of the systems in which such efforts were successful followed the criterion proposed by Cohen and Turnbull (1961; Turnbull, 1961). These authors suggested that glasses would be most easily formed by compositions lying near a deeply plunging eutectic. Such compositions would require a relatively smaller undercooling below the melting point to reach the glass transition temperature. This suggestion is supported by a kinetic analysis, as described below.

The critical cooling rates for glass formation in metallic systems have been calculated by Lewis and Davies (1977), using an approach due to Uhlmann (1972). The analysis involves the construction of time–temperature transformation curves based on calculated rates of crystal nucleation and growth. These permit an estimate of the cooling rate required to avoid the formation of a detectable degree of crystallinity. This approach yields order-of-magnitude agreement with experiment for a broad range of metallic glasses and indicates that the ratio of the glass transition temperature to the melting point is the dominant factor in determining the glass-forming tendency of an alloy. In transition metal–metalloid systems, the glass transition temperature T_g is weakly dependent on composition (Lewis and Davies, 1976), so that the melting point T_m is the dominant factor, and glass-forming regions tend to lie near deep eutectics. Structural models, which may account for the existence of these eutectics, have been proposed by Bennett et al. (1971), Chen and Park (1973), and Nagel and Tauc (1975). In metal–metal systems, such as Cu–Zr, the refractory component raises T_g substantially, so that low T_m is less important.

The structure of metallic glasses has been reviewed by Cargill (1975), and more recently by Waseda et al. (1977), who offer the following generalizations. The amorphous structure consists of tetrahedra as the fundamental units of a close-packed disordered atomic distribution. Unlike silica glass, in which the directional covalent bonding requires that the tetrahedra share only corners, the tetrahedra in metallic glasses share faces; this is illustrated in Fig. 3. In metal–metalloid glasses, the atomic radii differ by more than 15%, so that the metalloid atoms fill the spaces between the close-packed metal atoms. In metal–metal glasses, both species are close packed. Chen et al. have studied the influence of such atomic packing arrangements on the glass transition temperature (Chen, 1973), formation and thermal properties (Polk and Chen, 1974), thermal expansion and density (Chen et al., 1973), and elastic constants, hardness, and flow properties (Chen et al., 1975) of metallic glasses.

The mechanical properties of amorphous metals have been reviewed by Masumoto and Maddin (1975) and are summarized in Table I. The metal alloys have higher tensile strengths than their crystalline counterparts, and also have high values of toughness; that is, they can be cold-rolled with area reductions of 80%. Amorphous metal alloys do not show the rapid reduction in strength with increasing size that is typical of metal

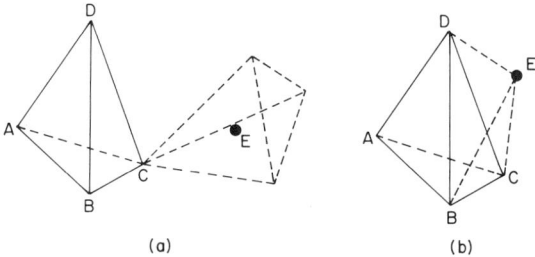

FIG. 3. Schematic diagrams of the difference between the near-neighbor distribution of atoms for (a) an oxide glass and (b) for a metallic glass. (From Waseda et al. (1977), with permission of Chapman and Hall Ltd.)

TABLE I

CHARACTERISTIC MECHANICAL PROPERTIES OF AMORPHOUS METALS[a]

Properties	Effect
Strength	Higher than any of the conventional metals; lower than metallic whiskers.
Rigidity	Lower than their crystalline counterparts by 20 ~ 40%.
Elasticity	Anelasticity is larger than that of crystalline metals.
Plasticity	Inhomogeneous deformation at low temperatures; viscous type deformation at T_g.
Work hardening	Almost negligible or some softening.
Mode of fracture	A large shear slip is observed at the shear fracture; the fracture surface consists of smooth sections produced by shear slip and veinlike sections produced by unstable rupture.
Size effect	Size effect in strength is small.
Strain rate effect	The apparent fracture strength is decreased as the strain rate is increased.
Temperature effect	Strength depends significantly upon temperature; appreciable softening is observed at around T_g.
Ductility, toughness	Toughness is high even though the strength is extremely high; ductility develops at around T_g.
Fatigue	There is a fatigue limit; the fatigue ratio is about 0.35.
Workability	Cold rolling of 30–50% is possible.

[a] From Masumoto and Maddin (1975).

whiskers. In a review of the structure and properties of metallic glasses, Chaudhari and Turnbull (1978) point out a number of interesting features of these materials. One of their most useful properties is high corrosion resistance, which is assumed to result from the dense random packing structure, which is free of dislocations and intercrystalline boundaries. Also important is the magnetic behavior of glass metals, which can be either ferromagnetic or ferrimagnetic. For example, magnetic bubbles have been observed in thin-film rare-earth–transition-metal alloys, and these may be applicable in information storage systems. The electrical resistivities of these materials tend to be high and to have weak temperature dependence; the sign of the temperature coefficient can be changed by alloying. Superconductivity has been reported in a number of amorphous metals, with transition temperatures that may be higher or lower than in their crystalline counterparts. Superconductivity and electronic properties of amorphous transition metal alloys have been reviewed by Johnson (1979), and soft magnetic properties have been reviewed by Hasegawa and O'Handley (1979). Thorough reviews of the properties of glassy metals have recently been provided by Chen (1980) and Cahn (1980).

Research on these materials is very active and has produced a substantial literature, as well as some commercial products, such as the Metglas alloys made by Allied Corporation.

Anantharaman and Suryanarayana (1971) have reviewed the progress in splat cooling from the time of its invention through 1971. They describe a number of devices and discuss a broad range of nonequilibrium metallic phases and their properties. Other good reviews have been prepared by Jones (1972) and by Giessen and Wagner (1971), and a comprehensive bibliography of work on splat cooling between 1958 and 1972 has been prepared by Jones and Suryanarayana (1973).

Moss et al. (1964) devised a procedure for melting droplets in a plasma and spraying them against a cold metal plate. Krauth and Meyer (1965) used this technique in the first application of splat cooling to produce oxide glasses. They found that ZrO_2–Al_2O_3 binaries containing 30 to 60 mole % ZrO_2 formed glasses. The density, refractive index, and crystallization behavior were studied as functions of composition. The systems ZrO_2–SiO_2 and ZrO_2–ZrN were also studied, but neither produced glass.

The next workers to apply splat cooling to oxide melts were Sarjeant and Roy (1967). Using the gun technique, they prepared amorphous flakes of MoO_3, TeO_2, WO_3, and V_2O_5. Unfortunately, the samples were too small and irregular for property measurements. These authors later extended their work (Sarjeant and Roy, 1969) to a number of binary systems, including alkaline earths plus TiO_2 or Al_2O_3, alkalis plus Ta_2O_5 or Nb_2O_5, MgO–SiO_2, and Al_2O_3–SiO_2, among others. Each of these systems exhibited glass formation over a substantial range of composition. The authors point out that those not forming glasses are characterized by close-packed crystal structures and low viscosities.

Kantor et al. (1973) used splat cooling by the gun technique to produce glasses in the following binary systems: Fe_2O_3 plus BaO, CaO, and PbO, Al_2O_3–PbO, and Ga_2O_3–PbO. They found that the glass-forming regions of these systems almost invariably included eutectic compositions. This work provided the first examples of glasses based on Fe_2O_3 and not including any of the usual glass formers.

Suzuki and Anthony (1974) examined 46 binary systems using a splat cooling device developed by Chen and Miller (1970). A molten drop falls between two rapidly rotating steel rollers that are held together by pressure from springs. The resulting foils were 1–40 μm thick; the cooling rate was estimated to be $\sim 10^5$ °C/sec. These authors concentrated on compositions near a eutectic in those systems for which a phase diagram was available. The results are summarized in Table II (NCS = noncrystalline solid). They found that the glass-forming (or NCS-forming) tendency increased with the number of oxygens per cation and with the radius ratio of the cations. The latter trend had also been recognized by Sargeant and Roy (1969).

Coutures *et al.* (1974) used two splat cooling devices to produce glasses from the refractory oxides Ln_2O_3–Al_2O_3 (Ln = La, Nd, Gd, Ho, Er, Yb, Lu, Y) and Ln_2O_3–Ga_2O_3 (Ln = Ga, Gd, Er). One device was of the piston-and-anvil type with a solar furnace for a heat source; the other was a variation on the gun technique. Only eutectic compositions were studied; the resulting glasses were examined by differential thermal analysis.

TABLE II

BINARY OXIDE QUENCHES[a]

System	Composition (mole %)	Phase	System	Composition (mole %)	Phase
ZrO_2–La_2O_3	33–40ZrO_2	Crystal	Al_2O_3–Ga_2O_3	50Al_2O_3	Crystal
CeO_2–ZrO_2	20CeO_2	Crystal	Y_2O_3–Ga_2O_3	35–50Y_2O_3	NCS + crystal
MgO–ZrO_2	50MgO	Crystal	ZrO_2–Fe_2O_3	30ZrO_2	Crystal
CaO–ZrO_2	40CaO	Crystal	CaO–Fe_2O_3	40CaO	NCS + crystal
MgO–Y_2O_3	50MgO	Crystal	Y_2O_3–Fe_2O_3	38Y_2O_3	NCS + crystal
MgO–La_2O_3	40–60MgO	Crystal	La_2O_3–Fe_2O_3	25La_2O_3	NCS + crystal
CeO_2–NiO	20–40CeO_2	Crystal	CeO_2–Nb_2O_5	20,30CeO_2	NCS + crystal
CaO–NiO	45CaO	Crystal	CeO_2–Nb_2O_5	40CeO_2	NCS
Y_2O_3–NiO	20–40Y_2O_3	Crystal	CeO_2–Ta_2O_5	20CeO_2	NSC + crystal
La_2O_3–NiO	30–50La_2O_3	Crystal	MgO–Ta_2O_5	20MgO	Crystal
CeO_2–Al_2O_3	30CeO_2	NCS	ZrO_2–Nb_2O_5	25,40ZrO_2	Crystal
CeO_2–Al_2O_3	50CeO_2	NCS + crystal	ZrO_2–Ta_2O_5	50ZrO_2	Crystal
ZrO_2–Al_2O_3	30–43ZrO_2	NCS + crystal	CaO–Nb_2O_5	25CaO	NCS + crystal
Y_2O_3–Al_2O_3	20Y_2O_3	NCS	Y_2O_3–Ta_2O_5	20–40Y_2O_3	NCS + crystal
ZrO_2–BaO	40ZrO_2	NCS	La_2O_3–Nb_2O_5	20La_2O_3	NCS
NiO–BaO	25NiO	NCS	Al_2O_3–Nb_2O_5	30–50Al_2O_3	NCS + crystal
CeO_2–TiO_2	20CeO_2	NCS	Al_2O_3–Ta_2O_5	50–70Al_2O_3	NCS
ZrO_2–TiO_2	20–50ZrO_2	Crystal	Y_2O_3–Nb_2O_5	15Y_2O_3	Crystal
Y_2O_3–TiO_2	15–50Y_2O_3	Crystal	Y_2O_3–Nb_2O_5	20–30Y_2O_3	NCS + crystal
La_2O_3–TiO_2	20La_2O_3	NCS	BaO–Nb_2O_5	30BaO	NCS
La_2O_3–TiO_2	45La_2O_3	NCS + crystal	TiO_2–Nb_2O_5	20–60TiO_2	Crystal
Al_2O_3–TiO_2	20Al_2O_3	Crystal	TiO_2–Ta_2O_5	50TiO_2	Crystal
BaO–TiO_2	30BaO	NCS	Ga_2O_3–Nb_2O_5	30Ga_2O_3	NCS
BaO–TiO_2	58BaO	NCS + crystal	Ga_2O_3–Ta_2O_5	20Ga_2O_3	NCS + crystal
CeO_2–Ga_2O_3	20,25CeO_2	NCS	Ta_2O_5–Nb_2O_5	30–70Ta_2O_5	Crystal
ZrO_2–Ga_2O_3	20–40ZrO_2	Crystal	HfO_2–La_2O_3	35HfO_2	Crystal

[a] Reprinted with permission from *Mat. Res. Bull.*, **9**, T. Suzuki and A. Anthony. Copyright 1974, Pergamon Press, Ltd.

Shishido et al. (1978) used a piston–anvil device with a laser heat source to make titanates, niobates, and tantalates of the lanthanide elements. Shishido (1979) also made glasses in the $Gd_2O_3 \cdot xAl_2O_3$ system.

To date, splat quenching studies of oxides have concentrated on establishing glass-forming compositions involving refractory materials. In contrast to the work on metals, which has a longer history and a larger following, little information has been provided concerning the properties of these unusual oxide glasses. As with metals, when such measurements are made, it is to be expected that many fascinating and useful materials will appear.

3. Laser Spin Melting

A novel technique called laser spin melting produces rapid cooling by dividing the melt into small droplets that quickly radiate their heat. Also, since the droplets touch no solid surfaces while cooling, the melt is not exposed to nucleating agents. The method (Topol et al., 1973) involves rotating a ceramic rod at a known speed of 8000 to 30,000 rpm with a high-speed drill motor. The end of the rod is melted with a 250-W continuous-wave CO_2 laser and molten droplets measuring 100–800 μm are spun off. Based on pyrometry data, the cooling rate of a 500-μm droplet was estimated to be ~4000°C/sec. Topol et al. produced glasses containing at least 80 wt % of the oxides Al_2O_3, Ga_2O_3, La_2O_3, In_2O_3, ZrO_2, HfO_2, Nb_2O_5, or Ta_2O, with 20 or less weight % of $CaO + SiO_2$. The best glass formers were in the Nb_2O- and Ta_2O-based systems. Using the same technique, Topol and Happe (1974) produced glasses from the pure lanthanide oxides Sc_2O_3, Y_2O_3, Sm_2O_3, Gd_2O_3, Yb_2O_3, and Lu_2O_3, as well as from binary and ternary mixtures based on these oxides. In both of these studies the indices of refraction and Abbé numbers were reported for most of the glasses.

4. Laser Film Melting

A new quenching technique, which takes advantage of solid conduction, as in splat cooling, and which provides the ultimate in thermal contact between melt and substrate, has been reported by United Technologies Research Center (1976). The beam from a CO_2 laser is passed across the surface of a metal block, raising the surface temperature to several thousand degrees. As the beam passes, the thin molten layer is rapidly cooled to form a glassy skin. The surface layer is reported to be harder and more corrosion resistant than the original alloy.

B. Melting under High Pressure

Melting under pressure has received little attention, but two noteworthy reports suggest the interesting possibilities offered by this approach. Datta

TABLE III

Glass Melted under Pressure[a]

Gas	Mole % dissolved	Glass host	Temp. (°C)	Pressure (bars)
Ar	1	(28.5)Na$_2$O–(28.5)B$_2$O$_3$–(43)SiO$_2$ (wt %)	750	5,000
Ar	2	K$_2$O · 4SiO$_2$	750	5,000
Ar	4	(20.7)Na$_2$O–(5.6)CaO–(73.7)SiO$_2$ (mole %)	750	10,000
H$_2$O	17	(20.7)Na$_2$O–(5.6)CaO–(73.7)SiO$_2$ (mole %)	900	1,000
H$_2$O	25	(20.7)Na$_2$O–(5.6)CaO–(73.7)SiO$_2$ (mole %)	800	1,000
CO$_2$	0.9	(20.7)Na$_2$O–(5.6)CaO–(73.7)SiO$_2$ (mole %)	900	1,000
CO$_2$	1.3	(20.7)Na$_2$O–(5.6)CaO–(73.7)SiO$_2$ (mole %)	800	1,000

[a] Based on Roy et al. (1964, p. 11).

et al. (1964) found that a number of glasses could be made in carbonate systems by melting under 1 kbar pressure. Glasses containing 40–60 mole % MgCO$_3$ in combination with K$_2$CO$_3$ were stable as long as no hydroxyl was present. Infrared absorption spectra indicated no change in the coordination of carbon; that is, CO$_3$ groups were still present. Glasses could also be formed by adding to the binary system less than 10 mole % of the carbonates of Mn, Pb, or Ca; the system CaF$_2$–Ca(OH)$_2$–CaCO$_3$–BaSO$_4$ also produced glasses. In the same laboratory, Roy et al. (1964) produced glasses containing extraordinary amounts of dissolved gases. Their results are summarized in Table III.

C. Metal Evaporation Technique

A novel method of making multicomponent glass was described by Nagel et al. (1976), who used it to produce an optical waveguide with a K$_2$O–SiO$_2$ core. The method involved evaporation of potassium metal and deposition of the vapor on a silica surface. The silica was then heated in vacuum to 600°C so that the metal would react to produce a potassium silicide layer; this product was then heated to ~1600°C in the presence of oxygen to produce K$_2$O–SiO$_2$ glass. In the course of heat treatment the potassium diffused into the silica, so that a concentration gradient resulted. This was desirable for the optical waveguide application.

This technique is suitable for making thin coatings of multicomponent glass (or perhaps ceramic or glass–ceramic) on glass or ceramic substrates. Nagel et al. suggest that the method is useful for making optical waveguides with any of the following dopants: Cs, Rb, K, Cd, Na, Zn, Mg, Li, Tl, P.

III. Deposition Methods

In this section we consider two classes of glass-forming methods: (1) An amorphous condensed phase is formed in a vapor and deposited on a substrate (e.g., flame oxidation); (2) an amorphous phase is formed on a substrate by chemical reaction (e.g., CVD or thermal oxidation) or by bombardment by atoms or ions that comprise the glass (e.g., sputtering). Several methods of the second class involve no chemical reaction, and they will be described separately as nonreactive processes. Processes of the second class are typically used to produce films with thicknesses less than a few hundred microns, and they are of greatest interest in the electronics industry. Table IV, adapted from Amick *et al.* (1977), compares a number of methods used to make dielectric films and illustrates the range of materials and applications that have been developed. Reactive deposition methods have the capacity to produce bulk glasses, as well as films, and this has led to the development of practical low-loss fiber optical waveguides. Methods aimed at this and other applications will be described in detail.

A. Nonreactive Deposition

1. Evaporation

A simple and relatively rapid method of producing an amorphous film is illustrated in Fig. 4. The material to be deposited is boiled off in a vacuum and condenses on the substrate; since most of the atoms travel in straight lines, a mask may be used to produce a pattern of deposition that is sharply defined. Evaporation methods have been described in detail by Holland (1956) and Berry *et al.* (1968). Evaporant may be produced by electrically heating in metal boats or coils, or by electron beam heating. The residual gas pressure is commonly reduced to 10^{-4} to 10^{-7} Torr in

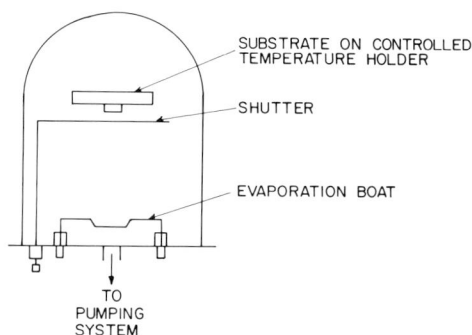

Fig. 4. Simple evaporation apparatus. (From Sinclair, 1968.)

TABLE IV

COMPARISON OF ALTERNATIVE DEPOSITION TECHNIQUES FOR DIELECTRIC FILMS[a]

Deposition process	Source material	Typical deposition rate	Typical substrate temperature (°C)	Crystalline nature of deposit	Sources of impurities	Typical films formed	Typical electronics applications
			Low-pressure depositions				
Evaporation	High-vapor-pressure solid (e.g., SiO)	1–2 μm/min	25–200	Amorphous	Filament/crucible, walls	SiO	Optical coatings
Reactive evaporation	Solids [e.g., Al in gas (e.g. O_2)]	1–2 μm/min	25–200	Amorphous	Filament/crucible, walls	Al_2O_3	Passivation layers
Sputtering; reactive sputtering	Low-vapor-pressure solids [e.g., Si, SiO_2 with gas (e.g., Ar)]	10–100 Å/min	25	Amorphous	Walls, gases used, sources	SiO_2,[b] Si_3N_4, TaN	Passivation layers
Plasma deposition	Gases (e.g., SiH_4–NH_3 mixtures)	10–100 Å/min	25–400	Amorphous	Walls, gases, electrodes	Si_3N_4, Si	Passivation layers, solar cells
Low-pressure CVD	Gases (e.g., SiH_4–NH_3 mixtures)	0.1–1 μm/min	600–1000	Amorphous, polycrystal	Walls, gases, susceptors	SiO_2, Si_3N_4, ORPS[c]	Passivation layers
			1-atm depositions				
Thermal oxidation	Substrate plus gas	Nonlinear ~1 μm for 1 h	800–1200	Amorphous	Walls, substrate surface	SiO_2, Al_2O_3	Channel oxides (MOS), diffusion masks, etch masks, passivation layers
Chemical vapor deposition	Gases (e.g., SiH_4–NH_3 mixtures, SiH_4–O_2 mixtures)	0.1–1 μm/min	800–1200, 250–500	Polycrystal, Amorphous	Source gases, walls	SiO_2,[b] Si_3N_4, PSG,[d] BSG[d]	Passivation layers, diffusion sources, dielectrics for multi-level metal
Anodization	Substrate plus electrolyte	Nonlinear 100–1000 Å in 1 min	25	Amorphous	Electrolyte	Al_2O_3, Ta_2O_5	Passivation layers
Ion implantation	High-purity ion source (mass spectrometer)		25	Amorphous	Substrate surfaces	Nitride layers	Etch resists

[a] From Amick et al. (1977).
[b] Stoichiometric composition usually not obtained.
[c] ORPS = oxygen-rich polycrystalline silicon.
[d] BSG = borosilicate glass; PSG = phosphosilicate glass.

order to avoid collisions between ambient and evaporant atoms. The temperature of the material to be deposited is then raised until its vapor pressure is $\sim 10^{-2}$ Torr; higher pressure would lead to atomic collisions, interfering with the desired straight-line paths. The primary sources of impurities are residual gases from the vacuum system and gases released from the evaporant and heater upon heating. If an active material is to be deposited, it will getter such impurities. In such a case, it is good practice to deposit on a shutter, until the contamination is reduced to an acceptable level, before depositing on the substrate.

Evaporation is a good technique for fabricating amorphous metal films, because the substrate can be held at very low temperatures to reduce the mobility of the deposited atoms. In this way, ferromagnetic films of iron (Grigson *et al.*, 1964) and Co–Au alloys (Mader and Nowick, 1965) have been formed. Evaporating alloys presents a problem, as the individual components generally have different vapor pressures, so that the composition of the film differs from that of the source. To combat this problem, flash evaporation may be used: Small particles of the alloy are dropped onto a surface that is so hot that evaporation is instantaneous. Platakis and Gatos (1976) devised a flash evaporation apparatus that gave film compositions indistinguishable from their sources in the system As–Se–Sb. A simple alternative is to use separate sources for each constitutent of the alloy and to rotate the substrate to obtain homogeneous coverage. A variant on this approach has been used to make continuous flexible glass films of SiO_2 and B_2O_3 (Hanlein, 1956).

Evaporation of oxides presents difficulties because the high temperatures required can lead to reaction between the source and heater, and to dissociation of the oxide. For instance, thermal evaporation of SiO_2 leads to films of SiO (Pliskin and Lehman, 1965), because of the much higher vapor pressure of the reduced species; similar behavior is observed with Al_2O_3. The original oxide can be obtained by oxidation of the film.

An important use for electron beam evaporation is the deposition of antireflection coatings of Ta_2O_5 and SnO_2 on photovoltaic cells (Amick *et al.*, 1977). They can be deposited directly or as metal films that are subsequently oxidized thermally (Revesz *et al.*, 1976). Other glasses that have been made by vacuum evaporation include such diverse materials as ZnS (Tien, 1971), As–S (Sinclair *et al.*, 1960), Si–As–Te (Tick, 1975), Bi–Se (Wood *et al.*, 1969), and alkali nitrates (Smyrl and Devlin, 1972). Zakharov *et al.* (1975) used a ruby laser to evaporate amorphous alloys of the systems Ch–Sb–S–I, where Ch = Ge, Sn, Pb, Cd, or Cu. The resulting amorphous films were very similar in composition and properties to the starting glasses. Mass-spectrometric analysis showed that the vaporized species included both atoms and complex molecular groups.

2. Sputtering

A simple sputtering system is illustrated in Figure 5. A high dc voltage is used to produce positive ions in a low-pressure gas and to drive the ions into the cathode; atoms of the cathode are ejected (sputtered) and they coat the entire apparatus, including the substrate. If the gas includes O_2 or N_2, the film will be the oxide or nitride of the sputtered material, and the process is called reactive sputtering. The physics of sputtering has been extensively studied and is thoroughly treated by Townsend *et al.* (1976).

There are a number of versions of sputtering, such as triode and tetrode dc sputtering, which produce more efficient ionization of the gas and thereby permit operation at lower pressures (Sinclair, 1968). This leads to less contamination of the deposited film by the ambient gas. Bias sputtering involves a dc system in which the substrate is weakly biased negative, so that some sputtering of the film occurs, which keeps the surface free of adsorbed impurities (Maissel and Schaible, 1965). A similar result is achieved with asymmetric ac sputtering (Frerichs, 1962), in which the deposit is formed during one half of the cycle and sputtered clean in the other. In order to sputter dielectric materials, it is necessary to use the rf technique of Anderson *et al.* (1962). Sputtering occurs in the first half of the cycle, and in the second, the plasma electrons neutralize the positive charge that has accumulated on the target. With this technique, a wide range of glasses can be sputtered to yield films of uniform thicknesses, which closely resemble the starting material in composition and properties.

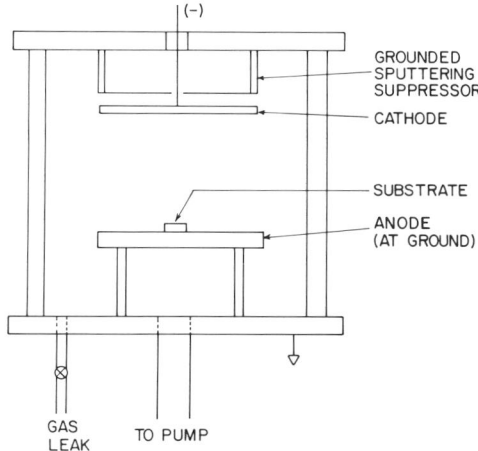

FIG. 5. Simple sputering (diode) apparatus. (From Sinclair, 1968.)

Davidse and Maissel (1966) applied rf sputtering to produce amorphous films from silica, calcium aluminosilicate, lead borosilicate, and borosilicate glasses. Goell (1973) sputtered (in oxygen) a hot-pressed disk of barium carbonate and silica to obtain barium silicate glass films for use in integrated optics; the refractive index of the film could be varied from 1.48 to 1.62 by varying the BaO content of the disk from 0 to 40 wt %. Oxide glass films have also been made from $GeO_2-P_2O_5-V_2O_5$ (Braski and Early, 1972) and $Bi_{12}GeO_{20}$ (Mitsuyu et al., 1976), among other systems. Amorphous Al_2O_3 films can be made by rf sputtering to produce films that are good barriers to moisture penetration; a summary of work on alumina films is provided by Kennedy (1974). Conducting glass films of cadmium stannate have been reported by Lloyd (1977).

Amorphous films of silicon nitride can be formed by rf (Hu, 1966) or dc reactive (Janus and Shirn, 1967) sputtering of a silicon cathode in pure nitrogen. Watts et al. (1974) formed As_2S_3, $Ge_{28}Sb_{12}Se_{60}$, and $Ge_{33}As_{12}Se_{55}$ glasses by rf sputtering; the resulting films were promising optical waveguides. Gallagher and Roy (1975) used rf sputtering to produce binary carbide glasses, containing Cd, Sn, Pt, Ir, Au, Fe, Al, Er, or Dy; some showed indications of phase separation on a scale of 2–5 nm. Buckley and Spalvins (1972) applied lubricating coatings of MoS_2 and Teflon by rf sputtering. Burt et al. (1980) used rf sputtering to prepare amorphous Be films up to 25 μm thick; these are of interest for the ablator layer in laser fusion targets. Sputtering has also been used to form amorphous metal alloys (Allen et al., 1976). Brimhall et al. (1980) have made 200-μm films of FeW and MoCo, the glass-forming ability of which is attributed to their complex crystal structure. Unlike eutectic alloys, these crystallize without the appearance of intermediate metastable phases. Extensive reviews of the recent literature on sputtering have been compiled by Hippler (1980a,b,c).

3. Ion Implantation

Ion implantation (or ion plating) is a physical vapor deposition process in which the substrate acts as the cathode. A dc arc is established in a low pressure of argon, so that the substrate is cleaned by sputtering. Then the coating material is evaporated, and as the atoms pass through the glow discharge area, some are ionized. The positively charged evaporant ions bombard the substrate, becoming implanted in its surface and sputtering the substrate. As a result, the deposited film makes intimate contact with the substrate, and the continuous sputtering action minimizes adsorbed impurities. Films made in this way can coat complex shapes, front and rear, because the evaporant ions follow field lines that surround the substrate. The technique is succinctly discussed by Chambers et al. (1972), and the physics of the process is treated in detail by Townsend *et al.* (1976).

Ion implantation may be used to inject oxygen or nitrogen into the surface of silicon, yielding an oxide or nitride layer after annealing (Wada and Ashikawa, 1976). Several types of planar optical waveguides have been made by ion implantation. Goell *et al.* (1972) produced a region 1 μm thick on fused silica that was uniformly bombarded with lithium ions. The layer had an estimated refractive index of 1.505 (3% greater than silica) and optical loss of 8–10 dB/cm. Webb *et al.* (1975) implanted singly charged ions of H, He, C, N, O, Ne, A, Kr, Xe, and Bi. They found that the resulting thin-film waveguides had lower refractive index than others made by ion exchange, but that the implanted specimens were less lossy. Another advantage of implantation is that the profile and index change of the guiding region can be controlled by the choice of ion and the bombarding energy.

Several features of the deposition methods described in this section are compared in Table V (adapted from Buckley *et al.*, 1972). A review of metastable alloy formation by ion implantation has been presented by Poate and Cullis (1980).

4. Film Structure

The structure of deposited films has been explored by Pliskin *et al.* (Pliskin, 1977; Pliskin and Lehman, 1965; Pliskin *et al.*, 1967) using a combination of methods, including infrared spectroscopy, etching, moisture exposure, and annealing. Some of their results will be briefly summarized.

Oxide films formed by thermal oxidation of silicon (Section II.B.1.b.) are the most stable and most similar to fused silica in physical properties. The nonreactive deposition methods discussed in this section produce films that are more porous and less dense and therefore have lower refractive indices. Films made by electron beam evaporation tend to have an oxygen deficiency that raises the refractive index, but this effect is offset by the very high porosity of these films. The latter property leads to high etch rates for this type of film. In general, for the same type of deposition process, films deposited on colder substrates are less dense, have more bond straining, have more hydrogen-bonded silanol groups, and are less similar in structure to thermal oxide than higher-temperature films. Heat treatment, generally at temperatures well below the annealing point of the glass, makes all of the films indistinguishable from one another. These generalizations have been shown to apply to oxide glasses other than SiO_2, including borosilicate and calcium aluminosilicate.

TABLE V

GENERALIZED COMPARISON OF VARIOUS COATING DEPOSITION TECHNIQUES

Characteristic	Coating technique		
	Vapor deposition	Ion plating	Sputtering
Materials capable of deposition	Elemental metals, some compounds	Metals, alloys, some compounds	Metals, alloys, compounds, cermets, ceramics, polymers
Nature of process	Vacuum evaporation	Vacuum plasma	Vacuum plasma
Nature of film	May not be uniform	Dense, pore free, does not blister	Dense, fairly pore free
Adhesion of film to surface	Poor	Excellent	Good
Type of interface	Sharp between coating and substrate unless thermally diffused	Diffuse or graded	Relatively sharp
Uniformity of film	May vary	Good uniformity	Fairly good uniformity
Deposition rates	Very fast	Very fast	Relatively slow
Film thickness control	Poor	Excellent	Excellent
Size of object that can be coated	Limited by size of vacuum chamber	Limited by size of vacuum chamber	Limited by size of vacuum chamber
Complex geometric surfaces	Only surfaces facing source are coated	Complete coverage of all surfaces	Complete coverage of all sources
Equipment required	Vacuum chamber and heat source	Vacuum chamber, heat source, and high voltage supply	Vacuum chamber and dc or rf power supply

[a] From Buckley et al. (1972).

B. REACTIVE DEPOSITION

Deposition processes described in this section are all characterized by the fact that, upon the introduction of sufficient activation energy (thermal, electrical, etc.), a chemical reaction is initiated somewhere during the process that leads to glass formation. For example, silicon tetrachloride vapor is oxidized, by application of heat or in an rf glow discharge, to produce solid vitreous silica; or a solid silicon metal surface is thermally oxidized to form a vitreous silica layer. Many processes discussed here involve a gas phase in the reaction. Some of the most important commercial applications of any unconventional glass-forming processes described in this chapter are found under this general heading, for here are the principal processes used for glass passivation of integrated circuits, for production of high-quality bulk fused silica (mirror blanks, tubing, crucibles, etc.) and for fabrication of high-purity optical waveguide fibers.

It is convenient to further categorize these reactive processes according to the final physical form of the deposited glass: thin film or bulk (large) solid. Some general statements and boundary conditions describing processes in each group can then be made, but these should be considered as *only guidelines* rather than hard-and-fast rules. Thin-film processes generally result in a glass coating less than 5 μm thick, deposited on a solid substrate. These processes involve reaction temperatures below ~1000°C (often below 500°C) and glass deposition rates below 1 μm/min, and they usually involve heterogeneous reactions (i.e., nucleation is surface catalyzed). On the other hand, bulk solid processes are typically used to fabricate much larger, often free-standing, glass bodies, whose size is more readily described by three dimensions rather than two; or by weight (sometimes tons) rather than by nearly invisible film thicknesses. These bulk glass processes involve high-temperature reactions (>1000°C) and deposition rates of 0.1–3.0 g/min (or 10–1000 μm/min). The initial glass formation is usually by a homogeneous reaction in the gas phase (i.e., nucleation in the vapor phase, without requiring surface catalysis).

1. Glass Film Formation

Processes for depositing thin glass films have received considerable attention in recent years, ever since these dielectric materials proved their worth as both active and passive components in semiconductor technology. Glass films are especially useful for encapsulating devices and integrated circuits, since they provide a reasonably hermetic seal, which improves component reliability. Published results on both reactive deposition processes and glass film properties are far too numerous to be fully presented here. Some excellent reviews of this work (including also nonreactive processes, Section III.A), citing many hundreds of references in total, have been compiled by Amick *et al.* (1977), Revesz (1973), Sinclair (1968), and Plishkin *et al.* (1967). The last is part of one volume in a serial publication entitled *Physics of Thin Films* (Academic Press) that treats all aspects of this important technology. The following subsections are limited to brief descriptions of the important reactive film deposition processes, along with examples of glass types and additional principal references.

The low deposition–reaction temperatures sometimes used here lead to low atom mobility and promote formation of disordered (i.e., noncrystalline) thin-film atomic structures. This often allows formation of amorphous solids that cannot readily be obtained by direct melt or bulk deposition (Section III.B.2.) techniques. Sinclair (1968) and Secrist and Mackenzie (1965) have concluded that such novel amorphous solid films should indeed be called glasses.

a. Reactive Sputtering As noted in Section III.A.2, if a gas is introduced to the main chamber (Fig. 5) that then reacts with the sputtered atoms prior to deposition on the cold substrate, the process is called reactive sputtering. Oxygen and nitrogen atmospheres (diluted in argon), combined with a high-purity silicon metal cathode, will lead to deposition of high-purity amorphous silicon dioxide (for example, Schreiber and Froschle, 1976) or silicon nitride films (Janus and Shirn, 1967), respectively. These films are usually purer than those formed from metal oxide cathodes, but deposition rates are generally lower (Amick *et al.*, 1977). Numerous glasses have been obtained by this general method, including $PbTeO_3$, GeO_2, SiO_2, $Al_2O_3-SiO_2$, $PbO-SiO_2$, SnO_2 (Sinclair and Peters, 1963), mullite (Williams *et al.*, 1963), aluminum borosilicate (Perri and Riseman, 1966), and phosphosilicates (Kennedy, 1973). In each case, appropriate cathode metals and reactive gases were chosen to obtain the desired glass. Compositions can be varied by controlling the exposed surface area of the various cathode materials.

b. Thermal Oxidation One of the oldest and commonly used methods of forming a primary passivating film of SiO_2 on silicon is to heat the metal in dry oxygen, in H_2O-containing oxygen, or in steam. An SiO_2 layer grows inward from the surface by a thermal oxidation mechanism. Although this process will occur slowly even at room temperature, yielding a layer ~10 Å thick after long exposures, it is more common to heat the silicon wafer to 600–1200°C to achieve 1-μm-thick films in ~1 h (Goodman and Breece, 1970). At these elevated temperatures, the film growth kinetics is controlled mainly by the inward diffusion of the oxidizing species (such as O_2, H_2O, or OH). Revesz (1973) reviewed fully the film growth mechanisms for various atmospheres and thermal cycles. Oxidation proceeds almost two orders of magnitude faster in a steam atmosphere compared to dry oxygen (Pliskin *et al.*, 1967), and impurities in the silicon (especially sodium and hydroxyl) can also greatly affect this rate (Revesz, 1973). The dramatic influence of water in the oxidizing ambient is thought due to the increased density of broken Si–O–Si bridges, which facilitates diffusion. One experimental arrangement used by Evitts *et al.* (1964) for thermal oxidation of silicon is shown in Fig. 6. In industrial practice, a "wet" oxygen cycle (controlled mixtures of O_2 and H_2, or steam) is first used, followed by a dry oxygen cycle to remove silanol groups.

This approach works only for materials that form coherent inert oxides, with most efforts concentrating on silicon. One-micron-thick films of 60 wt % $PbO-SiO_2$ glass have been obtained by suspending silicon wafers in air

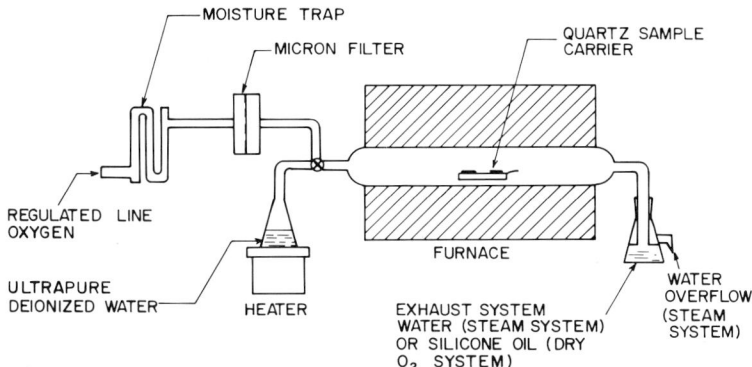

FIG. 6. Simple thermal oxidation equipment used with a silicon sample. (From Evitts et al. (1964); reprinted by permission of the publisher, The Electrochemical Society, Inc.)

over a lead oxide bed for 4.5 h at 650° C (Pliskin and Lehman, 1965). When thermally grown silica films are exposed to $POCl_3$ or P_2O_5 in an oxidizing atmosphere (for example, during the diffusion of phosphorus through the film into the silicon wafer), a thin phosphosilicate glass layer (~0.5 μm) forms on the surface (Pliskin and Gnall, 1964; Kooi, 1964). Similarly, a borosilicate glass film forms when a B_2O_3 diffusion source is used (Pliskin and Gnall, 1964). Boron and phosphorus levels exceeding 30 wt % have been obtained in these glasses. Surfaces of selected III–IV semiconductors have also been thermally oxidized (Wilmsen, 1976).

c. Anodic Oxidation An oxide layer can also be grown on a metal or semiconductor surface, such as silicon, by making it the anode in an electrolytic cell, immersing this anode in a suitable electrolyte (often aqueous), and passing a current through it. By applying a sufficient overvoltage, the metal surface will oxidize forming, in some cases, a glass film up to several thousand angstroms thick. The electrolyte acts primarily as a charge carrier, sometimes providing additives (and impurities) to the film. The process is usually carried out at room temperature to avoid developing a porous structure, but still results in films that are less dense than those formed by higher temperature methods. A simple anodization cell is shown in Fig. 7, and the general process is treated in detail by Young (1961). [This is the only reactive film-forming process discussed in this chapter that does not include a gas phase somewhere in the reaction (perhaps *liquid* deposition best describes it).]

Schmidt and Michel (1957) reported the first successful application of this technique to grow SiO_2 glass films on silicon substrates. Using a 0.04 N solution of KNO_3 in methylacetamide as the electrolyte (the KNO_3 is added to improve conductivity), they achieved film growth "rates" of 3.8 Å/V and used field strengths of 2.6×10^7 V/cm. Duffek et al. (1965) found

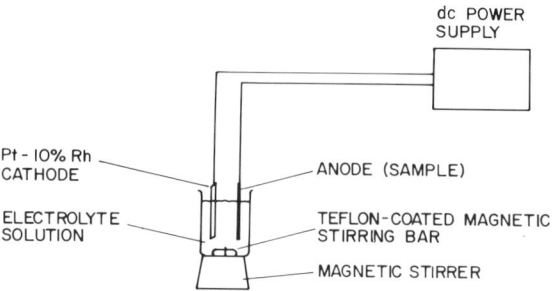

FIG. 7. An experimental anodization cell arrangement (From Van, 1976.)

improved SiO_2 film characteristics when trace amounts of water (<1%) were added to a 0.04 N KNO_3 in ethylene glycol electrolyte solution, and Croset et al. (1971) showed that 80% of the oxygen in the film actually comes from the water (the rest from the salt). Although low-temperature (600°C) heat treatment of these anodic films improves their properties, making them comparable to thermally oxidized silicon films (Schmidt and Ashner, 1971), the technique is not as widely used in the semiconductor industry as thermal oxidation. Doped silica films can be obtained by judicious choice of electrolyte. For example, Schmidt and Owen (1964) obtained SiO_2 glass films containing between 0 and 23 wt % P_2O_5 by using phosphoric acid in water solutions, and Croset and Dieumegard (1973) achieved fluorine-doped SiO_2 by adding potassium fluoride to an electrolyte solution consisting of diethylene glycol, potassium nitrate, and 0.4 wt % water. Anodic oxidation of GaAs has also been achieved (first by Revesz and Zaininger, 1963, and more recently by Hasegawa et al., 1979) to obtain a uniform glassy surface layer. Hasegawa used an electrolyte mixture of water, carboxylic acid, and ethylene glycol to produce a glass film several thousand angstroms thick (at 21 Å/V) in 15 min at room temperature. Anodic oxidation of InAs, InSb, GaP, and GaAsP has also been achieved, as reviewed by Schnable and Schmidt (1976). Transparent Ta_2O_5 glass films have been obtained on Ta metal anodes (Vermilyea, 1960).

d. Chemical Vapor Deposition (CVD) This technique involves primarily heterogeneous reaction and deposition of organometallic or metal halide vapors at a heated solid substrate surface. It is the most widely used and versatile method for forming dielectric films in the fabrication of semiconductor devices, and it is especially important for depositing secondary passivation layers of glass or crystalline oxides or nitrides over silicon and metallized interconnects. For example, when gas mixtures such as SiH_4, PH_3, and oxygen (or $SiCl_4$, $POCl_3$, and oxygen) are passed over a heated silicon wafer, the vapors will heterogeneously react at the water surface, depositing a thin layer of $P_2O_5-SiO_2$ glass over all heated exposed surfaces. Usually, temperatures and gas concentrations are kept at levels that minimize or completely avoid homogeneous reaction in the vapor phase. The main goal has been to achieve dense high-quality passivating layers at low enough temperatures to avoid causing thermal damage to the substrate. Reactor temperatures are typically below 1000°C, glass deposition rates <1 μm/min, and film thicknesses <5 μm.

CVD has been very actively studied during the past 10 to 15 years and many variations have evolved. This general approach for thin-film preparation has been fully described and reviewed by Feist *et al.* (1969, includes 304 references) and Bryant (1977, with 198 references), and its specific application to the deposition of glasses has been carefully reviewed by Kern *et al.* (Kern, 1975; Kern *et al.* 1976, with 70 references; Kern and Rosler, 1977, 100 references) and Amick *et al.* (1977). The important CVD methods used for glass deposition are summarized in Table VI, and the most common reactor designs are shown in Fig. 8. Although most work has concentrated on the glass systems listed in Table

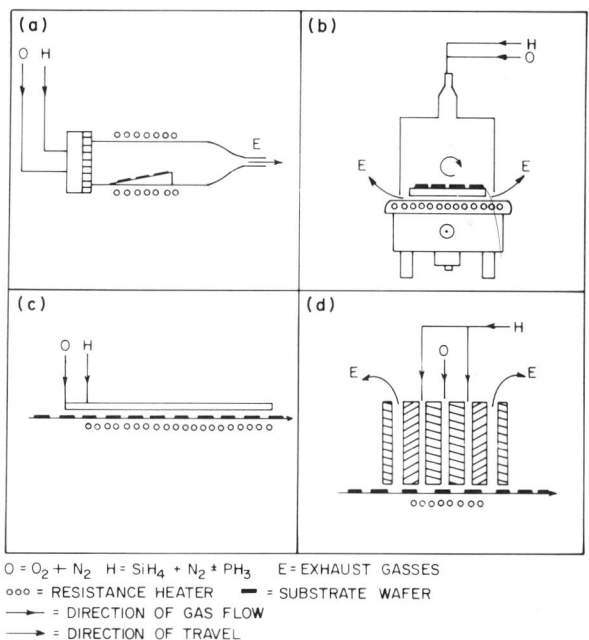

$O = O_2 + N_2$ $H = SiH_4 + N_2 \pm PH_3$ E = EXHAUST GASSES
ooo = RESISTANCE HEATER — = SUBSTRATE WAFER
⟶ = DIRECTION OF GAS FLOW
⟶ = DIRECTION OF TRAVEL

FIG. 8. Basic types of CVD reactors for preparing passivation overcoat layers. (a) Horizontal tube displacement flow reactors; (b) rotary vertical batch-type reactors; (c) continuous reactors employing premixed gas flow fed through an extended area slotted disperser plate; (d) continuous reactors employing separate, nitrogen-diluted, oxygen and hydride streams that are directed toward the substrate by laminar flow nozzles. (From Kern, 1975.)

TABLE VI

PRINCIPAL CVD METHODS FOR GLASS FILM DEPOSITION[a,b]

Process	Typical glass system	Advantages	Limitations	Conclusions
Normal pressure (760 Torr) Low Temperature CVD (450°C)	SiO_2, P_2O_5–SiO_2, B_2O_3–SiO_2	Good for <500°C; alkali gettering; good control; autoprocessors	Particles; pinholes; not conformal; poor Au adhesion	Best process for P_2O_5–SiO_2 final passivation, and for gettering/tapering/secondary passivation.
High Temperature CVD (850°C)	SiO_2, Si_3N_4, P_2O_5–SiO_2, B_2O_3–SiO_2, Al_2O_3, etc.	High quality; versatile; conformal	Poor economy; poor uniformity; >700°C; wall deposits	Conventional system for Si_3N_4, SiO_2, PSG[c], Al_2O_3. Good quality but poor economy.
Low Pressure (<1 Torr) Low Temperature CVD (450°C)	SiO_2, P_2O_5–SiO_2	Good for <500°C; conformal; economical	Poor uniformity; spacing sensitive; not established	Under development; potentially interesting.
Plasma-enhanced CVD (300°C)	$Si_xN_yH_z$	Alkali barrier; pinhole free; conformal; good for <350°C	Thermally unstable; variable properties; hard to control	Only system for "silicon nitride" final passivation. Good moisture and alkali barrier.
High-temperature (850°C)	Si_3N_4, P_2O_5–SiO_2	Economical; conformal; high quality; clean, uniform	Low deposition rates; >700°C; problems of mass control	Most economic process for Si_3N_4, SiO_2. Good-quality films.

[a] After Kern and Rosler (1977).
[b] As related to semiconductor passivation technology.
[c] PSG = Phosphosilicate glass

VI, many other glasses have been made by CVD using other elements available as vaporizable–decomposable compounds (both organometallics and halides). A few examples, showing the wide range possible, are zinc silicates and alumina silicates and borosilicates (Kern and Heim, 1970), arsenosilicates (Wong, 1972), lead silicates (Li and Tsang 1972), amorphous tantalum aluminates (Tsang *et al.*, 1976), and amorphous boron (Vandenbulcke and Vuillard, 1976).

The use of electrical energy through glow discharge plasmas (low pressure; rf or microwave) to partially or fully replace thermally supplied energy in activating the desired deposition reactions allows low-temperature operation of reactors (Table VI). One version of this, in which certain reactive vapor species are formed in a microwave plasma away from the deposition zone, was used by Secrist and MacKenzie (1966) to prepare glass films of SiO_2, GeO_2, B_2O_3, Ti_xO_y. More commonly, the plasma is generated directly above the substrate surface. Sterling and Alexander (1974) used this technique to prepare chalcogenide glass films, such as $Te_{48}As_{30}Ge_{10}Si_{12}$. This approach has also been adapted to prepare oxide glasses for optical waveguides with microwave plasma (see Section III.B.2.b., p. 84).

2. Bulk Glass Processes

Thermally activated *homogeneous* oxidation of starting mixtures of metal halide vapors is the principal reaction used to obtain large bulk glasses of high optical quality and purity by vapor deposition techniques. Glass deposition rates generally exceed 0.1 g/min and are often in the 1–2 g/min range for a single deposition device (such as a torch). Metal halides such as $SiCl_4$, $GeCl_4$, $TiCl_4$, BCl_3, $POCl_3$ are commonly used as starting compounds, but organometallics, such as SiH_4, $(CH_4)_3B$ can be also employed. Recent reaction kinetics studies by Powers (1978) and French and Pace (1977) for some of these metal halide vapors indicate that homogeneous oxidation is the principal reaction mechanism, especially at the elevated processing temperature (>1500°C) typically used to form bulk glasses by vapor deposition. For example, extrapolation of the Powers (1978) data for $SiCl_4$ indicates the reaction $SiCl_4 + O_2 \rightarrow SiO_2 + 2Cl_2$ is 99% complete in 0.09 sec at 1400°C and in 0.004 sec at 1600°C. When the rate constant $K \gtrsim 1$, the reaction is homogeneous (i.e., gas phase nucleated, no catalyzing surface required) and produces a finely divided particulate glass material commonly called "soot." For most metal halides used, this occurs quite rapidly above ~1300°C. Fused silica soot formed by passing a $SiCl_4$–oxygen mixture through a methane–oxygen flame is shown in Fig. 9. These high-purity, inclusion-free glass soots have high surface areas ($\gtrsim 20$ m^2/g), providing a strong driving force for rapid, thermally activated,

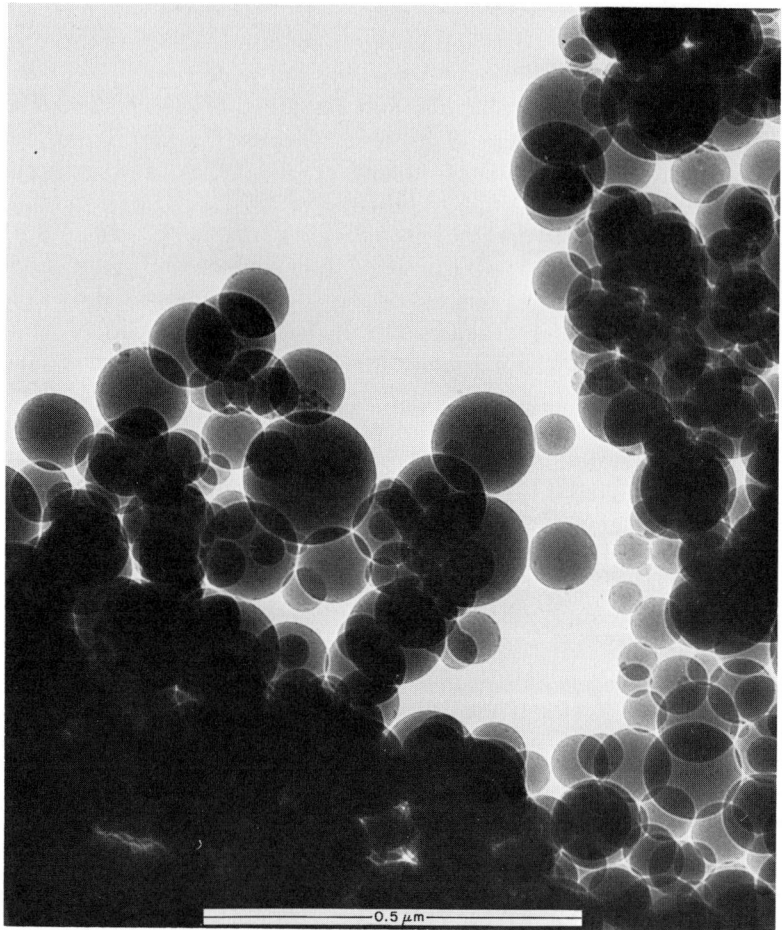

Fig. 9. Transmission electron micrograph of fused silica soot formed by homogeneous oxidation of $SiCl_4$ in a flame.

viscous sintering. The kinetics of this soot sintering were described by Scherer (1977). In practice, this allows one to convert glass soot, formed during the vapor oxidation step, into solid inclusion-free glass bodies. This approach has proved especially useful for making very high-quality (chemically and optically homogeneous) refractory glass compositions (such as pure and doped fused silica) that are extremely difficult to obtain in equivalent quality by conventional melting techniques. The following two sections detail how the generation and processing of glass soot has been used to make both bulk fused silicas and optical waveguide fibers.

a. Fused Silica and Silicates Bulk fused silica is commercially produced by a variety of techniques, including vapor deposition, and the physical properties of the resultant glass depend strongly on which method is used. This is extensively reviewed by Dumbaugh and Schultz (1969) and Brueckner (1970). Vapor-deposited bulk fused silica, often called synthetic fused silica, is classified according to the scheme of Hetherington (1966) as type 3 if it has a "high" hydroxyl content (typically 1000 ppm OH), or type 4 if the hydroxyl level is "low" ($\lesssim 50$ ppm OH). The metallic impurity level of type 3 and 4 fused silicas is generally much lower, and optical homogeneity higher, than for other types, resulting in improved optical transmission, radiation resistance, and increased electrical resistivity. This has led to numerous important applications for these synthetic fused silicas, including passive optics (lenses, windows, prisms, fibers), high-temperature lamp envelopes (tubing), contaminant-free crucibles for silicon crystal growing, and furnace muffles.

Hyde (1942) first described the general preparation of type 3 fused silica using a flame hydrolysis (oxidation) process. A vapor mixture of $SiCl_4$ and oxygen were fed to a torch where they reacted in a methane–oxygen or hydrogen–oxygen flame to produce a stream of fused silica soot, which was deposited onto a rotating target such as a nickel rod and sintered to a solid glass. Dalton and Nordberg (1941) and Nordberg (1943) prepared fused silicas containing additions of TiO_2, Al_2O_3, and B_2O_3 using this general flame reaction method with appropriate mixtures of metal halide vapors. If the target is kept above $\sim 1800°C$ during SiO_2 soot deposition, simultaneous sintering of the soot will occur, yielding in a single step a solid, bubble-free glass. One method for achieving this is shown in Fig. 10, where the heat from the soot-generating burners is also used to sinter the soot as it hits the hot fused sand target. Layer by layer, a boule of solid fused silica is deposited that, because hydrogen-containing fuels are used, contains ~ 1200 ppm OH. Very large boules, weighing over 1000 lb, can be

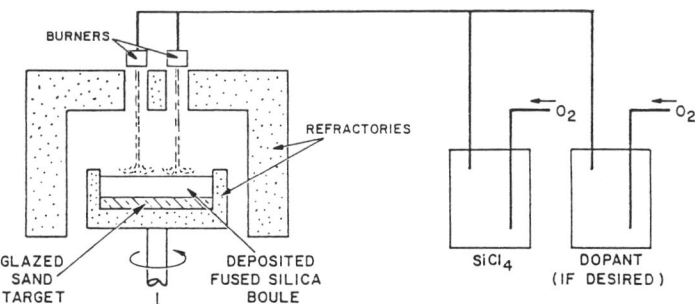

FIG. 10. Fused silica boule formed by simultaneous soot deposition and sintering in a flame oxidation furnace.

obtained by using many soot deposition burners and/or running furnaces for many days at a time. Typical deposition rate is 1.5 g/min burner. It is important to keep the target surface above ~1800°C to maintain a smooth surface that exhibits stable growth. At slightly lower temperatures, clear glass nodules form on the surface, trapping bubbles and debris between them as they grow.

This general boule approach is used to manufacture numerous commercial fused silicas as outlined in Table VII. Optical blanks, windows, crucibles, tubing, and so on, are all produced by further processing these boules using conventional cutting, grinding, polishing, and flameworking techniques, although Gray (1971a) has described a method of depositing tubing directly. One application, reflective optics for large telescopes (which takes advantage of the low thermal expansion of fused silica), clearly demonstrates the massive "bulk" nature of the glass that can be produced by this approach (see Fig. 11). This material is also used extensively for windows in manned spacecraft (and the space shuttle) because of its refractory nature, thermal shock resistance, and optical homogeneity. Addition of TiO_2 further lowers the expansion of fused silica (Nordberg, 1943; Schultz, 1976), making it especially attractive for mirror blanks. Corning Code 7971 TiO_2–SiO_2 glass, which contains ~7.5 wt % TiO_2, is made by this process for this purpose (Table VII).

Winterburn (1966) has shown that by using a hydrogen-free flame in this boule process, it is possible to produce bulk fused silica that is essentially water-free (type 4 glass). Mixtures of oxygen and fuels such as carbon disulfide, carbon monoxide, or cyanogen have been considered, but the

TABLE VII

COMMERCIAL FUSED SILICAS MADE BY VAPOR DEPOSITION BOULE PROCESSES

Manufacturer	Type 3 Hydrogen-containing flame	Type 4	
		Oxygen plasma	Soot preform
Corning Glass Works	Code 7940		Code 7943 (obsolete)
	Code 7971 (TiO_2–SiO_2)		
Dynasil	Dynasil		
General Electric	Type 151		
Heraeus (Amersil)	Suprasil	Suprasil W	
Quartz et Silice	Tetrasil		
Thermal Syndicate	Spectrosil	Spectrosil WF	
Westdeutsche Quarzsahmerlye	Synsil		

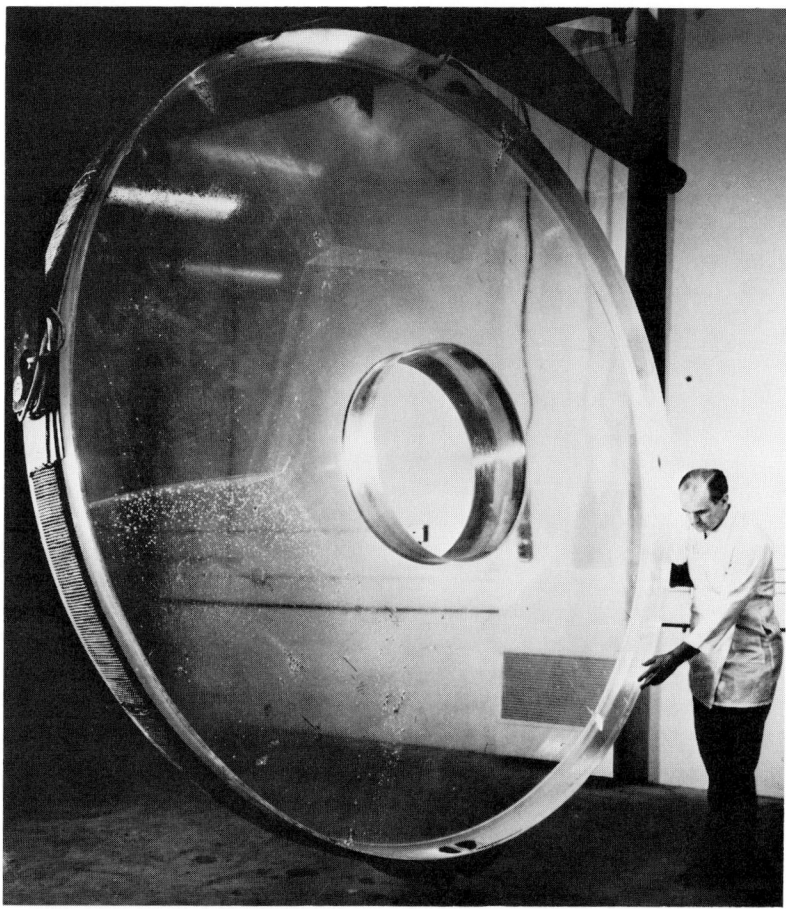

FIG. 11. 108-in.-diameter fused silica mirror blank for a University of Texas telescope; formed by fusing hexagonal blocks cut from individual boules. (Courtesy Corning Glass Works, Corning, New York.)

most widely used method involves an oxygen plasma torch. One arrangement is shown in Fig. 12. This process was described in detail by Nassau and Shiever (1975) and results in fused silica containing as low as 1 ppm OH. Glass deposition rate for a single torch is ~0.4 g/min. The chemical reactions occurring in the tail flame of this torch have been identified using spectrographic techniques by Wood et al. (1978). Commercial water-free fused silicas produced by this general plasma deposition approach are listed in Table VII. A variation on this process, where the plasma torch is replaced by a CO_2 laser, was used by Kobayashi et al. (1975) to grow a

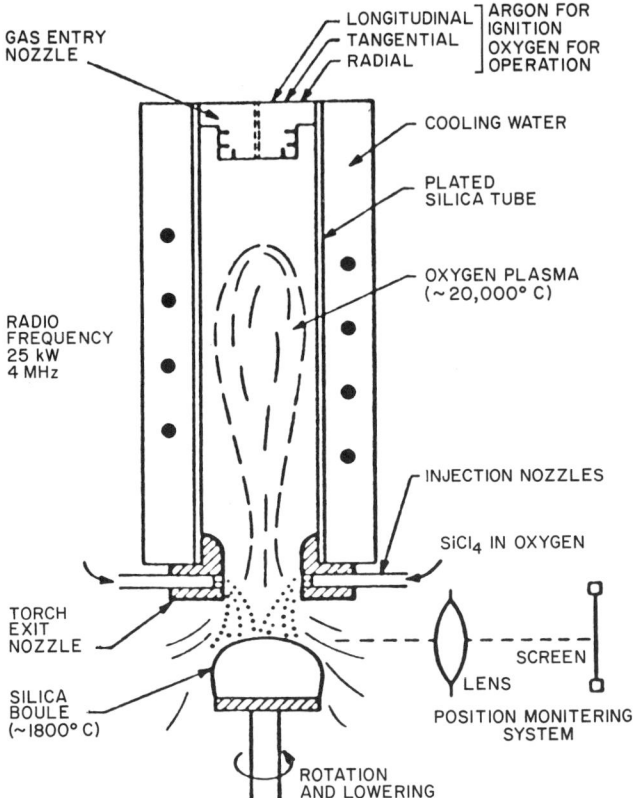

FIG. 12. Production of bulk fused silica boule by plasma torch method. (From Nassau and Shiever, 1975.)

small water-free fused silica rod (~1 cm diameter × 5 cm length), but the glass deposition rate was a low 0.06 g/min using a 200-W laser. This variation is not believed to be used commercially.

Additives such as Al_2O_3, ZrO_2, Nb_2O_5, Ta_2O_5, MoO_3, TiO_2, and selected 3d transition elements (Schultz, 1974a,b,c; Schultz and Voorhees, 1974, Flamenbaum et al., 1975), as well as Na_2O (Mattmuller, 1967), have all been incorporated in type 3 fused silica boules by vapor additions to the $SiCl_4$ stream. Rau (1977; Rau et al., 1977) achieved 3 wt % fluorine in type 4 fused silica using CCl_2F_2 in the oxygen plasma torch method. However, to avoid almost complete vaporization of such additives as GeO_2, P_2O_5, and B_2O_3, which have appreciable vapor pressures at preferred boule-forming temperatures, it is useful to separate the soot deposition and soot sintering steps for refractory glass compositions. Both can then be done at lower temperatures (<1500°C) to still obtain a high-quality glass product.

This is achieved in most optical waveguide processes, as described in the next section. Undoped fused silica has also been made by separating these two steps, as described by Nordberg (1943), Flamenbaum et al. (1974), and Gray (1971b), to obtain products ranging from tubing and crucibles to small optical blanks. These processes all involve soot deposition to obtain a porous, semisintered, preform that is then fully sintered to an inclusion-free bulk glass in a subsequent thermal treatment. This porous fused silica body can also be doped, prior to final sintering, by liquid and gaseous impregnation techniques (Schultz, 1975; Dumbaugh and Schultz, 1975).

b. Optical Waveguides The first 20-dB/km glass optical waveguide fiber (Kapron et al., 1970) was drawn from a glass blank or preform that had been fabricated by a vapor phase deposition process. This success ultimately led to extensive worldwide technical efforts to develop vapor phase processes capable of providing practical glass fibers for the now burgeoning optical communications industry. It has proved to be a very effective and versatile fabrication approach. Single-mode, step-index multimode, and graded-index multimode fibers have all been made by vapor phase techniques. Multimode fibers have been fabricated with bandwidths as high as 3 GHz (Keck and Bouillie, 1978) and with total attenuations as low as 0.5 dB/km, bordering on the intrinsic limits of the glasses used (Horiguchi et al., 1977). Numerous important variations on vapor phase processes have been developed during the past 8 years to achieve such impressive results. These have been reviewed extensively by Schultz (1979a, 1979b), including performance details of both process and fiber, process advantages and disadvantages, and areas of future work. A comprehensive review of optical communications technology is found in a special issue of the *Proceedings of IEEE,* edited by Miller (1980). A briefer review of key processes is presented here.

Mixtures of compounds such as $SiCl_4$, $GeCl_4$, BCl_3 (or BBr_3), and $POCl_3$ are generally used as the starting raw materials because from them physically compatible oxide glasses (typically silicate-based) can be formulated, for both the core and cladding parts of the waveguide, which exhibit excellent optical properties (attenuation, refractive index, dispersion) in the spectral region of interest. Reactive, high vapor pressure, organometallic compounds of these same elements, which are commonly employed in semiconductor vapor deposition technology, are generally avoided in optical waveguide processes, principally because a by-product of their decomposition/oxidation is —OH, which can be incorporated in the glass as a light-absorbing impurity.

Three principal types of vapor processes have been used to make optical waveguides: (1) outside, (2) inside, and (3) soot melt processes. Basic descriptions of each are presented below.

The term *outside* describes the general environment in which the glass soot is generated and deposited. The metal halide–oxygen vapors are reacted in an open heat source (such as a flame) to produce a hot glass soot stream that is deposited directly onto an external target surface (sometimes called a bait). Although this process is carried out "externally," the soot generation–deposition region can be enclosed in a chamber for atmospheric control. These processes are used mainly for multimode step- and graded-index fiber fabrication.

One important outside process, developed at Corning Glass Works and commonly called OVPO† (Outside Vapor Phase Oxidation), involves a flame heat source, a removable rod target, and a lateral orientation for soot deposition (Keck *et al.*, 1973, 1974). In practice, a hot soot stream of the desired glass composition is generated by passing the vapor stream through a methane–oxygen flame directed at a rotating and traversing ceramic target rod (Fig. 13a). The glass soot sticks to this rod in a partially sintered state and, layer by layer, a cylindrical porous glass peform is built up (~ 200 layers each of core and cladding glass). The average pore size is ~ 0.3 μm; overall porosity is $\sim 75\%$. Average soot collection efficiency is $\sim 50\%$ and effective deposition rate (i.e., for soot collected on preform) is ~ 2 g/min. When soot deposition is completed, the porous preform is slipped off the target rod. By properly controlling and sequencing the metal halide vapor stream composition during the soot deposition process, it is possible to build into this porous preform the desired glass compositions (and thus, refractive indices) for both the core and cladding regions (Fig. 13b). Multimode step-index as well as graded-index fiber designs can be achieved (Schultz, 1979a,b). This porous glass preform is then zone sintered (between ~ 1400 and $1600°C$, depending on composition) to a solid, bubble-free, glass blank by passing it through a furnace hot zone (Fig. 13c) in a controlled atmosphere, such as helium (see Scherer, 1977, 1979 for sintering kinetics). The gases evolved by the preform during zone sintering escape through the porous regions rather than becoming trapped as bubbles in the fully sintered glass. The central hole remains in this blank, but disappears when it is drawn into fiber at much higher temperatures ($\sim 1800-1900°C$) (see Fig. 13d).

An important alternative to this outside multistepped lateral deposition process is the use of an axial deposition orientation that can potentially allow continuous blank making. This concept is under extensive study in Japan, where it is called VAD (Vapor Axial Deposition). Initial results on VAD were published by workers at the Nippon Telephone and Telegraph Corporation (NTT) Ibaraki Laboratory (Izawa *et al.*, 1977a,b,c, 1978). As

---

† Conceptionally, this term actually applies to all outside-type processes.

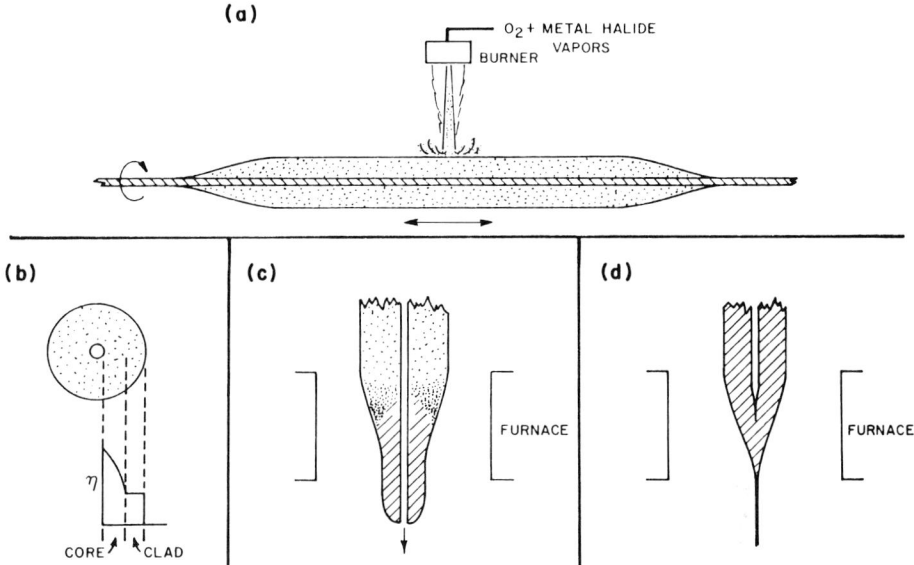

FIG. 13. An OVPO process for making optical waveguides. (a) Soot deposition, (b) soot preform cross section, (c) preform sintering, (d) fiber drawing. (From Schultz, 1979b.)

shown in Fig. 14, the process involves simultaneous flame deposition of both core and cladding glass soots onto the end (i.e., axially) of a rotating fused silica target rod. As the porous soot preform grows, at an overall soot deposition rate of ~0.5 g/min (Sudo *et al.*, 1978), it is slowly retracted through a graphite resistance furnace, where it is consolidated to a solid glass blank by zone sintering. In principle, this process should allow continuous blank making.

Another general family of vapor phase processes used to make optical waveguides involves reaction within a glass tube and glass deposition on the *inside* walls of this tube. This basic concept was first conceived at Corning Glass Works (Keck and Schultz, 1973) and was actively pursued and more fully developed at Bell Telephone Laboratories (MacChesney 1974a,b) resulting in a category of processes called IVPO (Inside Vapor Phase Oxidation). It has also been commonly called MCVD (Modified Chemical Vapor Deposition). This approach has received considerable attention throughout the world and given excellent low-attenuation step- and graded-index multimode fibers as well as single-mode fibers. The extensive effort on this approach, through 1976, has been reviewed by Pearson (1976), Rigterink (1976), and Küppers and Lydtin (1977).

The principal IVPO process to date involves externally heating a rotating fused silica or high silica glass tube (typically 25 mm o.d. × 1.5 mm

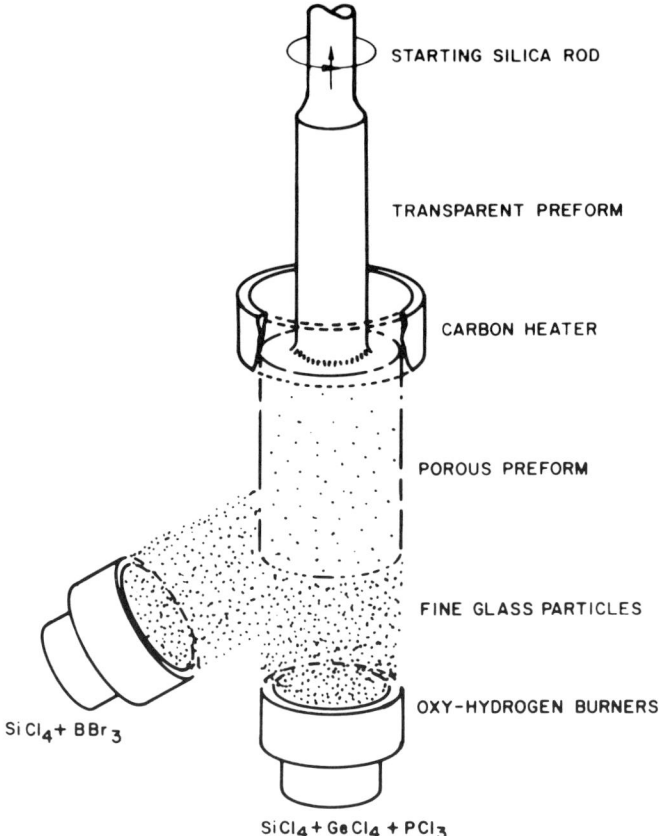

Fig. 14. Vapor phase axial deposition (VAD) process (From Izawa et al., 1978).

wall thickness × ~1 m long) with an oxy-hydrogen torch to thermally trigger vapor phase oxidation of the metal halide gases that are flowing inside (total flows ~1.5 l/min), and causing glass soot formation (see Fig. 15). Ultimately, the tube will become the fiber cladding, and the majority of the soot formed inside will become the fiber core. As the hot soot flows downstream, ~50–60% of it is attracted to the cold walls of the tube (not yet heated by the torch), where it is deposited as a thin porous layer (typically ~30 μm thick) not unlike the porous OVPO preform structure described earlier. The rest flows out the exhaust end of the tube. The torch is steadily moved toward this downstream portion of the tube (at ~15 cm/min), and as it passes over the soot deposit, it zone sinters the soot to a clear, bubble-free glass layer ~7 μm thick. When the torch reaches the exhaust end of the tube, it is quickly returned to the inlet end, and the process is repeated. After ~70 passes, enough glass is deposited (~0.5 mm) on the inside wall of the tube to provide the fiber core size desired for a multimode waveguide. The metal halide vapors are then stopped, and the flame intensity is increased, so that as the torch now traverses the tube it softens and collapses it to a solid rod blank, which is then drawn into fiber. Based on a recent study (Edahiro *et al.*, 1978), it may be possible to substitute a carbon resistance-heated furnace for the torch in this blank-making process.

By carefully controlling the metal halide vapor composition during each pass of the torch, it is possible to fabricate either step-index or graded-index multimode fibers. Often a series of barrier layers of B_2O_3-SiO_2 or SiO_2 are first deposited on the tube inside wall prior to core glass deposition. These low attenuation barrier layers act as part of the final fiber cladding and effectively minimize fiber attenuation related to tube impurities and core–clad interface irregularities. Single-mode fibers are readily achieved by decreasing the amount of core glass deposited.

IVPO glass deposition rates of ~0.1–0.3 g/min have been achieved (compared to earlier values of ~0.1 g/min) by using helium carrier gas and including small amounts of P_2O_5, both to improve soot sintering kinetics (Akamatsu *et al.*, 1977; O'Connor *et al.*, 1977). Since the starting tube, which is not deposited glass, actually becomes 0.75 volume fraction of the final fiber (for a 0.5 core–fiber diameter ratio) an effective deposition rate that is ~4 times higher than actual can be used for comparison with OVPO-type processes (where *all* of the glass is deposited). Thus, the 0.3 g/min translates into an effective (blank or fiber-making) rate of 1.2 g/min. A thermophoresis model has been successfully used to explain why soot

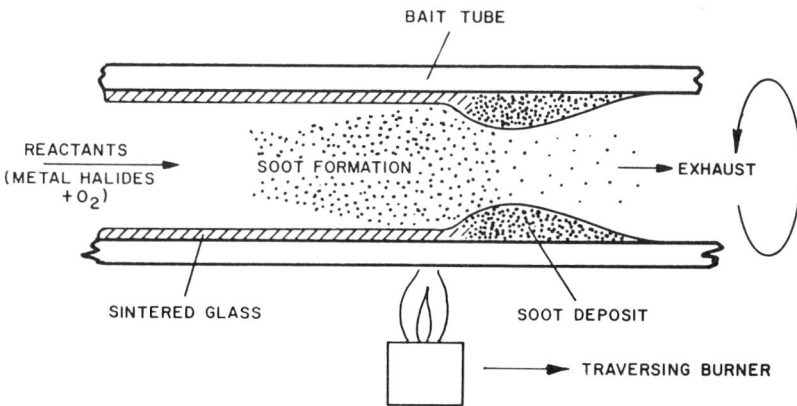

FIG. 15. IVPO process; cutaway view representing deposition on inside tube wall. (From Schultz, 1979b.)

actually collects on the tube wall in the IVPO process, and optimization of the hot zone based on this model may lead to further rate increases (Simpkins et al., 1978; Walker et al., 1978).

A high-temperature rf plasma technique has also been used to improve deposition rates of the IVPO process (Jaeger et al., 1978). In this approach the external torch is replaced by an internal high-temperature argon–oxygen plasma (4.5 MHz), which traverses the tube length during glass deposition, analogous to the torch. Actual rates of 0.3 g/min were achieved (with a 25-mm-i.d. tube), but potentially higher values may be possible, especially in larger-diameter tubes (since deposition efficiency approached 100%).

Another IVPO approach, which incorporates both internal and external heating during glass deposition, is the Philips microwave plasma process (see Fig. 16) (Küppers and Lydtin, 1977; Küppers et al., 1976, 1977; Geittner et al., 1976). Using a nonrotating 12-mm-i.d. SiO_2 tube placed in a furnace at ~1100°C, ~2000 glass layers are deposited on the inner wall by the heterogeneous reaction of metal halides (no soot formed) at ~10–30 Torr pressure using a nonisothermal plasma (200 W/2.45 GHz) to fabricate a graded index preform. Deposition efficiency is 100% and the deposition rate is ~0.3 g/min. This increased rate over earlier values of ~0.1 g/min was achieved using larger-diameter tubes (12 mm i.d. versus ~6 mm) to allow a higher vapor flux.

Typical glass compositions and multimode fiber designs for the OVPO and IVPO processes are shown in Fig. 17.

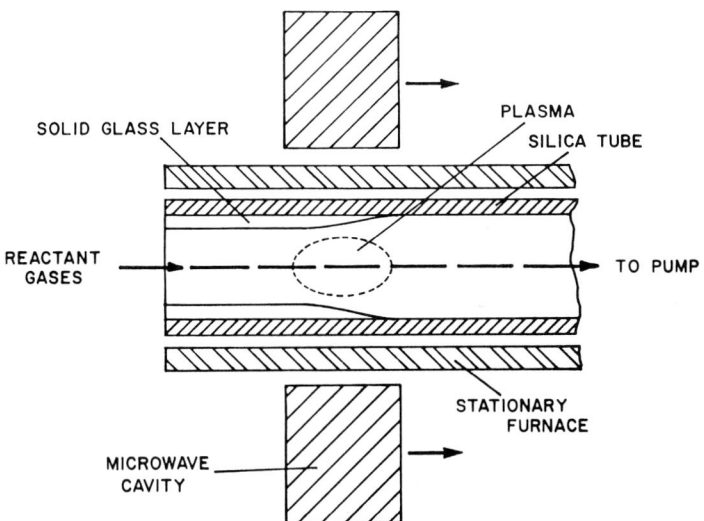

FIG. 16. Microwave plasma IVPO process. (From Küppers et al., 1977.)

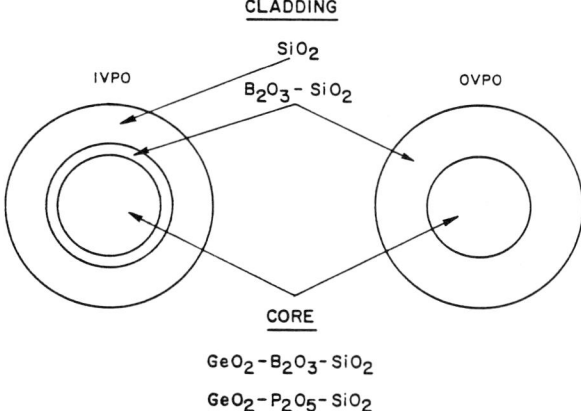

FIG. 17. Typical multimode glass optical waveguide cross sections. Core diameter, ~55–65 μm; cladding thickness, ~25–35 μm.

Finally, a third general approach, called soot melts, begins to bridge the gap between vapor phase deposition processes and the conventional direct melt methods that use high-purity batch-melted glasses and double-crucible fiber drawing (a potentially continuous process). High-purity soots made by flame oxidation are used as starting batch in this approach. Initial studies reported (Inoue et al., 1977) for low melting (~1400°C) Ga_2O_3–GeO_2–P_2O_5 glasses used the process shown in Fig. 18 and resulted in step-index fibers (~0.2 numerical aperture) with minimum attenuation of 9.8 dB/km at 820 nm. By bubbling oxygen through the melt, the hydroxyl level was reduced to ~40 ppm, and in 1978 an attenuation of 8.5 dB/im was reported (Nishinari et al., 1978). A variation (Akamatsu

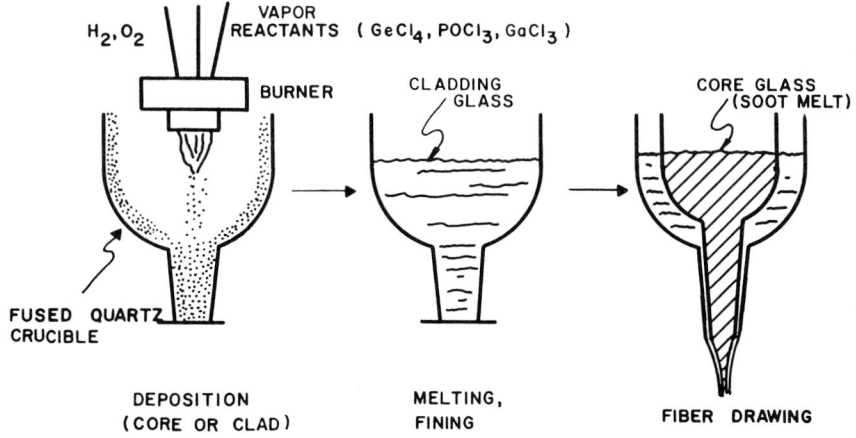

FIG. 18. Soot melt process (description based on Inoue et al., 1977).

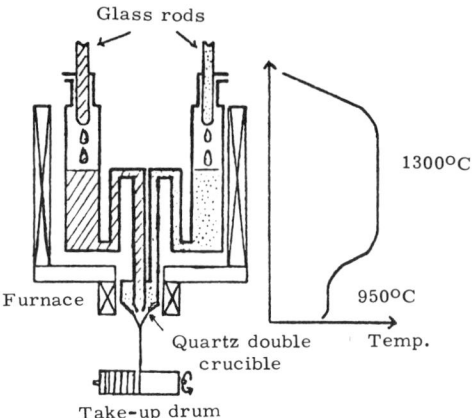

FIG. 19. Soot melt process alternative method. (From Akamatsu *et al.*, 1978).

et al., 1978) on this process, shown in Fig. 19, is analogous to the standard double-crucible method. The starting rods of Ga_2O_3–GeO_2–P_2O_5 core and cladding glasses are prepared from initial soot melts and can be continuously fed into the double-crucible fiber drawing equipment. Attenuation of 8.5 dB/km at 820 nm was achieved for step-index fibers with numerical aperture of 0.3.

IV. Solution Methods

Amorphous SiO_2 coatings produced by solution methods were developed by German optical firms during World War II (see, for instance, patent issued to Schott, 1939). Methods of application and properties of such coatings have been described in a good review by Schroeder (1969). The formation of silica gel from an alkoxide solution can be visualized as follows (Kingery, 1960):

$$Si(OR)_4 \xrightarrow[\text{catalyst}]{H_2O} Si(OH)_4 \xrightarrow{-H_2O} SiO_2,$$

where R is an alkyl group. The silica resulting from condensation of the silicic acid is an amorphous gel containing numerous fine pores. The gel can be sintered to clear glass, in the form of thin films, at low temperatures. Thicker bodies are more difficult to consolidate because gaseous products of the decomposition of precursor materials, and water from the hydrolysis, tend to produce bubbles. McCarthy *et al.* (1971) confirmed the observation made by Heany (1937, 1938) that vitreous silica bodies can be formed at low temperatures from various synthetic silicas having very large specific surface area. McCarthy *et al.* found that SiO_2 made by hydrolysis of tetraethyl orthosilicate could be hot-pressed to produce a

2. UNUSUAL METHODS OF PRODUCING GLASSES

translucent disk with some transparent regions at 1050°C under 272 atm, and at 500°C under 4000 atm.

A novel method of making porous glasses was introduced by Kistler (1931), who dried a gel by heating in an autoclave above the critical temperature and pressure of the solvent. The resulting material is called an aerogel. Nicolaon and Teichner (1968) extensively studied silica aerogels, obtaining pore volumes up to 18 cm^3/g; the pores were stable upon heating up to 700–900°C. The preparation and properties of inorganic oxide aerogels have been reviewed by Teichner *et al.* (1976).

The difficulty of making bulk glass directly from gels, i.e., without crushing and melting, is illustrated by the work of Hansen and Hood (1970). These workers describe a method for casting an article from silica gel, dehydrating, and sintering at 1200–1300°C in vacuum to a clear, theoretically dense body. Unfortunately, the drying step required more than 2 months under conditions of controlled hymidity. This excessively slow dehydration procedure was required, according to Shoup (1976), because the very small (<300 Å) pore size of silica gel produces high capillary pressures that cause splitting of the gel. Shoup has developed a method of making silica gel with controlled, uniform pore size from mixtures of silica sol and alkali silicates. As long as the pore size exceeds ~600 Å, the gel can be processed in hours, rather than weeks, and can be sintered at 1450°C to produce clear, dense bodies centimeters thick. The sintering kinetics of such gels have been analyzed by Scherer (1977). In order to make equivalent bodies by melting quartz, temperatures at least 300° higher would be required.

Several workers have explored solution methods of making multicomponent glasses. Roy (1956) suggested three methods for preparing glass batch: spray drying solutions of nitrates, coprecipitation of salts, and hydrolysis of alkoxide–salt solutions. In a patent filled in 1967, Schroeder and Gliemeroth (1971) described a method of gelling mixtures of alkoxides and halides of glass constituents, then crushing, drying, and sintering to a bubble-free mass. In this way they could make glasses at temperatures several hundred degrees lower than would be required for melting standard batch, and could obtain excellent homogeneity. Leven and Thomas (1972) described a similar process involving alkoxides and organic salts. Roy (1969) pointed out that methods such as that described by Luth and Ingamells (1965), of gelling inorganic salts and silica sol, which had long been in use for preparing homogeneous crystalline oxide powders, could be used for making glass batch. McCarthy and Roy (1971) showed that such methods could be used to make soda-lime, flint, and other standard glass types, which could be melted at temperatures 100–200°C lower than would be required for ordinary batch materials. Dislich (1971a) showed how one could obtain homogeneity on an atomic scale by polymerization of alkoxides; the resulting gels were used to prepare multicomponent glasses, ceramics (MgAl spinel), and glass ceramics. Dislich (1971b) hot-pressed granules of dehydrated gel containing $86.85SiO_2-5.91B_2O_3-2.62Al_2O_3-3.91Na_2O-0.66K_2O$ to form clear molded articles at low temperatures (650–700°C). Konijnendijk et al. (1973) made clear borosilicate glasses by hot-pressing or melting gels. Gorlich et al. (1976) made amorphous powders of SiO_2-TiO_2 and SiO_2-GeO_2 from ethanol solutions of $Si(OEt)_4$, $TiCl_4$, and $GeCl_4$.

A method of making multicomponent glasses directly from the gel, i.e., without crushing, drying, and remelting, was developed by Shoup (1972). He found that he could coprecipitate a number of oxides with silica to form a gel with the oxides intimately bound to the silica network. In this way he produced SiO_2-K_2O glasses doped with about 1–9 wt % of a number of transition metal oxides, Al_2O_3 or P_2O_5. Most of the fired bodies (maximum firing temperature was 650°C) were opaque or translucent when fired in air, but vacuum firing produced some clear, seed-free glasses. Yoldas (1977) has made glass and ceramic articles, with dimensions of a few centimeters, entirely from hydrolysis of mixtures of alkoxides. The gels produced by hydrolysis were heated to 400–500°C to drive off water and organic groups, and transparent, monolithic bodies resulted. The crystalline articles described by Yoldas are highly porous, with pore diameters in the range of 3–10 nm. There has been considerable recent interest in the preparation of monolithic glasses from gels. Yamane et al. (1978) hydrolyzed $Si(OCH_3)_4$ with additions of ammonia; the latter slowed gelling, lowered the density, and raised the surface area of the gel. Nogami and Moriya (1980) compared the effect of NH_4OH and HCl on the hydrolysis of $Si(OC_2H_5)_4$. With additions of base, the gel consisted of 100-Å particles, had a bulk density of 0.8 g/cm³, and sintered above 1050°C by viscous flow; acid produced a continuous solid phase (i.e., no particles) with a density of 1.8 g/cm³, which sintered at 700°C. Yamane et al. (1979) found that when silica gel, made from $Si(OCH_3)_4$, was heated above ~300°C, residual organics burned off, creating micropores and increased surface area. Using refractive index measurements, Brinker (1980) showed that the solid phase of the gel was less dense than the corresponding glass. Shrinkage upon firing apparently involves densification of the solid phase, as well as collapse of the pores.

Some special applications of solution methods of commercial significance have appeared. For instance, amorphous SiO_2 fibers have been produced by dissolving sodium silicate in an alkaline organic solution, then spinning fibers according to the process used to make rayon yarn (Wizon and Robertson, 1967). The fiber or fabric can by pyrolyzed to produce pure silica, or a carbon–silica mixture with each phase having a continuous microstructure. Kanichi et al. (1976) drew fibers of TiO_2-SiO_2

and $Al_2O_3-SiO_2$ from liquid mixtures of the metal alcoholates. Heating the drawn fibers at 500°C led to the formation of oxide glass fibers. The preparation of films, fibers, and monoliths from alkoxides has been ably reviewed by Sakka and Kamiya (1980).

V. Solid State Transformations

A. GLASS FORMATION BY RADIATION

Radiation damage in solids results from the collision of an energetic particle with an atom in the lattice. The recoiling atom generally has sufficient energy to leave its site and collide with neighboring atoms, displacing them. Fast neutrons have a relatively low incidence of collision, but each collision causes a large number of displacements; in contrast, charged particles have a much higher collision probability, but less energy is transferred to the target atom and fewer displacements occur. Another effect of irradiation is the "thermal spike" that occurs when an energetic particle causes large atomic vibrations and this kinetic energy is rapidly transferred to neighboring atoms. The result is sudden heating, sometimes sufficient to cause local melting, and fast quenching. According to Kircher (1964), the duration of temperatures of ~ 1000 K is $\sim 10^{-10}-10^{-11}$ sec over a region of thousands of atoms.

The extent of damage done to a crystal structure by radiation depends not only on the ease of displacing an atom from its site, but also on the ease with which it can return. Crawford and Wittels (1955) have pointed out that the radiation stability of the cation–oxygen bonds decreases in the order Be–O, Al–O, Zr–O, Si–O, and this is the order of decreasingly ionic bonding. Thus as the directionality of the bonding increases, the ability of the collision-damaged bonds to heal diminishes. Such reasoning could explain why diamond becomes amorphous after neutron bombardment (Levy and Kammerer, 1955).

Several ceramic materials containing silica can be rendered amorphous by exposure to neutron bombardment. The structural disorder, and attendant decreases in density and thermal conductivity, increases with dosage. For instance, beryl ($3BeO \cdot Al_2O_3 \cdot 6SiO_2$) is merely distorted by exposure to 2.8×10^{20} (fast) neutrons/cm², but becomes amorphous when the dose is increased by 29%. Similarly, zircon ($ZrSiO_4$), garnet ($Ca_3Al_2Si_3O_{12}$), topaz [$Al_2(OH)_2SiO_4$], and tourmaline [$NaMgB_3Si_6O_{27}(OH)_4$] all become amorphous after sufficient neutron bombardment according to Crawford and Wittels (1955). Radiation effects on these and other ceramic materials are described by Wullaert et al. (1964).

Primak and Bohmann (1962) emphasize the importance of the kinetic stability of the disordered phase. For example, UO_2 is very stable on irradiation, but naturally occurring uranium oxides of complex composition are metamictized (i.e., disordered by their own radioactivity). This implies that the rate of recrystallization in these materials is slower than the rate of atomic displacement by radiation. Similarly, pure Al_2O_3 shows little structural change after irradiation, but $Al_2O_3-UO_2$ compacts, in which the UO_2 is finely dispersed, are completely metamictized.

Of the materials that become glassy upon irradiation, silica is by far the most thoroughly studied. Wittels and Sherrill (1954) showed that quartz, low cristobalite, low tridymite, and silica glass were all reduced to the same phase by heavy neutron bombardment. The resulting material appears amorphous to x rays, is optically isotropic, and has a density of 2.30 g/cm^3, versus 2.20 g/cm^3 for unirradiated silica. Simon (1957) studied the structure of glasses made by irradiating quartz and vitreous silica with 1.4×10^{20} (fast) neutrons/cm^2. The two glasses were identical within experimental error, but differed from unirradiated vitreous silica. The average Si–O distance remained 1.61 a.u. as in vitreous silica, but the average Si–O–Si bond angle decreased from 142° to 138°, so that the neighboring Si atoms were closer together. The radial distances between the O–O and Si-2nd-O atoms, and of the more distant neighbor atoms show a wider distribution in the irradiated samples than in vitreous silica. Primak (1960) reviewed the work on radiation effects on silica, and, based on the change in physical properties with dosage and with subsequent annealing, offered the following interpretation. The region of the thermal spike [$\sim 4 \times 10^4$ atoms reach 2500°C for $\sim 10^{-12}$ sec (Primak, 1958)] in quartz does not become vitrified; if it did, disordering should be complete at a much lower dosage than is actually observed. The development of disorder within the thermal spike results from a cooperative rearrangement of neighboring tetrahedra much like the displacive phase transformations of quartz and cristobalite, to accommodate the displaced atoms. The material does not become fully isotropic until dosages sufficient to displace every atom in the solid.

Glass formation by irradiation with ^{235}U fission fragments has been reported in a Pd–20% Si alloy (Lesueur, 1973), resulting in a material that previously has been formed only by splat cooling. The disordering was attributed to the partially covalent bonding, which hindered recrystallization within the thermal spike.

B. SHOCK-INDUCED GLASS FORMATION

The formation of partially and fully disordered phases by shock events has been of interest in geology since amorphous feldspars† were first found

---

† For the convenience of nongeologists, mineral names used in this article are defined in Table VIII.

at meteorite impact sites (Gumbel, 1870, cited in Stoffler and Hornemann, 1972). Extensive study of naturally and experimentally shocked materials has recently been stimulated by the lunar exploration (Apollo) program. Much of the surface of the moon is covered with glassy material, and it is of great interest to determine whether this reflects volcanic activity or the result of meteoritic bombardment. Such information is critical in the formulation of theories concerning the evolution of the solar system. Recently, shock-produced glasses have proved to be of commercial as well as cosmic interest.

There are about forty sites on earth that contain rocks exhibiting shock deformation, and, although some controversy exists concerning the possible volcanic origin of some of these, most may confidently be identified as meteorite impacts (Dence, 1971). Petrographic examination of materials from these locations indicates that a continuum of disordered states exists, including deformed crystals, high-pressure phases, and amorphous phases produced by solid state transformation or quenching of a melt (Chao, 1967). Glasses produced by shock, without melting, are called diaplectic or, synonymously, thetomorphic. Such glasses retain the grain boundaries of the original crystal, showing no indication of flow; Chao (1967) shows examples of amorphous silica grains in a matrix of biotite grains that have remained crystalline.

The first report of diaplectic glass formed in the laboratory was by DeCarli and Jamieson (1959), who produced amorphous silica at pressures

TABLE VIII

Mineral Names

Name	Composition
Feldspars:	
Albite (Ab)	$NaAlSi_3O_8$
Anorthite (An)	$CaAl_2Si_2O_8$
Orthoclase (Or)	$KAlSi_3O_8$
Microcline	$KAlSi_3O_8$
Sanidine	$KAlSi_3O_8$
Alkali Feldspar	Ab–Or solid solution
Plagioclase	Ab–An solid solution
Jadeite	$NaAl(SiO_3)_2$
Biotite	$K(Mg, Fe^{2+})_3Si_3AlO_{10}(OH)_2$
Olivine	$(Mg, Fe^{2+})_2SiO_4$
Forsterite	Mg_2SiO_4
Zircon	$ZrSiO_4$
Oligoclase	$([NaSi]_{0.9-0.7} [CaAl]_{0.1-0.3})AlSi_2O_8$
Granodiorite	Quartz + alkali Feldspar + plagioclase

exceeding 350 kbar in a shock wave produced by an explosively driven aluminum plate. Next, Milton and DeCarli (1963) showed that plagioclase was rendered noncrystalline by a shock of 250–300 kbar, but that the glassy regions retained the grain outlines and cleavage cracks of the original sample. The refractive index of this diaplectic glass was intermediate between that of the ordinary crystalline phase and a glass of the same composition produced by quenching a melt. The same plagioclase, when shocked to 600–800 kbar, exhibited vesicles (i.e., bubbles) and evidence of flow. At this pressure the shock causes the temperature to rise to ~1500°C, exceeding the melting point of the mineral and leading to the formation of a melt glass with properties closer to that of a conventionally melted glass. The progression from diaplectic to melt glasses with increasing shock pressure has now been studied in detail in a number of minerals. Much of this work has been reviewed by Stoffler (1972, 1974).

Shock pressures are produced in the laboratory using devices similar to that shown in Fig. 20, from Kleeman (1971). The impact is produced by striking the sample with a projectile that is propelled by high explosive; the duration of the pressure pulse is a few microseconds. Methods of calculating the magnitudes of the pressures and temperatures achieved during a shock event are described by Wackerle (1962) and by Gibbons and Ahrens (1971).

Stoffler and Hornemann (1972) reviewed the work on quartz and feldspar glasses produced naturally and experimentally. Transformation of single crystals of quartz to diaplectic glass requires pressures of at least 360 kbar, at which the temperature rises only to ~500°C. As the shock

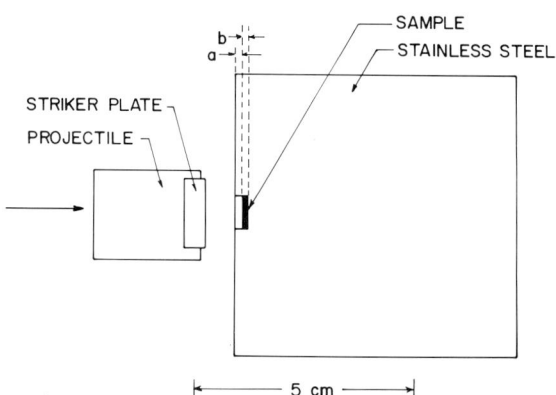

FIG. 20. Schematic of the configuration of sample, holder, and projectile at the instant prior to impact. (From J. D. Kleeman, *J. Geophys.* Res. **76,** 5499–5503 (1971), copyrighted by the American Geophysical Union.)

2. UNUSUAL METHODS OF PRODUCING GLASSES

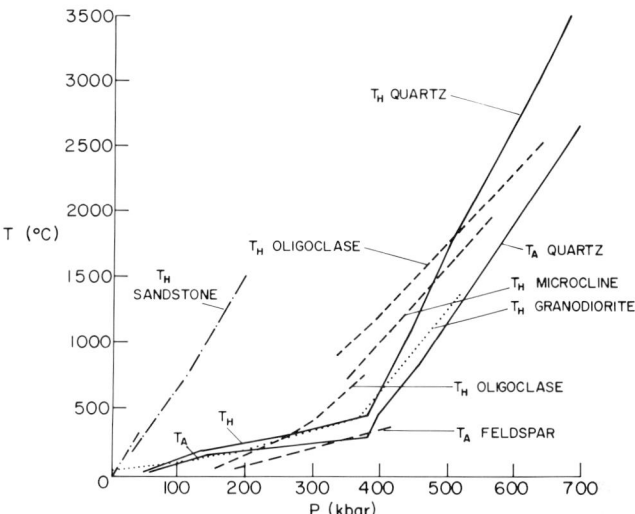

FIG. 21. Shock pressure versus shock temperature T_H and postshock temperature T_A for framework silicates. Data for feldspars from Ahrens *et al.* (1969), data for quartz from Wackerle (1962). For comparison, T_H of granodiorite (Borg, 1972) and of sandstone (Ahrens and Gregson, 1964) are plotted. (From Stoffler and Hornemann, 1972.)

pressure rises, the shock temperature increases rapidly, so that at ~500 kbar the softening point of the glass is exceeded. At that stage, vesicles and flow are evident, and the refractive index of the glass approaches that of ordinary fused silica. The variation of shock temperature with pressure is illustrated in Fig. 21. Minerals with lower melting points are transformed at lower temperatures: feldspars form melt glasses at pressures of 425–450 kbar. Alkali feldspars are anomalous in that the vesiculated shock-melted glasses are higher in refractive index than the corresponding conventionally melted glasses. Stoffler and Hornemann suggest that this behavior may be attributable to the high viscosity of these melts, which could inhibit structural rearrangement after pressure release.

The minerals that form diaplectic glasses most readily are quartz and network (or framework) silicates, such as plagioclase. Orthosilicates, such as zircon and forsterite, can be disordered, but do not become amorphous. The similarity in the ease of glass formation by shock and by irradiation has been pointed out by Dobretsov *et al.* (1970) and is illustrated in Tables IX and X, from that paper. Arndt *et al.* (1971) have shown another correlation between shock- and radiation-produced silica glasses, which is that the density–refractive-index relations are the same. This is illustrated in Fig. 22, from Stoffler and Hornemann (1972), which shows that SiO_2 glasses produced by static compression, shock compression, and

TABLE IX

TRANSFORMATIONS OF SILICATE AND OXIDE POWDERS UNDER SHOCK WAVE TREATMENT IN THE CYLINDRICAL CASE[a]

Type		Characteristic features	Examples	
			Initial minerals	New phases[b]
I	1.	Indistinct or absent axial zone, i.e., unstable three-shock configuration	SiO_2 (quartz, glass)	Destroyed "quartz," s.r.o. phase of high density, traces of stishovite, rarely of coesite
	2.	Formation of glasslike phases of variable density (without fusion)	Framework minerals: $KAlSi_3O_8$ (orthoclase) $NaAlSi_3O_8$ (albite) $CaAl_2Si_2O_8$ (anorthite)	Destroyed "orthoclase," s.r.o. phase of high density, traces of high-pressure phase ?, s.r.o. phase, jadeite + SiO_2, s.r.o. phase (maskelinite)
IIa	1.	Distinct axial zone, corresponding to Mach's three-shock configuration	Silicates: $ZrSiO_4$ (zircon)	Destroyed (metamictic) zircone, SiO_2 + ZrO_2 (monoclin.), glass
	2.	No glasslike phases, with partial or complete lattice deformation	$MgSiO_3$ (enstatite) $K(M,Fe)_3AlSi_3O_{10}(OH)_2$	SiO_2(s.r.o.) + Mg_2SiO_4, glass; destroyed Mg-mica + $FeFe_2O_4$ or Fe + SiO_2 + glass
IIb	a.	With decomposition to constituents	Noncomplex silicates and oxides: Mg_2SiO_4 (forsterite)	Fine-grade fracturing and partial deformation of the lattice
	b.	With polymorphic transformations	Al_2O_3 (α and γ)	Traces of new phase high pressure (?) a-Al_2O_3
III		No phase transformation; partial lattice deformation	TiO_2	No new phases

[a] From Dobretsov et al. (1970).
[b] Destroyed: phase with partially or completely destroyed lattice; s.r.o.: short-range-order (glasslike) phase.

neutron irradiation show similar behavior. Curiously, naturally occurring diaplectic glasses follow a distinctly different curve. This may result from contamination with water or other impurities; other possible explanations are discussed by Stoffler (1974). The apparent similarity of the synthetic silica glasses is remarkable, considering that the duration of the thermal spike in irradiated silica is $\sim 10^{-12}$ sec (Primak, 1958), the pressure pulse in explosively shocked silica lasts $\sim 10^{-6}$ sec (Dobretsov, 1970), and static pressure is applied for >1 sec. These vastly different kinetic processes seem to lead to glasses with the same properties. Whether the glasses are actually identical in structure should be investigated by x-ray analysis, but it is difficult to obtain diaplectic glass with uniform properties in sufficient quantity for such a study.

When diaplectic glasses are annealed, some show a decrease in refractive index toward the value characteristic of melt glass, and others increase in refractive index and revert to single crystals or polycrystalline aggregates (Stoffler and Hornemann, 1972). This is directly analogous to the behavior of neutron-irradiated quartz as described by Primak and

TABLE X

COMPARISON OF THE SHOCK WAVE TRANSFORMATIONS AND TRANSFORMATIONS CAUSED BY IRRADIATIVE TREATMENT[a]

Minerals	Type of irradiative treatment	Type of transformation caused by shock wave treatment	Analogy with shock treatment
SiO_2(quartz, glass)	Fast neutrons up to $2 \cdot 10^{20}$ neutrons/cm²	1	Line broadening, decrease in SiO_2 density up to glass formation. Increase in glass density
$NaAlSi_3O_8$ $CaAl_2Si_2O_8$	Fast neutrons	1	Decrease in density up to formation of glass
ZrO_2(monoclin.)	Fast neutrons up to $6 \cdot 10^{19}$ neutrons/cm²	IIb	Decrease in the density and transformation to a new modification (to high temperature, cubic one with irradiative treatment; high pressure, rhombic one with the shock wave treatment)
$ZrSiO_4$	α-particles up to $3 \cdot 10^{-4}$ α-particles/atom	IIa	Lattice deformation up to the x-ray amorphous state with decrease in density, decomposition to SiO_2 (x-ray amorphous) and ZrO_2 (various modifications)
	Fast neutrons up to $3 \cdot 10^{20}$ neutrons/cm²	IIa	Decrease in density, disappearance of the far-order lines
Mg_2SiO_4	Fast neutrons	IIb	No observable change (except for disappearance of the weak lines)

[a] From Dobretsov et al. (1970), as modified from Lastman (1963).

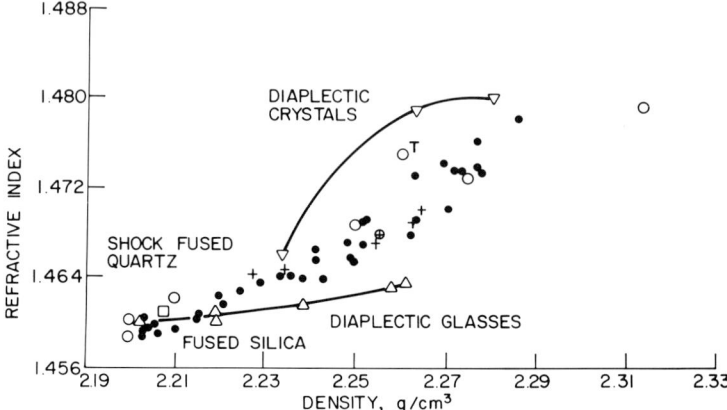

FIG. 22. Density–refractive index plot of disordered quartz and quartz glasses of various origin: △, diaplectic quartz glasses, Ries (Engelhardt *et al.*, 1967); ▽, diaplectic quartz, Ries (Engelhardt and Bertsch, 1969); ●, silica glass densified by static compression (Arndt and Stoffler, 1969); ⊙, silica glass densified by shock compression (Arndt *et al.*, 1971); +, silica glass densified by neutron irradiation (Primak, 1958); ⊕, quartz neutron irradiated (Primak, 1958); T, tridymite. (From Stoffler and Hornemann, 1972.)

Bohmann (1962). After low doses of fast neutron irradiation, the material consists of amorphous regions scattered throughout the crystal; annealing at this stage can reproduce the original single crystal. At higher doses, the crystal nuclei are widely separated by the amorphous silica, and annealing leads to the growth of a polycrystalline microstructure. Finally, at the highest doses, no effective nuclei remain, and annealing causes the amorphous structure to adjust in the direction of ordinary fused silica. In view of the fact that the degree of disorder in shock specimens varies continuously, from fully crystalline to fully amorphous, with increasing shock intensity, the same sort of behavior would be expected.

As mentioned earlier, the impetus for shock research has been the need to understand phenomena associated with meteorite impacts on earth and on the moon. Obviously, these natural events involve much greater forces than can be employed in the laboratory. In major impact events, pressures of several megabars are expected: material shocked above ~2 Mbar will vaporize, and most of that shocked between 0.5 and 2 Mbar will be partially or totally fused (Dence, 1971). This is consistent with recent experimental studies in the range up to 1 Mbar (Schaal and Horz, 1977). The tendency of such melts to remain glassy or to crystallize under various cooling conditions can be analyzed using sophisticated kinetic models, together with experimental data for viscosity and crystallization rates (Uhlmann *et al.*, 1977).

TABLE XI
Shock-Produced Optical Glasses

Oxide glass		Fluoride glass	
Component	W%	Component	W%
B_2O_3	13	CaF_2	23
La_2O_3	38	SrF_2	32
ThO_2	24	LaF_3	2
Nb_2O_3	8	AlF_3	22
Ta_2O_5	17	$NaPO_3$	21

Some commercial applications for shock-produced glasses have been reported. Schott Glaswerk has patented (Schott, 1968) a process for making diaplectic glass for optical applications, taking advantage of the elevated refractive index so obtained. Using shock pressures of 5–10 kbar, they fabricated the two compositions shown in Table XI. The oxide glass cannot be made by conventional melting; the fluoride glass has a much higher refractive index than a melted glass of the same composition (1.5527 versus 1.4644). A very interesting development has been reported by Cline (1978). Explosive compaction has been used at Lawrence Livermore Labs to fuse glassy metallic particles into amorphous cylinders (~1-cm diameter, ~15-cm length) and disks (~5-cm diameter, 3-mm thick). This is significant because ordinary powder processing leads to crystallization of metallic glasses, but shock welding preserves the amorphous structure.

References

Ahrens, T. J., and Gregson, V. G., Jr. (1964). *J. Geophys. Res.* **69**, 4839–4874.
Ahrens, T. J., Petersen, C. F., and Rosenberg, J. T. (1969). *J. Geophys. Res.* **74**, 2727–2746.
Akamatsu, T., Okamura, K., and Ueda, Y. (1977). *Proc. Topical Meeting Opt. Fiber Transmission, 2nd, Williamsburg, Virginia* Paper Tu C3-1.
Akamatsu, T., Nakamura, O., Goto, J., and Ueda, Y. (1978). *Eur. Conf. Opt. Commun., 4th, Genoa, Italy.*
Allen, R. P., Dahlgren, S. D., and Mery, M. D. (1976). *Int. Conf. Rapidly Quenched Met., 2nd,* pp. 37–44.
Amick, J. A., Schnable, G. L., and Vossen, J. L. (1977). *J. Vac. Sci. Technol.* **14**, 1053–1063.
Anantharaman, T. R., and Suryanarayana, C. (1971) *J. Mater. Sci.* **6**, 1111–1135. See also discussion between these authors and H. Jones, *J. Mater. Sci.* **7**, 349–354 (1972).
Anderson, G. S., Mayer, W. N., and Wehner, G. K. (1962). *J. Appl. Phys.* **33**, 2991.
Anthony, T. R., and Cline, H. E. (1979). *J. Appl. Phys.* **50**(1), 239–254.

Arndt, J., and Stoffler, D. (1969). *Phys. Chem. Glasses* **10**, 117–124.
Arndt, J., Hornemann, U., and Muller, W. F. (1971). *Phys. Chem. Glasses* **12**, 1–7.
Bennett, C. H., Polk, D. E., and Turnbull, D. (1971). *Acta Metall.* **19**, 1259.
Berry, R. W., Hall, P. M., and Harris, M. T. (1968). "Thin Film Technology." Van Nostrand Reinhold, New York.
Bletry, J. (1973). *J. Phys. D: Appl. Phys.* **6**, 256–275.
Borg, I. Y. (1972). *In* "Flow and Fracture of Rock" (H. C. Heard, I. Y. Borg, N. L. Carter, and C. B. Raleigh, eds.). *American Geophysics Union Monogr.*
Braski, D. N., and Early, B. F. (1972). Rep. 1972, ORNL-TM-3830.
Brimhall, J. L., Wang, R., and Kissinger, H. E. (1980). *J. Mat. Sci.* **15**, 2605–2611.
Brinker, C. J. (1980). *Am. Ceram. Soc. Bull.* **60**(3), 364.
Brueckner, R. (1970). *J. Non-Cryst. Solids* **5**, 123–177.
Bryant, W. A. (1977). *J. Mat. Sci.* **12**, 1285.
Buckel, W. (1954). *Z. Phys.* **138**, 136.
Buckley, D. H., and Spalvins, T. (1972). *In* Sputtering and Ion Plating. NASA SP-5111, pp. 71–87.
Buckley, D. H., Johnson, R. L., and Bisson, E. E. (1972). *In* Sputtering and Ion Plating. NASA SP-5111.
Burt, R. J., Meyer, S. F., and Hsieh, E. J. (1980). *J. Vac. Sci. Technol.*, **17** (1), 407–410.
Cahn, R. W. (1980). *Contemp. Phys.* **21**(1), 43–75.
Cargill, G. S. III (1975). *Solid State Phys.* **30**, 227–320.
Chambers, D. L., Carmichael, D. C., and Wan, C. T. (1972). *In* Sputtering and Ion Plating, NASA SP-5111.
Chao, E. C. T. (1967). *Science* **156**, 192–202.
Chaudhari, P., and Turnbull, D. (1978). *Science* **199**, 11–21.
Chen, H. S. (1973). *J. Non-Cryst. Solids* **12**, 333–338.
Chen, H. S. (1980). *Rep. Prog. Phys.* **43**(4), 353–432.
Chen, H. S., and Miller, C. E. (1970). *Rev. Sci. Instrum.* **41**, 1237.
Chen, H. S., and Miller, C. E. (1976). *Mat. Res. Bull.* **11**, 49–54.
Chen, H. S., and Park, B. K. (1973). *Acta Metall.* **21**, 395.
Chen, H. S., Krause, J. T., and Coleman, E. (1975). *J. Non-Cryst. Solids* **18**, 157–171.
Chen, H. S., Krause, J. T., and Sigety, E. A. (1973). *J. Non-Cryst. Solids* **13**, 321–327.
Cline, C. F. (1978). Reported at the *Gordon Res. Conf. Glass, Tilton, New Hampshire.*
Cohen, M. H., and Turnbull, D. (1961). *Nature (London)* **189**, 131.
Coutures, J., Sibieude, F., Rouanet, A., Foëx, M., Revcolevschi, A., and Collongues, R. (1974). *Rev. Int. Htes. Temp. Réfract.* **11**, 263–268.
Crawford, J. H. Jr., and Wittels, M. C. (1955). *U. N. Int. Conf. Peaceful Uses At. Energy, 1st* **7**, 654.
Croset, M., and Dieumegard, D. (1973). *J. Electrochem. Soc.* **120**, 526.
Croset, M., Petreanu, E., Samuel, D., Amsel, G., and Nadai, J. P. (1971). *J. Electrochem. Soc.* **118**, 717.
Dalton, R. H., and Nordberg, M. E. (1941). U. S. Patent 2,239,551, April 21.
Datta, R. K., Roy, D. M., Faile, S. P., and Tuttle, O. F. (1964). *J. Am. Ceram. Soc.* **47**, 153.
Davidse, P. D., and Maissel, L. I. (1966). *J. Appl. Phys.* **37**, 574–579.
De Carli, P. S., and Jamieson, J. C. (1959). *J. Chem. Phys.* **31**, 1675–1676.
Dence, M. R. (1971). *J. Geophys. Res.* **76**, 5552–5565.
Dislich, H. (1971a). *Glastech. Ber.* **44**, 1–8.
Dislich, H. (1971b). *Angew. Chem. Int. Ed.* **10**(6), 363–370.
Dobretsov, N. L., Deribas, A. A., and Maly, V. I. (1970). *Phys. Earth Planet Interiors* **3**, 348–355.

Duffek, E. F., Benjamin, E. A., and Mylroie, C. (1965). *Electrochem. Technol.* **3**, 75.
Dumbaugh, W. H., and Schultz, P. C. (1969). *Kirk Othmer, Ency. Chem. Technol.* **18**, 73–103.
Dumbaugh, W. H., and Schultz, P. C. (1975). U. S. Patent 3,864,113, February 4.
Duwez, P., and Willens, R. H. (1963). *Trans. Metall. Soc. AIME* **227**, 362–365.
Duwez, P., Willens, R. H., and Klement, W., Jr. (1960), *J. Appl. Phys.* **31**, 1136–1173.
Edahiro, T., Chiyoda, K., and Nakahara, T. (1978). *Jpn. Natl. Conv. Inst. Elec. Comm. Eng.*, Paper 906.
Engelhardt, W. von, and Bertsch, W. (1969). *Contr. Mineral. Petrol.* **20**, 203–234.
Englehardt, W. von, *et al.* (1967). *Contr. Mineral. Petrol.* **15**, 91–100.
Evitts, H. C., Cooper, H. W., and Flaschen, S. S. (1964). *J. Electrochem. Soc.* **111**, 688.
Feist, W. M., Steele, S. R., and Readey, D. W. (1969). *Phys. Thin Films* **5**, 237.
Flamenbaum, J. S., Schultz, P. C., and Voorhees, F. W. (1974). U. S. Patent 3,806,570, April 23.
French, W. G., and Pace, L. J. (1977). *Int. Conf. Integrated Opt. Opt. Fiber Commun., Tokyo, Japan* Paper B 9-1, pp. 319–322.
Frerichs, R. (1962). *J. Appl. Phys.* **33**, 1898.
Gallagher, S. A., and Roy, R. (1975). *J. Am. Ceram. Soc.* **58**, 255–256.
Geittner, P., Küppers, D., and Lydtin, H. (1976). *Appl. Phys. Lett.* **28** (11), 645–646.
Gibbons, R. V., and Ahrens, T. J. (1971). *J. Geophys. Res.* **76**, 5489–5498.
Giessen, B. C., and Wagner, C. N. J. (1971). *In* "Liquid Metals" (S. Beer, ed.).
Goell, J. E. (1973). *Appl. Opt.* **12**, 737.
Goell, J. E. *et al.* (1972). *Appl. Phys. Lett.* **2**(15), 72–73.
Goodman, A. M., and Breece, J. M. (1970). *J. Electrochem. Soc.* **117**, 982.
Gorlich, E. *et al.* (1976). *Roczniki Chem.* **50**, 1673–1679.
Gray, F. L. (1971a). U. S. Patent 3,620,704, November 16.
Gray, F. L. (1971b). U. S. Patent 3,619,440, November 9.
Grigson, C. W. B., Dove, D. B., and Stilwell, G. R. (1964). *Nature (London)* **204**, 173.
Gumbel, C. W. von (1970). *Sitz. Ber. Kg. Bsyer. Akad. Wise. Munchen* **1**, 153–200.
Hanlein, W. (1956). *Proc. Int. Glass Cong., 4th* **VIII-2**, 419–423.
Hansen, K. W., and Hood, H. P. (1970). U. S. Pat. 3,535,890, October 27.
Hasegawa, R., and O'Handley, R. C. (1979). *J. Appl. Phys.* **50**(3), 1551–1556.
Heany, J. A. (1937), French Patent 819,710, October 26.
Heany, J. A. (1938). Canadian Patent 372,557, March 15.
Hetherington, G. (1966). *J. Br. Ceram. Soc.* **3**, 595–598.
Hippler, R. (1980a). NTIS Rep. No. PB80-809163.
Hippler, R. (1980b). NTIS Rep. No. PB80-809171.
Hippler, R. (1980c). NTIS Rep. No. PB80-809189.
Holland, J. (1956). "Vacuum Deposition of Thin Films." Wiley, New York.
Horiguchi, S. *et al.* (1977). *Nat. Conv. IECE J.* 821.
Hu, S. M. (1966). *J. Electrochem. Soc.* **113**, 693.
Hyde, J. F. (1942). February 10. U. S. Patent 2,272,342,
Inoue, K., Goto, J., Arima, T., Nakamura, O., and Akamatsu, T. (1977). *Int. Conf. Integrated Opt. Opt. Fiber Commun., Tokyo, Japan* pp. 387–390.
Izawa, T. *et al.* (1977a). Japanese Natl. Conv. of Inst. Electron. Commun. Engineers Paper, p. 792.
Izawa, T., Kobayashi, S., Sudo, S., and Hanawa, F. (1977b). *Int. Conf. Integrated Opt. Opt. Fiber Commun. Tokyo, Japan* p. 375.
Izawa, T., Miyashita, T., and Hanawa, F. (1977c). U. S. Patent 4,062,665, December 13.
Izawa, T., Kobayashi, S., Sudo, S., Taka, F., Shibata, N., and Nakahara, M. (1978).

Japanese Natl. Conv. of Inst. Electron Commun. Engineers, Technical Digest, Paper 909, March meeting.
Jaeger, R. E., MacChesney, J. B., and Miller, T. J. (1978). *Bell Syst. Tech. J.* **57**(1), 205–210.
Janus, A. R., and Shirn, G. A. (1967). *J. Vacuum Sci. Technol.* **4**, 37.
Johnson, W. L. (1979). *J. Appl. Phys.* **50**(3), 1557–1563.
Jones, H. (1972). *J. Sheffield Univ. Metall. Soc.* **11**, 50–57.
Jones, H., and Suryanarayana, C. (1973). *J. Mat. Sci.* **8**, 705–753.
Kanichi, K., Sumio, S., and Tashino, N. (1976). *Yogyo Kyokai Shi* **84**, 614–618.
Kantor, P., Revcolevschi, A., and Collongues, R. (1973). *J. Mat. Sci.* **8**, 1360–1361.
Kapron, F. P., Keck, D. B., and Maurer, R. D. (1970). *Appl. Phys. Lett.* **17**, (10), 423–425.
Keck, D. B., and Bouillie, R. (1978). *Opt. Commun.* **25**(1), 43–48.
Keck, D. B., and Schultz, P. C. (1973). U. S. Patent 3,711,262, January 16.
Keck, D. B., Schultz, P. C., and Zimar, F. (1973). U. S. Patent 3,737,292, June 5.
Keck, D. B., Schultz, P. C., and Zimar, F. (1974). U.S. Patent Re 28,029, June 4.
Kennedy, T. N., (1973). U. S. Patent 3,743,587, July 3.
Kennedy, T. N. (1974). *Electron. Packag. Prod.* **14**, 136.
Kern, W. (1975). *Solid State Technol.* **18**, 25.
Kern, W., and Heim, R. C. (1970). *J. Electrochem. Soc.* **117**, 562.
Kern, W., and Rosler, R. S. (1977). *J. Vac. Sci. Technol.* **14**, 1082.
Kern, W., Schnable, G. L., and Fisher, A. W. (1976). *RCA Review,* **37**, 3.
Kingery, W. D. (1960). "Introduction to Ceramics," p. 145. Wiley, New York.
Kircher, J. F. (1964). *In* "Effects of Radiation on Materials and Components," p. 35. Van Nostrand Reinhold, New York.
Kistler, S. S. (1931). *Nature (London)* **127**, 741.
Kleeman, J. D. (1971). *J. Geophys. Res.* **76**, 5499–5503.
Klement, W. Jr., Willens, R. H., and Duwez, P. (1960). *Nature (London)* **187**, 869–870.
Kobayashi, S., Sudo, S., Miyashita, T., and Izawa, T. (1975). *Appl. Opt.* **14**(12), 2817.
Konijnendijk, W. L., van Duuren, M., and Groenendijk, H. (1973). *Verres Refract.* **27**, 11–14.
Kooi, E. (1964). *J. Electrochem. Soc.* **111**, 1383.
Krauth, Von A., and Meyer, H. (1965). *Ber. Deut. Ker. Ges.* **42**, 61–72.
Krepski, R. *et al.* (1975). *J. Mat. Sci.* **10**, 1452.
Küppers, D., Koenings, J., and Wilson, H. (1976). *J. Electrochem. Soc.* **123**(7), 1079–1083.
Küppers, D., and Lydtin, H. (1977). *Int. Conf. Chem. Vapor Dep. Proc., 6th* pp. 461–476. Electrochemical Society.
Küppers, D., Lydtin, H., and Meyer, F., (1977). *Int. Conf. Integrated Opt. Opt. Fiber Commun., Tokyo, Japan* Paper B 9-1, pp. 319–322.
Lastman, B. (1963). *In* "Uranium Dioxide, Properties and Nuclear Application" (I. Belle, ed.). U. S. Atomic Energy Commission.
Lesueur, D. (1973), C. E. N., Commis. At Emerg., Fontenay-aux-Roses, France, Rep. CEA-R-4502.
Levene, L., and Thomas, I. M. (1972). U. S. Pat. 3,640,093, February 8; (1969). Appl. 805, 841, March 10.
Levy, P. W., and Kammerer, O. F. (1955). *Phys. Rev.* **100**, 1787–1788.
Lewis, B. G., and Davies, H. A. (1976). *Mat. Sci. Eng.* **23**, 179.
Lewis, B. G., and Davies, H. A. (1977). Inst. Phys. Conf. Ser. No. 30, Chapter 2, Part 1, pp. 274–282.
Li, P. C., and Tsang, P. J. A. (1972). German Patent 2,148,120, May 25.
Lloyd, P. (1977). *R.S.R.E. Newsl. Res. Rev.* **1**, 9/1–9/4.
Luth, W. C., and Ingamells, C. O. (1965). *Am. Mineral.* **50**, 255–258.

MacChesney, J. B., O'Connor, P. B., DiMarcello, F. V., Simpson, J. R., and Lazay, P. D. (1974a). *Proc. Int. Cong. Glass, 10th, Kyoto, Japan* Paper 6-40.
MacChesney, J. B., O'Connor, P. B., and Presby, H. M. (1974b). *Proc. IEEE* **62**(9), 1278–1279.
McCarthy, G. J., and Roy, R. (1971). *J. Am. Ceram. Soc.* **54**, 639–640.
McCarthy, G. J., Roy, R., and McKay, J. M. (1971). *J. Am. Ceram. Soc.* **54**, 637–638.
Mader, S., and Nowick, A. S. (1965). *Appl. Phys. Lett.* **7**, 57–59.
Maissel, L. I., and Schaible, P. M. (1965). *J. Appl. Phys.* **35**, 237.
Masumoto, T., and Maddin, R. (1975). *Mat. Sci. Eng.* **19**, 1–24.
Mattmuller, R. (1967). U. S. Patent 3,334,982, August 8.
Miller, S. E. (1980). *Proc. IEEE* **68**(10), 1169–1360.
Milton, D. J., and DeCarli, P. S. (1963). *Science* **140**, 670–671.
Mitsuyu, T., Wasa, K., and Hayakawa, S. (1976). *J. Electrochem. Soc. Solid State Sci. Technol.* **123**, 94–96.
Moss, M., Smith, D. L., and Lefever, R. A. (1964). *Appl. Phys. Lett.* **5**, 120.
Nagel, S. R., and Tauc, J. (1975). *Phys. Rev. Lett.* **35**, 380.
Nagel, S. R., Pearson, A. D., and Tynes, A. R. (1976). *J. Am. Ceram. Soc.* **59**, 47–49.
Nassau, K., and Shiever, J. W. (1975). *Am. Ceram. Soc. Bull.* **54**(11), 1004–1011.
Nicolaon, G. A., and Teichner, S. J. (1968). *Bull. Soc. Chim. Fr.* **5**, 1900–1911; **8**, 3107–3113; **9**, 3555–3561; **11**, 4343–4347.
Nishinari, Y., Goto, J., Nakamura, R., and Akamatsu, T. (1978). *Jpn. Nat. Conv. Inst. Electron. Commun. Eng.* Paper 920.
Nogami, M., and Moriya, Y. (1980). *J. Non-Cryst. Solids* **37**, 191–201.
Nordberg, M. E. (1943). U. S. Patent 2,326,059, August 3.
O'Connor, P. B., MacChesney, J. B., and Melliar-Smith, C. M. (1977). *Electron. Lett.* **13**(7), 170–171.
Pearson, A. D. (1976). In "Applied Solid State Science," Vol. 6. Academic Press, New York.
Perri, J. A., and Riseman, J. (1966). French Patent 1,438,826, May 13.
Pietrokowsky, P. (1963). *Rev. Sci. Instr.* **34**, 445.
Platakis, N. S., and Gatos, H. C. (1976). *J. Electrochem. Soc. Solid-State Sci. Technol.* **123**, 1409–1410.
Pliskin, W. A. (1977). *J. Vac. Sci. Technol.* **14**(5), 1064–1081.
Pliskin, W. A., and Gnall, R. P. (1964). *J. Electrochem. Soc.* **111**, 872.
Pliskin, W. A., and Lehman, H. S. (1965). *J. Electrochem. Soc.* **112**, 1013.
Pliskin, W. A., Kerr, D. R., and Perri, J. A. (1967). *Phys. Thin Films* **4**, 257–324.
Poate, J. M., and Cullis, A. G. (1980). In "Treatise on Materials Science and Technology," Vol. 18, pp. 85–133. Academic Press, New York.
Polk, D. E., and Chen, H. S. (1974). *J. Non-Cryst. Solids* **15**, 165–173.
Powers, D. R. (1978). *J. Am. Ceram. Soc.* **61**(7-8), 295–297.
Primak, W. (1958). *Phys. Rev.* **110**, 1240–1254.
Primak, W. (1960). *J. Phys. Chem. Solids* **13**, 279–286.
Primak, W., and Bohmann, M. (1962). *Prog. Ceram. Sci.* **2**, 103–177.
Rao, Bh. V. J. (1963). *Phys. Chem. Glasses* **4**, 22–34.
Rau, K. (1977). Heraeus Quarzachmelze, British Patent Specification 1,492,920 November 23.
Rau, K., Muhlich, A., and Treber, N. (1977). Topical Meeting on Opt. Fiber Transmission II, Williamsburg, Virginia, Paper TuC4-1.
Revesz, A. G. (1973). *J. Non-Cryst. Solids* **11**, 309–330.
Revesz, A. G., and Zaininger, K. H. (1963). *J. Am. Ceram. Soc.* **16**, 606.

Revesz, A. G, Allison, J. F., and Reynolds, J. H., (1976). *COMSAT Tech. Rev.* **6,** 57.
Rigterink, M. D. (1976). *Am. Ceram. Soc. Bull.* **55**(9), 775–779.
Roy, R. (1956). *J. Am. Ceram. Soc.* **39**(4), 145–146.
Roy, R. (1969). *J. Am. Ceram. Soc.* **52,** 344–345.
Roy, D. M., Faile, S. P., and Tuttle, O. F. (1964). *Am. Ceram. Soc. Bull.* **43,** 291 (abstract only).
Ruhl, R. C. (1967). *Mat. Sci. Eng.* **1,** 313–320.
Sakka, S., and Kanichi, K. (1980). *J. Non-Cryst. Solids* **42,** 403–422.
Sarjeant, P. T., and Roy, R. (1967). *J. Am. Ceram. Soc.* **50,** 500–503.
Sarjeant, P. T., and Roy, R. (1969). *In* "Reactivity of Solids" (J. W. Mitchell *et al.*, eds.), pp. 725–733. Wiley, New York.
Schaal, R. B., and Horz, F. (1977). *Proc. Lunar Sci. Conf., 8th,* 1697–1729.
Scherer, G. W. (1977). *J. Am. Ceram. Soc.* **60**(5-6), 236–246.
Scherer, G. W. (1979). *J. Non-Cryst. Solids* **34,** 239–256.
Schmidt, P. F., and Ashner, J. D. (1971). *J. Electrochem. Soc.* **118,** 325.
Schmidt, P. F., and Michel, W. (1957). *J. Electrochem. Soc.* **104,** 230.
Schmidt, P. F., and Owen, A. E. (1964). *J. Electrochem. Soc.* **111,** 682.
Schnable, G. L., and Schmidt, P. F. (1976). *J. Electrochem. Soc.* **123,** 310 C.
Schott, Jenaer Glaswerk (1939). German Patent 736,411, U. S. Patent 2,366,516.
Schott, Jenaer Glaswerk (1968). French Patent 1,537,617.
Schreiber, H. U., and Froschle, E. (1976). *J. Electrochem. Soc.* **123,** 30.
Schroeder, H. (1969). *Phys. Thin Films* **5,** 87–142.
Schroeder, H. and Gliemeroth, G. (1971). U. S. Patent 3,597,252.
Schultz, P. C. (1974a). U. S. Patent 3,785,722, January 15.
Schultz, P. C. (1974b). *J. Am. Ceram. Soc.* **57,** 309.
Schultz, P. C. (1974c). U. S. Patent 3,848,152, November 12.
Schultz, P. C. (1975). U. S. Patent 3,859,073, January 7.
Schultz, P. C. (1976). *J. Am. Ceram. Soc.* **59**(5-6), 214–219.
Schultz, P. C. (1979a). *In* "Fiber Optics: Advances in Research and Development" (B. Bendow and S. Mitra, eds.), pp. 3–31. Plenum Press, New York.
Schultz, P. C. (1979b). *Appl. Opt.* **18**(21), 3684–3993.
Schultz, P. C., and Voorhees, F. W. (1974). U. S. Patent 3,801,294, April 2.
Scott, M. G. (1974). *J. Mat. Sci.* **9,** 1372–1374.
Secrist, D. R., and Mackenzie, J. D. (1965). *J. Am. Ceram. Soc.* **48**(9), 487–491.
Secrist, D. R., and Mackenzie, J. D. (1966). *Ceram. Bull.* **45,** 784.
Shishido, T. (1979). *J. Mat. Sci.* **14,** 823–830.
Shishido, T., Okamura, J., and Yajima, S. (1978). *J. Mat. Sci.* **13,** 1006–1014.
Shoup, R. (1972). U. S. Patent 3,678,144.
Shoup, R. (1976). *Colloid Interface Sci.* **3,** 63–69.
Simon, I. (1957). *J. Am. Ceram. Soc.* **40**(5), 150–153.
Simpkins, P. G., Greenburg-Kosinski, S. E., and MacChesney, J. B. (1978). *Fiber Opt. Symp. Abstr.* 137. Electrochem Society Meeting, Pittsburgh, Pennsylvania.
Sinclair, W. R. (1968). *Glass Ind.* **49,** 22–28.
Sinclair, W. R., and Peters, F. G. (1963). *J. Am. Ceram. Soc.* **46,** 20–23.
Sinclair, W. R., Flashen, S. S., and Peters, F. G. (1960). *J. Am. Ceram. Soc.* **43,** 168.
Smyrl, N., and Devlin, J. P. (1972). *J. Phys. Chem.* **76,** 3093–3094.
Sterling, H. F., and Alexander, J. H. (1974). U. S. Patent 3,843,392. October 22.
Stoffler, D. (1972). *Fortschr. Mineral.* **49,** 50–113.
Stoffler, D. (1974). *Fortschr. Mineral.* **51,** 256–289.
Stoffler, D., and Hornemann, U. (1972). *Meteoritics* **7,** 371–394.

Sudo, S. et al. (1978). *Elec. Lett.* **14**(17), 534.
Suzuki, T., and Anthony, A. (1974). *Mat. Res. Bull.* **9**, 745–754.
Teichner, S. J., et al. (1976). *Adv. Colloid Interface Sci.* **5**, 245–273.
Tick, P. A. (1975). U. S. Patent 3,858,548.
Tien, P. K. (1971). *Appl Opt.* **10**, 2395.
Tien, P. K., and Ballman, A. A. (1975). *J. Vac. Sci. Technol.* **12**, 892–904.
Topol, L. E., and Happe, R. A. (1974). *J. Non-Cryst. Solids* **15**, 116–124.
Topol, L. E., Hengstenberg, D. H., Blander, M., Hoppe, R. A., Richardson, N. L., and Nelson, L. S. (1973). *J. Non-Cryst. Solids* **12**, 377–390.
Townsend, P. D., Kelly, J. C., and Hartley, N. E. W. (1976). "Ion Implantation, Sputtering and Their Applications." Academic Press, New York.
Tsang, P. J., Anderson, R. M., and Cvikevich, S. (1976). *J. Electrochem. Soc.* **123**, 57.
Turnbull, D. (1961). *Trans. Metall. Soc. AIME* **221**, 422.
Uhlmann, D. R. (1972). *J. Non-Cryst. Solids* **7**, 337.
Uhlmann, D. R., Klein, L. C., and Handwerker, C. A. (1977). *Proc. Lunar Sci. Conf., 8th* 2067–2078.
United Technologies Research Center (1976). Reported in "Machine Design," p. 6, April 18, and "Optical Spectra," p. 22, April.
Van, T. B. (1976). PhD Thesis Ceramic Eng., Univ. of Illinois, Urbana-Champaign, Illinois.
Vandenbulcke, L., and Vuillard, G. (1976). *J. Electrochem. Soc.* **123**, 278.
Veltri, R. D. et al. (1979). *J. Mat. Sci.* **14**, 3000–3002.
Vermilyea, D. A. (1960). In "Non Crystalline Solids," Chapter 14. Wiley, New York.
Wackerle, J. (1962). *J. Appl. Phys.* **33**, 922–937.
Wada, Y., and Ashikawa, M. (1976). *Jpn. J. Appl. Phys.* **15**, 1725.
Walker, K. L., Homsy, G. M., Nagel, S. R., and Geyling, F. T. (1978). *Electrochem Soc. Meeting Pittsburgh, Pennsylvania Fiber Opt. Symp.*, Abstr. 137.
Waseda, Y., Okazaki, H., and Masumoto, T. (1977). *J. Mat.Sci.* **12**, 1927–1949.
Watts, R. K., DeWit, M., and Holton, W. C. (1974). *Appl.Opt.* **13**, 2329–2332.
Webb, A. P. et al. (1975). *J. Phys. D* **8**, 1567–1574.
Williams, J. C., Sinclair, W. R., and Koonce, S. E. (1963). *J. Am. Ceram. Soc.* **46**, 161.
Wilmsen, C. W. (1976). *Thin Solid Films* **39**, 105.
Winterburn, J. A. (1966). U. S. Patent 3,275,408, September 27.
Wittels, M., and Sherrill, F. A. (1954). *Phys. Rev.* **93**, 1117–1118.
Wizon, I., and Robertson, J. A. (1967). *J. Polym. Sci. Part C* **19**, 267–281.
Wong, J. (1972). *J. Electrochem. Soc.* **119**, 1071.
Wood, C., Schottmiller, J. C., and Ryan, F. W. (1969). German Patent 1,801,636.
Wood, D. L., MacChesney, J. B., Miller, T. J., and Fleming, J. W. (1978). Electrochemical Society Fall Meeting, Pittsburgh, Pennsylvania, Abstract No. 140.
Wullaert, R. A. et al. (1964). In "Effects of Radiation on Materials and Components," pp. 277–402. Van Nostrand Reinhold, New York.
Yamane, M. et al. (1978). *J. Mat. Sci.* **13**, 865–870.
Yamane, M. et al. (1979). *J. Mat. Sci.* **14**, 607–611.
Yoldas, B. E. (1977). *J. Mat. Sci.* **12**, 1203–1208.
Young, L. (1961). "Anodic Oxide Films." Academic Press, New York.
Zakharov, V. P. et al. (1975). *Izv. Akad. Nauk SSSR, Neorg. Mater.* **11**, 626–628.

CHAPTER 3

Inorganic Glass-Forming Systems*

N. J. Kreidl

DEPARTMENT OF CHEMICAL ENGINEERING
UNIVERSITY OF NEW MEXICO
ALBUQUERQUE, NEW MEXICO

I. Vitreous Silica	107
A. General Structural Principles	107
B. Structure and Properties	113
C. Diffusion	119
D. Preparation and Characterization	119
E. High-Density Forms of Noncrystalline Silica	122
II. Alkali Silicate Glasses	122
A. Structural Principles and Distribution of Alkali	122
B. Phase Separation	131
C. Structure and Properties	135
D. Properties of Mixed Alkali Silicate Glasses	142
E. Diffusion	144
F. The Systems $H_2O-R_2O-SiO_2$	145
III. Soda-Lime Glasses	147
A. General Considerations	147
B. Conventional Soda-Lime Glasses	149
C. Phase Separation	150
D. Additives to $Na_2O-CaO-SiO_2$ Glasses to Promote Fining	151
IV. Other Cations in Silicate Glasses	153
A. Calcium, Magnesium, and Beryllium	153
B. Barium and Strontium	154
C. Zinc Silicate Glasses	154
D. Lead, Thallium, and Bismuth	155
E. Other Constituents in Silicate Glasses	158
V. Borate Glasses	160
A. B_2O_3 Glass	160
B. Alkali Borate Glasses	163
C. Other Borate Glasses	168
VI. Borosilicate Glasses	171
A. Systems and Structures	171
B. Phase Separation in Alkali Borosilicate Glasses	174

* Section X, Ionic Salt and Solution Glasses, is by C. A. Angell.

	C. "Reconstructed" Glass	175
	D. Other Technical Alkali Borosilicate Glasses	178
	E. Alkali-Free Borosilicate Glasses	180
VII.	Aluminosilicate Glasses	181
	A. The Role of Al in Silicates	181
	B. The System Al_2O_3–SiO_2	181
	C. Alkali Aluminosilicate Glasses	183
	D. Lacy's Model	187
	E. Aluminosilicate Glasses Containing F and OH	188
	F. Alkali-Free Aluminosilicate Glasses	189
	G. Boroaluminosilicate Glasses	190
	H. Galloaluminosilicate Glasses	191
VIII.	Phosphate Glasses	191
	A. Structure	191
	B. Properties and Applications	195
IX.	Other Oxide Glasses	196
	A. Germanate Glasses	196
	B. Aluminate (Gallate and Beryllate) Glasses	200
	C. Antimonate and Arsenate Glasses	201
	D. Vanadate Glasses	202
	E. Titanate Glasses	203
	F. Tellurite Glasses	204
	G. Niobate and Tantalate Glasses	207
	H. Tungstate and Molybdate Glasses	208
	I. Other Oxide Glasses	208
X.	Ionic Salt and Solution Glasses	209
	A. General Considerations	209
	B. Nitrate and Nitrite Glasses	210
	C. Carbonate Glasses	212
	D. Formate and Acetate Glasses	213
	E. Thiocyanate Glasses	215
	F. Sulfate Glasses	215
	G. Dichromate Glasses	216
	H. Aqueous Hydrate Glasses	216
	I. Zinc-Chloride-Based Halide and Related Glasses	223
	J. Bismuth-Chloride-Based Glasses	225
XI.	Fluoride and Oxyhalide Glasses	226
	A. Fluoride Glasses	226
	B. Halides in Oxide Glasses	228
XII.	Chalcogenide Glasses	231
	A. Introduction	231
	B. Sulfur	236
	C. Selenium	237
	D. Tellurium	240
	E. Arsenic and Antimony Chalcogenide Glasses	241
	F. Chalcogenide Glasses Containing Germanium and Silicon	249
	G. Silver, Copper, and Thallium in Chalcogenide Glasses	256
	H. Chalcogens in Oxide Glasses	258
	References	260

I. Vitreous Silica

A. GENERAL STRUCTURAL PRINCIPLES

Because of the apparent simplicity of its chemical constitution and the technological importance of its properties, vitreous silica (SiO_2) has been the subject of extensive investigation [see particularly the texts by Sosman (1965), Scholze (1965), Dumbaugh and Schultz (1969), Brückner (1970, 1971), and Doremus (1973)]. However, neither the structure nor the properties and their relation to structure are simple.

In the vitreous and crystalline forms of silica, most crystalline silicates, and silicate glasses, the tetrahedral arrangement of O and Si is of fundamental significance (Goldschmidt, 1926; Zachariasen, 1932; Warren et al., 1936; Mozzi and Warren, 1969). This arrangement is to a large extent determined by the radius ratios of the constituent ions. The ratio of R_{Si} to R_O is about 0.30. Such a radius ratio demands, according to Table I, the tetrahedral arrangement generally found in silicate structures. In the tetrahedral arrangement of SiO_2, the valence of 2 of each of the surrounding four O's is obviously satisfied only if each of the O's belongs to two tetrahedra.

The tetrahedral arrangement is also favored by the bonding of electrons in SiO_2. Silicon may form from its ground state $Ne3s^23p^2$ an sp^3 hybrid with four directed bonds, which may interact with the p orbitals of O. This covalent bonding may be in partial equilibrium with the two other possibilities, primarily the pure ionic $Si^{4+} + (O^{2-})_4$, and the double-bonded (π bond) $O{=}Si{=}O$ (Pauling, 1952; Noll, 1963; Yip and Fowler, 1974). More detailed electronic models were presented by Griscom (1977). The designations Si^{4-}, O^{2-} frequently used in this and other texts in the description of silicate glasses are merely conventional.

The SiO_4 tetrahedra are almost always linked only at O corners. This is the case in quartz, cristobalite, tridymite, and vitreous silica. The excep-

TABLE I

RADIUS RATIO AND COORDINATION REQUIREMENTS

Ratio	Coordination	Polyhedron
0.155–0.225	3	Triangle
0.225–0.414	4	Tetrahedron
0.414–0.732	6	Octahedron
0.732–0.904	8	Hexahedron, cube
>0.904	12	Icosahedron

tional linkage along O–O edges is found in the fibrous polymorph of crystalline SiO_2 (Weiss, 1954); and a more striking exception of octahedral SiO_6 is found in the high-pressure crystalline polymorph stishovite (Presinger, 1962). Recent x-ray-diffraction studies (Mozzi and Warren, 1969, Da Silva *et al.*, 1975; Uhlmann and Wicks, 1979) have established in detail the tetrahedral arrangement in vitreous SiO_2. The three basic distances are Si–O 1.62 Å, O–O 2.65 Å, Si–Si 3.12 Å. The maximum in the distribution of Si–O–Si angles is at 144°, with most angles being within 10% of this maximum. A two-dimensional illustration corresponding to these data is given (Doremus, 1973) in Fig. 1. The resonance character of bonding becomes evident when one considers that pure σ bonding would require a 90° angle, and pure π bonding a 180° angle (Scholze, 1965). Positron annihilation work (Prjanishnikov *et al.*, 1971; Bartenev *et al.*, 1970; Abarenkov *et al.*, 1970) also supports the experimental distribution of bond angles in relation to electronic configuration. Various arrangements in space of corner-sharing SiO_4 tetrahedra are possible; hence the abundance of polymorphs (over 20!) of SiO_2 (Table II). For details refer to Sosman (1965).

Although these structural principles are the same for almost all crystal-

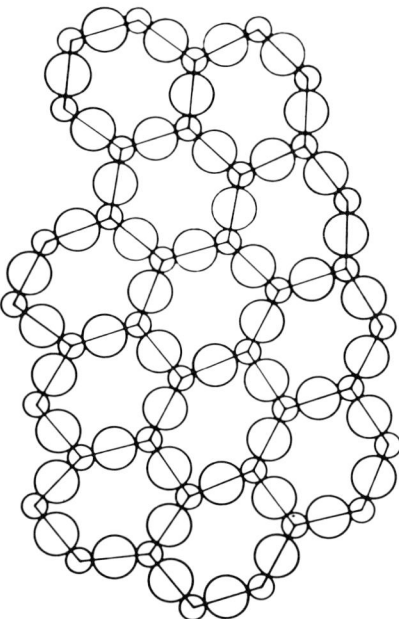

FIG. 1. Two-dimensional schematic of the structure of vitreous SiO_2. Si diameters somewhat enlarged (Doremus, 1973).

TABLE II

Twenty-Two Phases of Silica

Modification	Temperature range[a] (°C)		Modification	Temperature range[a] (°C)	
Crystalline Silica					
Quartz[b]			Cristobalite[b]		
Low	−273–573	s	Low	−273–272	m
High	573–867	s	High	272–1470	m
				1470–1723	s
			Silica W	From SiO + O in vapor	
Tridymite[b]			Coesite[c]	~300–1700	
S-I	−273–64	m	Keatite[d]		
S-II	64–117	m	Stishovite[e]		
S-III	117–163	m			
S-IV	163–210	m	Amorphous Silica		
S-V	210–475	m			
S-VI	475–867	m	Liquid	Above 1723	s
	867–1470	s			
M-I	−273–117	m	Vitreous	−273–1723	m
M-II	117–163	m	Supra-piezo-vitreous[f]		m
M-III	Above 163	m	Compacted vitreous[g]		m
			Silica M	Indefinite	n

[a] s = stable; m = metastable; n = produced from all other phases by high-speed neutrons. Pressure = 0 except as noted by footnotes c–g.

[b] To avoid any possible confusion that their use might cause, the early Greek-letter designations for high-temperature modifications are not used.

[c] Pressure ~15–40 kbar.

[d] Pressure ~0.8–1.3 kbar.

[e] Pressure ~160 kbar.

[f] Pressure ~35 kbar.

[g] Pressure ~100 kbar and lower.

line polymorphs of SiO_2 and vitreous SiO_2, the structure of vitreous SiO_2 differs from that of the crystalline forms in the lack of systematic orientation of the tetrahedra. Deviations from the cristobalite structure, the one most similar among crystalline SiO_2 structures to that of vitreous SiO_2, do not necessarily signify randomization. The entropy of melting of cristobalite to vitreous SiO_2 is remarkably small, of the order of 1 cal/mole (Rossini, 1952), as is the volume change (Pauling, 1960). From these and other considerations, the presence of substructures has been assumed, e.g, by Hicks (1967), Evans and King (1966), King (1967).

Konnert and Karle (1972, 1973) derived from their radial distribution function the possibility of a tridymite-like structure of vitreous SiO_2. Roy and Cohen (1974) dispute the justification of a definite assignment, emphasizing that various complex tetrahedral network structures resembling somewhat cristobalite or tridymite features are possible. In their view the experimental fact that vitreous SiO_2 invariably crystallizes to cristobalite* suggests at least a nonuniform structure containing disordered cristobalite-like regions. In fact, Konnert et al. (1974) insist only that their data demand a model showing some order far from one involving randomly connected tetrahedra. Nukui et al. (1978) claim that cooling the cristobalite-like structure in the melt may form a more quartz-like one.

It has been suggested that groupings differing only by their distribution of Si–O–Si angles corresponding to α and β cristobalite may exist and transform in vitrous SiO_2 (Vukcevich, 1972). Random network models constructed by, e.g., Bell (1970), Bell and Dean (1966, 1968), Bell and Hibbins-Butler (1975), Duering (1966), Polk (1971), Turnbull and Polk (1972), Ordway (1964, 1969) are becoming increasingly successful in the more detailed elucidation of possible disordered structures.

Analysis of possible ring features often shows a statistical predominance of five-membered rings; however, this should not lead one to accept too easily the pentagonal structures that have been proposed (Tilton, 1957; Robinson, 1965; Boganov et al., 1966). In liquids one frequently finds 5-coordination, but such arrangements do not permit repetition. Too much "glue" is required to connect pentagonal features to be meaningful (Turnbull and Polk, 1972; Hicks, 1967).

If and when these details in disordering a cristobalite-like three-dimensional network of SiO_4 tetrahedra are established, they will be found to underlie the remarkable time–temperature dependence of important properties, particularly density, as treated extensively by Brückner (1970, 1971). Conventional x-ray diffraction cannot determine unequivocally the intermediate-range order, but small-angle scattering (SAXS) strongly suggests that the asymptotic scattering found for SiO_2 glass is restricted to thermal fluctuations, with the remaining inhomogeneities almost certainly being "technological."

The electronic structure of vitreous SiO_2 has been the subject of extensive study (Loh, 1964; Abarenkov et al., 1970; Reilly, 1970; Bennett and Roth, 1971a,b; Ruffa, 1973; Rowe, 1974) and has been reviewed by Sigel (1973, 1974, 1977) and Griscom (1977). An energy level scheme like that proposed by Rowe (1974) seems consistent with experimental evidence from x-ray emission and absorption, photoconductivity, UV photoelec-

---

* This crystallization on the surface is suppressed by the addition of small amounts of $MoSi_2$, SiO, Si (Schlichting and Schubert, 1978).

tron spectroscopy, electron energy loss spectroscopy (ELS), etc. (Sigel, 1977) (Fig. 2). Optical absorption and reflection data suggest electronic transitions within tetrahedra. This accounts for the great similarity between the spectra of all crystalline modifications and of vitreous SiO_2. Empty states may correspond with excitons from a conduction band involving Si–Si as well as Si–O interactions, or localized Si–O antibonding (Sigel, 1977).

The band gap of fused silica is about 10 eV, in agreement with earlier estimates by Nagel (1970), Distefano and Eastman (1971), Sigel (1974), Yip and Fowler (1974). The valence band appears to be about 10–11 eV wide, and conduction bands may extend at least 8 eV beyond the gap (Griscom, 1977). Oxygen hybridization is limited. A residual charge of -1.2 on

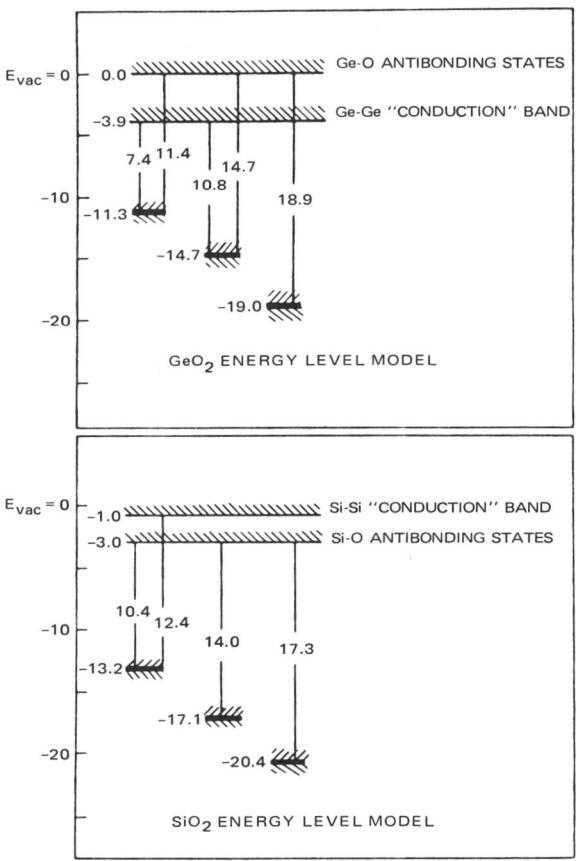

FIG. 2. Energy-level models for SiO_2 and GeO_2 consistent with experimental data. Transitions from various valence state levels to antibonding and conduction band states are shown (after Rowe, 1974).

oxygen estimated by Yip and Fowler (1974) indicates partly ionic, partly covalent bonding, in general agreement with Pauling's classification as 50% ionic–50% covalent (Griscom, 1977).

At least five defect centers can be identified by γ irradiation (Friebele *et al.*, 1977). "Wet" (high OH) SiO_2 is more stoichiometric than "dry" SiO_2, in which Si–Si bonds may be able to trap electrons. The effect of high-energy radiation on hydrogen sites in SiO_2 was studied by means of doping with deuterons (Shelby *et al.*, 1979). Other defects include nonbridging oxygens with or without alkali neighbors. Such NBO defects may have alternating + and − charges like those in chalcogenide glasses (Mott, 1977), in which case a lone-pair O orbital of one NBO may bond another O. These defects must play a role in thermally grown SiO_2 films in which electron mobility is much higher than expected from the angle disorder established by Mozzi *et al.* (Mott, 1977).

As in chalcogenide glasses (Kastner, 1978; Kastner and Hudgens, 1978), over and under coordination will exist in vitreous SiO_2^- where, however, more ionic bonding will involve positively charged trivalent Si^{3+} and negatively charged monovalent O^- (or D^-) defects (Greaves, 1979). These under radiation will trap holes or electrons to form O^0, Si^0 defects:

Si^+ Defect $\qquad\qquad\qquad$ Si^0 Defect

O^- Defect $\qquad\qquad\qquad$ O^0 Defect

Equivalent dipole centers exist, corresponding to Kastner's intimate alternate valence pairs (see Section XII):

trapping holes and electrons to form

$$\overset{\bullet}{\underset{\bullet}{O}} === \overset{O}{\underset{O}{\bullet\ Si\ \uparrow}} O$$

i.e., Si⁰ defects, and

$$\overset{\bullet}{\underset{\bullet}{O}} === \overset{\overset{\uparrow}{O}}{\underset{O}{\bullet\ Si^+}}$$

i.e., O^0 defects in the dipole center.

In the presence of OH, Si^+ and O^- convert to nonbridging OH

$$-\underset{|}{\overset{|}{Si}}{}^+ + H_2O + O^- \longrightarrow -\underset{|}{\overset{|}{Si}}OH \quad HO$$

For these models Greaves (1979) derives band levels

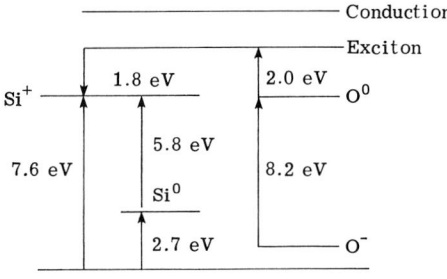

B. Structure and Properties

1. *Specific Volume*

The specific volume of fused silica has a minimum around 1550°C (Brückner, 1970, 1971) as shown in Fig. 3. Therefore, below the temperature of minimum specific volume, SiO_2 glass when cooled rapidly has a *smaller* specific volume than when cooled slowly. Most other glasses show the reverse behavior since a looser high-temperature state is frozen in on rapid cooling. In OH-rich SiO_2 glasses, the effect is much less pronounced and occurs at lower temperatures (Brückner 1970, 1971). Possibly OH

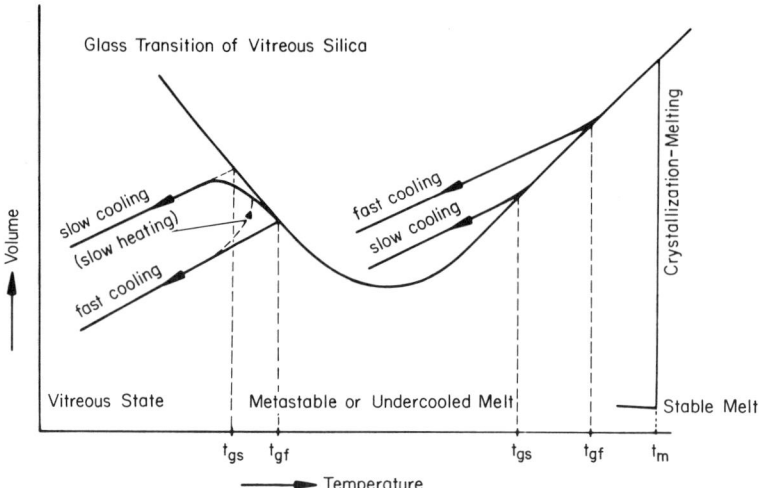

FIG. 3. Volume versus temperature for low-OH SiO_2 glass. t_{gf}, t_{gs} = glass temperatures for fast and slow cooling (Brückner, 1970).

locations are paired. A description as Si–OH and Si–H groupings was based on Raman spectra (van der Steen, 1977; van der Steen and Van den Boom, 1977). Typical values of density are 2.21–2.22 g/cm³ (Dumbaugh and Schultz, 1969). The density increases permanently by up to ~10% after subjection to high pressure at intermediate temperatures (Roy and Cohen, 1961), and by ~3% after 10^{19} n/cm⁵ sec neutron irradiation (Primak et al., 1953; Primak, 1962). The *refractive index* is affected by these volume effects by as much as 4×10^{-4} (Nassau et al., 1979). Refractive index data were provided by Waxler and Cleek (1971) and Gibbons and Kleeman (1970).

2. Coefficient of Expansion

The *coefficient of expansion* of vitreous silica is very small (5×10^{-7}) over important temperature ranges, in part because of the volume changes described above. It is negative below about $-100°C$. The expansion of crystalline SiO_2 can be considered as composed of (a) bond extension and (b) configurational expansion—reorientation and repacking of tetrahedra. The low expansion of vitreous SiO_2 might be related to groupings resembling those in high-temperature cristobalite in which configurational expansion is limited by the absence of the zigzag features of the low-temperature modification. In any case, the low-temperature negative thermal expansion coefficient seems to be associated with weak transverse vibrational modes involving Si–O–Si bridges.

3. Compressibility

The *compressibility* shows an abnormal increase with increasing pressure below about 35,000 kp/cm² (Hillig, 1962; Krause *et al.*, 1979).

4. Elastic Moduli

The *elastic moduli* (Spinner, 1962) increase instead of decreasing as in most solids, with increasing temperature above about 200°C, to a maximum at about 1100°C. The room temperature Young's modulus is 730 kbar, the shear modulus 311 kbar, the Poisson's ratio 0.17. At the maximum, the Young's and shear moduli are about 10% higher than at room temperature (see Chapter 1, Volume 5). The bulk modulus has a negative pressure dependence.

5. Strength

When exceptional care is taken to avoid surface flaws and corrosion (at about $-200°C$), the strength of vitreous SiO_2 may approach the theoretical strength (2 compared to 3.5×10^6 psi) (Hillig, 1962). For the flaw and atmosphere dependence of strength consult Volume 5. In design practice, one can only expect an impact strength of the order of 10^4 psi, still higher than that of conventional silicate glasses. The decrease of strength with temperature is interrupted by a maximum around 400°C due to the removal of OH.

6. Sound and Ultrasound

The temperature coefficient of the *sound velocity* is *positive* from 0 to 800°C (Morey, 1954). The shear wave velocity is 3.76×10^5 cm/sec. The high-frequency loss is unusually low (0.08 dB × frequency and 0.05 dB × frequency/ft MH), respectively, for shear and compressive waves (Fagan, 1951). This led to an early anticipation of the two-site-levels model of noncrystalline solids (Anderson and Bommel, 1955) by postulating two equilibrium positions for the bridging O.

Ultrasonic absorption spectra show a main peak at about 50 K, and a smaller one at lower temperature (near 5 K) (Brawer, 1975). The main peak is insensitive to, whereas the smaller peak depends on, OH concentration. The main peak may be explained by the two-site-levels (TSL) model.

7. Hardness

Knoop and Vickers *hardness* values are about 550 and 700 kP/mm², respectively (Dumbaugh and Schultz, 1969; Mackenzie, 1960a,b; Westbrook, 1960).

8. Thermal Properties

The *thermal conductivity* at 400 K is 0.00361 cal/cm sec K (Carwile and Hage, 1966). It increases to 0.00415 cal/cm sec K at 600 K, and decreases to 0.00161 cal/cm sec K at 100 K. At 10 K it is three to four orders smaller than that of crystalline quartz. This observation has been ascribed at times to residual crystallinity (Rawson, 1967), but the similar behavior of quite different amorphous solids suggests that it is related to a similar mean free path of dominant phonons (Kittel, 1967; Brawer, 1975). The deviation of the low-temperature specific heat from the Debye CT^3 law for crystals—a deviation usually described by $CT + C_3T^3$, where C_3 is up to twice the value predicted by Debye's model—has not been explained conclusively. But the deviation appears to be associated with defects in bonding characterized by phonons representing well-defined excitations (Phillips, 1972; Anderson *et al.*, 1972; Love, 1973; Schroeder, 1977—who gives a critical review mobilizing the evidence of Brillouin scattering). Scattering is governed by density fluctuations frozen in at the fictive temperature. The low depolarization ratio is in line with an isotropic SiO_4 tetrahedral network.

9. Electronic Properties

The dc conductivity of vitreous silica is almost entirely dependent on alkali and OH impurities. The change in slope of log resistivity versus $1/T$ at 225°C is caused in a manner not well explained by OH impurities (Dumbaugh and Schultz, 1969). The *dielectric constant* and *dielectric loss* are much lower than in most oxide glasses (see the volume on electrical properties). The hole mobility in thermally grown (MOS) SiO_2 films (Hughes, 1975) is $\mu \simeq 20 \exp(0.6/kT)$ eV cm²/V sec. The energy to form pairs is field dependent, becoming 18 eV/pair, and T independent. Most holes go to the Si–SiO_2 interface. Holes do not stay long enough to form definite traps, but as in other amorphous insulators move through a wide distribution of shallow traps (Gill, 1972).

The dielectric relaxation of SiO_2 at 40 K (1000 Hz) appears to be intrinsic, attributable to the lateral O motion in Si–O–Si (Anderson and Bommel, 1955; Fontanella *et al.*, 1979). It is observed in all types of vitreous SiO_2's independent of, e.g., OH or Al impurities. At very low temperatures a relaxation region is associated with H–O bonds (Jaeger, 1968; Fontanella *et al.*, 1979). The 240-K relaxation is associated with Al (Mahle and McCannon, 1969; Fontanella *et al.*, 1979). The disappearance of this relaxation after γ irradiation is attributed to the well-known hole trapping at the oxygen neighbor of Al, leading to the diffusion of alkali away from this site (de Vos and Volger, 1967).

10. Optical Properties

Silica glass is valued for its high transmission, particularly at low wavelengths in the *ultraviolet*. The UV absorption edge is largely influenced by impurities; in the purest types of SiO_2 it lies below 160 nm at 1-cm thickness (Dumbaugh and Schultz, 1969). Its shape compared to that of quartz reveals the distribution of Si–O–Si angles. A band at 242 nm is found in various types of fused silica and is most likely associated with electron traps, e.g., at Ge^{4+}, Si^{4+}, Al^{3+}, or a vacancy. Ultraviolet reflectance spectra arise essentially within the tetrahedra and are thus remarkably similar for crystal and glass, while sputtered films reveal a lack of close-range order by the absence of resolved reflectance spectra, which only begin to appear on vacuum annealing (Sigel, 1974).

In the infrared, OH causes the absorption bands at 2.73 μm and 1.38 μm. The overtone of the Si–O vibration at 8.83 μm lies at 4.45 μm, which band practically cuts off transmission for most uses of vitreous silica. A band at 2.22 μm is ascribed to OH interaction with Si–O vibrations by Adams and Douglas (1959). Vibrational spectra at very low temperatures support double-well potential models (Brawer, 1975). The similarity to cristobalite and the high degree of short-range order appear to be evidenced by the analysis of Gaskell and Ward (1976). The entire transmission behavior of the types discussed below is best summarized in Fig. 4 and Table III (Dumbaugh and Schultz, 1969). In the figure, curve A corresponds to

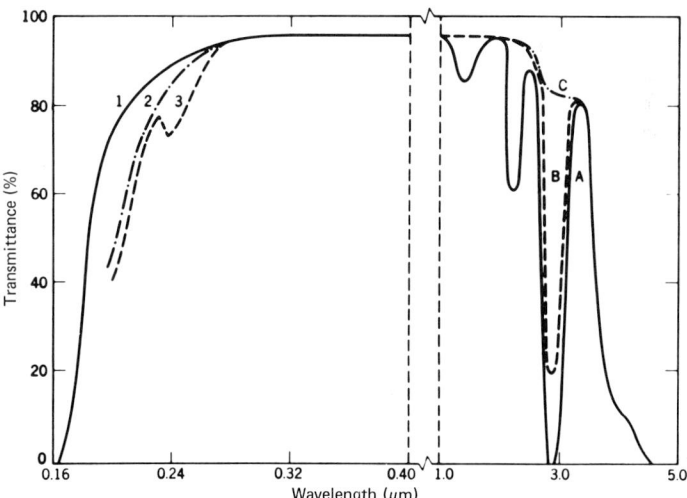

FIG. 4. Transmissions of some silica glasses. Identification of curves is given in Table III.

TABLE III

Manufacturer	Type	UV curve (Fig. 4)	IR curve (Fig. 4)
Amersil, Inc. (Heraeus)	Optosil (Herasil)	3	B
	Homosil	3	B
	Ultrasil	2	B
	Infrasil	2	C
	Suprasil	1	A
Corning Glass Works	Code 7940	1	A
	Code 7943	3	C
Dynasil Corp. of America	Dynasil	1	A
General Electric Company	Types 102, 104	3	B
	Types 105, 106, 204	3	C
	Type 125	2	C
	Type 151	1	A
Quartz et Silice	Pursil 453	3	C
	Pursil Ultra	2	C
	Tetrasil	1	A
Thermal American Fused Quartz Co.	I.R. Vitreosil	3	C
Thermal Syndicate Ltd.	Vitreosil 055, 066, 077	3	B
	Spectrosil	1	A
	Spectrosil WF	1	C
Westdeutsche Quarzschmelze GmbH	Synsil	1	A

glasses prepared by hydrolysis, thus having high OH contents, curve C to glasses prepared in dry atmospheres containing little OH, but more impurity. In H_2-impregnated SiO_2, *Raman* spectroscopy shows (a) OH stretching at 3685 cm^{-1}, (b) SiH stretching at 2254 cm^{-1}, and (c) physically dissolved H_2 at 4135 cm^{-1} (van der Steen, 1977). On drying, (a) and (b) go into (c) and H_2 diffuses.

Fluorescence is generally due to impurities.

The *refractive indexes* for the wavelengths generally considered in optical design are listed in Table IV (Dumbaugh and Schultz, 1969).

The Brewster constant relating *birefringence* to stress is 3.40 nm/cm kP (Dumbaugh and Schultz, 1969).

TABLE IV

REFRACTIVE INDEXES FOR SOME WAVELENGTHS

$\lambda(\mu m)$	0.21386	0.48613	0.58926	0.65627	3.7007
n_λ	1.53429	1.46313	1.45841	1.45637	1.39936

C. Diffusion

Silica is permeable to gases such as He, D_2, H_2, Ne, Ar, N_2, and O_2. Diffusivity (D), permeability (P), and solubility (S) for 1 atm and a given temperature are related to each other by $P = DS$.

The high permeability of He is best illustrated by the comparison of the activation energy of diffusion E_D (in kilocalories per mole) in the relation $D = D_0 \exp(-E_D/RT)$ with that of other gases:

	He	H_2	D_2	D_2
E_D	6	10	11	29

The diffusion of alkali has an activation energy decreasing with temperature from about 28 to 21 kcal. For details see Dumbaugh and Schultz (1969) and Doremus (1973).

D. Preparation and Characterization

1. General Considerations

The most important types of vitreous silica and their designations after Brückner (1970, 1971) are the following:

I. Electromelted from quartz in vacuum or an inert medium at low pressure.
II. Flame fused from quartz.
III. Hydrolyzed from $SiCl_4$ vapor.
IV. Dry (e.g., plasma flame) processed from $SiCl_4$ (also for lowest loss) in O_2 or inert gas at very high T (Kurosaki and Usui, 1978), silanes, SiH_4, etc.

Table V lists the dominant characteristics of brands corresponding to these and some other types.

The kinetics of the synthesis of type IV has been explored in detail by Powers (1978), who finds the reaction complete in minutes at 1600°C with a tenfold oxygen excess.

Silica glass may be obtained at low temperatures by hot-pressing the gel (Decuttignigs *et al.*, 1977). To give an example, one low-T synthesis leading to monolithic shapes is accomplished as follows (Yamane and Okand, 1979): Gel is prepared by hydrolyzing $Si(OMe)_4$. The gel is shaped while still porous (50–80 Å), then vacuum treated at 250°C to remove H_2O, next exposed to air to decompose residual organics, then vacuum heated at 800°C, and finally heated near 1000°C to collapse pores. Another typical starting material for gel preparation is $Si(OC_2H_5)_4$ (Nogami and Moriya, 1980; see also the chapter by Zarzycki in the volume on processing).

TABLE VA

Preparation and Characteristics (Q = Quartz) of Four Types of SiO_2 Glass

	Type			
	I	II	III	IV
Process	Electromelted Q	Flame-fused Q	Hydrolyzed $SiCl_4$	Dry from $SiCl_4$ (plasma)
Example (Brands)	IR-Vitreosil[a] Infrasil[b] Pursil[b]	Vitreosil OG, OH[a] Herasil,[b] Homosil[b] Ultrasil,[b] Optosil[b]	7940[c] Dynasil[d] Spectrosil,[a] Suprasil[b]	7963[c] Spectrosil[a] WF
GE[++]	105, 201, 204, 124, 125, 104			
OH	<5 ppm	0.15–0.04 wt %	~0.1 wt %	~0(<0.4 ppm)
Al (ppm)	30–100	<1	<0.2	<0.2
Na (ppm)	4	<1	<0.2	<0.2
Cl (ppm)	—	—	100	Up to 200
IR bands	—	—	2.73, 2.6, 2.22, 1.38 μm	—
UV cutoff (nm)	240	240	190	190
Fluorescence (nm)	280, 390 (except GE 124)	280, 390	—	—

[a] Thermal Syndicate England.
[b] Amersil, Heraeus.
[c] Corning.
[d] Dynasil Corp.
[e] GE = General Electric.

TABLE VB

Other Forms of Preparation

Type	Preparation	Type	Preparation
V	Reactive sputtering (Peters, 1966; Revesz, 1972)	VIII	Hot-pressed multiform (Vasilov, 1960, Pivinskii, 1970)
VI	Oxidation of Si (Atalla et al., 1959, Rudenberg, 1962)	IX	Thermal oxidation (Revesz, 1972)
VII	Glow discharge	X	Anodic oxidation (Revesz, 1972)

Note: For a nontransparent variety, see Dumbaugh and Schultz (1969). Lower viscosity (activation energy 122 comp. to 170).

2. Doped Silicas

SiO_2 accepts significant amounts of additives with more or less difficulty. Cooxidation or cohydrolysis is possible by feeding another chloride with $SiCl_4$ to the combustion system. The most important application is to optical fibers to achieve step or gradient cladding (Gossink, 1977; McChesny et al., 1974; Keck et al., 1973).

Another important example is the joint hydrolysis of $SiCl_4$ and $TiCl_4$, resulting in a glass containing about 7% TiO_2 and having less than one-tenth the coefficient of expansion of pure vitreous silica ("ultra-low-expansion" glass) (Rathmann et al., 1968; Kozlova et al., 1968, 1969). Clad fibers are produced on the same principle (Carpenter, 1973). Still lower expansions can be achieved for useful temperature ranges (Schultz and Smyth, 1974). Ti may also be incorporated from alcoholates (Kamiya et al., 1977).

More recalcitrant additives can be accommodated using the so-called soot process (Flamenbaum and Schultz, 1973; Schultz and Smyth, 1974). Fused SiO_2 is deposited from $SiCl_4$ by hydrolysis on a rotating nonmetallic mandrel in the outer cone of the flame. The layer of the resulting soot is introduced into an induction furnace at 1000–1700°C at about 0.25–2.5 cm/min.

Doping can be accomplished by addition to the $SiCl_4$ vapor of, e.g., $ZrCl_4$, $SnCl_4$, $GeCl_4$ (Hammond and Norman, 1977; Hammond, 1978; Huang et al., 1978), PCl_3 (Hammond and Norman, 1977), and also from ozone and organic Si and P compounds at low temperatures (Maeda and Sato, 1977), and by implantation (Haack, 1977) as with $TaCl_5$, $TiCl_4$, $AlCl_3$ (Kueppers, 1976). The addition of GeO_2 (Huang et al., 1978) and $GeO_2 + B_2O_3$ (McChesney et al., 1974; Gossink, 1977) for the purpose of obtaining gradient index fibers has led to increased knowledge of these binaries and ternaries. The maintenance of the typical open structure of SiO_2 glass in the binary SiO_2–GeO_2 is documented by the fact that the density and n_D conform to invariance of effective ionic volume and to the Lorenz–Lorenz relation $(n^2 - 1)/(n^2 + 2) = 4\pi/3 \, \Sigma N_e \alpha_e$ (N_e is the number of electrons, α_e the polarizability). The addition of P_2O_5 suppresses the OH content (Gossink, 1977). Other reports on doped SiO_2 include

B_2O_3	(in gradient fibers) (Carpenter, 1973; Hammond and Norman, 1977, Mita, 1979)
CuO	(Schultz and Smyth, 1974)
FeO	(Schultz and Smyth, 1974)
Cr_2O_3	(Schultz and Smyth, 1974)
N	from $SiCl_4$ and NH_3 (Kato et al., 1972)
Nd_2O_3	(Amosov et al., 1969)

Nb_2O_5 (in gradient fibers) (Carpenter, 1973)
V_2O_5 (Settarova et al., 1973; Schultz, 1976)
CoO (Settarova et al., 1973)
NiO (Schultz and Smyth, 1974)
Al_2O_3 (in gradient films) (Carpenter, 1973; Leko, 1977)
Ga_2O_3 (Leko et al., 1977)
Eu_2O_3 (Settarova et al., 1973)
WO_3 (Zhilova et al., 1972) (to control heat capacity)
SiF_4 (Abe, 1970; Kueppers, 1976; Mita, 1979)

E. High-Density Forms of Noncrystalline Silica

Silica glass can be permanently compressed to a higher density (2.61) (+18%, almost that of quartz) by applying pressures up to 100 kbar (Roy et al., 1961; Simon, 1953; Kennedy, 1960). Suprapiezo vitreous SiO_2 was first reported forming abruptly at about 30 kbar by Bridgeman (1939). The compressibility increases with pressure (Andersen, 1958; Bogardus, 1965). Densification is enhanced by OH content (Arndt et al., 1969). Yet Infrasil shows the largest densification in spite of its low OH content because of a high impurity and vacancy content (Arndt et al., 1969).

At much higher pressures (400 kbar) permanent densification decreases according to Arndt et al. (1969). Yet Kawai et al. (1971) report densities above 5.9 (refractive index above 1.95) at about 2 Mbar. Densification is removed by annealing at 300°C, at rates that increase with temperature. The activation energy of annealing is remarkably small (1–10 kcal) (Brückner, 1970; DeCarli and Jamieson, 1959; DeCarli et al., 1964). The form obtained by shock loading (Arndt et al., 1971) is comparable to that obtained under static pressure. The shock form still contains most Si in 4-coordination, unlike the crystalline compressed form stishovite. Yet a 6-coordinated noncrystalline form analogous to a stishovite may exist (DeCarli and Jamieson, 1959; DeCarli et al., 1964).

Pi bonding seems to decrease with pressure, and refractive index data suggest a certain "remembrance" of crystalline polymorphs according to Revesz (1972) and Arndt et al. (1969). Densified SiO_2 glass expands on irradiation (Revesz, 1972). Enthalpy, as determined by solubility changes in HF, increases with densification to 80 kbar. The energy change corresponds to that of densification (Couty and Gabatier, 1978).

II. Alkali Silicate Glasses

A. Structural Principles and Distribution of Alkali

The vast majority of industrially important glasses are silicate glasses containing, besides SiO_2, two or more major oxides. Binary alkali silicate glasses, technologically limited mostly because of their reactivity with H_2O, have nevertheless been the subject of extensive studies as simple systems providing insight into structure–property relations. On adding alkali oxides to SiO_2, the extra oxygen provided by Na_2O to the continuous $(SiO_4)^{4-}$ network increases the O/Si ratio in the network above its value of 2 in SiO_2, resulting in the appearance of oxygens bonded to only one Si atom. These oxygens are termed "nonbridging" and are charge-compensated by the monovalent cations, while their position in the structure remains relatively unchanged in the SiO_4 tetrahedron. But there is some weakening of the Si–O bond with increasing alkali, as indicated by the decrease of the Si K_β x-ray emission wavelength (Sakka and Kamiya, 1976) and by time-of-flight n-scattering (Misawa *et al.*, 1980; Sakka 1976). This weakening is associated with the fact that the coordination sphere of oxygen around alkali includes bridging as well as nonbridging oxygens. The weakening can be evaluated by the chemical shift of the peak assigned to bridging oxygens in the XPS (x-ray photoelectron spectrum) toward that of nonbridging oxygen (Brückner *et al.*, 1978, 1980). The effect increases with increasing alkali field strength from Cs^+ to Li^+.

The weaker bonding of nonbridging oxygen is manifest in the shift toward the visible of the UV absorption edge (Stroud *et al.*, 1965; Mott, 1977; Greaves, 1979). In Greaves' (1979) scheme of defects in SiO_2, the nonbridging O^- and the compensating R^+ can be conceived as defect centers. The well-known hole trap in SiO_2 can be considered shifted below the intrinsic state; the electron trap is modified by the near R^+, giving rise to a 5.3-eV absorption, compared to 5.5 eV for the E′ center in SiO_2. This absorption corresponds with the ESR signal observed by Sigel (1973/1974) and others.

In alkali (e.g., sodium silicate) glasses, among many possibilities of detailed arrangement, that of a random distribution of the sodium among the $(SiO_4)^{4-}$ units is one extreme. The other extreme would be the appearance of definite sheets, chains, rings, and isolated tetrahedra as in crystalline silicates, depending on their O/Si ratio. Some early evidence (see Mackenzie, 1960a,b; Bockries and Lowe, 1954; Brockries and Mackenzie, 1955; Bockries *et al.*, 1956, 1958) pointed at a distribution of some anionic groupings in the melt and glass rather than an entirely random structure. A more significant than conspicuous difference is illustrated in Fig. 5. In the lower alkali range, the linked-sheet systems stable in crystalline forms do not arrange easily in the cooling melt. Consequently, solid glasses in this range tend to exhibit two-phase structures starting at low Na_2O concentrations with islands of sodium silicate groups in the SiO_2-rich matrix, changing with increasing Na_2O content to SiO_2-rich islands in sodium silicate matrices (Ohlberg *et al.*, 1965).

In these structures, sodium is essentially bonded ionically and is mobile

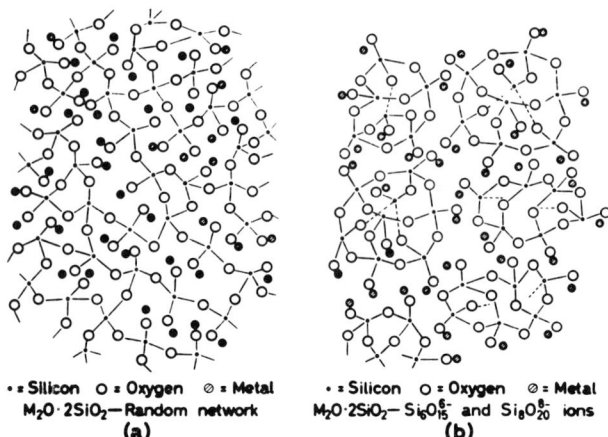

FIG. 5. (a) Random structure versus (b) anionic grouping in alkali silicate glasses (Mackenzie, 1960).

at relatively low temperatures, an important property reaching into more complex technological glasses. This behavior may be less prominent both toward Cs^+ and Rb^+ via K^+, because of their larger size, and toward Li^+ because of its higher field strength or (Kolesova, 1971) more covalent bonding to oxygen.

Although Warren's classical study (Warren *et al.*, 1938) was often used in support of a random structure, Warren himself had pointed out—as noted by Weyl and Marboe (1964)—that his x-ray data are also consistent with SiO_2 island groups in a Na_2O-rich silicate structure, the other extreme approached, as we now know, in certain sodium silicate glasses heat-treated to complete phase separation. Warren *et al.* (1938) had established

(1) Si–O peaks near 1.65 Å,
(2) O coordination of Si 4.2–4.5,
(3) irresolvable O–O and Na–O peaks at 2.3–2.6 Å,
(4) an Si–Si peak near 3.2 Å, vanishing at 35 Na_2O, and
(5) O coordination of Na: 6, K: 10.

Warren's own cautious conclusions were limited to

(1) *average* coordination of Si is 4,
(2) discrete molecules as postulated in the past by Preston *et al.* (1936) are unlikely, and
(3) neither random distribution nor a mixture of SiO_2 and sodium silicate type groupings is contradicted by the x-ray evidence.

Recent electron radial distribution (Urnes, 1969) and n-diffraction data (Hoffman et al., 1966; Anderson and Urnes, 1975) generally confirm that Si–Si, Si–2nd-O, and O–O coordination remain the same as in pure vitreous SiO_2. Porai-Koshits (1955, 1965; Porai-Koshits and Averyanov, 1969), after modifying the controversial crystallite hypothesis of SiO_2 glass structure, maintained the original (Valenkov and Porai-Koshits, 1936) postulate of volume elements of different groupings in alkali silicate glasses. Hartleif (1938), immediately following Warren in his study of potassium silicate glasses, believed he had sufficient evidence for nonrandom distribution of K^+ ions. The case for grouping was supported by Domenici et al. (1970) (in lithium silicate glasses), Blau (1951), Dietzel (1966), Milberg and Peters (1969) (in thallium glasses), and Brosset (1958), the last demonstrating clustering in barium silicate glasses.

The difficulty for straightforward x-ray evidence lies in the small difference in the electronic densities of Na, Ca, Si, and O. Ohlberg and Parsons (1964) attempted to demonstrate the distribution of alkali by the elegant approach of replacing sodium by the high-electron-density species silver in a diffusion process conducted at temperatures low enough (300°C) to preclude structural rearrangement and subjecting the substituted glass to x-ray-diffraction studies. Na–Na distances in random distribution should average near 7 Å. Although Ohlberg's lowest peak corresponding to 3.4 Å was, it seems with justification, disputed by Prins (1965) and Urnes (1969), the distance of 4.9 Å can still be considered evidence for nonrandom grouping. Unfortunately, the study of Guaker and Urnes (1973) on silver-substituted Na_2O–SiO_2 glasses shed severe doubt on the possibility of demonstrating shortest Na–Na distances convincingly. They attribute a very large peak between 2.3 and 4.3 Å with a maximum at 3.25 Å to Na–O and Na–Si distances (with Ag substituting for Na) on the basis of equivalent distances in crystalline meta- and disilicates. If Na–Na distances had made a major contribution, a shift to higher radius would have to be expected. Thus Guaker and Urnes (1973) reject the evidence for short Na–Na distances and prefer the reasonable assumption that the glasses have a wide Na–Na distribution as in silicate crystals. It should be noted that this does not exclude considerable order; i.e., it does not confirm a random distribution (Breskhovskikh et al., 1971). ^{29}Si NMR data in Na_2O–SiO_2 glasses also seem to indicate a nonrandom Na^+ distribution (Mosel et al., 1974).

Porai-Koshits' (1955) differential method indicates definite distances (Fig. 6). As a matter of fact, initial groupings of Na_2O–SiO_2 and SiO_2 can be considered indicated in his analysis of a $Na_2O \cdot 2SiO_2$ glass (Fig. 7). Breskhovskikh et al. (1971) seem to be able to conclude from NMR

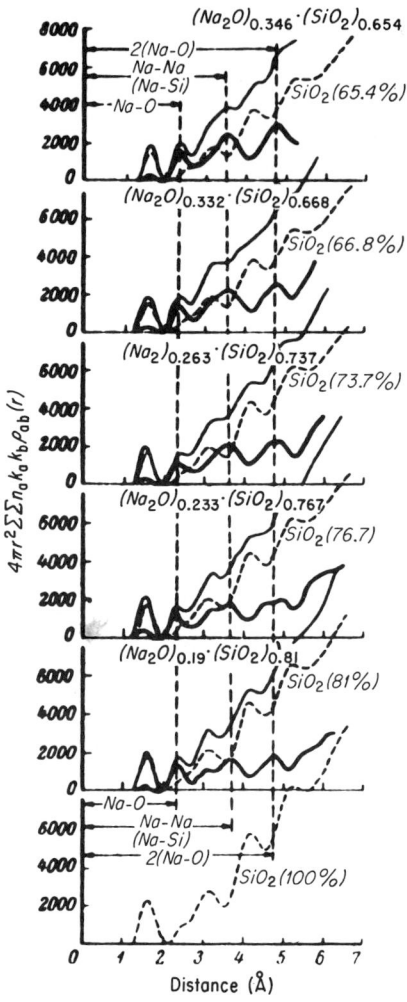

FIG. 6. X-ray diffraction of sodium silicate glasses (Porai-Koshits, 1955).

spectra some ordering of alkali, with a preference for mixed neighboring when two alkalis are present.

Positron annihilation experiments combined with x-ray spectroscopy appear capable of revealing pairing of alkali ions and a tendency to microheterogeneity involving lower and higher alkali regions (Gorbachev et al., 1978). It seems also possible to "count" groupings of two or more nonbridging oxygens, as was tried in the case of the system K_2O–SiO_2 using NMR signals based on the spin $\frac{1}{2}$ of 4%-abundance ^{29}Si (Harris and

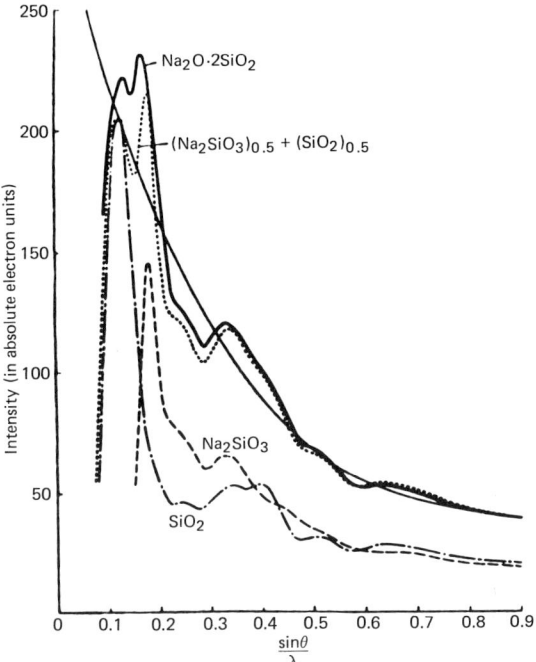

FIG. 7. Indication of Na_2O–SiO_2 and SiO_2 grouping from x-ray diffraction analysis of sodium silicate glasses (Porai-Koshits, 1955).

Bray, 1980). Ordering of K^+ is also indicated for $K_2O \cdot SiO_2$ glass where RDF from x-ray data suggest close similarities to the structure of the crystalline potassium metasilicate (Misawa *et al.*, 1976–1980).

With increasing alkali content, the concentration of nonbridging oxygens and the isolation of tetrahedra increases, saturating formally at Na_4SiO_4, where all four oxygens are nonbridging and all tetrahedra are isolated. With increased alkali content, glass formation becomes less pronounced as nucleation and growth rates become more rapid.

Infrared bands are broadened and shifted increasingly by alkali in the order K > Na > Li (Simon, 1953). A quantitative interpretation of IR and Raman spectra has been attempted by Gaskell and Johnson (1976).

The *phase diagrams* of the systems (a) Li_2O–SiO_2, (b) Na_2O–SiO_2, and (c) K_2O–SiO_2 are shown in Figs. 8a–c and 9. One observes (see also Table VI):

(a) The liquidus termperature decreases steeply from SiO_2, in most concentration ranges in the order Rb^+ (most) to Li^+ (least). The

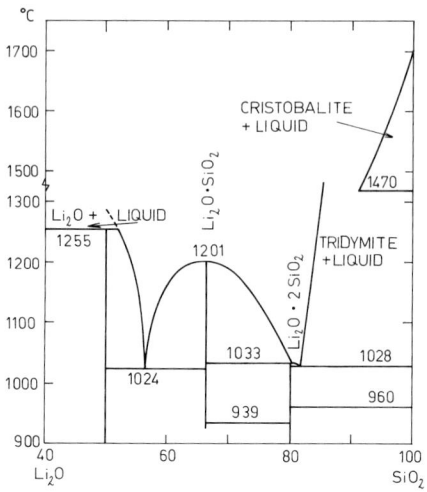

FIG. 8a. Schematic of the phase diagram of the system Li_2O–SiO_2.

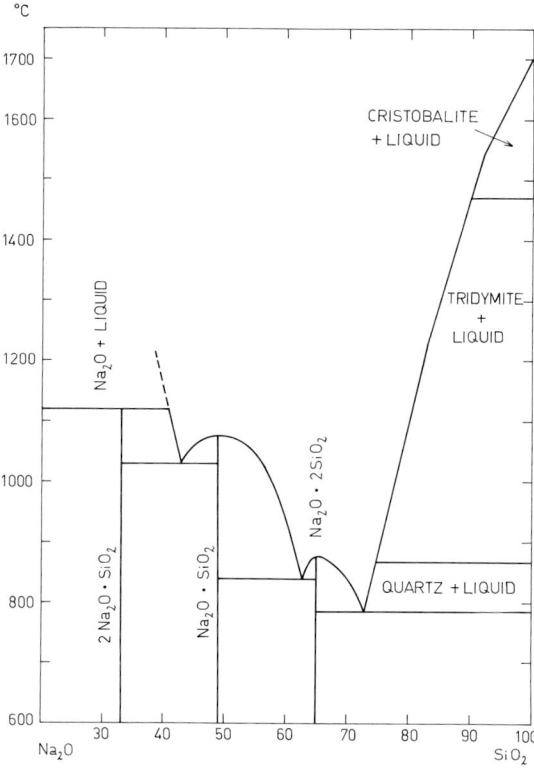

FIG. 8b. Schematic of the phase diagram of the system Na_2O–SiO_2.

3. INORGANIC GLASS-FORMING SYSTEMS

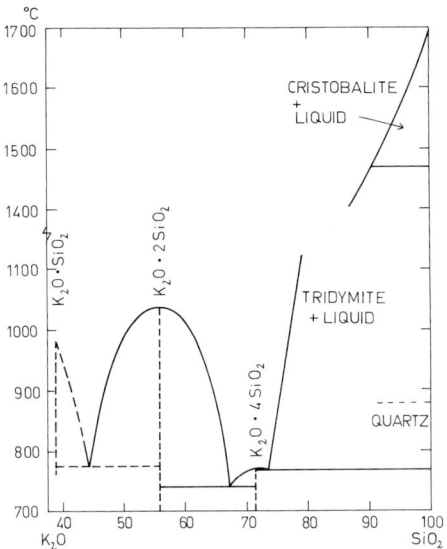

FIG. 8c. Schematic of the phase diagram of the system K_2O-SiO_2.

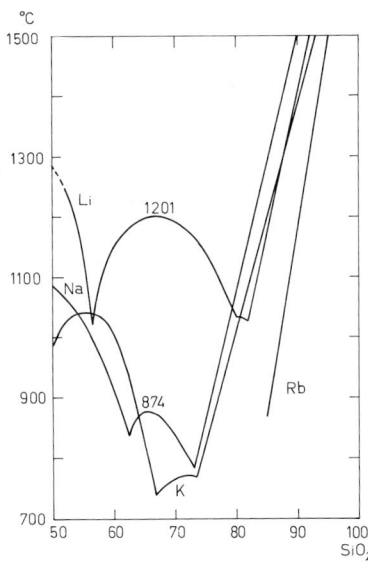

FIG. 9. Comparison of the phase diagrams of the systems Li_2O-SiO_2, Na_2O-SiO_2, and K_2O-SiO_2.

interaction of R^+ with nonbridging oxygen may be expected to increase in this order of decreasing size (increasing field strength).

(b) Compound formation tendency decreases in the same order. Tetrasilicates exist only starting with $K_2O \cdot 4SiO_2$.

(c) In the range near the disilicate ratio, the more pronounced compound formation of $K_2O \cdot 2SiO_2$ reverses the order of liquidus temperatures (highest for potassium silicate melts).

Details are illustrated in Table VI. One sees that to decrease the melting point of SiO_2 (1713°C) to 1500°C one requires least Rb and K, most Li.

Rawson (1967) makes the instructive remark that the glass-forming ability for equal amounts of nonbridging oxygen seems to be less in silicate than phosphate glasses because of the lower melting points of crystalline phosphate systems. This, according to him, clearly shows the unsatisfactory nature of considerations of glass stability based solely on structural arrangement, a view supporting concepts of glass formation expounded by Rawson (1967) and in this volume (Chapter 1).

Similarly, the range of glass formation in the higher melting lithium silicate glasses is smaller (25–37% mole) than in Na_2O (0–50), K_2O, Rb_2O, and Cs_2O (0–60) glasses. Also the kinetics of crystallization of the secondary phase $Li_2O \cdot SiO_2$ is more important in this system (Rawson, 1967; Mazurin *et al.*, 1973; Imaoka and Mazurin, 1966; Blair and Urnes, 1961; Moore and Carey, 1957; Marinov and Dimitriev, 1964).

More recently a new phase $(Na_2O)_3(SiO_2)_8$ was reported (Williamson and Glasser, 1965) in a small area of the phase diagram. At about 700°C it decomposes into quartz plus $Na_2O \cdot 2SiO_2$.

The first addition of alkali, i.e., that of the order of 1%, does not lead to lower melting glasses, as might be expected considering the strong decrease in liquidus temperature. Rather, cristobalite precipitates easily, most likely via the formation of asymmetrical R^+-nonbridging oxygen defects (Weyl and Marboe, 1964).

In the phase diagram Na_2O–SiO_2 the first binary compound is

TABLE VI

CHARACTERISTICS OF ALKALI INTRODUCTION INTO SiO_2 STRUCTURE

	Rb_2O	K_2O	Na_2O	Li_2O
To decrease MP to 1500°C (mole %)	5	7	10	16
To decrease MP to 870°C (mole %)	15			
Liquidus for metasilicate ($R_2O \cdot SiO_2$) (°C)		976	1089	1201
Liquidus for disilicate		Highest	Lowest	Intermediate
Tetrasilicate exists	Yes	Yes	No	No

$Na_2O \cdot 2SiO_2$. It has a sheet structure (Pant and Cruikshank, 1968) with Si–O distances 1.643 Å, 1.609 Å (bridging), and 1.578 Å (nonbridging); Si–O–Si angles 138.9° and 160.0°. Each Na has five links to O at distances of 2.29–2.60 Å. According to Zarzycki (oral communication), nonbridging oxygens tend to have a 10% shorter bond length.

The establishment of an *energy level diagram* for alkali silicate glasses does not seem advisable at this time. Alkalis seems to leave the interband peak fairly unaltered while perturbing the excitation peak (Sigel, 1977). Rowe (1974) had indicated how impurities should affect the antibonding level of SiO_2. The general situation has been reviewed by Tauc (1975).

B. Phase Separation

1. General Considerations. Na_2O-SiO_2

The phase diagrams of the R_2O-SiO_2 systems do not show areas of stable liquid–liquid immiscibility. However, glass–glass immiscibility below the liquidus may always occur as a nonequilibrium phenomenon if crystallization is avoided. The inevitable critical range for separation lies at a temperature at which the viscosity is not too high for a sufficient rate of separation. The system Na_2O-SiO_2 does indeed show liquid–liquid phase separation. The thermodynamic and kinetic conditions for this phenomenon are described in detail in the volume on submicrostructure.

Phase separation in the Na_2O-SiO_2 and Li_2O-SiO_2 systems was established, at times after hot dispute, once tools such as light scattering, small-angle x-ray scattering, and electron microscopy had been utilized with confidence (Prebus and Michener, 1952, 1954; Poraï-Koshits, 1955; Vogel, 1959, 1960, 1966; Vogel and Byhan, 1964; Vogel and Gerth, 1958; Vogel *et al.*, 1958, 1969, 1974; Zarzycki and Mezard, 1962; Sella *et al.*, 1964; Oberlies, 1964; Ohlberg *et al.*, 1965).

In the system Na_2O-SiO_2 at 600°C the gap ranges from less than 1 to about 20 Na_2O (Hammel, 1965; Hammel and Ohlberg, 1965). The dome of immiscibility is indicated in Fig. 10. In accord with theory, spherical SiO_2-rich phases are observed in the low-Na_2O region, spherical Na_2O-rich phases in the high-Na_2O area, while the center region exhibits interconnected microstructures. This behavior was investigated by many workers, such as Tran (1965), Charles (1966), Tomazawa *et al.* (1968), Tomazawa and Obara (1973), and Redwine and Field (1968, 1969).

Under realistic conditions the evolution of phases is far more complex than it would appear from theory and the established immiscibility dome. First, sequential phase separation must be expected and has been observed and interpreted in step-cooling experiments (Porai-Koshits and Averyanov, 1969, Senard *et al.*, 1968a). As sketched in Fig. 11, a phase R_I

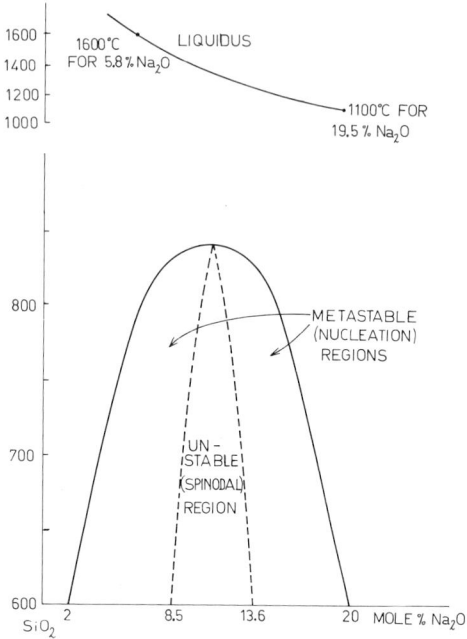

FIG. 10. Below-liquidus immiscibility in the system Na_2O–SiO_2 (schematic after Hammel, 1965, 1967).

separated at a higher temperature from composition A will, once more, when held at a lower temperature separate into S_{II} and R_{II}, which later, in turn, will separate into S_{III} and R_{III}, with Na_2O content increasing from R_I to R_{III}. On continuous cooling, the condition becomes quite complex since separation first increases in rate for thermodynamic reasons, then decreases in rate for kinetic reasons. It was an important contribution of Neilson (1969) to show that holding in the spinodal region does not guarantee interconnected microstructures. In a 12.5% Na_2O glass the following sequences were observed in a given regime (Neilson, 1972):

(a) spinodal heterogeneity,
(b) reversion toward homogeneity,
(c) growth of SiO_2-rich particles leading to a matrix richer in Na_2O than predicted from the diagram, and
(d) coarsening of particles.

It was demonstrated by Seward et al. (1968b) that a discrete particle structure can develop into an interconnected structure and vice versa on heat treatment. Some similar results were noted by Haller and Macedo,

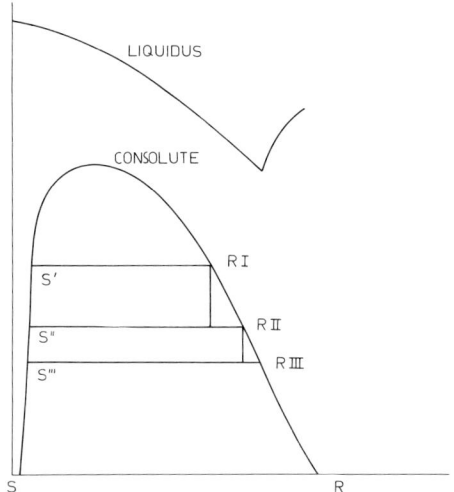

FIG. 11. Sequential phase separation in the system alkali oxide(R)–SiO$_2$(S) (schematic): A = original composition; R_I, R_{II}, R_{III}-alkali contents of alkali-rich phase after first, second, third hold; S', S", S"'-corresponding SiO$_2$ contents; consolute = boundary of immiscibility dome (after Porai-Koshits and Averyanov, 1969).

(1968). The complexity of initial phase separation can be demonstrated by small-angle neutron studies (Roth and Zarzycki, 1974). It becomes increasingly evident that composition and density fluctuations start above the critical temperatures of the immiscibility dome (Porai-Koshits *et al.*, 1971; Ohlberg and Golob, 1973; Golubkov *et al.*, 1975; Schroeder, 1977). In the binary Na$_2$O–SiO$_2$ system, the following effects of third components in small quantities have been observed.

Hydroxyl accelerates phase separation without affecting the equilibrium conditions much (Maklad and Kreidl, 1971; Kreidl and Maklad, 1969). The critical temperature is lowered and thus immiscibility practically decreased in the order given by small amounts of Ca < Ba < Cd < Zn < Mg < Pb (Moriya, 1970). The large effect in decreasing immiscibility of Al (Kreidl and Maklad, 1969; Moriya, 1970b) and P is ascribed to their high field strength (Moriya, 1970b). Taking into account the distribution of bridging and nonbridging oxygens, a parallel between reduction of liquidus and immiscibility temperatures can be seen (Tomazawa and Obara, 1973). The effect of Ga is similar to that of Al (Topping and Murthy, 1973). Density changes were used to indicate phase separation by Shaw *et al.* (1969), electrical conductivity by Charles (1966) and Hayami and Terai (1972), and magnetic spin resonance by Johnstone (1967).

The behavior of viscosity and T_g in phase-separated glasses was studied by Li et al. (1970), Mazurin and Tret'yakova (1970), and Maklad and Kreidl (1971). The connected phase may dominate a property. The elastic modulus of a phase-separated sodium silicate glass may decrease less or even increase with temperature (Redwine and Field, 1968, 1969). The coefficient of expansion may also be affected in this manner.

2. Atomistic Models

Atomistic models for immiscibility of glass have been proposed by Warren and Pincus (1964), Levin and Block (1957), and Schmitt (1962). They are based on geometrical postulates. Other, more chemical approaches were advanced by Charles (1968, 1969).

3. Role of Alkali Species

The deviation of the melting-point depression indicates the tendency toward phase separation. Mg, Ca, and Sr show definite stable immiscibility. The S shapes of the liquidus curves of Ba, Li, and Na correlate with the tendency to subliquidus glass–glass immiscibility; the straight lines for K, Rb, and Cs correlate with the effective lack of observed immiscibility in the binaries K_2O (Rb_2O, Cs_2O)–SiO_2 (Moriya, 1970). In K_2O–SiO_2, there is a mere indication of separation (Gupta and Mishira, 1969; Charles, 1966, 1967, 1968) but no immiscibility boundary has been established. Rates of separation are too slow at the temperature where a critical immiscibility temperature may be required thermodynamically (perhaps 530°C) (Moriya, 1970a,c). Monophasic fluctuations smaller than 200 Å have, however, been observed between 7 and 17 K_2O by Andreyev et al. (1971). In Li_2O, as would be expected, the phase separation range is larger, the critical temperature higher (Fig. 12) (James and McMillan, 1970; Moriya, 1970a,c; Doremus, 1970; Averyanov and Porai-Koshits, 1966) (Table VII) (Tomazawa et al., 1972). At the lowest temperatures for which observations are possible, the boundary approaches but does not

TABLE VII

CONSOLUTE TEMPERATURES FOR ALKALI–SILICA BINARIES

Alkali	Li_2O	Na_2O	K_2O
T_c	~1200	850	(530) not observed, estimated by Moriya (1970a,c)
Approximate come position (mole %)	~10	~8	—
Approximate range of immiscibility at 0.8 T_c (mole %)	25	18	(Scholze, 1977)

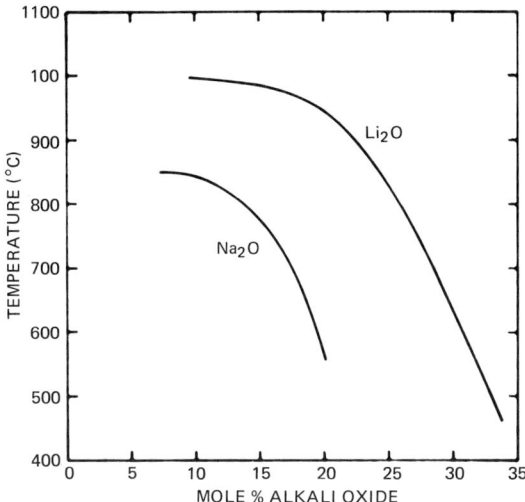

FIG. 12. Phase separation in Li_2O-SiO_2 and Na_2O-SiO_2 glasses (Doremus, 1973).

exceed the composition of $Li_2O \cdot 2SiO_2$ (Tomazawa et al., 1972). Crystallization at interfaces is frequent.

The approximate location of the critical temperature (apex of the immiscibility dome) is 1230°C for Li_2O, 850° for Na_2O. From chemical activities a quite rough preestimate of these temperatures had been made by Charles (1967, 1968).

C. STRUCTURE AND PROPERTIES

The *density* of binary alkali silicate glasses increases generally with increasing alkali contents as the cavities in the loose $(SiO_4)^{4-}$ network are filled. The influence of the kind of alkali (Table VIII) becomes more evident when the molar volume is considered. While Na contributes mostly to the filling of cavities, K additionally *expands* and Li *contracts* the network (Fig. 13). This trend can still be recognized at high temperatures (Fig. 14). The changes of density with thermal history cause the "secular rise" and "ice-point depression" of thermometers, particularly in the case of the presence of two alkalis.

The increase in the asymmetry of SiO_4 tetrahedra by the introduction of nonbridging oxygens is a major cause for the large increase in the *coefficient of expansion* with increasing alkali content in binary alkali silicate glasses. The influence is larger the smaller the field strength or larger the ionic radius, so that the coefficient of expansion generally increases in the

TABLE VIII

Density[a] at Room Temperature

Li_2O		Na_2O		K_2O		Rb_2O		Cs_2O	
Mole %	Density	Mole %	Density	Mole %	Density	Mole %	Density	Mole %	Density
10	2.231	(9.7)	2.291	(11.6)	2.3172	(2.7)	2.498	(12.2)	2.926
15	2.258	(14.5)	2.334	(15.3)	2.3523		2.708	(15)	2.9[c]
20	2.277	(19.1)	2.383	(20.1)	2.3923		2.89[b]	(20)	3.268
25	2.302	(24.4)	2.431	(24.8)	2.4242			(26.7)	3.529
30	2.334	(28.3)	2.459	(29.3)	2.4522	(31.5)	3.040		
35	2.354	(34.6)	2.498						

[a] For Li_2O, Na_2O, K_2O, and Rb_2O: Shaw and Uhlmann (1971a,b) Morey and Merivin (1932), Young et al. (1939), and Charles (1966).
[b] For Rb_2O and Cs_2O. Evetrop'ev and Davlovskii (1967), Shmidt and Alekseeva (1964).
[c] Hakim and Uhlmann (1967).

series Li–Cs. Some data are given in Table IX and Fig. 15 (Shartis et al., 1952).

The *refractive index,* depending on both atomic polarizability and molar volume, exhibits particularly large values for the low-molar-volume lithium silicate glasses. Otherwise the refractive index increases with alkali content because of both the higher polarizability of the alkali compared to oxygen and silicon and the greater polarizability of nonbridging oxygen. Data are given in Table X.

In relation to *elastic properties,* the introduction of alkali generally decreases the modulus, to a greater extent with increasing ionic radius. In fact, the introduction of Li_2O may cause a small increase in modulus.

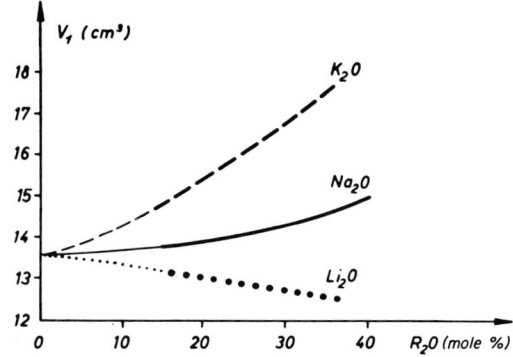

FIG. 13. Molar volumes of binary alkali silicate glasses (Scholze, 1965).

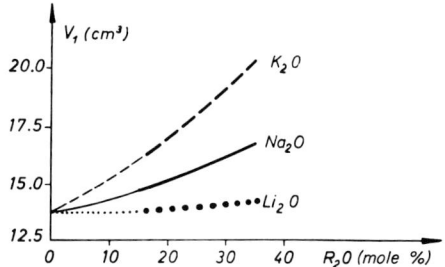

FIG. 14. Molar volumes of binary alkali silicate melts at 1400°C (Scholze, 1965).

Poisson's ratio is lowest in pure SiO_2 and increases as alkali is introduced (Table XI).

The *viscosity* decreases rapidly by several powers of 10 as the first addition of alkali increases the concentration of nonbridging oxygens (Fig. 16) (Scholze, 1965). The influence of the kind of alkali is relatively small (Table XII). Only for high concentrations and at low temperatures, the stronger bonding of Li^+ to oxygen causes lithium silicate glasses to have the highest viscosity, reversing the order K > Na > Li to Li > Na > K (Endell, 1940). A subcritical viscosity anomaly was ascribed to droplets with fluctuations related to a variable-order parameter (Schnaus *et al.*, 1976).

Hardness decreases rapidly with increased alkali, generally to a greater extent as the ionic radius increases (Table XIII).

Electrical conductivity depends on alkali content under the contrasting

TABLE IX

Coefficient of Expansion α $(\times 10^7)^{a,b}$

Li_2O		Na_2O		K_2O		Rb_2O		Cs_2O	
Mole %	α (°C^{-1})	Mole %	α (°C^{-1})	Mole %	α (°C^{-1})	Mole %	α (°C^{-1})	Mole %	α (°C^{-1})
10.3		61.2						12.2	74.4
16.10	53.6	14.8	83.9						
		18.4	101.5		100[b]	20	125[b]	26.7	118.4
26.4	85.6			26	160				152.6
32.30	110.0	30	155						

[a] Values for T from room temperature to T_g.

[b] Dietzel and Sheybany (1948) for Li_2O; Redwine and Field (1968) for Na_2O; Shmidt and Alexeeva (1964) for Rb_2O and Cs_2O.

[c] Amrhein (1963).

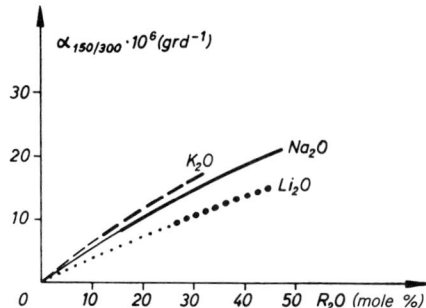

FIG. 15. Expansion coefficients of binary alkali silicate glasses (Scholze, 1965, after Shartsis *et al.*, 1952).

influence of size and bond strength. Therefore, Na_2O glasses tend to show a higher conductivity than either K_2O glasses (large, less mobile ion) or Li_2O glasses (ion more tightly bonded) (Table XIV). The much lower conductivity of mixed alkali glasses is discussed below. Nonbridging oxygen uncompensated by alkali may simulate extravagantly large jump distances.

TABLE X

REFRACTIVE INDEX

Li_2O		Na_2O[e]		K_2O[g]		Rb_2O[f]		Cs_2O[f]	
Mole %	$n_D - 1$	Mole %	$n_D - 1$	Mole %	$n_D - 1$	Mole %	$n_D - 1$	Mole %	$n_D - 1$
10		9.7	0.4740	10.1	0.482	11.8	0.487	12.2	0.510
15		14.5	0.4810	13.8	0.486	15.3	0.495		
20	+0.503	19.1	0.4900	21.6	0.496	20	0.504	20	0.530
25	0.513[a]	24.4	0.4975	25.6	1.500	26.6	0.514	26.7	0.548
30	0.524[b]	28.3	0.5020	29.9	1.504				
	0.529[d]								
35	0.5392[c]	34.6	0.5075	34.3	1.508	33.1	0.521		
45	1.555[b]	45.9	0.5150						
50	1.559[b]	49.2	0.5170	48.9	1.520				
	1.557[d]								

[a] Shelby and Day (1969).
[b] Dubrovo and Shmidt (1959).
[c] Hubbard and Cleek (1952).
[d] Kracek (1930).
[e] Morey and Merwin (1932).
[f] Shmidt and Alexeeva (1964).
[g] Rao (1963).

TABLE XI
Moduli and Poisson's Ratio

Mole %	$G(\times 10^{11})$ (dyn/cm²)		$E(\times 10^{11})$ (dyn/cm²)	μ	E(kg/mm²)
	Li_2O^a	Na_2O^a	Li_2O^a	Li_2O	Na_2O
10	3.048		7.426	0.218	
20	3.140	7.699	7.699	0.226	
25		2.54[b]			
30	3.173	7.881		0.242	
14.6					6335
24.4					5746
33.3					5710

[a] Shaw and Uhlmann (1971).
[b] Shelby and Day (1969).
[c] Jagdt (1960).

The *dielectric constant* increases in the series Li–K (Lacharme and Isard, 1978), as nonbridging oxygens become more polarized and the specific effect of a higher number alkali becomes manifest (Table XV). The effect of frequency, generally a decrease, shows most for the more mobile Li^+. The mobility of alkali generally causes large dielectric losses, which can be estimated semiquantitatively as a function of alkali content (Stevels, 1946).

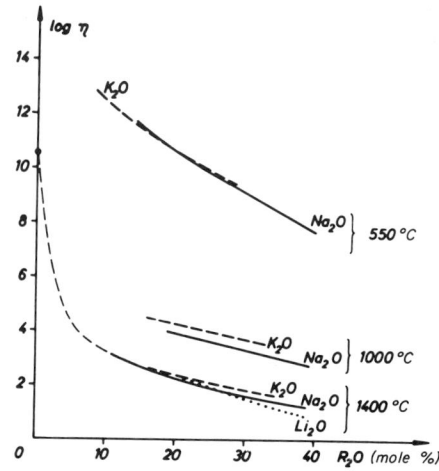

FIG. 16. Viscosity of binary alkali silicate melts (Scholze, 1965).

TABLE XII

Viscosity[a]

C	Mole %	η	η
Li$_2$O	13	426	438
	12	449	459
	10	500	506
	8	565	575
Na$_2$O	13	473	464[b]
	12	502	478[b]
	10	571	520[b]
	8	652	585[b]
K$_2$O	13	491	448
	12	518	567
	10	583	607
	8	622	658
Rb$_2$O	see Li and Uhlmann (1970)		

[a] Nemilov (1969). For higher viscosities, see Li and Uhlmann (1970). For low-temperature viscosities, see Meiling and Uhlmann (1967).
[b] Mazurin et al. (1970).

The *chemical resistance* is influenced decisively by the addition of alkali to silica. Resistance to acidic media is decreased as increasing concentration of alkali favors the exchange of R$^+$ for H$^+$ from the environment. This results in a weak hydrated layer that may, if sufficiently thin and coherent,

TABLE XIII

Microhardness and Surface Tension

	Mole % Na$_2$O		Ref.
		10^7 Nm^{-2}	
Microhardness	34.4	460	(Mazurin et al. 1973)
	23.4	415	
		10^{-3} Nm^{-1}	
Surface tension[a]	34.6	364	(Ainsworth, 1954)
(at 1300°C)	23.3	357	

[a] The surface tension is increased when Li$_2$O, decreased when K$_2$O replaces Na$_2$O.

TABLE XIV

Electrical Conductivity ($\log \sigma$) (σ in Ω^{-1} cm^{-1})

Mole %	Li$_2$O	Na$_2$O		K$_2$O		Rb$_2$O	Cs$_2$O
5	0	10.45[b]	7.33[b]	11.29	8.33	13.72	
10	7.5			8.68	6.24	9.68	
20, 21	5.9	6.45[b]	4.36[b]			7.53	
25		6.3	4.10				
27							6.42
30	5.4	5.4	3.57				
40, 36	4.5	4.4	3.16				
Temperature (°C)	150	150	300	150	300	150	150
Reference	(Kondratyev and Smirnova, 1970)	(Petrovskii, 1959)		(Evstrop'ev and Davevlovskii, 1967)			(Parfenov et al., 1959)

[a] For high temperatures, see Urnes (1959).
[b] Mazurin (1963).

dry to a resistant silica layer. Resistance to alkaline and neutral media is also decreased as, at nonbridging oxygens, the silicate network starts to dissolve. The role of alkali remains as decisive in more complex silicate glasses.

TABLE XV

Dielectric Constant[a]

Li$_2$O		Na$_2$O[b]		K$_2$O[c]	
Mole %	D at 10^5 Hz	Mole %	D at 4 × 10^{10} Hz	Mole %	D at 10^6 Hz
6.7	4.4			10	5.9
14.7	4.97			17.5	6.9
		20	6.5[d]	4.5	7.3
				25.6	8
30.7	6.57			29.9	9.4
39.8	8.03				

[a] Charles (1963). For more detail see Mazurin (1973). For Cs$_2$O and Rb$_2$O see Amrhein (1963).
[b] Prasad and Isard (1967).
[c] Rao (1963).
[d] Data for varying frequencies; see Provenzano et al. (1971).

Mechanical strength is also influenced decisively by the alkali in silicate glasses, since mechanical failure in these glasses is predominantly one of stress corrosion starting at flaws. Mechanical strength is a major topic of Vol. 5 of this treatise.

D. PROPERTIES OF MIXED ALKALI SILICATE GLASSES

Silicate glasses (and other oxide glasses as well) containing more than one alkali exhibit characteristic minima or maxima in properties ("mixed alkali" or "polyalkali" effect) (Isard, 1969; Day, 1972, 1976). Minima are observed for hardness (Gehlhoff and Thomas, 1926), high-temperature elastic moduli (Jagdt, 1960), expansivity (Terai, 1971), electrical conductivity, dielectric constant and dielectric loss (Mazurin and Borisovski, 1957; Hakim and Uhlmann, 1967; Lengyel and Boksay, 1954; Hayami and Terai, 1972; etc.), and viscosity (Poole, 1949; Matusita *et al.*, 1975). Maxima are observed for the low-temperature moduli (Gehlhoff and Thomas, 1926), chemical durability (Peddle, 1920; Sen and Tooley, 1950), and the activation energy of electrical conduction (Terai, 1971). The electrical resistivity ρ may rise more than two orders of magnitude, e.g., to $\log \rho = 12.6$ for $Na_2O \cdot K_2O \cdot 4SiO_2$ from 8.1 and 9.1 for $Na_2O \cdot SiO_2$ and $K_2O \cdot 2SiO_2$, respectively (Lengyel and Boksay, 1954, 1955). Additional representative data are given in Fig. 17 and all data are summarized in

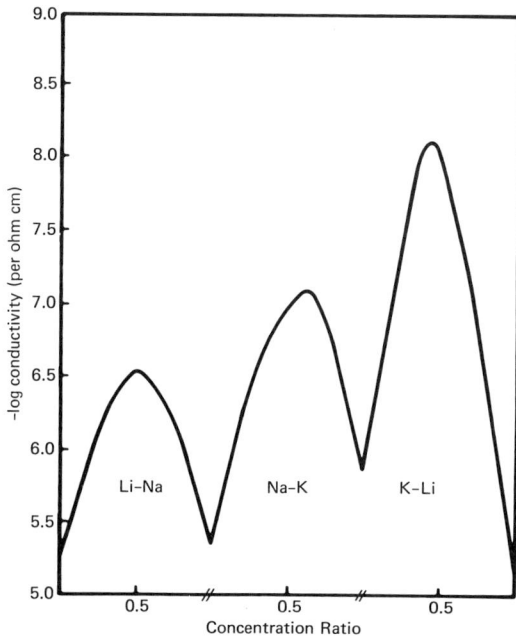

FIG. 17. Ionic conductivity in mixed alkali silicate glasses (Doremus, 1973).

TABLE XVI

SUMMARY OF PROPERTIES FOR MIXED ALKALI GLASSES[a]

Property	Deviation from linearity (additivity) with addition of second alkali
Physical	
Density	Small, ~±10%
Refractive index	Small, ~±10%
Molar volume	Slight, <±5%
Hardness	Small, ~±10%
Thermal expansion	Small, <±10%, usually positive dev.
Chemical durability	Moderately higher, alkali extraction lowered 4–6 times
Thermal conductivity	No reported data
Mechanical	
Strength	No reported data
Static fatigue	No reported data
Elastic modulus	Small, ~±20%, deviation-temperature dependent
Compressibility	No reported data
Internal friction	Major, large new peaks
Electrical	
Conductivity	Major, negative deviation, lower by factor of 10^2–10^6
Loss	Major, negative deviation, lower by factor of 10^1–10^3
Dielectric constant	Small (~25%) to moderate (~50%) usually negative deviation, but temperature and frequency dependent
Mass transport	
Alkali diffusion coefficient	Major, lower by factor of 10^2–10^4
Viscosity	Major, negative temperature-dependent deviation, lower by factor of 10^1–10^2
Gas permeability	Small, <10% negative deviation

[a] Day (1976).

Table XVI (from Day, 1976). One may attribute maximum acidity to the mixed silicate glass (Forland and Tashiro, 1951); i.e., the potential of $(Na,K)_2O \cdot 3SiO_2$ is equal to that of $Na_2O\ 4SiO_2$, and a current flows from $(Na,K)_2O \cdot 3SiO_2$ to $Na_2O \cdot 3SiO_2$. Hydrogen bonding is lower than average in a mixed K_2O–Li_2O silicate glass (Scholze, 1959). Nonlinear relations extend, as expected, to optical refractivity. But the effect on volume and refraction is much less spectacular, indicating that packing is only one of several causes for the phenomenon.

The absorption of mechanical energy ("internal friction"), on the other hand, shows very large changes with the addition of small amounts of a second alkali, the losses now being much higher, in the mixed alkali glass (Roetger, 1958; Jagdt, 1960; Steinkamp et al., 1967) and manifested in a new large "mixed peak." Day (1972, 1976) suggests that large effects are associated with properties involving alkali movements. The diffusion coefficient of the slower moving ion and its size are involved critically in a play of elastic dipoles that are electrically inactive. Higgins *et al.* (1973) believe that the electrical field just decays faster. Others, such as Hendrickson and Bray (1972a,b), postulate a mass effect. At least in borate glasses, some such effect may be present. This was shown with the isotope pair ^{6}Li–^{7}Li (concentration 1:1) for the activation energy of electrical conduction (maximum) and dielectric constant and loss (minima) (Abou-el-Leil *et al.*, 1978). An extensive treatment of the mixed alkali effect will be found in the volume on viscosity and relaxation and in the review by Day (1976).

No single mechanism limited to the interacting species can explain all mixed alkali effects. The structural change in the total system will have to be properly considered (Doremus, 1974). Mixed alkali effects can also be observed in ion-exchanged alkali silicate glasses (Day, 1972). Heat capacity data support (Moynihan *et al.*, 1976) the notion of Shelby and Day (1969) that the term mixed alkali effect should be reserved for ionic transport phenomena. More recently, the behavior of mixed alkali glasses with a small concentration of the second cation in ionic conduction experiments seems to be explained best by assuming that an alkali silicate glass is a weak electrolyte (Moynihan *et al.*, 1980).

Technologically, the mixed alkali effect has considerable importance. Examples are

(1) Thermometer glasses in which the presence of a second alkali must be strictly excluded in order to avoid instability due to volume relaxation. A typical thermometer glass is composed of (all numbers are wt %):

Jena 16 III: SiO_2, 67.5; B_2O_3, 2; Al_2O_3, 2.5; CaO, 7; ZnO, 7; Na_2O, 14.

(2) Low-dielectric-loss glasses in which a second alkali is desirable, e.g.,

Corning 7070: SiO_2, 71; B_2O_3, 26; Li_2O, 0.5; K_2O, 1

Jena 3079 III: PbO, 30; B_2O_3, 52.7; Al_2O_3, 5; Li_2O, 4; K_2O, 8

In these mixed alkali glasses the dielectric loss (tan δ) is of the order of 10^{-3}.

E. DIFFUSION

There is a growing need for understanding diffusion mechanisms in alkali silicate glasses, not only from a conceptual but also from a technological viewpoint. Since alkali ions are held much more loosely in the silicate structure than the other constituents, their diffusion is generally much more rapid. In binary alkali silicates the immediate neighbor is normally a nonbridging oxygen represented in the grouping $O_3SiΘNa^+$. However, alkali diffusion does not depend on this; on the contrary, it will be shown in other sections that diffusion is enhanced when the introduction of Al^{3+} converts nonbridging oxygens into bridging oxygens.

It does not seem that the conventional Schottky (vacancy) defects are available in glass, and, indeed, the correlation between diffusion and conductivity data appears to exclude the vacancy mechanism, as well as the mechanism of movement through interstitial positions. The most likely mechanism is the one termed "interstitialcy mechanism," in which an interstitial moves by jumping to a "lattice" site, pushing its occupant to another interstitial site (Barr et al., 1970). In mixed alkali glasses the situation is more complicated since individual moves of unlike species have no correlation (Terai, 1971; Isard 1972). A more complicated situation also exists in phase-separated alkali silicate glasses (Doremus, 1969, 1973, 1974) since a species may be absorbed preferentially in the low alkali phase. Ion mobility ratios were tabulated by Doremus (1969, 1973) for Li, Na, K, Rb, and Ag. Diffusion is also affected by small amounts of OH (Maklad and Kreidl, 1971; Boulos and Kreidl, 1972; McVay and Farnum, 1972).

F. THE SYSTEMS $H_2O-R_2O-SiO_2$

The heating of alkali silicate batches in an autoclave (80–200°C, 50–100% steam) results in rubberlike glasses containing up to 30% H_2O (Stookey, 1970), which can be used as a molding powder, hydraulic binder, or insulator. Glass ceramics containing 15% H_2O can be obtained similarly. Applications demand more complex glasses, such as (mole %) 76.9 SiO_2; 16.5 Na_2O; 2.1 Al_2O_3; 4.5 MgO; 18.8% H_2O (Wu, 1980). The relation of hydrogen bonding to the number of nonbridging oxygens is of great interest in the development of such compositions. Two types become manifest through dehydration peaks. The removal of the less tightly bonded species by dehydration at the lower dehydration peak temperature leads to particularly stable, high-OH, moldable glasses (Wu, 1980a,b).

1. Soluble Silicate Glasses[*]

High-alkali silicate glasses containing OH are soluble in H_2O and are used in soaps, detergents, adhesives, cements, coatings, soil and water treatments, etc. They are classified by SiO_2/Na_2O ratios (from 0.15 to 4) and particle size. They are used as such or in aqueous solution. Usually Na_2O is the alkali. Lithium silicate glasses are practically insoluble. Soluble silicate glasses are synthesized at 1450°C from soda ash and sand. The

---

[*] See Wills (1969).

melt is dissolved as drawn in rotary dissolvers or after cooling and crushing under pressure with steam. Recently, NaOH has been used. When lithium glass is desirable, one obtains a product by dissolving silica gel in LiOH solution. The solubility of water in such melts increases (Scholze, 1959) from 20% at 1250°C to 40% at 1480°C. Under steam, water progresses into the glass along a sharp front and the outer layer dissolves.

The silicate solutions consist of colloidal micelles charged negatively and stabilized by alkali ions. Thermodynamic phase equilibria can be constructed, but become established only after finite times. Polymer ion sizes can be established by chromatography. There are 18 crystalline compounds in the ternary $Na_2O-SiO_2-H_2O$, 6 in the K_2O system.

2. "Water" in Oxide Glasses

With H_2O in the batch or melting atmosphere, protons are accommodated in hydrogen-bonded positions characterized by three IR absorption bands (Scholze, 1959, 1966; Franz and Kelen, 1967) in increasing order of hydrogen bonding:

(1) 2.75–2.95 μm, least hydrogen-bonding character, for positions near bridging oxygens;
(2) 3.35–3.85 μm, for positions near one nonbridging oxygen; and
(3) 4.25 μm, in high-alkali glasses involving more than one nonbridging oxygen in the neighborhood of the proton.

Type (1) is sometimes rather loosely termed a "free hydroxyl grouping." It is found predominantly in pure SiO_2 and is obtained by the reaction $\equiv Si-O-Si\equiv + H_2O \rightarrow 2\equiv Si-OH$. It is, in fact, difficult to obtain silicate glasses free of significant amounts of OH. A normal OH content is 0.03 wt % (generally increasing with alkali concentration). This seemingly small concentration is capable of decreasing the glass transition temperature by as much as 40°C. Viscosities in the working range are also affected. Automatic feedback batch corrections must take account of OH content. Other properties of oxide glasses are correspondingly influenced by OH contents of the order of 0.01% (Boulos and Kreidl, 1972), and resarch results should always be reported giving the OH content of samples used. An often sufficient assessment of the OH content is provided by measuring their absorption bands (Goetz and Vosahlova, 1968).

A dramatic effect of OH in oxide glasses is the appearance of an intensive high-temperature peak in internal friction measurements (Day, 1976). The effect is largest in phosphate and smallest in borate glasses, probably because of its association with nonbridging oxygens. Other important effects include crystallization, phase separation, diffusion of alkali and gases, expansion, and electrical conductivity.

It is interesting that backward-wave phonon echoes (Shiren et al., 1977) have amplitudes proportional to the unassociated hydroxyl content in several glasses. This suggests a distribution of inhomogeneously broadened resonances sensitive to ultrasonic and electromagnetic fields.

Relatively large OH concentrations protect SiO_2 against radiation-induced transmission losses and can be achieved under high-pressure impregnation (Faile et al., 1967; Faile and Roy, 1970). Above 7 wt % water, the structure begins to contain molecular water while the Si–OH content remains constant (Bartholemew et al., 1980). The band at 1.91 μm was assigned to molecular water in experiments conducted with Na_2O–K_2O–ZnO–Al_2O_3–SiO_2 glass. Nucleation and growth rates are increased by OH. Parallelism of this effect and a dramatic decrease in viscosity suggest their kinetic nature (Gonzalez-Oliver et al., 1979).

III. Soda-Lime Glasses*

A. GENERAL CONSIDERATIONS

Most conventional glasses are based on the compositional system Na_2O–CaO–SiO_2 (N–C–S). The historical development of these glasses conforms to the rationale: SiO_2, the excellent glass former, is made to fuse more easily by Na_2O, the resultant chemical vulnerability being remedied by CaO. In the binary system CaO–SiO_2, glass formation is difficult. Typical ternary compositions of conventional container, flat, and other commercial glasses are near the ratio $Na_2O : CaO : 6SiO_2$, at about 15–18 Na_2O, 10 CaO, 72–75 SiO_2 (Fig. 18). Raman spectra indicate that chain-like silicate anions change little when CaO enters the sodium oxide–silica system and that the resulting structures resemble those of crystalline devitrite (Na_2O, $3CaO$, $6SiO_2$) (Brawer and White, 1977).

The introduction of CaO into binary Na_2O–SiO_2 glasses introduces additional nonbridging oxygens. At the same time, the bonds of the bridging oxygens in the coordination sphere of Ca^{2+} are affected. The XPS (x-ray photoelectron spectrum) peak corresponding to bridging oxygen suffers a chemical shift toward that of nonbridging oxygen even more than in the case of alkali (Brückner et al., 1980).

The phase diagram, first worked out by Morey and Bowen (1925, 1931), is illustrated in Fig. 19 (from Shadid and Glasser, 1971). Interesting points are shown in Table XVII.

The compound with the composition nearest to that of commercial soda-lime glasses is devitrite, $Na_2O \cdot 3CaO \cdot 6SiO_2$, which melts, decom-

---

* Na_2O–CaO–SiO_2 glasses.

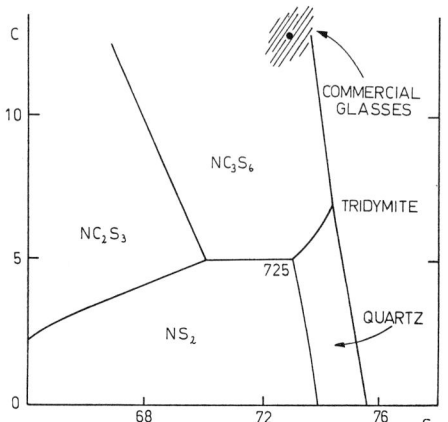

FIG. 18. Commercial glass region located in the phase diagram $Na_2O-CaO-SiO_2$ (N = Na_2O, C = CaO, S = SiO_2).

posing into $CaO \cdot SiO_2$ and liquid at 1060°C. Na_2O introduces enough nonbridging oxygens into SiO_2 to provide meltability and compatibility for Ca, which interacts with ionic oxygen to suppress solubility.

The ternary is a complex system. The most important results of more recent investigations (Shadid and Glasser, 1971; Muir and Glasser, 1974) of the ternary $Na_2O-CaO-SiO_2$ (N–C–S) may be summarized as follows (Q = quartz, L = liquid, E = eutectic) (Figs. 19 and 20):

(1) There are primary phase fields N_3S_8 and NCS_5.
(2) In the S-rich field two eutectics, both at 755 ± 5°C, exist: (a) $NCS_5-Q-N_3S_8$ and (b) $NCS_5-NS_2-N_3S_8$.
(3) The eutectic $Q-NS_2-NC_3S_6-L$ is metastable.

TABLE XVII

DATA FOR POINTS ON PHASE DIAGRAM OF FIG. 19

Compound or eutectic (E)	CaO (C)	Na_2O (N)	SiO_2 (S)	Melting T or liquidus (°C)
S	—	—	100	1723
E (Quartz (Q)–NS_2)	—	—	—	790
NS (Metasilicate)	—	50	50	1088
E (NS_2–NS–N_2CS_3)	1.8	37.5	60.7	821
NS_2 (disilicate)	—	$\frac{1}{3}$	$\frac{2}{3}$	874
E(NC_3S_6–NS_2–Q)	5.2	21.3	73.5	725
NC_3S_6 (devitrite)	10	30	60	1060 (decomp.)

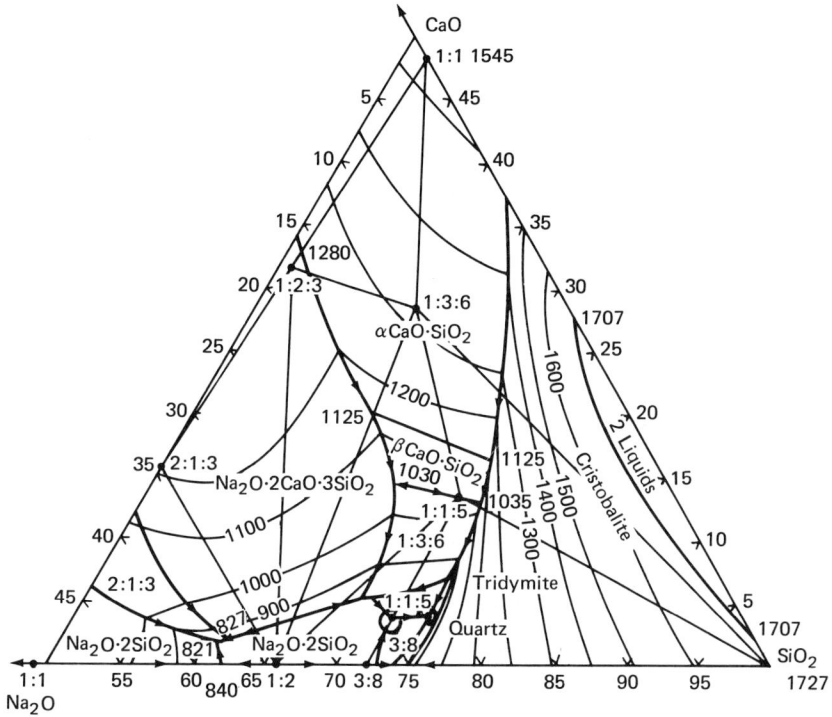

FIG. 19. Phase diagram of the ternary system Na$_2$O–CaO–SiO$_2$ (N–C–S) (Shadid and Glasser, 1971).

(4) NC$_2$S$_3$ is not a phase of fixed stoichiometry but varies from 28.5 to 53.5 mole % NS with several polymorphic forms, one of which may be the new compound NCS$_2$.

(5) N$_2$CS$_3$ has much smaller variability in stoichiometry, and disproportionates into NS and N-rich solid solution NC$_2$S$_3$. In the high-T and low-S regions, new work confirms classical data.

B. CONVENTIONAL SODA-LIME GLASSES

Deliberate or, historically, accidental addition of small amounts of Al$_2$O$_3$ improves conventional glasses, particularly by suppressing devitrification and increasing durability. Other important variants in commercial soda-lime glasses are MgO (0–4%), controlling ("shortening") setting rate ($d\eta/dT$, η = viscosity); SO$_4$ (less than 1%); BaO (about 0.4%); B$_2$O$_3$ (less than 1%); and TiO$_2$ (less than 1%) as nonalkaline fluxes and fining agents. To improve durability, the alkali content has been decreased con-

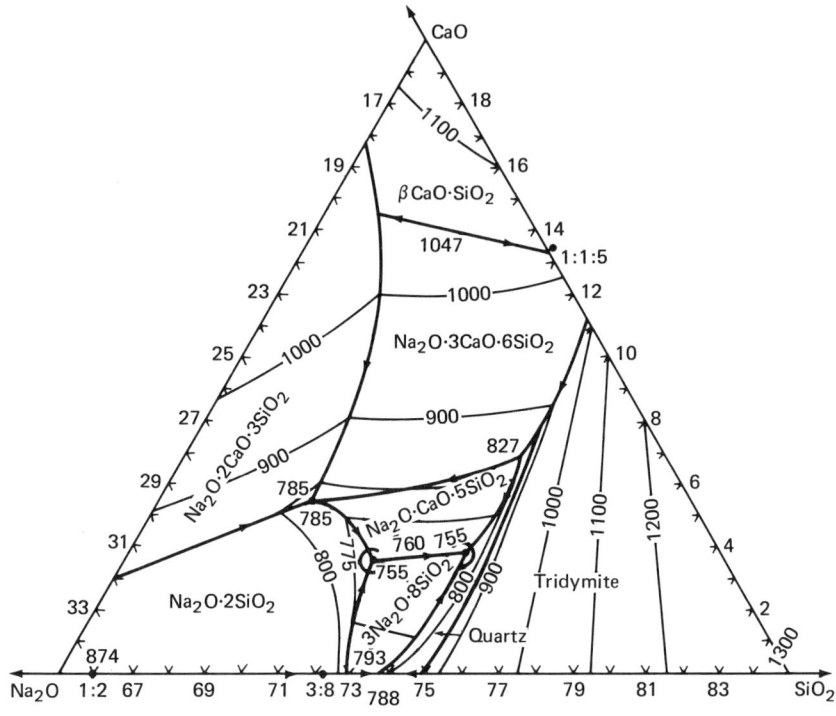

FIG. 20. Detail of Fig. 19.

sistently during the past decades, from about 17% before 1935, to less than 14% since 1950 (Sharp, 1933, 1940; Moore et al., 1947). These decreases necessitated increases in the fusion temperature, addition of the more expensive nonalkaline fluxes plus Al_2O_3, and some decrease in SiO_2. For ecological reasons, however, the addition of fluorine and B_2O_3 has been discouraged since 1970, and sulfate has been used more frequently. A typical modern composition (wt %) (Stadler and Ladue, 1978) is shown in Table XVIII.

C. Phase Separation

The metastable region of liquid–liquid phase separation in the binary Na_2O–SiO_2 extends into the ternary and connects with the stable immiscibility region in the CaO–SiO_2 binary (Ohlberg et al., 1965; Burnett and Douglas, 1970). The ternary immiscibility diagram (Burnett and Douglas, 1970) is illustrated in Fig. 21. Nucleation and growth of these phases are strongly affected by melting history and atmosphere. Vacuum enhances and excess oxygen discourages nucleation and growth (Kumar and

TABLE XVIII

Typical Modern Compositions (in wt %) of Soda-Lime Glasses

	1972	1977	Variation
SiO_2	72.19	72.15	66.2–74.7
Al_2O_3	1.81	2.13	1.25–2.5
Fe_2O_3	0.12	0.11	9.16–13.40
CaO	9.55	10.66	9.16–13.40
MgO	1.51	0.91	9.16–13.40
BaO	0.17	0.08	0.0–0.47
Na_2O	13.96	13.83	12.88–17.30
K_2O	0.59	0.57	12.88–17.30
SO_3	0.16	0.14	0.08–0.22

Rindone, 1979a,b). Nucleation and growth of crystalline phases increase with OH (and F) content as the viscosity decreases (Gonzalez-Oliver et al., 1979).

D. Additives to Na_2O–CaO–SiO_2 Glasses to Promote Fining

Additives to commercial soda-lime glasses intended to promote the removal of bubbles are termed *fining agents*. The most important fining processes are

(1) Buoyancy, increasing with increasing bubble size (mm/sec for 1-mm diameter, mm/day for 0.01-mm diameter), and decreasing viscosity

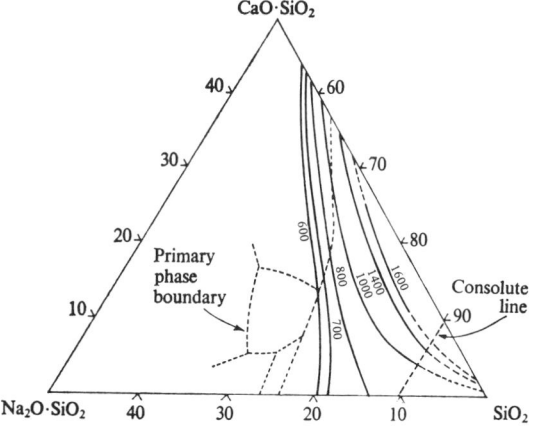

FIG. 21. Subliquidus immiscibility in the ternary Na_2O–CaO–SiO_2 (Burnett and Douglas, 1970).

(1400°C required in conventional melts). The desired increase in size is affected by (a) diffusion of gas from glass into bubble, and (b) coalescence.

(2) Dissolution of gases from bubbles into the glass. The increase of solubility with decreasing temperature is most significant for O_2, less for CO_2, absent for N_2.

Fining agents provide specific reactions to promote these processes, often complex and understood increasingly only through recent studies involving careful gas analysis. The most frequently used reaction is that of accepting O_2 at low temperatures and releasing it at high temperatures.

Because of their polyvalence, As and Sb are capable of such action. Oxidized to As_2O_5 and Sb_2O_5 at low temperature, they release oxygen at high temperature:

$$As_2O_5 \rightarrow As_2O_3 + O_2$$
$$Sb_2O_5 \rightarrow Sb_2O_3 + O_2$$

These reactions are simplifications of more reactions taking place when As and Sb are built into the silicate structure. The efficiency of the process is increased in the presence of $NaNO_3$, which aids the original oxidation to As_2O_5 and Sb_2O_5 during glass formation. Sulfates also are capable of releasing O_2 on decomposition (Nemec, 1977; Konijnendijk and Buster, 1977). For details see, in particular, Cable and Haroon (1970), Cable and Martlew (1971), Nemec (1977), and Mulfinger (1966).

In the release of oxygen at elevated temperatures, the alkali concentration plays an important role. In a model glass, $70SiO_2$–$30K_2O$–$1As_2O_3$, the following reactions were suggested by the interpretation of laser Raman data for As^{3+} and As^{5+} (Verweij, 1979): The original batch contains much CO_2 and some N_2. The first liquid forming is rich in K_2O and CO_2. As more SiO_2 is dissolved, the relative concentration of K_2O decreases, and CO_2 is released forming small bubbles. In the first liquid phase below 700°C, when most SiO_2 remains unreacted,

$$As_2^{III}O_3 + 3K_2CO_3 \rightarrow 2K_3As^VO_4 + 2CO_2$$

The K_3AsO_4 is in the form of grains on the surface of the K_2CO_3. Between 700 and 850°C, K_2CO_3 reacts with SiO_2 forming two layers (a) crystalline $K_2O \cdot 2SiO_2$ (b) liquid $K_2O \cdot SiO_2 \cdot CO_2$ (containing metasilicate chains and CO_3^{2-} anions plus $(As^VO_4)^{3-}$ anions). Between 850°C and 1100°C, the crystalline $K_2O \cdot 2SiO_2$ layer grows at the expense of both SiO_2 and the $K_2O \cdot SiO_2 \cdot CO_2[(AsO_4)^{3-}]$ liquid. This crystalline $K_2O \cdot 2SiO_2$ layer melts at 1015°C, forming the final glass. In it, CO_2 is poorly soluble and escapes. The reaction of the arsenate groups leads to oxygen in two reactions:

$$2K_3As^VO_4(liq) + 2SiO_2 \rightarrow K_4As_2^yO_7 + K_2O \cdot 2SiO_2 \quad (1)$$

About 10% of this compound dissociates at 1100°C releasing oxygen:

$$K_4As_2O_7 + 2SiO_2 \rightarrow K_2O \cdot 2SiO_2(liq) + 2KAsO_2(liq) + O_2(gas) \qquad (2)$$

Gradually, less CO_2 and more O_2 evolve (Nemec, 1977). At 1400°C most bubbles contain oxygen. On cooling, the oxygen bubbles dissolve since the solubility of oxygen increases with decreasing temperature (Greene and Lee, 1965; Greene and Platts, 1969).

IV. Other Cations in Silicate Glasses

A. CALCIUM, MAGNESIUM, AND BERYLLIUM

In the system $CaO-SiO_2$, glass formation is extremely difficult. There is a large immiscibility gap from about 35 (Tewhey and Hess, 1979) to 45 mole % (Shelby et al., 1979) down to about 2% CaO. The gap is skewed; the consolute temperature is ~1871°C near 90% SiO_2 (Tewhey et al., 1979), although earlier workers reported values above 2000°C (Toropov et al., 1956). The lower T limit is 1695°C. The eutectic is above 1400°C. In both phases, secondary phases exsolve. In the CaO-rich droplets, a SiO_2-rich phase crystallizes to cristobalite.

Magnesium oxide reinforces the structure of complex silicate glasses, revealing the formal character of the network-modifying concept, which fits only larger or monovalent cations. Crystalline magnesium silicates are better described by oxygen packing with small Si^{4+} and Mg^{2+} in interstices than by disrupted SiO_4 networks. MgO thus is best suited to impart a high modulus of elasticity to special glasses such as fiber glasses (see, e.g., Aerojet, 1969). An example (wt %) is SiO_2 65%, Al_2O_3 19%, MgO 15%, Li_2O less than 1%, etc. Most likely some of the Mg is 4-coordinated in some glasses. Raman spectra indicate that silicate anions in $Na_2O-MgO-SiO_2$ glasses resemble those in the binary Na_2O-SiO_2 as well as the ternary $Na_2O-CaO-SiO_2$, with the appearance, however, of more disorder (clustering of silicate groups) and a related widening of the immiscibility dome (Brawer and White, 1977). MgO replacing CaO shortens the working range of soda-lime glasses.

Tetrahedra of BeO_4 have been identified by IR spectroscopy (Semin and Kenton, 1970). Addition of BeO imparts the highest modulus of elasticity $(15.7-20.3 \times 10^6$ psi) to complex silicate glasses (Tiede, 1964; Bacon, 1971, McMarlin, 1971). Two composition ranges (wt %) (Tiede, 1964) are given in the accompanying tabulation. An yttrium beryllium aluminosilicate glass (Bacon, 1971) contained (wt %) BeO 5%, Y_2O_3 30.24%. In such glasses a modulus of 20×10^{-6} psi was attained or approached.

SiO_2	Al_2O_3	CaO	MgO	BeO	ZrO_2	TiO_2	CeO_2	Li_2O
45–60	0–3	9–19	6–10	7–12	1–3	2–10	0–4	0–4
49–57	0–3	11–14	6–9	8–11	2–3	4–8	4–8	4–8

B. Barium and Strontium

The S-shaped liquidus curve near SiO_2 in the binary $BaO-SiO_2$ (Kracek, 1930) was taken as an indication of incipient liquid–liquid immiscibility. Extrapolation from ternary fields ($BaO-B_2O_3-SiO_2$—Levin, 1965; Toropov et al., 1956a,b) suggested metastable immiscibility near BaO 8 from 1430°C (compared to Charles' (1968) calculation from activities at BaO 8, 1600°C). A metastable miscibility gap has indeed been found by Argyle and Hummel (1963) and Seward et al. (1968) between BaO 2 and 28 with an apex at approximately BaO 10 and 1460°C. Recently, Shelby et al. (1979) studied extensively the immiscibility regions in the binaries CaO, SrO, $BaO-SiO_2$, with the following results:

	CaO	SrO	BaO
Upper limit of immiscibility (mole %)	46	42	40
Upper limit of connected SiO_2-rich phase (mole %)	33	25	22

The upper limit of immiscibility in the $BaO-SiO_2$ system inferred here is undoubtedly too high in light of the results of Seward et al. (1968a). In any case the resulting morphology significantly determines density, thermal expansion, refractive index, and helium permeability of these binary glasses. Structural changes in the system can be followed by radiation-induced defect analysis (Bishay and Goma, 1967).

The ternary $BaO-Na_2O-SiO_2$ containing a large region of immiscibility was described by Burnett and Douglas (1971). A review on Sr in silicate glasses was published by Zagar (1971) when strontium glasses were considered for fission waste storage.

C. Zinc Silicate Glasses

The phase diagram $Na_2O-ZnO-SiO_2$ was worked out by Holland and Segnit (1966). It contains fewer compounds but an extensive area of solid solution. Although it is divalent, zinc may be 4-coordinated by oxygen in alkali silicate glasses, reinforcing the polymer (Hurt and Philips, 1970; Vargin and Slepanov, 1964; Hinz and Mitsch, 1968). An $Li_2O-ZnO-SiO_2$ glass ceramic is usable as a seal for Pd–Ag pins to stainless steel in a 60 N_2–40 air oxidizing atmosphere (Leedecke and Baca, 1978, 1979). Alkali (Li_2O, K_2O) borosilicate glasses were also suggested by the same author, and glass ceramics by Hinz and Mitsch (1968) and McMillan

(1964). See also Section VII of this chapter. In the latter application, the most important system is $Li_2O-ZnO-SiO_2$ (Stewart et al., 1967; McMillan et al., 1964). In it, phase separation precedes crystallization of $Li_2O \cdot 2SiO_2$. A fine dispersion of this phase yields strong glass ceramics. Other systems comprise $CaO-ZnO-SiO_2$ (Segnit, 1954), $MgO-ZnO-SiO_2$ (Segnit et al., 1965), and $SrO-ZnO-SiO_2$ (Smakota et al., 1974) glasses.

Zinc oxide has semiconductor properties. This may carry into glass systems, as indicated by observations of Ohno (1965).

D. LEAD, THALLIUM, AND BISMUTH

Very large amounts of Pb, Tl, and Bi (e.g., to 90 wt % PbO) can be accommodated in silicate glasses (see, e.g., Fajans and Kreidl, 1948) in apparent contradiction to the postulate for a continuous network of $(SiO_4)_n^{4-}$ (at 88 wt % PbO all SiO_4 tetrahedra would be isolated). The ease of glass formation is related to the electronic structure of Pb^{2+}, the species found in these glasses ($6s^2$), which is characterized by its tendency toward asymmetric bonding. This asymmetry is found in crystalline lead compounds, for instance, in PbO, in which four Pb–O distances are 2.3 Å and four 4.3 Å (Bordovski and Izvochikov, 1967). This asymmetry is also related to the complex structure in a large unit cell of various lead-containing crystalline silicates, which appears to be maintained in the short-range order of high-lead silicate glasses (Mydlar et al., 1970; Imaoka and Hasegawa, 1980a,b): Fig. 22.

The low liquidus temperature and the high-lead phase that the high-lead glasses resemble in short-range order are exhibited in the phase diagram studied in considerable detail by Billhardt (1969) and Smart and Glasser (1974). The strong relation of high-PbO glass structures to those of crystals is also demonstrated by the similarity in the triplet fundamental UV

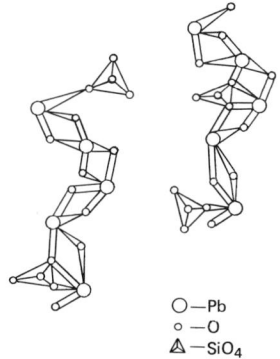

○ —Pb
○ —O
△ —SiO_4

FIG. 22. Model of lead silicate glass structure (Mydlar et al., 1970).

absorption edge (Smirnova, 1966), which is associated at higher energy with excitation of oxygen ions and at lower energy with transitions within Pb^{2+} ions (Stroud and Lell, 1971), and by Raman spectroscopy (Morozov et al., 1969; Verweij and Konijnendijk, 1976).

Nemilov (1968) considers this relation on the basis of various properties, particularly viscosity. The methods of polymer chemistry were used successfully by Masson et al. (1970) and Goetz et al. (1976) to establish models of grouping in high lead glasses. Verweij and Konijnendijk (1976) point to the close similarity of the structure of glasses with close to 50 mole % PbO with that of the crystalline compound $PbO \cdot SiO_2$ (alamosite), in which a spiral of Pb^{2+} ions and their coordinating O^{2-} is perpendicular to SiO_4 tetrahedral chains. The 30 mole % PbO glass appears to show, in addition, evidence (1015-cm^{-1} peak) for the nonbridging oxygen ion manifest in the spectrum of K_2O-SiO_2 glasses. Although the structures of ternary $K_2O-PbO-SiO_2$ compounds have not been determined to date, glasses once more resemble crystals of similar compositions in their Raman spectra, such as $K_2O \cdot PbO \cdot 4SiO_2$.

Metastable phase separation has been reported over large areas of the system by Pavlushkin et al. (1968) and Calvert and Shaw (1970). The S shape of the liquidus as well as electron microscopy indicate a range to about 35 mole % PbO; but immiscibility also appears to be possible at high-PbO contents.

The presence of fundamental UV absorption bands at relatively low energies approaching the visible range (Kim and Yamane, 1980) due to the easily polarized electronic structure are associated with the high refraction and high dispersion of all lead oxide glasses (the refractive index must rise steeply on the low-energy side of an absorption band). This behavior has made high-PbO glasses useful as optical materials for several centuries. The refractive properties and compositions of three typical optical (flint) glasses are given in Table XIX, in comparison with a soda-lime (crown) glass. The UV absorption band is shifted to shorter wavelength when B_2O_3, and to longer wavelength when GeO_2, substitute for SiO_2 (Kim and Yamane, 1980).

The high refractive index is also responsible for the brilliance of art glass, called crystal in the trade. In many countries this trade name must cover 30 wt % PbO in the composition.

Other consequences of the high polarizability of lead in silicate glasses are the reversion of the stress-optical coefficient at a characteristic PbO content, allowing the design of zero-stress-optical-coefficient glasses (Pockels, 1902); the polarizability of lead glasses in electrical fields (Quincke, 1880; Wuellner and Wien, 1902, 1903), and their high Verdet constant relating optical rotation to the magnetic field (Kreidl and Rood,

TABLE XIX

Optical Flint Glasses[a]

Name	Composition (mole %)					n_D	V^b
	PbO	R$_2$O	SiO$_2$	CaO	Other		
Light flint	12	11	76			1.572	42.5
Dense flint	19.5	6.5	73.5			1.617	36.6
Extra dense flint	34.5	2	62.5			1.751	27.7
Crown		15	73	10	3	1.523	58.6

[a] Kreidl and Rood (1965), Kreidl (1974).
[b] Dispersion is conventionally expressed by the V value comparing the differences of indexes at two wavelengths (F, C) to that at the standard sodium D line wavelength. Low V indicates high dispersion; $V = (n_D - 1)/(n_F - n_C)$.

1965). Lead can be reduced chemically at the surface of high-lead glasses, causing blackening and electroconductivity. The most recent application is to channel multipliers (Gerhardt, 1973).

Lead is excluded from glasses resistant to damage by reduction. Lead is an important component in the low-loss glasses investigated and developed by Armistead (1946a,b,c), which demand exacting compromise properties. In a typical composition (wt %)

SiO$_2$, 40; PbO, 50; K$_2$O, 6.3; Na$_2$O, 2.7; Li$_2$O, 1.0

SiO$_2$ should be less than 50, K$_2$O:Na$_2$O should be 1.5 to 4.1, (K$_2$O + Na$_2$O):Li$_2$O should be 4 to 19.1.

Shielding against x rays depends on the absorption generally increasing with atomic number. Because of its high atomic number and low valence in oxide glasses, lead is a preferred component of shielding glasses (Kreidl and Hensler, 1958). Up to 30 wt % PbO was used in many glasses requiring low sealing temperatures, e.g., in the lamp and television industries. However, TV tubes proper no longer contain lead because of weight limitations. PbO and BaO are blended for dispersions intermediate between those of high-index flint and barium crown optical glasses; e.g.,

Barium flint mole %: 66 SiO$_2$; 7 R$_2$O; 9 BaO; 9 PbO

Under a thermal gradient, anisotropic crystallization can be obtained in a sodium lead silicate glass (Melling and Duncan, 1980).

Bismuth (Shabanova, 1967) and *thallium* (Otto *et al.*, 1969; Otto and Milberg, 1967) glasses are similar in many respects to lead glasses. In all three cases the electronic species stable in silicate glasses is characterized by electrons above a completed shell (one in Tl^+, two in Pb^{2+}, three in Bi^{3+}).

In contrast to alkali silicate glasses, in thallium silicate glasses there is a tendency to substructure as with PbO in lead glasses (Blair and Milberg, 1974; Panek and Bray, 1977).

The high refractive index of high-Bi glasses (close to 2.0) has made them useful for special optical problems, e.g., the fabrication of beads for reflecting highway paints. Glasses in the system Bi_2O_3–SiO_2, the phase diagram of which was established by Speranskaya et al. (1968), form to about 65 Bi_2O_3 under conventional conditions. The relation to the structure of equivalent crystalline compounds is similar to PbO glasses (Shabanova, 1967). Semiconductivity in Bi glasses is described by Watanbe et al. (1970). In glasses containing substantial amounts of Bi_2O_3, chemically reduced surfaces are applied to channel multiplier glasses (Gerhardt, 1973).

E. OTHER CONSTITUENTS IN SILICATE GLASSES

Tin (Bartney et al., 1970; Zorina and Vakhramev, 1969; Rogozhin et al., 1971; Itoh and Mori, 1977) seems to be present as Sn^{4+} with a coordination number of 6, but some Sn^{2+} can be obtained (Min'ko, 1973; Dannheim et al., 1976). The ionic character of the Sn–O bond increases from Li to K (Bartenev et al., 1970).

Titanium dioxide can be introduced into alkali silicate glasses in large quantities. Low liquidus temperatures can be achieved for high SiO_2 contents in line with a very low melting ternary compound. The structure of ternary glasses was investigated by ESR spectroscopy (Sidorov and Tyul'kin, 1968). For $Na_2O : TiO_2 = 3$, Ti has 6-coordination; for less Na_2O, the coordination decreases toward 4 at $Na_2O : TiO_2 = 1$. In between, nonbridging O's are introduced. Some Ti^{3+} can be obtained (Kurkjian and Peterson, 1974). Titanosilicate glasses may show a positive temperature coefficient of the moduli. The introduction of TiO_2 yields glasses with high refractive index, having much lower density than lead glasses (Schott, 1976). TiO_2 is an important crystallization catalyst (see Section VII). Alkali-free binary TiO_2–SiO_2 glasses can be obtained by hydrolysis of alkoxides (Kamiya et al., 1974).

Zirconium dioxide. Glasses in the SiO_2–ZrO_2 binary have been obtained as fibers for reinforcement of cement from spinnable viscous liquids formed from metal alcoholates reheated at 500°C. The ZrO_2 oxide content can be up to 26% (Kamiya et al., 1977). In sodium silicate glasses, replacement of some SiO_2 by ZrO_2 increases chemical durability (Weyl and Marboe, 1964), particularly in alkaline media. *Thorium dioxide* acts similarly.

Cadmium oxide. Up to 45–55 mole % can be introduced into silicate glasses, which are photochromic in the absence of GeO_2, B_2O_3, P_2O_5, and transition elements (Kumato et al., 1977). They are stabilized by Al, alkalis, and alkaline earths.

Transition elements can be accommodated in large amounts in silicate glasses. (For important effects, see the volume on optical properties.) McMillan (1962) gives details on increasing accommodation in aluminosilicate glasses in the order Ni < Fe < Co < Mn. See also Kuzentsova and Estrop'ev (1972) and Kutateladze et al. (1974). Iron silicate glasses are the subject of metallurgical slag research (Gaskell and Ward, 1967; Pavlushkin et al., 1968). An iron aventurine (the old name for glasses containing lustrous metallic flakes) contained 44.8% Fe_2O_3 (Hoskikawa and Nambu, 1978). Supermagnetic behavior is associated with precipitation of ferrite from such glasses (Shaw and Heasley, 1967; Collins and Mulay, 1970). Uranium in large amounts has been considered in the study of reactor problems (Cojocaru, 1966). Twenty percent plutonium is contained in a sodium silicate glass from which single crystals of plutonium trioxide were crystallized (Phipps et al., 1964).

Ferric iron structurally replaces Al (Coleman et al., 1970). Sites were determined by Mossbauer spectroscopy (Levy et al., 1976). Semiconduction is observed in high-Fe glasses and glazes (Russak and McLaren, 1973; Rusetskaya and Ermolenko, 1970; Ballard and Pye, 1976). Chemically resistant glass for reinforcing cement was based on the system Na_2O–CaO–FeO-Fe_2O_3–MnO–SiO_2 (70%) (Paul and Youseff, 1978). Iron silicate glass with strong adherence to steel improves sintered steel by filling its voids. On cooling, it crystallizes to a glass ceramic (Santt, 1978).

Copper is mobile enough in silicate glasses to cause phenomena related to those caused by alkali ions; but effects are complicated by its multiple valency states; Cu^0, Cu^+, Cu^{2+} (White, 1971; Hoffman and Weyl, 1957; Day and Steinkamp, 1969). Cu^+ acts like Na^+, Cu^{2+} like Mg^{2+}, Cu^+ particularly decreases the moduli. Up to 20 wt % Cu_2O can be introduced into lead silicate glasses, which by suitable heat treatment produce the so-called cuprous oxide aventurine glass (Ahmed and Ashour, 1981). The transition elements are to be discussed more extensively in connection with the color of glasses (see the volume on optical properties).

The *rare earths* are to be discussed in connection with laser glasses in the volume on optical properties. The binaries and ternaries with SiO_2 were investigated by Margaryan et al. (1971a,b) and those with CeO_2, Y_2O_3, and La_2O_3 by Bacon (1971). Lanthanum silicate glasses were produced from gels by Mukherjie et al. (1976). Hosts for Nd^{3+} other than silicate glasses have become more important in laser applications (see Sections VIII and XI on phosphate and fluoride glasses). High elastic moduli were obtained with Y_2O, La_2O_3, and TiO_2 (Makishima et al., 1978).

Nitrogen may not only be present in the form of dissolved nitrates (NO_3 grouping) but may also substitute for O (nitride grouping) (Elmer, 1965). This bonding is revealed by the 2.96-μm band after ammonia treatment preceding vacuum consolidation. The annealing point increases, surpassing that of SiO_2. Acid treatment of nitrided "porous SiO_2" leaches additional boron, leading to 99 SiO_2 glass of high resistance. It is assumed that N in this state can occur in other silicate glasses as well.

Ceramics and glasses may be built up from doubly substituted $Si_{1-x}Al_xO_{4(1-x)}N_x$ tetrahedra in wide ranges allowing numerous further substitutions (sialon ceramics and sialon glasses) (see Jack, 1972, 1978). The Y–Al–Si–O–N glasses were first found by Jack (1976) and studied extensively by Loehman (1979). The simultaneous substitution Y, Al:Si, and N:O produces high average cross-linking. The glasses are prepared by melting AlN, Y_2O_3, and SiO_2 in boron nitride crucibles at 1550–1700°C under argon and quenching (5–30 g in 2–3 min to <800°C). Typical compositions include (at. %) 0–6.5 N; >50 O; 5–17 Y; 15–21 Si; 3–12 Al. Glasses are gray to black. T_g increases with N, expansion decreases, hardness and toughness increase. Electrical resistance is high (see also Ukyo et al., 1979), dielectric loss is low. Infrared spectra indicate Si–N bonds (Leedecke and Baca, 1980). Crossbreeds with B can be obtained (Jankowski and Risbod, 1980). Nitrogen can diffuse through sialon glass (Frischat and Schrimpf, 1980). Glass ceramics can be obtained (Leedecke and Loehman, 1980). They may have an expansion coefficient of 21 and Knoop hardness 636. Composition ranges (mole %) are 40–85 SiO_2; 2–17 N; 13 R_2O + RO; >15 B_2O_3, Al_2O_3; e.g., 55 SiO_2; 21 Al_2O_3; 12 Si_3N_4; 2 AlF_3; 7 MgO; 3 Li_2O.

Silver is only sparingly soluble in silicate glasses. It is generally introduced by diffusion at relatively low temperatures. Silver diffusion has been applied to the production of yellow "silver stain" in commercial silicate glasses. Silver (and gold) have been introduced by implantation (Arnold and Borders, 1976).

V. Borate Glasses

A. B_2O_3 Glass

The structure of both crystalline and vitreous B_2O_3 consists of planar O triangles centered by B (Warren et al., 1936; Bray et al., 1963; Mozzi and Warren, 1970; Strong and Kaplow, 1968). In crystalline B_2O_3 the B–O distance is 1.35 Å, the O–O distance 2.40 Å, the angle B–O–B 140°. In vitreous B_2O_3, x-ray diffraction (Mozzi and Warren, 1970) established a

B–O distance of 1.37 Å, B–O–B angles between 120° and 130°, with most triangles arranged in boroxol 3-rings with three oxygens in, three oxygens outside the ring (Fig. 23a), as in some borate crystals (Krogh-Moe, 1965; Borelli, 1963; Hanst *et al.*, 1965; Mozzi and Warren, 1970; Bril, 1976); and NMR (Jellison, 1977; Jellison *et al.*, 1977) as well as SAX (Zarzycki, 1978) studies seem to support this model. The strongest Raman line (806 cm^{-1}) of crystalline boroxol structures is found in vitreous B_2O_3 (Bronswijk, 1977). The boroxol structure persists in the melt according to Raman spectroscopy (Furukawa and White, 1980).

Among other structural models for B_2O_3, one involving two BO_4 types (Berger, 1953) used by Weyl and Marboe, (1964) to draw conclusions on structure property relations now appears incorrect, particularly in the light of the predominance of the coupling constant of 2.76 MHz^{-1} corresponding to BO_3 in NMR data (Bray, 1958, 1964; Bray and O'Keefe, 1963; Bray *et al.*, 1963; 1960). In the arrangement of BO_3 as described, space is not filled: every boroxol ring has neighbor groups not connected by closed B–O bonds. Bonding is prevalently covalent. Ito *et al.* (1951) have postulated irregular tetrahedra with one faraway O as in boracite, which cannot be excluded entirely (Krogh-Moe, 1960; Bray, 1966). B_2O_3 exhibits a very large depolarization (0.3) of light scattering as a consequence of anisotropy in the boroxol structure (Schroeder, 1977).

The viscosity of B_2O_3 at 1260°C is as low as $10^{-11.6}$ times that of SiO_2, even though the B–O bond is stronger than the Si–O bond (Mackenzie, 1956; Araujo, 1966). The relation in bond strength is revealed by the molar refraction of O as listed in Table XX. Although Barber and Fajans' (1954) molecular model has been generally discarded, his postulate of a group structure remains justified. All other *peculiarities* of B_2O_3 glass are ultimately associated with the more directed B–O bond: high expansion, low melting point of crystalline B_2O_3, extremely low surface tension (58.2 erg/cm^2 at 300°C; Al_2O_3 has 2050 erg/cm^{-2} at 580°C), low temperature specific heat, change of molar refraction with thermal history, and the relationship of borate glasses to organic polymers (Verebeichik and Odeleskii, 1960) in which they are soluble (Irany, 1943) and from which they can be extracted in molecular-weight fractions (Weyl and Barber, 1952). But the activation energy of viscosity E differs as much between

TABLE XX

Molecular Refraction of O in Some Oxides

B_2O_3 cryst.	B_2O_3 vitr.	Al_2O_3	SiO_2
3.28	3.50	3.48	3.54–3.66

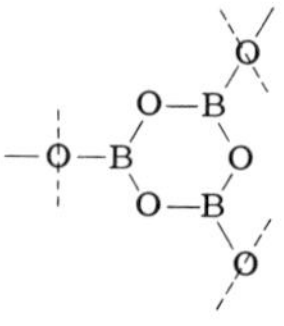

(a) The boroxol ring, observed in vitreous B_2O_3.

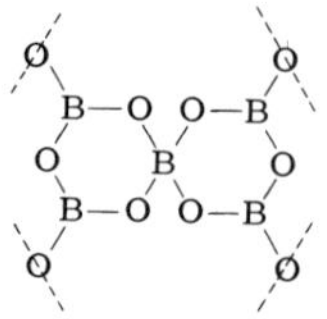

(b) The pentaborate group, observed in the compounds α-$K_2O \cdot 5B_2O_3$ and β-$K_2O \cdot 5B_2O_3$.

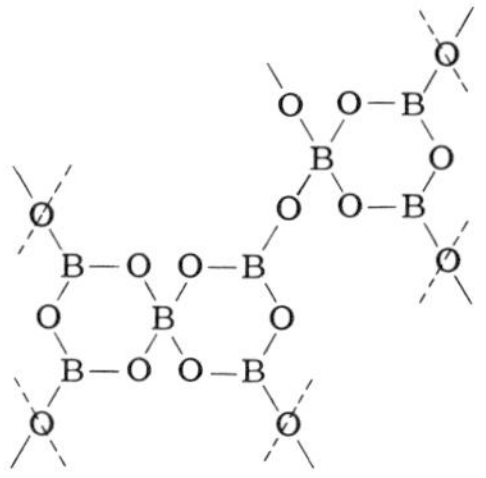

(c) The tetraborate group, observed in the compound $Na_2O \cdot 4B_2O_3$.

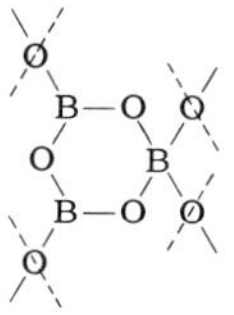

(d) The triborate group, observed in the compound $Cs_2O \cdot 3B_2O_3$.

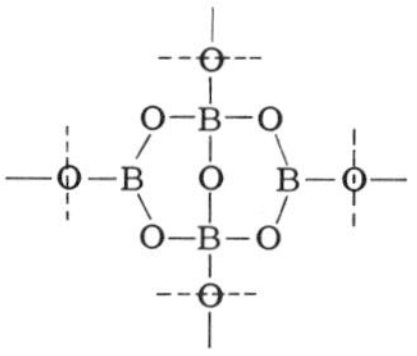

(e) The diborate group, observed in the compound $Li_2O \cdot 2B_2O_3$.

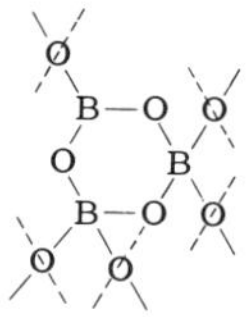

(f) The di-triborate group observed in the compound $K_2O \cdot 2B_2O_3$.

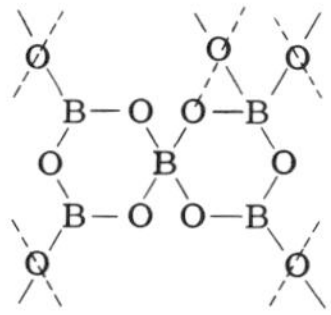

(g) The di-pentaborate group, observed in the compound $Na_2O \cdot 2B_2O_3$.

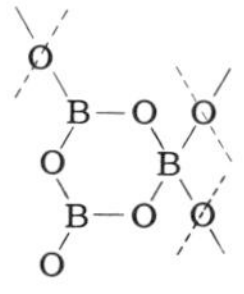

(h) The triborate group with one nonbridging oxygen ion, observed in the compound $Na_2O \cdot 2B_2O_3$.

FIG. 23. Borate groups observed in several borate compounds. Dotted line through the oxygen ions indicate that these are of the bridging type.

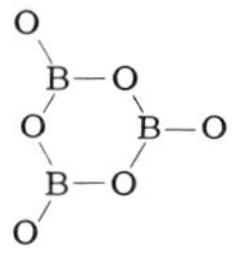

3. INORGANIC GLASS-FORMING SYSTEMS

(i) The ring-type metaborate group, observed in the compounds $Na_2O \cdot B_2O_3$ and $K_2O \cdot B_2O_3$.

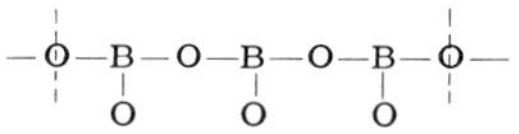

(j) The chain-type metaborate group, observed in the compounds $Li_2O \cdot B_2O_3$ and $CaO \cdot B_2O_3$.

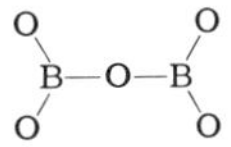

(k) The pyroborate group, observed in the compounds $2MgO \cdot B_2O_3$ and $2CaO \cdot B_2O_3$.

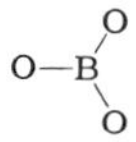

(l) The orthoborate group, observed in the compounds $3MgO \cdot B_2O_3$ and $3CaO \cdot B_2O_3$.

FIG. 23. (*continued*)

organics and B_2O_3 as between B_2O_3 and network liquids such as SiO_2 (Table XXI).

The behavior of B_2O_3 under high pressure, exhibiting the significant role of H bonds, appears to add to objections against any simple free volume concept (Corsaro, 1976).

B. ALKALI BORATE GLASSES

The addition of alkali (R_2O) to B_2O_3 produces $(BO_4)^{5-}$ groupings until about 40 R_2O (Krogh-Moe, 1962; Svanson *et al.*, 1962; Bray *et al.*, 1963; Bray, 1966, 1978; Silver and Bray, 1958). Figures 24 and 25 demonstrate the formation and decay of $(BO_4)^{5-}$ as indicated by nuclear magnetic resonance. The principle of the conversion by alkali of $(BO_3)^{3-}$ to $(BO_4)^{5-}$ groups had been recognized early by Warren (1941), Abe (1952), Huggins and Abe (1957), Everstein *et al.* (1960), and others; however, NMR data disposed of the notion that a maximum in $(BO_4)^{5-}$ formation at about 16 R_2O caused the so-called boron oxide anomaly (property maxima and minima). Rather, $(BO_4)^{5-}$ formation continues above 15% and the boron

TABLE XXI

ACTIVATION ENERGY OF VISCOSITY (E_η) OF SOME MATERIALS

	Organics	B_2O_3	SiO_2
E_η	1.5–13	40	100

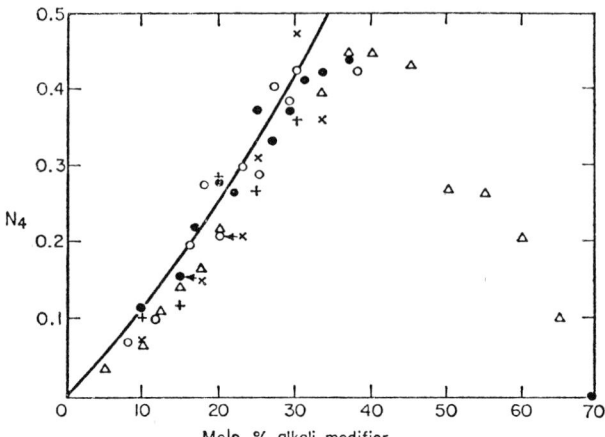

FIG. 24. The fraction N_4 of boron atoms in 4-coordination in alkali borate glasses: ●, Na_2O; ○, K_2O; △, Li_2O; +, Rb_2O; ×, Cs_2O (Bray and O'Keefe, 1963).

oxide anomaly has a more complex origin (Uhlmann and Shaw, 1969). In the case of the minimum of expansivity, BO_4 buildup may indeed decrease expansivity, but the increasing alkali concentration tends to increase it. At 30% R_2O the further increase in expansion is in line with the decay of BO_4 groups and the initiation of nonbridging oxygens indicated by NMR. Moreover, Uhlmann and Shaw's (1969) new expansivity data for all five alkali systems showed a wide variation in position and intensity of the minima. The attempt to explain viscosity minima by BO_4 maxima had found an early critic in Krogh-Moe (1962). Griscom (1978) concluded that the expansion characteristics of alkali borate glasses are not anomalous at all but vary predictably with complex structural changes. However, the sharp extremes found for several properties in well-annealed glasses by Mader and Lorentz (1978) remain of technological significance.

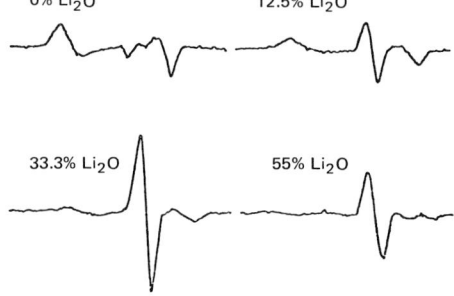

FIG. 25. Signal of BO_4 as in 33.3% Li_2O as indicated by nuclear magnetic resonance (Bray and O'Keefe, 1963).

Between 40 and 70 mole % alkali, the $(BO_4)^{5-}$ concentration decreases to zero (Bray and O'Keefe, 1963; Bray, 1966). Nonbridging oxygens in 3-coordinated BO_3 grouping will appear in increasing numbers (Bray and O'Keefe, 1963; Bray, 1966). A model for the first appearance and increase of nonbridging oxygens at BO_3 groups was offered by Beekenkamp (1965). Nonbridging oxygens may be detected by x-ray photoelectron spectroscopy (Brückner et al., 1980), by the hyperfine splitting of irradiated Ag centers introduced by doping (Assabghi et al., 1977), by quadruple interaction analysis (Kriz et al., 1971; Park et al., 1972), or by Mössbauer signals of Fe^{3+} (Nishida and Takashima, 1980). NMR spectra conclusively identify five sites corresponding to models by Krogh-Moe (1965) and Konijnendijk and Stevels (1975) (Table XXII). Although Peterson et al. (1974, 1975) warn about limitations in the assignment of spin resonance signals to $(BO_4)^{5-}$, Taylor and Friebele (1974) maintain confidence in the uniqueness of the postulated borate glass structure.

Snyder (1978) suggests and amplifies the application of quantum chemistry to the structural interpretation of borate systems. Cooper (1978) convincingly calls for topological considerations for obtaining important details beyond first-neighbor arrangements, using crystallization kinetics and gas permeation (Shackleford, 1978) data. Raman evidence supports the following models in detail (see Fig. 23) (Konijnendijk and Stevels, 1975, 1978; White et al., 1978).

(1) From *0 to 20%* R_2O, tetraborate groups (Fig. 23c) containing $(BO_4)^{5-}$ groups in the definite relations to $(BO_3)^{3-}$ groups shown in the figure are formed, while boroxol groups (Fig. 23a) (signal at 806 cm^{-1}) decay. A few "loose" BO_3 triangles and BO_4 tetrahedra are present. Data

TABLE XXII

Percentages of Sites versus R_2O Content in Borate Glasses[a,b]

Designation	Site	% R_2O						
		5	10	15	20	25	30[c]	35[c]
T⁴	1	5	10	15	17	14	11	8
D⁴	2	—	1	3	8	19	28	34
B³	3	79	58	38	16	4	—	—
T³	4	16	30	44	51	44	33	24
D	5	—	1	3	8	19	28	34

[a] Jellison and Bray (1978a,b).
[b] T = tetraborate, D = diborate, B = boroxol. Superscripts refer to coordination. 3-coordinated B may be next to 0, 2 or 3 4-coordinated B; 4-coordinated B may be next to all or 3 3-coordinated B.
[c] Nonbridging oxygen ignored.

are similar to those obtained with crystalline tetraborate glasses and the interpretation agrees with that based on NMR, x-ray, and melting-point depression.

(2) From *20 to 35%* R_2O, diborate (Fig. 23e) and di-triborate (Fig. 23f) groups form. No signals for ring metaborate (Fig. 23i) dipentaborate (Fig. 23g), and triborate with nonbridging O (Fig. 23h) are obtained. The pairing of B^{IV} in diborate is confirmed by NMR (Rhee and Bray, 1971) and Krogh-Moe's (1962) melting-point depression data as well as by studies by Nagel and Bergeron (1974) and Kumar (1963).

(3) At *40 to 50%* R_2O, diborate, and orthoborate (Fig. 23l), pyroborate (Fig. 23k), and meta-ring-borate groups (Fig. 23i) occur revealed by the 755, 820, and 630-cm^{-1} signals, respectively. Also, nonbridging oxygens, earliest for Cs^+, occur. Raman assignments can be reinforced by comparison with crystalline systems, e.g., B_2O_3, $Cs_2O \cdot 9B_2O_3$ (Bronswijk, 1977). The latter is the only crystal system having both boroxol and triborate rings, giving signals at 806 and 770 cm^{-1}, respectively.

The BO_4 tetrahedra in such structures persist in the melt, although some rearrangement to less dense groups, as triborate to pentaborate, is indicated by band broadening (Furukawa and White, 1980). At higher temperature, less alkali is needed to attain the maximum BO_4 concentration and to form BO_3 groups with nonbridging oxygens. For this reason, different authors may obtain different results at low temperature when the fictive temperature varies (Araujo, 1979). The pursuance of these studies leads Araujo (1980) to stress the classical Abe postulate that BO_4 groups should not combine, and must be separated by BO_3 groups. This explains the early (in terms of alkali concentration) appearance of nonbridging oxygens and promotes the dissociation of diborate groups.

Structural work in general agreement with these models includes the study of viscosity (Mackenzie, 1956, 1960; Shartsis and Shermer, 1953, 1954) and absorption and emission of transition and rare earth elements (Kurkjian, 1965; Berkes and White, 1966; Dunicz, 1966; and many more).

Alkali ions in borate glasses are mobile (Rhee and Bray, 1971; Boulos and Kreidl, 1971a,b) and show the mixed alkali effect in internal friction (Terai *et al.*, 1973; Boulos and Kreidl, 1971a,b). The mixed alkali effect appears to be affected by the mass ratio as evidenced by a maximum in electrical resistivity (and minima in dielectric constant and loss) obtained at a 1 : 1 ratio of ^6Li and ^7Li isotopes used in lithium borate glasses (Abou el Leil *et al.*, 1978). Heat of solution data on 30 Li_2O– 30 K_2O– 10 Al_2O_3– 60 B_2O_3 glasses appear to support the views of Terai and Sugita (1978) on the role of increased attraction of unlike alkali sites as a cause of the mixed alkali effect.

Alkali borate glasses tend to phase-separate (Vogel, 1959, 1960; Shaw and Uhlmann, 1970). In the system $Na_2O-B_2O_3$, for example, the consolute was given as near 614°C at about 15 Na_2O (Charles, 1966). Golubkov et al. (1977), however, found no evidence for phase separation meeting their criteria for the binaries Na_2O, K_2O, Rb_2O, and $Cs_2O-B_2O_3$. They consider inhomogeneities found as thermal and concentration fluctuations. The topic remained the subject of spirited argumentation at the Alfred Conference (Porai-Koshits et al., 1978; Seward, 1978) and the Prague International Congress in 1977.

Alkali borate glasses are much densified under >40 kbar pressure above 250°C (Bridgman and Simon, 1953; Stewart et al., 1967).

As in silicate glasses, the electronic structure of borate glasses is the subject of an increasing amount of investigation, including the use of transition-metal (Armstrong et al., 1977), rare earth (Weber et al., 1978), or silver (Bishay et al., 1978) impurities as probes. In particular, after γ irradiation, ESR signals may permit identification of nonbridging oxygens. In the undoped glass, it appears that for predominantly BO_4 groupings, charge transfer from O to B prevails; for BO_3 groupings with nonbridging oxygens the charge transfer is from nonbridging oxygens to alkali cations. It has become possible to make quite satisfactory calculations of molecular orbitals corresponding with the models incorporating boroxol, pentaborate, triborate, tetraborate, dipentaborate, and diborate groupings (Kawazoe et al., 1978a,b,c). Basicity was related to two quite different orbital situations. Delocalization of the out-of-plane π-type nonbonding levels decreases sharply in the 15–20% alkali region. In-plane basicity is nearly constant over the entire 0–35% range. Curiously, π-type basicity does not differ much for NBO and bridging O in tetra- and diborate groups. It seems that the properties are controlled by π-electron distributions. In this situation, at present, orbital calculations do not seem to allow the determination of nonbridging oxygen formation, which is believed to occur much before BO_4 saturation according to a lever rule.

A copper [$Cu^{(II)}$] probe was used (Kawazoe et al., 1978b) to demonstrate sharp rises in π basicity at 17 and 50% K_2O, much larger than in phosphate or sulfate systems, while σ basicity varied little within borates in comparison to other systems.

The incorporation of OH will affect the *density* of alkali borate glasses, mostly of $Li_2O-B_2O_3$ glasses of low Li_2O content where open space is available and a slight contraction is caused by weak hydrogen bonding to adjacent bridging oxygens (Franz, 1978). *Refraction* is mostly affected at low K^+ content where B–OH replacing B–O–B is most effective. The influence on T_g is small, with the increase associated with a relaxation with increasing OH of the expanding framework of alkali borates of high alkali content. T_g can be evaluated from viscosity–temperature relations (Mazurin and Potselueva, 1978). B_2O_3 losses and OH content vary strongly with the use of raw materials (Hartung and Heide, 1978; Cable, 1978).

Acoustic losses differentiate between BO_4 and OH influences. As in germanate, and in contrast to silicate, glasses, the sound velocity increases with alkali content in borate glasses (Krause and Kurkjian, 1978).

C. OTHER BORATE GLASSES

1. Alkaline Earth Borate Glasses

The structures frequently resemble those of the crystalline compounds (Block and Piermarini, 1964), with ring formation an important feature. For example, monoclinic $BaO \cdot 2B_2O_3$ (Block and Perloff, 1966) consists of a three-dimensional linkage of alternating single and double 6-rings, each ring being linked to a 5-ring through a nonring oxygen and each containing both 3- and 4-coordinated B. Rhombic $SrO \cdot 2B_2O_3$ is completely different, containing only 4-coordinated B (Block, 1967; Park and Bray, 1972). Its structure resembles that of $PbO \cdot 2B_2O_3$ (Wittmann and Beulich, 1965). The calcium borate system was described by Carlson (1932). The conductivity of iron-doped calcium borate glasses is electronic ($Fe^{2+} \rightleftarrows Fe^{3+} + e^-$) (Omar and Stevels, 1978). Immiscibility in the ternary K_2O–CaO–B_2O_3 extends to high K_2O contents (Kawazoe *et al.*, 1978c). Finally, in R_2O–MgO–B_2O_3 glasses, Mg promotes BO_4 formation (Toyuki, 1980).

2. Zinc Borate

In contrast, zinc borate melts and glasses differ from the crystalline structures (Ring, 1962; Krogh-Moe, 1962; Hummel and Harrison, 1964). Partial replacement of ZnO by La_2O_3 favors BO_3 triangles over BO_4 groups, as evidenced by IR spectra (Efimov and Mikhailov, 1979). Complex systems containing ZnO and B_2O_3, e.g., with PbO, R_2O, TiO_2, Sb_2O_5, are the basis of some solder glasses (Dale and Stanworth, 1949; Levand *et al.*, 1971, 1972) and glass ceramics (McMillan, 1964; Zhdanets and Kheifeis, 1967; Berezhnoi *et al.*, 1977).

3. Cadmium Borate Glasses

These glasses, also containing S, are described by Takahashi and Goto (1971). Photoelectric effects were observed by Lonsdale *et al.* (1978).

4. Lead Borate Glasses

Nuclear magnetic resonance indicates a shift from ionic, at low PbO content, to covalent bonding at high PbO content (Bray *et al.*, 1963).

Above 20 PbO, PbO_4 pyramids tied to BO_3 triangles are beginning to be preferred over configurations containing $(BO_4)^{5-}$ groupings. In crystalline lead borates, ring structures exist resembling those in strontium borates (Svanson et al., 1962).

At low PbO contents (about 0.5 to 20 mole %), subsolidus immiscibility occurs (Liedberg et al., 1965; Shaw and Uhlmann, 1971; Gough et al., 1969; Simmons, 1973). Lead borate glasses with a small amount of SiO_2 and other additives for stabilization serve as low-melting solders. In low-temperature solder glass for encapsulating diodes, up to (wt %) SiO_2 15, Al_2O_3, ZnO 8, TiO_2 7 are added (Davis and Greeson, 1979). To prevent sticking to the graphite boat used in sealing, F is added. [Example: (wt %) SiO_2 13, Al_2O_3 9, PbO 67, PbF_2 5, B_2O_3 6.

The addition of Cu^+ to such glasses results in the lowest softening ranges without affecting expansion (Pirooz, 1963; Francel and Hagedorn, 1964; Ellis, 1971). It is important in this application to assure a high Cu^+/Cu^{2+} ratio without, however, exceeding the low-T solubility of Cu^+ and thus causing precipitation (Takamori and Tomazawa, 1976). An example of an experimental glass of this sort is PbO, 64, B_2O_3 16, SiO_2 6, ZnO 9, CuO 5. Electrical conduction and sites in lead borate glasses containing Fe and Cu were studied by Hiroshima and Yoshida, (1977). From $PbO-TiO_2-B_2O_3$ glasses, $PbTiO_3$ can be grown (Bergeron and Russell, 1965a,b; Brown and Ginell, 1962).

5. $Bi_2O_3-B_2O_3$ and $PbO-BiO_3-B_2O_3$

A large area of glass formation exists in the systems $Bi_2O_3-B_2O_3$ and $PbO-Bi_2O_3-B_2O_3$ (Hirayama and Subbarad, 1962; Levin, 1965; Ioffe and Patrina, 1970). Conduction at high Bi_2O_3 content becomes electronic (Gough et al., 1969). The structure at vitreous $PbO \cdot 2B_2O_3$ differs considerably from that of the crystal in which all B is in unusual 4-coordination (Konijnendijk and Stevels, 1975; Konijnenkijk and Buster, 1976).

6. *Silver Borate*

Silver borate glasses may contain above 30 Ag_2O and are structured like alkali borate glasses (Kreidl and Maklad, 1971). The so-called boron oxide anomaly (around 16% for Na_2O) is not observed for ϵ and ϵ'' in Ag_2O glasses. Log conductivity is higher in Ag_2O glasses than in Na_2O glasses, with a lower temperature coefficient. ϵ' and ϵ'' are larger, increasing rapidly with Ag concentration. At 30%, clusters of Ag^+ and colloidal Ag^0 contribute (Tsuchiya et al., 1979). Gold (Au) is very sparingly soluble (~ 0.001 wt %) in borate glasses (Paul et al., 1979). Its solubility increases with alkali and, by complexing, with Cl.

7. Cabal Glasses

Cabal glasses is the name given glasses in the ternary $CaO-B_2O_3-Al_2O_3$ system. These are important because of their high electrical resistance, surpassing even that of SiO_2 glasses (Owen, 1961). In this system, Al and B compete for $(AlO_4)^{5-}$ and $(BO_4)^{5-}$ groups (Bray, 1966; Bishop, 1964; De Waal, 1969; Gough *et al.*, 1969; Sakka, 1977). Conduction is most likely by O^{2-}, and depends on the amount, not the kind, of alkaline earth that may substitute for Ca.

8. $PbO-B_2O_3-Al_2O_3$

The conductivities of $PbO-B_2O_3-Al_2O_3$ glasses (Gough *et al.*, 1969; Zarzycki and Naudin, 1967; Naudin and Zarzycki, 1968) show lower activation energies. A Ga_2O_3 analogue was described by Polukhin *et al.*, (1968).

9. Nabal Glasses

Nabal and similar glasses [$Na_2O(K_2O,Li_2O,Cs_2O)-B_2O_3-Al_2O_3$ glasses) behave similarly to cabal glasses (De Waal, 1969; Gresch *et al.*, 1976). They contain more nonbridging oxygens and exhibit the mixed alkali effect (Terai *et al.*, 1973).

10. Glasses Analogous to Cabal Glasses

These form when Fe_2O_3 replaces Al_2O_3 (Gdula and Tompkins, 1970; Moon *et al.*, 1975; Bishop *et al.*, 1977). Large $-Fe^{3+}-O-Fe^{3+}$ clusters (close to 5 Fe per cluster) with stable magnetic properties have been observed (Kornilova and Petrovskii, 1980). From them ferrites can be precipitated (Kerr and Jorgensen, 1971; Gdula and Tompkins, 1970; Bishop *et al.*, 1977). High concentrations of Fe_2O_3 (65 mole %) can be achieved in lead borate glasses, the FeO portion decreasing at higher total Fe content (Burro and Ardelean, 1979).

11. Fluoroborate Glasses

In fluoroborate glasses $B^{IV}O_3F$ and $B^{IV}O_2F_2$ groups are observed (Bray, 1964).

12. Lanthanum Borate Glasses

These glasses (Brewster *et al.*, 1947) are the basis of the low dispersion, high refraction glasses often termed rare earth glasses, more properly lanthanum crowns and lanthanum flints. They were discovered and first developed by Morey (1939), and contain Ba, Ta, and, to improve chemical resistance, Zr or Th. The objection to the latter's radioactivity led to the design of the Th-free varieties. Devitrification is minimized by Mg, Ca,

Zn, etc. (Faulstich, 1961; Katzschmann, 1965). Examples are given in the accompanying tabulation (compositions in weight percent).

	B_2O_3	BaO	Ta_2O_5	TiO_2	ZrO_2	La_2O_3	ThO_2	Nb_2O_5
(1)	12	—	28	12	6	42	—	—
(2)	40	20	—	—	—	20	20	—
(3)	24–36	—	—	—	—	45–60	—	8–20

13. Tl_2O and Cu_2O

Large amounts of Tl_2O and Cu_2O can be introduced (Heindorf and Vogel, 1970) in binary and polynary borate glasses: Tl (Baugher and Bray, 1969; Krogh-Moe, 1965; Momii and Nachtrieb, 1968); Cu (Rza *et al.*, 1971). With increasing Tl^+ content, bonds become more covalent and complex structures, as in $PbO-B_2O_3$ glasses, appear in the region of 20–35 mole % Tl_2O (Panek and Bray, 1977).

14. Tin Borate Glasses

These glasses were described by Eissa *et al.* (1974) and Paul *et al.* (1977). With increasing SnO content, BO_4 groups form. A maximum in chemical shift at 20% suggests a change from ionic to covalent bonding as in PbO glasses.

15. Nitrogen

Nitrogen is accepted in alkali borate structures (Alexandre-Ferraris *et al.*, 1977).

VI. Borosilicate Glasses

A. SYSTEMS AND STRUCTURES

1. $B_2O_3-SiO_2$

Because of experimental difficulties, a useful phase diagram has become available only quite recently (Rockett *et al.*, 1965) (Fig. 26). No compounds appear; the eutectic is near B_2O_3; there is no stable immiscibility; but there is metastable immiscibility with a flat symmetrical boundary and a consolute temperature of 520°C (Charles, 1968; Vasilevskaya *et al.*, 1980). In such phase-separated $B_2O_3-SiO_2$ glass, microstructural birefringence occurs (Tomazawa and Takamori, 1980). This immiscibility connects in the pseudobinary $Na_2O \cdot 4B_2O_3-SiO_2$ and extends into the ternary immiscibility dome in which Vycor glass occurs. Small amounts of strong cations such as Li^+ or Ca^{2+} accentuate immiscibility (Sastry and Hummel,

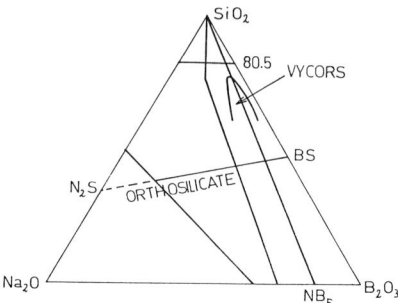

FIG. 26. Location of phase-separated "reconstructable" (Vycor) glass compositions in the ternary $Na_2O(N)-B_2O_3(B)-SiO_2(S)$ (after Volf, 1961).

1959). The structure of $B_2O_3-SiO_2$ glasses was investigated by Warren et al. (1939). Radial distribution data show a first unresolved peak from which positions were calculated as shown in the accompanying tabulation. The second (O–O) peak gives 2.40 Å, the third (Si–Si) 3.2 Å (decreasing with increasing B). Vapor-deposited glasses in this system are important for optical communication fibers (see Section I). Index of refraction versus composition data were tabulated by Hammond and Norman (1977).

Composition (mole %)	First peak (Å)	Composition (mole %)	First peak (Å)
100 B_2O_3	1.39	45 SiO_2	1.62
15 SiO_2	1.46	60 SiO_2	1.62
30 SiO_2	1.55	100 SiO_2	1.62

Amorphous $SiO_2-B_2O_3$ is obtained by the hydrolysis of ethyl silicate and borates and may be used as anticorrosive coatings (Sakurai and Mochizuki, 1978); or from gels (Antonova and Dyakova, 1979). Mixed $SiO_2-B_2O_3$ gels from sol–gel processing have been converted to glass by hot pressing. This material has a very low OH content and does not devitrify for ≥ 10 mole % B_2O_3. The refractive index has a minimum at 15% B_2O_3 (Jabra et al., 1979).

Infrared spectra of vapor-deposited $B_2O_3-SiO_2$ films allow the identification of B–O–Si bonds, using a bond-stretching vibration band at 1130 cm^{-1} (Wong, 1978; Jabra et al., 1979). The similarity to alkali borosilicate (Pyrex) spectra is instructive. For the important case of small B additions to SiO_2 glass, trigonal BO_3 groups appear to form in the SiO_4 network (Smith, 1978).

2. $R_2O-B_2O-SiO_2$

This system includes important technical glass compositions (Volf, 1961) with R = Na. The structural arrangement of Si and O in borosilicate glasses is similar to that in SiO_2 and silicate glasses (Warren et al., 1939), but phase separation is found over a wide range in this system (see Section VI.B). Because of the technological importance of this system, various techniques have been applied to obtain as much information as possible on the grouping of B, Si, O, and alkali.

Viscosity and density studies (Mackenzie, 1956, 1959; Li 1962; Riebling, 1964) suggest that below 20 SiO_2, B is arranged as in borate glasses, while $(SiO_4)^{4-}$ groups remain widely separated. At around 45 SiO_2, moderate departures from borate structures were suggested, and SiO_2 clustering was assessed by chemical extraction (Kolb and Hansen, 1965) based on a chromatographic technique (Lentz, 1964). Above 60 SiO_2, extensive clustering was found. Note that using mole %, Si corresponds to 2B, so that equal amounts of Si and B appear only at $66\frac{2}{3}$ SiO_2. Time–viscosity data suggest an ordering process (Huggins and Abe, 1957).

Raman spectra (Konijnendijk et al., 1976) lead to the populations of various species shown in Table XXIII.

Below R/B = 0.5 nearly all alkali ions cause the formation of six-member borate rings with one or two $(BO_4)^{5-}$ units arranged in tetraborate or diborate groups. Above R/B = 0.5 nonbridging oxygens form at SiO_4 tetrahedra, while below 0.5, SiO_4 tetrahedra connect as in SiO_2. Above

TABLE XXIII

BORATE AND SILICATE GROUPS PRESENT IN ALKALI BOROSILICATE GLASSES AS INDICATED BY THE RAMAN SPECTRA[a]

	Alkali/boron ratio <0.5	Alkali/boron ratio >0.5
15 mole % SiO_2	Boroxol	Diborate
	Tetraborate	SiO_4 (4 bridging O ions)
	Diborate	SiO_4 (1 nonbridging O ion)
	SiO_4 (4 bridging O ions)	
35 mole % SiO_2	Boroxol	Diborate
	Tetraborate	Metaborate
	Diborate	SiO_4 (4 bridging O ions)
	SiO_4 (4 bridging O ions)	SiO_4 (1 nonbridging O ion)
65 mole % SiO_2	Boroxol	Metaborate
	Tetraborate	SiO_4 (4 bridging O ions)
	(Diborate)	
	Metaborate	SiO_4 (1 nonbridging O ion)
	SiO_4 (4 bridging O ions)	

[a] Konijnendijk and Buster (1976).

R/B = 0.5, ring metaborate and pyroborate groups and orthoborate units appear to form as in alkali borate glasses. At R/B ~ 1, Na and K encourage the formation of ring-type metaborates at increasing SiO_2 contents. BO_4 units are *not* incorporated in the SiO_4 network as AlO_4 units are. As in binary alkali borate glasses some "loose" $(BO_3)^{3-}$ and $(BO_4)^{5-}$ units exist.

NMR studies appear to provide increasingly concrete and detailed structural models for this important ternary glass system. The fraction (N_4) of boron atoms in BO_4 units as a function of Na_2O/B_2O_3 (R) and SiO_2/B_2O_3 (K) have been the subject of extensive investigation (Milberg *et al.*, 1972; Scheerer *et al.* 1973; Zhdanov, 1974; Brungs and Cartney, 1975), indicating that for $R > 0.5$, Na generally converts BO_3 to BO_4 groups just as in the binary system. For $R > 0.5$, more recent work (Yun and Bray, 1978) permits identification of the fraction of asymmetric BO_3 units with one or two nonbridging oxygens (N_{3s}). N_4 rises to a maximum above $R = 0.5$. N_4 as well as N_3 are clearly, and for obvious reasons, functions also of K. As SiO_2 increases, high Na_2O contents convert the diborate grouping to the structural unit $(BSi_4O_{10})^{-1}$ (Bray, 1978) found in the mineral $Na_2O \cdot B_2O_3 \cdot 6SiO_2$ (Reedmergnerite) (Appleman and Clark, 1965). Bray (1978) suggests the shift of N_{4max} to $\frac{1}{16}K + 0.5$. For $SiO_2/B_2O_3 \geq 8$ ($K \geq 8$), $N_4 = 1$, which means that all boron is 4-coordinated, and the structure is like that of reedmergnerite embedded in SiO_4 groupings, which, if Na_2O exceeds the amount needed for $R_2O/B_2O_3 = 1$, will form nonbridging oxygens at SiO_4 tetrahedra. With all this evidence, the existence of a region of phase separation larger than that manifest in electron microscopy is not surprising.

B. Phase Separation in Alkali Borosilicate Glasses

The metastable immiscibility region in the ternary system Na_2O–B_2O_3–SiO_2 was mapped by Haller *et al.* (1970) (Fig. 27) and revised to show a lower consolute at 20–40% SiO_2 near $Na_2O = 0$ (Alekseeva *et al.*, 1977). The highly regular and connective character of the resulting microstructure is often associated with the spinodal mechanism of phase separation (Cahn and Hillard, 1959) and was considered by Charles (1964, 1968a,b). However, this microstructure may also result from a nucleation–growth–coalescence process (Haller and Macedo, 1968; Haller *et al.*, 1970; Srinivasan *et al.*, 1971; Seward *et al.*, 1968b). Indeed, in a specific case, at 100°C, within the spinodal region, a nucleation mechanism was observed. Although tie lines might suggest three-phase immiscibility, the complex microstructure is more likely the partial result of secondary-phase separation in which one of the two first-formed phases continues to

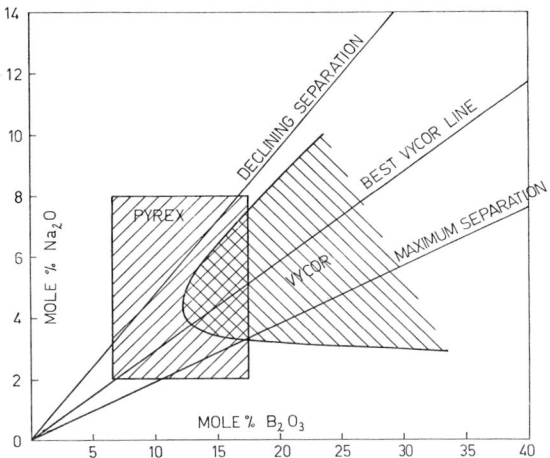

FIG. 27. Detail of the system $Na_2O_3-B_2O_3-SiO$ with location and limitation of Pyrex and Vycor borosilicate glass compositions (after Volf, 1961).

separate once more on cooling. Certain features of the tie line may still have to be considered unexplained until electron probe analysis furnishes additional evidence (Scholes, 1973).

Above the immiscibility boundary, some inhomogeneity persists even in the one-phase glass because of compositional fluctuations (Porai-Koshits et al., 1971). The radii of such inhomogeneities are about 25 Å; the mean square electron density differences are two orders smaller than in the immiscibility region. Biphasic alkali borosilicate glasses after stretching in the T_g range become negatively birefringent as the distribution of spherical minor phase particles becomes anisotropic through the flow of the lower viscosity matrix (Takamori et al., 1976).

C. "Reconstructed" Glass

1. Range

The ternary system $Na_2O-B_2O_3-SiO_2$ contains within its general immiscibility region a subregion in which, at characteristic temperatures (500–650°C), a boron- and sodium-rich phase is formed. This phase can be leached by mineral acids leaving behind a silica-rich skeleton that, at about 1000°C, shrinks to a homogeneous glass (Turner and Winks, 1926; Charles, 1964). The leachable phase may contain, e.g., 1 Na_2O–2.4 B_2O_3–0.15 SiO_2 (Charles, 1964), the shrinkable phase 96% SiO_2. The desired product is formed from the easily molten and shaped parent glass before phase separation, and obtained after phase separation, leaching, and shrinking as the reconstructed (Vycor) 96% SiO_2 glass with a very low coefficient of thermal expansion.

The region in which this process is feasible is bound (Figs. 26 and 27) approximately on the low-SiO_2 side by the line $(Na_2O)_2 \cdot SiO_2 - B_2O_3 \cdot SiO_2$, below which the coherence of SiO_2 begins to collapse (Skatula et al., 1958; Kuehne, 1955; Porai-Koshits, 1955; Molchanova, 1958) and on the high-SiO_2 side by the 80% SiO_2 line, above which the continuity of the boron-rich phase is impaired. Below 4% Na_2O and at less than $2B_2O_3/Na_2O$, the solubility of the B_2O_3 phase might be too low; at above 16% Na_2O and 10% B_2O_3/Na_2O, corrosion may prove too high. In simple terms, the region may be delineated (Hood and Nordberg, 1940) by 55 to 70 SiO_2, $10 - (0.1(SiO_2 - 55))$ Na_2O, balance B_2O_3. In this formula a higher, workable SiO_2 region is excluded in which strain in the leached product was found excessive (Hood and Nordberg, 1940).

The phase separation takes place well below the liquidus, around 750°C. The preferred compositions have high glass temperatures (Charles, 1964; Volf, 1961) and are not too far from those of stable Kovar sealing glasses. In fact, near the $Na_2O \cdot 2B_2O_3$ line where leachability decreases, the compositional area of Pyrex and sealing glasses starts (Pyrex compositions are more complex but near the ternary point 83 SiO_2–12 B_2O_3–5 Na_2O). In this area is found the lowest liquidus for highest silica content.

2. Influence of Other Constituents

The replacement of Na by K narrows the separation zone. Li widens and accentuates spontaneous separation (Sastry and Hummel, 1959). F favors separation. In the ternary $K_2O-B_2O_3-SiO_2$ immiscibility region, Kawazoe et al. (1979) observed three subregions: (1) B_2O_3, (2) $17K_2O \cdot 83B_2O_3$, and (3) $yK_2O \cdot (100 - y)B_2O_3$. Doping by P_2O_5, V_2O_5, MoO_3, and WO_3 shifts the immiscibility region so that a more soluble phase contains less SiO_2, thus accelerating the leaching process (Tomozawa and Takamori, 1980). The immiscibility regions of the quaternaries $X_2O-MO-B_2O_3-SiO_2$ (X = Na, K; M = Ca, Sr, Zn) were determined by Taylor et al. (1980). As in the ternaries $MO-B_2O_3-SiO_2$, the gap is larger in the order Zn > Ca > Sr, but the ZnO quaternary extends to lower Na contents. In this respect, in the presence of alkali, Zn^{2+} exhibits some network-forming character at Mg^{2+} and Zn^{2+} do in many other instances. Alumina somewhat suppresses the consolute (Dgebuadze and Averyanov, 1970) but also suppresses unwanted crystallization. For these reasons, both Vycor and Pyrex-type glasses contain of the order of 2 to 3 Al_2O_3. More Al_2O_3 would interfere with the process.

Vycor can be generalized as: $(55 - 1.25Al_2O_3)$ to 70 SiO_2; 0.1 to 4

Al_2O_3; $(10 - 0.1(SiO_2 - 55) - 0.17Al_2O_3)$ Na_2O; remainder B_2O_3. A typical composition would then be:

$$\begin{array}{rl} 60 & SiO_2 \\ 3.5 & Al_2O_3 \\ 10 - 0.1(60 - 55) - 0.17(3.5) = 10 - 0.5 - 0.6 = 8.9 & Na_2O \\ 28.6 & B_2O_3 \end{array}$$

GeO_2 suppresses phase separation in $Li_2O-B_2O_3-SiO_2$ glasses (Evstrop'ev et al., 1970). Rare earths were added by Wachtel (1970); Nd_2O_3 goes into the borate phase (Burkat et al., 1971).

3. The Leaching and Sintering Processes

a. Leaching Parameters. Typically, the soluble high-B phase is removed by $3 N$ HCl or $5 N$ H_2SO_4 at about 100°C. At low temperature and high acid concentration, the more soluble sodium is leached out first. The glass swells as the larger H_3O^+ ions replace Na^+. Then, when more B_2O_3 is leached, the SiO_4 skeleton shrinks (Eguchi et al., 1979). Maximum solubility is near $5Na_2O$ (Molchanova, 1957). The procedure may take 24 h/mm. Leaching is followed by more than 12 h washing, and gradual drying at 100°C, avoiding steam pressure. Normal thicknesses are kept below 3 mm. Leaching is accelerated by S, Sb_2O_3, ZrO_2, and TiO_2, and inhibited by F (Mitra et al., 1971). Influences of other constituents were described by Burkat et al. (1971) and by Molchanova (1957). Since iron accumulates in the soluble phase, the reconstructed glass is quite pure and transmits well in the ultraviolet (Macedo and Litovitz, 1976).

b. "Thirsty Vycor." The uncompacted leached glass is very hygroscopic: its surface sticks to the tongue. It is used for filters, catalysts, and membranes. A brand name is Vycor 7930 (nicknamed thirsty Vycor), whose pore size is normally 20–40 Å (Brunauer et al., 1938) although larger (50–220 Å) pores can be obtained.

c. Technological Aspects. Thick-walled product is made from sintered powder. The formed article is leached (Multiform 7900 brand glass) (Greene, 1943). High-Na_2O glass may swell under gel formation; low-Na_2O glass shrinks, causing strain. The latter is more objectionable. This led to an upper limit for the SiO_2 content. Special products can be obtained by *impregnating* the thirsty glass with solutions (Kuehne, 1955; Low et al., 1969). By impregnating with furfuryl alcohol and heat treating, a carbon-containing reconstructed glass is obtained (Elmer, 1976) whose conductivity can be controlled. High-SiO_2 fibers are made by leaching 62 SiO_2–28 Na_2O–2 Al_2O_3–8 B_2O_3 glass (Nordberg, 1950a,b; Labino, 1958).

Filters with controlled porosity are used in the separation of biological materials ("bioglass") (Haller and Macedo, 1968; Dobychin, 1962; Zhdanov et al., 1973; McMillan and Matthews, 1976). The glass was also used for salt rejection (Hill, 1966). By dipping a porous rod into $NaNO_3$, drying, and sintering, a gradient index rod can be produced from which gradient index fibers can be drawn (Tanaka, 1977). An SiO_2-rich internal surface has been obtained by leaching with a solution containing water, oxides, and an organic solvent (Macedo et al., 1978). High-temperature (~1000°C) treatment (dehydration) of the pores with hydrocarbons (e.g., natural gas in a nonoxidizing atmosphere), followed by heating in an oxidizing atmosphere around 1000°C to oxidize the formed C film, results in precursors for sintering to a reconstructed glass of higher T_g (Elmer, 1976). A water vapor treatment can increase the SiO_2 content of the resulting reconstructed (Vycor-type) glass to 99% (Eguchi et al., 1980).

d. Sintering. Typically, leached Vycor is sintered at 900–1200°C. Typical shrinkage is 35%. For improved purity (UV transmission), heating in hydrogen, then in a vacuum, precedes sintering. This treatment causes evaporation of As and reduction of Fe^{3+} to Fe^{2+} (Nordberg, 1950a,b; Li and Bray, 1962). Special treatments to minimize devitrification have been proposed (Jenaer, 1965).

4. Properties of Reconstructed Glass

Properties are listed and compared in Table XXIV. The structural difference from pure SiO_2 is large enough to give a different expansion curve characterized by a lower coefficient at high temperatures (Saunders, 1942). A typical final composition (Volf, 1961) is 95 SiO_2, 0.3–2 Al_2O_3, 2–5 B_2O_3, 0.02–0.25 Na_2O. The chief uses of Vycor glass are for low-expansion, shockproof articles such as high-grade laboratory ware, and for insulation.

D. OTHER TECHNICAL ALKALI BOROSILICATE GLASSES

At the high-SiO_2 end of the Vycor region and beyond, glasses are found with compositions selected for use as apparatus glasses, which represent a compromise between the high melting temperature of pure SiO_2 and the ranges of corrosion and phase separation. In this region also, the relatively low expansion of 35×10^{-7} can be attained, and the liquidus is the lowest for any high-SiO_2 glasses (Morey, 1931). These are generally known as Pyrex-type glasses, and were first developed by Sullivan and Taylor (1915).

An example of a typical composition is (wt %):

SiO_2, 80.6; B_2O_3, 12.0; Al_2O_3, 2.25; CaO, 0.3; MgO, 0.3; Na_2O, 4.3; K_2O, 0.6

TABLE XXIV

COMPOSITION AND SOME PROPERTIES OF LOW-EXPANSION GLASSES

	Composition (wt %)							Softening point (°C)	Anneal. temp. (°C)	Strain point (°C)	$\alpha \times 10^{-7}$ (K^{-1})	d (g cm^{-3})	ϵ	n_D
	SiO$_2$	B$_2$O$_3$	Na$_2$O + K$_2$O	Al$_2$O$_3$	CaO	MgO	K$_2$O							
Vycor	96.6	2.9	0.04	0.4				1510	890	790	8	2.18	3.8	1.458
Silica	100							1650	1140	1070	5.5	2.2	4.0	1.458
Pyrex	80.6	11.9	4.2	2.1	0.3	0.3	0.6	820	600	570	33	2.73	4.6	1.474

^{a} From Phillips (1960).

The addition of about 1–2% F has been suggested (Li, 1962; Hood and Nordberg, 1934a,b). Pyrex cannot be fined successfully by conventional means; NaCl is preferred (Riebling, 1964, 1966).

Alumina reduces and controls phase separation as well as crystallization. Ta_2O_5 was suggested as promoting increased acid resistance (Rogozhin *et al.*, 1971a,b). It is now generally recognized that Pyrex-type glass, although outside the leachable Vycor range, is within the ternary immiscibility region (Charles, 1964; Skatula *et al.*, 1958; and particularly Andreyev *et al.*, 1958).

Earlier apparatus glasses were in the lower SiO_2, higher R_2O extension of this range, often with the addition of ZnO for relatively high chemical resistivity. These glasses, often called Jena glasses from their origination by O. Schott in Jena around 1888, contained first 5 wt %, then 12 wt % B_2O_3, coresponding to expansion coefficients of 70 and 50 × 10^{-7}, respectively (soda-lime = 100, Pyrex = 30).

Optical glasses, termed borosilicate crowns, have higher alkali content (16 wt %), thus a higher expansion coefficient (80 × 10^{-7}). They were designed to provide low refractive index, lower dispersion, a conventional melting range, and good physical and chemical stability. A typical composition is (wt %):

SiO_2, 72.5; B_2O_3, 12.5; $Na_2(K_2)O$, 12; BaO, 1; ZnO, 2; $Sb_2(As_2)O_3$, 0.2

Other applications include (low-melting) *sealing glasses* and low-dielectric-loss glasses. Details on compositions of various borosilicate glasses and their properties are found in Volf (1961). Semiconductive steel enamels might be obtained from alkali borosilicate glasses containing Cu^+ (Dzevuskaya *et al.*, 1977). A borosilicate glass has been prepared from bentonite (Corning, 1977).

E. ALKALI-FREE BOROSILICATE GLASSES

Optical glasses termed *barium crowns* contain less or no alkali and large amounts of barium. Glasses in this composition range have high (1.6) refractive index, and dispersions that are low relative to that of equal-index lead (optical flint) glasses. An example is (wt %):

R_2O, 0; SiO_2, 57.5; B_2O_3, 6; Al_2O_3, 4.5;
BaO, 25; PbO, 0.5; ZnO, 6.0; $ZrO_2(TiO_2)$ 0.5

In the ternary $BaO-B_2O_3-SiO_2$ system, immiscibility boundaries were determined by Averyanov *et al.* (1979).

Doping of $BaO-B_2O_3-SiO_2$ glasses with TiO_2 causes dc and ac conduction, possibly due to the formation of ion chains (Rawal *et al.*, 1977). Alkali-free solder glasses with low expansion and low softening points contain CuO; e.g., (wt %): PbO, 66; B_2O_3 14; Al_2O_3 3; ZnO 10; Bi_2O_3 1.5; CuO. 2.5–5. The undesirable precipitation of Cu_2O is prevented by separate melting of high and low melting constituents (Powell and Frieser, 1979). Ag_2O and AgI, with GeO_2, V_2O_5 etc., were introduced into borosilicate glasses by Minami et al. (1979).

VII. Aluminosilicate Glasses

A. The Role of Al in Silicates

The radius of Al^{3+} is 0.57 Å; the radius ratio of aluminum to oxygen is 0.43, i.e., at the boundary between conditions for 4- and 6-coordination (Table I). In crystalline silicates, Al^{3+} is found in 4-coordinated (tetrahedral) as well as in 6-coordinated (octahedral) sites. If an $(AlO_4)^{-5}$ tetrahedron replaces a $(SiO_4)^{4-}$ tetrahedron, the valency 3 of Al causes a charge unbalance that is best compensated by monovalent alkali (R^I). This arrangement is typical for many silicate minerals such as the feldspars or zeolites. Two alkali ions may be replaced by one alkaline earth ion (R^{II}) in these structures. Excess Al^{3+} over $R^I + R^{II}$ may be found in 6-coordinated (octahedral, interstitial) sites. Stable, technologically valuable aluminosilicate glasses frequently contain comparable amounts of Al_2O_3, R_2O, and $R^{II}O$. In the absence of R^I and R^{II}, compatibility of Al and Si in the vitreous state is limited.

B. The System Al_2O_3–SiO_2

Alumina itself cannot form a glass under conventional conditions. Al_2O_3–SiO_2 glasses can be obtained by hydrolizing organic compounds, followed by drying and calcining at temperatures as low as 500°C (Kamiya et al., 1974), or by flame spraying and hot pressing (Gani and McPherson, 1977). SiO_2–Al_2O_3 36.6 wt % glass prepared by condensation from a high-frequency plasma showed phase separation on sintering, followed by crystallization of mullite (Lee and McPherson, 1980). The phase diagram of the binary system Al_2O_3–SiO_2 is illustrated in Fig. 28. It contains the compound mullite ($3Al_2O_3 \cdot 2SiO_2$). The glass scientist is particularly interested in the large area of metastable liquid–liquid immiscibility (MacDowell, 1966; McDowell and Beall, 1969; Aramaki and Roy, 1962; Risbud and Pask, 1977, 1978, 1979). At low temperatures, it extends from 7 to 55 mole % Al_2O_3. Above 1000°C, mullite crystallizes from the Al_2O_3-rich phase. Above 1200°C, the siliceous matrix crystallizes cristobalite. The miscibility gap can be suppressed by rapid cooling, which explains existing controversy about the extension of the gap. From an atomistic viewpoint, immiscibility above 5% Al_2O_3 might be associated with the entry of

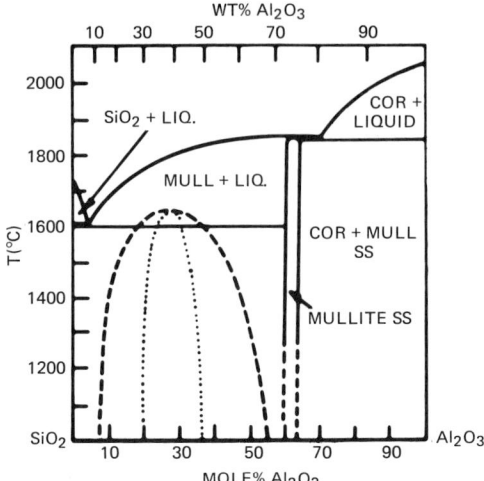

FIG. 28. Phase diagram Al_2O_3–SiO_2. COR = corundum; MULL = mullite (McDowell and Beall, 1969).

Al_2O_3 into the tetrahedral $(SiO_4)^{4-}$ network in the form of the tricluster—one O tribridging three tetrahedra per Al/Si substitution—which under normal pressure creates a high-energy state. The tricluster after Lacy (1963) is sketched in Fig. 29. Alkali and alkaline earth suppress immiscibil-

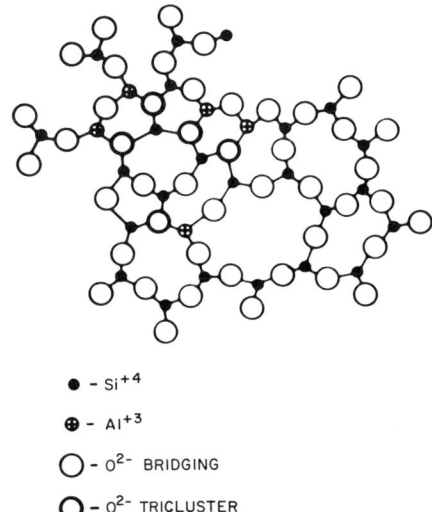

● – Si^{+4}
⊕ – Al^{+3}
○ – O^{2-} BRIDGING
○ – O^{2-} TRICLUSTER

FIG. 29. Lacy's (1963) tricluster model—a feature in the structure of alkali aluminosilicate glasses.

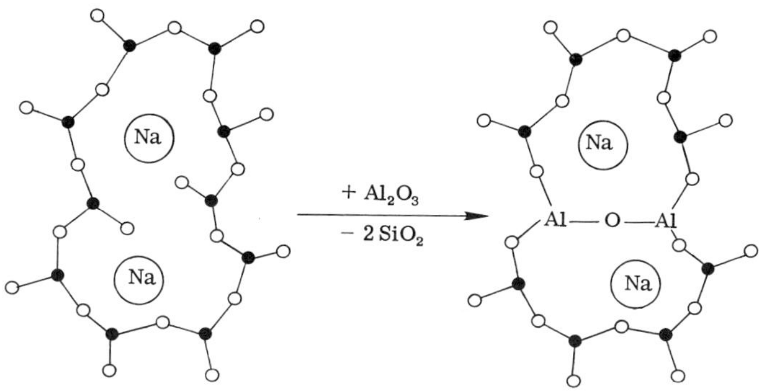

FIG. 30. Elimination of nonbridging oxygen in alkali silicate glasses by introduction of alumina.

ity as well as mullite crystallization and the numerous glass systems thus facilitated and described below are of increasing technological significance.

C. ALKALI ALUMINOSILICATE GLASSES

*1. General Considerations**

The simplest and generally satisfactory model for the structure of sodium silicate glasses with increasing additions of Al_2O_3 up to $Al/Na = 1$ is: (1) Substitution of 4-coordinated Al^{3+} for Si^{4+} in each tetrahedron; (2) elimination of one nonbridging oxygen per Al^{3+} introduced, since each Al^{3+} contributes only $\frac{3}{2}O^{2-}$ and thus ties half of a nonbridging oxygen as supplied per Na^+ (Fig. 30); (3) decreased preference of specific positions for Na^+ because of (2), and increased mobility of Na^+.

X-ray photoelectron spectra (XPS), however, reveal that there is in aluminosilicate glasses a significant chemical shift of the peak associated with bridging oxygen, indicating increasing similarity of the bridging oxygen's binding energy with that of the nonbridging oxygen with increasing Al content (Brückner *et al.*, 1980). Lam *et al.* (1980) take the more extreme position that nonbridging oxygens persist. For the technologist the important structural feature is the integrity of a rigorous network of 4-coordinated Al and the mobility of the alkali ion less tied to a specific nonbridging oxygen, as in Al-free silicate glasses. It can be agreed that the positive charge deficiency of the AlO_4 tetrahedra is compensated by a uniform electron charge density transfer from sodium ions not localized

---

* See, in particular, Beall *et al.* (1967) and McMillan (1964).

near specific AlO_4 tetrahedra. This is comparable (Lam et al., 1980) to impurity levels in a wide-band-gap semiconductor and suggests low-T conductivity measurements in such glasses.

In the following, the description of specific properties is based on the three simple features described above. Aluminosilicate glasses are the most important bases for glass ceramics (see Chapter 7). More complex models such as that of Lacy (1963) are more suitable for explaining certain features of alkali aluminosilicate glasses. These will be discussed in Section VII.D.

In aluminosilicate glasses, Si–O distances are generally enlarged, except near $R_2O/Al_2O_3 = 1$, suggestive of the absence of nonbridging oxygens (Sakka and Matsuita, 1976; Sakka and Kamiya, 1978; Sakka and Senga, 1978). As Al_2O_3 is added to Na_2O–SiO_2 glasses, light scattering is affected in two ways, depending on the ternary composition (Schroeder, 1977): (a) it decreases because of the decrease in spinodal area where there is practically no change in the fictive temperature, or (b) it increases because of an increase in the fictive temperature. Optimized compromises on this foundation may represent first-principle approaches for the low-scattering optical fiber candidate glasses studied by Pinnow et al. (1975). (After cleaning up the glass to reduce absorption losses, scattering remains the critical feature.)

If one compares $Na^+O^{2-}Si^{4+}(O_{1/2})_3^{4-}$ in a sodium silicate glass with $Na^+Al^{3+}(O_{1/2})_4^{4-}$ and $Si^{4+}(O_{1/2})_4^{4-}$ in a sodium aluminosilicate glass, the free energy of dissociation in the latter is smaller. Diffusion and conductivity (Gehlhoff and Thomas, 1926; Tsekhomski et al., 1963) are enhanced at the maxima (activation energy minima). With increased structural symmetry, expansivity decreases. The always-present OH decreases its hydrogen bonding character to attain, as shown by its spectra, the so-called free OH character of Al/Na = 1 (Scholze, 1959); that is, the H is most closely associated with a particular tetrahedral oxygen.

The substitution of Al for Si in glasses finds its analogue in the isostructural character of cristobalite (SiO_2) and carnegieite ($NaAlSiO_4$). Above Al/Na = 1.1, Al becomes 6-coordinated, that is, the center of an oxygen octahedron, or interstitial (Day and Rindone, 1962). However, independent of the acceptance of the Lacy model, there has been some indication of some Al in higher coordination even below Al/Na = 1 (Day et al., 1962, Li and Bray, 1962). The coordination of Al, first postulated from volume data, can be substantiated by ESR (Li and Bray, 1962), microprobe (White et al., 1958), x-ray data (Moori et al., 1970), including those with increasing resolution by diffusional substitution of Ag for Na (Prins, 1965; Urnes, 1972). These Ag substitutions prove that the 3.4-Å diffraction peak, at times considered to indicate grouping of Na ions, is in fact an Na(Ag)–Si or —O peak.

The stability of alkali aluminosilicate glasses is in agreement also with the low liquidus temperatures. The eutectic Na_2O 8; Al_2O_3 15; SiO_2 67 (wt %) is as low as 1062°C (Rawson, 1967), and even the compound albite, $Na_2O \cdot Al_2O_3 \cdot 6SiO_2$, melts as low as 1120°C. Albite melts are extremely viscous (log η = 8.5 at that temperature) and crystallize slowly.

A "mixed-site" effect (Lacourse et al., 1976) can be considered related to the mixed alkali effect. In it the same alkali species occurs in two different sites, for example, at a nonbridging oxygen near $(SiO_4)^{4-}$ and at a bridging oxygen at an $(AlO_4)^{5-}$ group. Raman spectra were obtained and interpreted by Moriya et al. (1978), including glasses containing TiO_2.

2. Complex Alkali Aluminum Silicate Glasses

Since Al_2O_3 is an important additive to soda-lime glasses, the quarternary Na_2O-CaO-SiO_2-Al_2O_3 is of interest (Brawer and White, 1977). One of the most efficient laser host glass systems is the system SiO_2 (~50); CaO ($\leq$20); Li_2O(~27); Al_2O_3(~2.5). The efficiency increases with CaO content; however, devitrification tendency limits this possibility. It has been proposed to synthesize higher CaO glasses in this system in a microgravity environment to avoid inhomogeneous nucleation (Neilson and Weinberg, 1977).

High OH contents in glasses of systems such as Na_2O-MgO-Al_2O_3-SiO_2 have been obtained by exposure to steam. Some of these glasses are moldable at low temperature and remain homogeneous and stable against either dehydration or attack by water (Wu, 1980a,b). See also Section II.F.

3. The Role of Alkali Species

a. General Considerations. The effect of changes in the alkali species may be considered less complex than in pure silicate glasses because of the virtual absence of nonbridging oxygens. The primary correlation is the increased weakening of Si–O–Al bonds with decreasing size, that is, increasing field strength of alkali, in the series (Ca)–K–Na–Li. The melting points of crystalline alkali aluminosilicates show this correlation (Table XXV).

b. Potassium Sodium Silicate Glasses. Sodium aluminosilicate glasses are chiefly considered as precursors of nepheline glass ceramics. Nepheline is* $K_xNa_yCa_z\square_{8-(x+y+z)}Al_{x+y+2z}Si_{16-(x+y+2z)}O_{32}$ and is based on $K_2Na_6Al_8Si_8O_{32}$, which may be considered "stuffed" Al-substituted SiO_2 [i.e., $Si_{16}(Al + R)O_{32}$]. Additionally, Mg^{2+}, Fe^{2+}, Ti^{4+} (the latter for Al^{3+}) may be substituted. The structure of nepheline may be regarded as a

---

* The square ($\square$) in the formula represents vacancies.

TABLE XXV

MELTING POINTS (MP) OF ALKALI ALUMINOSILICATES $(RAS_x)^a$

Compound	RAS_2	MP (°C)	Compound	RAS_4	MP (°C)	RAS_6	MP (°C)
Eucryptite	LAS_2	1400	Spodumene	LAS_4	1473		
Carnegieite	NAS_2	1526				NAS_6	1118
			Leucite	KAS_4	1680	KAS_6	1530

^{a} L = Li, N = Na, K = K, R = R_2O, A = Al_2O_3, S = SiO_2. Lithium aluminosilicates have the lowest melting points.

distorted tridymite structure (Buerger et al., 1954; Donnay et al., 1959). Compositions of glasses crystallized to nepheline glass ceramics may have the following ranges (Duke et al., 1967) (wt %): SiO_2, 40–50; TiO_2, 5–7.5; K_2O, 0–9.5; Na_2O, 10–20; Al_2O_3, 26–33, with some CaO, MgO, and BaO. Excess of SiO_2 and Al_2O_3 over the composition of nepheline assures a residual aluminosilicate matrix of low deformability. The heat treatment of the nepheline glass around 850°C produces (a) phase separation [when such an interim product is acid-leached, porous mullite fibers may be obtained (Aslanova and Kostareva, 1974)], (b) primary crystallization of carnegieite, $NaAlSiO_4$, from the high-alumina phase, which expels K, (c) crystallization of nepheline into which carnegieite dissolves, and (d) after long treatment and at higher temperatures, exsolution of anatase. (TiO_2 has been added as a crystallization catalyst.)

Crystallization results in an increase of the coefficient of expansion from 97.10^{-7} to as much as 130×10^{-7} and an increase in density from 2.5 to 2.7. Nepheline glass ceramics can be further strengthened by surface exchange of K for Na (Duke et al., 1967). Fe, Mn, etc., cluster in Na_2O–Al_2O_3–SiO_2 glasses, which are precursors of crystallized ferrite glasses (Stephanov and Novikov, 1976).

Neither the melt of the composition of albite ($NaAlSi_3O_8$), nor the glass quenched from 1800°C, nor the slowly cooled glass show any significant difference in x-ray radial distribution function from each other (Taylor and Brown, 1979b). But their x-ray radial distribution function does not resemble at all that of crystalline albite, whose structure is characterized by four-member rings. It appears that, on melting, the albite structure converts to a six-member ring structure. This agrees with the strong barrier both to melting and crystallization and the strong effect of OH content. The six-member ring of the liquid structure is maintained in the glass and resembles that in SiO_2 structures, especially tridymite (Konnert and Karle, 1973; Konnert et al., 1973). Taylor and Brown, (1979a,b) believe that this structural principle extends to orthoclase glass ($KAlSi_3O_8$) and to the complex systems $SiO_2-NaAlSi_3O_8-KAlSi_3O_8$, including nepheline and Kalsilite. The Si(Al)–O–Si(Al) angle in $Na(K)AlSi_3O_8$ is 146°, compared to 151° in Konnert's data on SiO_2.

c. Lithium Aluminosilicate Glasses. Lithium aluminosilicate glasses are technologically important as a base for the crystallization to glass ceramics of extremely low expansion coefficient. (15×10^{-7} to nearly 0) (McMillan, 1964). They range in composition as follows (wt %): SiO_2, 53–75; TiO_2, 3–7; Li_2O, 2–15; Al_2O_3, 12–36 (see Chapter 7 and McMillan, 1964). Commercial Cervit and Pyroceram brand products are examples. The structure of crystalline spodumene ($Li_2O \cdot Al_2O_3 \cdot 4SiO_2$) has been established by Li (1968). It has four Li sites where, in one close pair, only one site can be occupied. Lithium aluminosilicate glasses are important laser hosts. Crystallized spodumene glasses containing 2% Nd_2O_3 were described as a glass–ceramic laser material (Kasimova and Milyukov, 1977). In vitreous systems the addition of CaO was found advantageous (Section II.C.2). Glass formation in the ternary extends to 40 wt % Li_2O, with glasses up to 23% forming with ease (Eppler, 1963). The eutectic lies at $15.1 Li_2O \cdot 9.04 Al_2O_3 \cdot 75.35 SiO_2$ (Roy and Osborn, 1949).

According to Kanbara (1969) the addition of F can be used to control crystallization sequences and improve properties. Phase separation was studied extensively by Moriya *et al.* (1968). The critical temperature of metastable immiscibility in the binary Li_2O-SiO_2 system is reduced by the addition of Al_2O_3 (Marinov and Radenkova-Janova, 1967). The ternary was explored by Savva and Newns (1971), with the results illustrated in Fig. 31. The consolute temperature is reduced from about 1000°C for the binary to 870°C at $Li_2O/Al_2O_3 = 10.1$ and to 760°C at $Li_2O/Al_2O_3 = 5.1$. The SiO_2-poor phase may (Nakagawa and Izumitani, 1972) crystallize to β-quartz solid solution, which, above the consolute, may dissolve the droplets of the SiO_2-rich phase.

D. LACY'S MODEL

Briefly, Lacy's model (Lacy, 1963) postulates the formation of "triclusters," the unorthodox meeting of three AlO_4 groups at one oxygen. Triclusters form increasingly at high Al/Na ratios from an equilibrium distribution of (a) Na^+, (b) SiO_4 with one nonbridging O, (c) AlO_4 groups, and (d) AlO_6 groups tied to SiO_4 groups.

The existence of three regions relating increases in Al to property changes exemplified here by those of the diffusion coefficient D for Na is in accord with this model: I. Small additions of Al, (Al/Na $<<$ 1) lead to a *decrease* in D as AlO_6 forms. II. AlO_4 groups form increasingly as in the simpler model. *D increases* to a maximum. III. Triclusters form (Fig. 29) and *D decreases* once more.

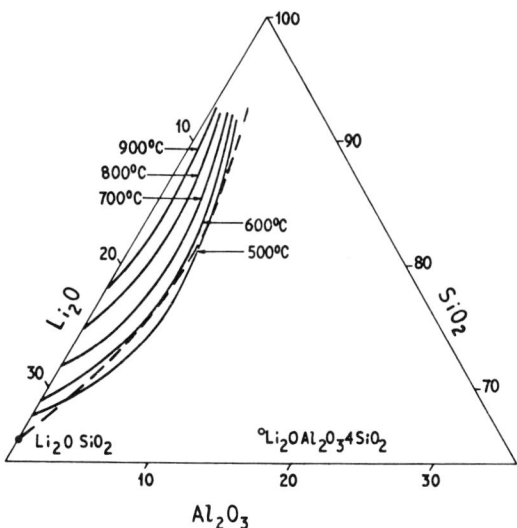

FIG. 31. Immiscibility in the system $Li_2O-Al_2O_3-SiO_2$: ——, experimental isotherms; – – – estimated limit of immiscibility (after Marinov and Radenkova-Janeva, 1967).

Another interesting complex model is by Yoldas (1971). It postulates Al in 6-coordination at low concentrations (0.25%) in accord with properties. In *sodium calcium aluminosilicate glasses* much more 6-coordinated Al and nonbridging oxygen is created on substituting Al for Si when compensating by substituting Ca for Na than in the ternary $Na_2O-Al_2O_3-SiO_2$ (Brawer and White, 1977). It should be noted that wollastonite, the primary phase at $x = 0.2$ in the quaternary $(1 - x)Na_2O \cdot (3 + 2x)CaO \cdot xAl_2O_3 \cdot (6 - x)SiO_2$ has two nonbridging O per Si.

The models of Lacy and Yoldas are strongly supported by conductivity data (Hayward, 1977), especially for glasses near Al/Na = 1 and near the Na_2O-SiO_2 immiscibility region. In other regions, nonbridging oxygens versus AlO_4 can explain the data. NBO can be determined by ESCA (Brückner *et al.*, 1978).

E. ALUMINOSILICATE GLASSES CONTAINING F AND OH

The most important modifications are those by F (Kanbara, 1969) and by OH (Ernsberger, 1974). In the latter case, surface OH introduced by steam permits the surface to crystallize first, resulting in a permanent compression layer. Crystallization catalysts include TiO_2, ZrO_2, and P_2O_5, as well as metals.

The lower valency of Al favors the stability of F in silicate glasses. This is the basis for the compositions of optical fluorine crowns in which the addition of a large amount of F without crystallization results in a reasonably low refractive index in a material more conventional than the fluoride crystals and glasses (Fraser and Upton, 1944).

A typical fluor crown is composed as follows (wt %):

SiO_2, 47; B_2O_3, 13.5; Al_2O_3, 15; R_2O, 16.5;   F, 8

and has a refractive index n_D = 1.46550, with a reciprocal relative dispersion of V = 65.8. Complex $MgO-Al_2O_3-SiO_2$ glasses containing alkali, fluorine, and other constituents can be converted to crystalline ceramics with fluorophlagopite (mica) as the major phase. These are machineable because of crack propagation perpendicular to certain lattice planes (Beall et al., 1967; Heidenreich et al., 1977; Grossman, 1972; Daniels, 1974).

F. ALKALI-FREE ALUMINOSILICATE GLASSES

A larger number of glasses with high softening ranges (>900°C), high electrical resistivity, relatively low expansion, and high low-temperature viscosity are found near ternary eutectics $R^{II}O-Al_2O_3-SiO_2$. These eutectics are listed in Table XXVI. Modifications of value are Ba, Zn, B, R^I (not much), to 7%F. Synthesis by solid reaction is possible. Polynary eutectics with TiO_2 are advantageous; they favor AlO_4 grouping (Toropov and Khatimchenko, 1969).

The glasses are used in fibers, discharge lamps, combustion tubes, cooking ware (because of their high annealing temperature). A typical "top-of-the-stove" composition is as follows (wt %):

B_2O_3, 10; SiO_2, 50; Al_2O_3, 13.5; C_aO, 16.5;   MgO, 4; Na_2O, 4.5; K_2O, 0.5

Cordierite magnesium aluminosilicate glasses crystallized to cordierite glass ceramics contain 54–80 SiO_2, 21–13.7 Al_2O_3, 25–4.5 MgO, and have expansion coefficients of 51 to 20 × 10^{-7} (Hammel and Ohlberg, 1965; Maurer, 1962; Ohlberg et al., 1962; Gates and Lent, 1967). Their dielectric losses are low (Mashkovich and Udenko, 1965).

TABLE XXVI

Some $R^{II}O-Al_2O_3-SiO_2$ Eutectics

System	Composition of (mole %)				Liquidus (°C)
	SiO_2	Al_2O_3	CaO	MgO	
$CaO-Al_2O_3-SiO_2$	62	15	23	—	1170
$MgO-Al_2O_3-SiO_2$	62	18	—	20	1345
$CaO-MgO-Al_2O_3-SiO_2$	62	18.5	10	9.5	1222

Other aluminosilicate glasses include those with the principal third constituent being La (Karlsson, 1970), Ce (Litvinov and Zhuraleva, 1965; Aleksandrov *et al.*, 1980), Be (Miller and Mercer, 1960; Ganguli and Saha, 1965; Riebling and Puie, 1966), Mn (Fulrath and Hollar, 1968), Pb (Naudin, 1968; Stookey, 1950), and Zn (Vargin and Milyukov, 1968; McMillan, 1964).

Copper aluminosilicate glasses are of interest because of the unusual combination of low expansion coefficient and low glass temperature. T_g is much lower than for SiO_2 or SiO_2-TiO_2 glasses. The glasses contain both Cu^+ and Cu^{2+}, but no correlation of their use characteristics with the Cu^{2+}/Cu^+ ratio exists. No correlation exists with the properties of glasses in which Cu is replaced by ions of similar charge and size (Zn, Ni, Na).

G. BOROALUMINOSILICATE GLASSES

The ternary system $B_2O_3-Al_2O_3-SiO_2$ contains wide areas of phase separation (Galakhov *et al.*, 1977). The fact that alkaliborosilicates have much weaker (lower melting) structures (Weyl and Marboe, 1964) (Table XXVII) permits fluxing of aluminosilicate systems of desirable properties by borosilicate systems.

In such boroaluminosilicate glasses, Al and B compete for the tetrahedral grouping (Kobayashi and Okuma, 1976), with Al^{IV} apparently being favored (Appen, 1959; Appen and Fu-Si, 1959; Appen and Hsi, 1959). The grouping changes with heat treatment (Stepanov and Novikov, 1972). Because the reactivity of alkali is particularly obnoxioux when surface areas are large, aluminoborosilicate glasses free of alkali, or with very low alkali content, have been developed. These are based on eutectic compositions in the systems $RO-B_2O_3-Al_2O_3-SiO_2$ involving more than one RO. A standard fiber glass widely produced in the industry is called E glass. The compositional field is (wt %)

SiO_2, 52–56;   CaO, 16–24; Al_2O_3, 12–16; B_2O_3, 8–13;   MgO, 1–6

For the alkali resistance required of fibers used to reinforce cement, the boron content might be decreased or eliminated, and ZrO_2 may be introduced.

TABLE XXVII

COMPARISON OF MELTING POINTS OF ALUMINOSILICATES AND BOROSILICATES

Compound	MP (°C)	Compound	MP (°C)
$Na_2O \cdot B_2O_3 \cdot 2SiO_2$	766	$CaO \cdot B_2O_3 \cdot SiO_2$	1000
$Na_2O \cdot Al_2O_3 \cdot 2SiO_2$	1576	$CaO \cdot Al_2O_3 \cdot SiO_2$	1550

As in the case of Vycor, E glass can be phase-separated and the non-silica constituents leached by acids. In this manner, nearly pure SiO_2 textiles can be obtained commercially from E-glass weaves.

In $Na_2O(Ag_2O, CuO(-Al_2O_3-B_2O_3-SiO_2$ glass systems, Ag shifts the Cu^+/Cu^{2+} equilibrium toward Cu^+ (Artamonova et al., 1978).

H. GALLOALUMINOSILICATE GLASSES

The introduction of Ga has structural effects similar to those of Al and combinations exist over wide ranges, for instance in the complex systems $CaO \cdot Al_2O_3-CaO \cdot Ga_2O_3-SiO_2$ (Lisenkov and Vasilev, 1979) and CaNa(Al–Fe–Ga–Si–O) (Virgo et al., 1978; Seifert et al., 1978).

VIII. Phosphate Glasses

A. STRUCTURE

1. General Considerations

The structures of phosphate glasses, particularly those low in alkali and/or high in aluminum content, have close relationships to those of silicate glasses because the sizes of P^{5+} and Si^{4+} are comparable, and excesses in charge can be compensated by trivalent Al^{3+}, and even by divalent Mg^{2+}, Ca^{2+} etc.:

$$-O-\underset{\underset{O}{|}}{\overset{\overset{O}{|}}{P^{5+}}}-O-\underset{\underset{O}{|}}{\overset{\overset{O}{|}}{Al^{3+}}}-O-$$

Consequently, PO_4 tetrahedra with a P–O bond length of about 1.6 Å are as basic as, and similar to, SiO_4 tetrahedra in silicate structures. The isostructures $AlPO_4$ and $SiSiO_4$ (SiO_2) (Huettenlocher, 1935) are basic to wide fields of ceramics. X-ray patterns for glasses based on them (Biscoe et al., 1941; Wignall et al., 1977) resemble those of silicate glasses (P–O = Al–O = 1.6 O–Si–O). The chief and significant difference is that the pentavalence of P introduces nonbridging oxygens in the absence of alkali (network modifiers):

$$-O-\underset{\underset{O}{|}}{\overset{\overset{O}{|}}{P}}=O$$

Thus the weakest, low melting, pure P_2O_5 glass is structurally strengthened rather than weakened by the addition of RO or even R_2O:

$$\begin{array}{ccc} | & | & | \\ O & O & O \\ | & & | \\ -O-P=O & \begin{array}{c} Na \\ Na \end{array} & O=P-O- \\ | & & | \\ O & O & O \\ | & | & | \end{array}$$

$$\begin{array}{cc} | & | \\ O & O \\ | & | \\ -O-P=O & Ca-O-P-O- \\ | & | \\ O & O \\ | & | \end{array}$$

Therefore, even alkali phosphates have higher melting points than pure P_2O_5.

As in silicate glasses (Section II), the bridging oxygen is strongly affected by the introduction of alkalis and nonbridging oxygens (Brückner *et al.*, 1980), as evidenced by the shift of the peak associated with bridging oxygens in the x-ray photoelectron spectrum (XPS). The double-bonded P=O occurring in the PO_4 network because of the pentavalency of P is also affected. Table XXVIII shows the relation of nonbridging oxygen and alkali concentration for simple phosphate and silicate glasses.

Various P–O groupings in glasses can be demonstrated by the NMR spectra of ^{31}P as a probe (Bray, 1966; Brandenberger, 1964), and the Al–O

TABLE XXVIII

Nonbridging Oxygen and Alkali Ratios in Simple Phosphates and Silicates[a]

Silicate	R	B	N	Silicate glass	Phosphate glass	Phosphate
				A		
SiO_2	2	4	0	0	0	$AlPO_4$
$Na_2Si_4O_9$	2.25	3.5	0.5	0.5	0	AlP_3O_9
$Na_2Si_2O_5$	2.5	3	1	1	0	P_2O_5
Na_2SiO_3 (meta)	3	2	2	2	1	$Na_2P_2O_6$
Na_4SiO_4 (ortho)	4	0	4	4	3	$Na_6P_2O_8$

[a] R = oxygen per Si, P, or Al; B = bridging oxygens, per tetrahedron; N = nonbridging oxygens per tetrahedron; A = alkali (e.g., Na) per Si, P, or Al. B + N = 4.

groupings, mostly AlO_4, in aluminophosphate glasses by x-ray fluorescence (Sakka, 1977).

From the viewpoint of phosphorus chemistry, condensed phosphates can be classified into three groups (Thilo, 1965):

(1) Metaphosphates $M_n^I(PO_3)_n$, which may form ring structures, e.g., 3-rings $(P_3O_9)^{3-}$ or 4-rings $(P_4O_{12})^{4-}$. Sodium hydrophosphates when heated yield glasses that anneal to metaphosphate glasses.

(2) Polyphosphates $(M_{n+2}^I)^{(n+2)} + (_nO_{3n+1})^{(n+2)-}$, which will form open chains, e.g.,

$$\begin{matrix} & O & O & & O \\ HO & P & O & P & O & P & O \cdots & P & OH \\ & O & O & & O \end{matrix}$$

with six characteristic types: two 2-chains with metal ions R, a 3-chain with RH, a 4-chain with Pb, Ca, and two "helical" chains with Ag, and so-called "Kurrol."

Glasses and crystals are known with polymerization n up to at least 1000. The value of n increases with temperature, decreases with water pressure, and depends on the cation. For Na/P $\geq$ 1, only polyanions exist, which can be analyzed by paper chromatography (Westman, 1960; Westman and Beatty, 1966; Westman et al., 1959; Murthy and Westman, 1966).

(3) Ultraphosphates, with R/P < 1, which are the cross-linked phosphates, including glasses with networks more similar to those in conventional silicate glasses.

2. Phosphorus Pentoxide

Crystalline P_2O_5 is a molecular compound (P_4O_{10}); it is hexagonal and melts at 420°C. The vapor contains P_4O_{10} molecules. Like the somewhat similar sulfur, it polymerizes when heated to higher temperatures, from which condition it quenches to a glass. The vapor pressure of this glass is lower than that of one monomeric crystal. The polymerized rhombic crystal melts at 500°C. The exact structure is not too well understood (Wells, 1950); there seem to be some P–O distances much shorter than those in the bridging tetrahedral P–O bonds.

3. Alkali-Free and Low-Alkali Phosphate Glasses

Numerous glass forming systems exist. $Al(PO_3)_3$, $Mg(PO_3)_2$, $Ca(PO_3)_2$, $Sr(PO_3)_2$, $Ba(PO_3)_2$, $Zn(PO_3)_2$, alkali metaphosphates, and metaphosphates containing Ag, Cu, and the transition and rare earth elements are good glass formers and can be combined to obtain desired properties for many applications. In BaO–(SrO)–P_2O_5 glasses, P–O distances are 1.4 Å, smaller than in the crystalline metaphosphates (Strugach *et al.*, 1979). They have structures based on PO_4 networks (ultraphosphates), possibly containing meta-rings.

The most important component of stable phosphate glasses is $Al(PO_3)_3$. Glasses containing large amounts are comparable to glasses high in SiO_2 because of the analogy of $AlPO_4$ and SiO_2 (= $SiSiO_4$). $AlPO_4$ crystallizes in three modifications [quartz (berlinite), tridymite, and cristobalite] like SiO_2, with similar displacive transformations. Again the tridymite form is obtained in the presence of alkali or ammonium, since with increasing purity the tetrahedral linkage is unfavorable (Floerke, 1967). In $AlPO_4$–SiO_2 mixtures, tridymite formation is enhanced and stable even at high temperatures. In the quartz structure, on the contrary, SiO_2 and $AlPO_4$ are immiscible, particularly on the low-SiO_2 side. The replacement, when possible, is substantial: Al by Si^I, P by Si^{II}.

Large areas of borophosphate glasses exist (Kreidl and Weyl, 1941; Beekenkamp, 1966). Large concentrations of P_2O_5 in alkali borate glasses enhance the formation of BO_4 groups by charge compensation ($B^{3+}OP^{5+}$), and of BPO_4 grouping resembling that in crystalline BPO_4 (Kreidl and Weyl, 1941), as evidenced by NMR resonance (Beekenkamp, 1966; Syritskaya, 1971; and Bray, 1978). FeO–Fe_2O_3–P_2O_5 glasses are good conductors with a maximum near Fe^{2+}/Fe^{3+} = 1/1. More than 50% FeO + Fe_2O_3 can be accommodated (Vaughn and Kinser, 1975; Hansen, 1965). Other more complex phosphate glass systems were described by (Shcheglova (1971) (Ga_2O_3), Syritskaya (1971) (La_2O_3,CeO_2), Matveev *et al.* (1971) (TiO_2, CuO, Nb_2O_3), and Lagzdons *et al.* (1980) (WO_3, also with reduced W).

In the system TeO_2–P_2O_5, much TeO_2 is accommodated in the PO_4 structure, while Te–O distances change rapidly when some P_2O_5 is added to TeO_2 as evidenced by *n*-diffraction (Neov *et al.*, 1980). Rare-earth phosphates were studied in connection with the development of laser glasses (Levenberg and Lunter, 1979; Valters and Lunter, 1979 [Eu^{3+}], Denker *et al.*, 1979; Litvin *et al.*, 1979). However, the most important laser host glasses are in the systems containing F (Section XI).

4. High-Alkali Phosphate Glasses

These resemble organic polymers with chain structures containing polyphosphoric acid ions (Brady, 1958; Murthy and Westman, 1967; Takahashi, 1962). Glasses can be dissolved and analyzed for chain length by conventional chromatography (van Wazer, 1958; Westman, 1960; Westman and Beatty, 1966; Westman *et al.*, 1959; Schulz and Hinz, 1956). Mercury can be introduced into such glasses (Weyl and Marboe, 1964).

5. Silver Phosphate

Silver phosphate glasses with high Ag content contain polymeric chains with some covalent Ag–O–P bonds (Bartholomew, 1972, 1973). Very high conductivities are observed in silver phosphate glasses containing Ag_2SO_4 and Li^+ (Malugani et al., 1978) or AgI (Malugani et al., 1978; Reggiani et al., 1978). The replacement of some PO_4 by SO_4 hardly changes the Raman spectra, indicating structural comparability in the series I, Br, Cl; the conductivity could be increased from 10^{-2}–10^{-3} (I) to 10^{-3}–10^{-7} (Br) to 10^{-5}–10^{-7} (Cl) Ω/cm (Nambu and Minami, 1977; Katsuda and Minami, 1979; Minami, 1980). Chromatography and IR spectra show that for $AgO/P_2O_5 = 3$ to 2, Ag^+I (Cl^-, Br^-), PO_4^{3-}, and $P_2O_7^{4-}$ ions persist, while for lower ratios, polyphosphate units are revealed.

A very large area of glass formation extends between phosphate and fluoride glasses (see also Section XI). The replacement of O by F destroys bridges while maintaining the general structure. Margaaryan and Arutyunan (1972) believe that in specific cases new network types such as $BaPO_3F$ appear (Artyushkina et al., 1972).

Two alkalis, Ag and Na, as well as H and Na, etc., exhibit "mixed alkali effects" as in silicate glasses. The effect of H is much more pronounced, probably because of the higher concentration in nonbridging oxygen sites (Day and Stevels, 1974). Random distribution of two alkalis is indicated by IR and Raman Spectra (Rouse et al., 1978).

B. Properties and Applications

Among characteristic properties and applications are the following:

1. Chemical Resistance

Some high-$Al(PO_3)_3$ glasses resist HF to a remarkable extent (American Optical, 1947; Morgan, 1947). Calcium barium aluminophosphate glasses resist alkali vapors (Dale and Stanworth, 1951). Chemical reactivity plays a role in glassy fertilizers of low T_g (Ray et al., 1973). For equal alkali contents, phosphate glasses sinter at lower temperatures, suggesting their use in enamels for certain metals like Al (Godron, 1966) or binders (Lyon et al., 1966). Low melting Li_2O–ZnO–P_2O_5 glasses have high chemical resistivity (Tindyala et al., 1978).

2. Optical Properties

While pure SiO_2 excels for stable ultraviolet transmission, more easily meltable aluminophosphate glasses are quite satisfactory (Hood, 1926; Duffy, 1972). The specific environment for Nd^{3+} (De Paolis and Mauer, 1947; Deutschbein 1967) and merit factors involving high gain coefficient and low nonlinear index have induced considerable development work with phosphate and fluorophosphate systems for laser hosts (Weber, 1976; Voronko et al., 1976; Section XI). Dispersion–refraction relations also differ enough to have encouraged the development of a small number of optical phosphate glasses. The high amounts of Ag^+ accepted in phosphate systems have led to the design of solid state dosimetry phosphate glasses in which the reaction $Ag^+ + h\nu \rightarrow Ag^{2+}$ (hole center formation) permits permanent dose readout using the characteristic fluorescence of Ag^{2+} (Schulmann et al., 1951; Schulmann and Etzel, 1953; Kreidl and Blair, 1956). An example of the basic composition is

$$50 \text{ Al }(PO_3)_3, \quad 25 \text{ Ba}(PO_3)_3, \quad 25 \text{ KPO}_3 + 8 \text{ AgPO}_3$$

It is, however, advantageous for minimizing energy dependence to replace K by Li, and Ba by Mg. Aluminum metaphosphate is an advantageous base, also, for fast-recovering photochromic glasses (Chance-Pilkington, 1977).

3. Electrical Properties

Pyroelectric effects were observed on cooling a zinc phosphate glass in electrical fields above 100 V/cm (Drake and Scanlan, 1970). Sodium trapping and polarized control were discussed by Balk and Eldrige (1969) for a silicate–phosphate glass. Semiconduction—quite generally observed and applied in nonoxide glasses—also occurs in oxide glasses containing transition elements. In these glasses, the glass-former is often P_2O_5 (Kinser, 1970). Ag_2O, 58.3; P_2O_5, 41.7 glass showed bistable switching at 110 V after forming (Kaes et al., 1972, 1973).

4. Other Properties and Uses

Aluminum alloys have been extended in phosphate glass (Gordon, 1966). Iron phosphate glasses are magnetic and may be truly noncrystalline (Logan and Yung, 1976; Wedgwood and Wright, 1976).

IX. Other Oxide Glasses

A. Germanate Glasses

Germanium oxide forms a glass with a tetrahedral network structure resembling that of α quartz (Laubengeyer and Morton, 1932; Zarzycki, 1957), differing from the more open cristobalite-like structure of SiO_2 glass (Rawson, 1967). The radius ratio Ge^{4+}/O^{2-} (0.43) is close to the top value for fourfold coordination (0.414) (Table I). There is therefore a high-temperature (>1033°C) crystalline form of GeO_2 that is isostructural with α quartz and differs from the low-temperature 6-coordinated rutile form by, for example, higher density and acid resistance. GeO_2 glass has a much higher coefficient of expansion (77 × 10^{-7}) (Krishna-Murthy and Scroogie, 1965; Mackenzie, 1959) than SiO_2 glass (5 × 10^{-7}), indicating more assymetry. The Ge–O bond (~1.74 Å) is about 8% longer than the Si–O bond (Smith and Isaacs, 1964; Leadbetter and Wright, 1972a,b). The best data have been obtained with n-diffraction (Ferguson and Hass, 1970; Leadbetter and Wright, 1972a,b): Ge–O, 1.73A (average coordination, 3.9); O–O, 2.83 (5.9); Ge–Ge, 3.45 (4.3); average angle Ge–O–Ge, 133°. Small-angle scattering data (Pierre *et al.*, 1972) and IR data (Borelli, 1969) are consistent with a random structure.

The electron affinity of Ge^{4+} is greater than that of Si^{4+}. Hence, particularly under reducing conditions or under high-energy radiation, $Ge^{4+,e}$ centers resembling Ge^{3+} and pairing with oxygen vacancies may appear in larger concentrations (Galimov *et al.*, 1971). Glasses exist between $GeO_{1.9}$ and GeO_2, and diphasic glasses below $GeO_{1.9}$ (Deneufville and Turnbull, 1970). GeO_2 glasses may be photoconductive (Boehm, 1972).

The energy-level model after Rowe (1974) is illustrated in Fig. 2. Sigel (1977) provides substantial evidence for this model. The level scheme is characterized by a broader band than in SiO_2, with a lower band edge (Cohen and Smith, 1958).

Light scattering data exhibit a large depolarization ratio (0.31), which would be in contradiction to a GeO_4–tetrahedral model analogous to that of SiO_2. It may indicate features including some GeO_6 groups as postulated by Wemple (1973; Schroeder, 1977).

The great difference between the structures of SiO_2 and GeO_2 accounts for their nonideal mixing properties (Riebling, 1968, 1971, 1972a,b, 1973; Riebling and Dalton, 1970). SiO_2–GeO_2 glasses are used as Si coats (Riebling and Dalton, 1970).

As in borate glasses, the addition of alkali causes a change in coordination (here IV to VI), which is most noticeable at about 15 R_2O (Krishna-Murthy and Scroogie, 1965, Shaw and Uhlmann, 1971a,b). [For details and other concepts, see also Imaoka (1962) and Tyulkin and Shalunenko, (1971).]

More recent x-ray diffraction studies (Kamiya *et al.*, 1979) confirm these observations showing (Fig. 32) an increase of GeO_6 groups to about 35% (less than formerly believed) for alkali additions up to about 30%, followed by a sharp decrease (more than formerly believed).

More detailed conclusions were made by Verweij *et al.* (1979; Verweij, 1979; Verweij and Buster, 1979) for the binary alkali glasses $XR_2O(1-x)$-GeO_2:

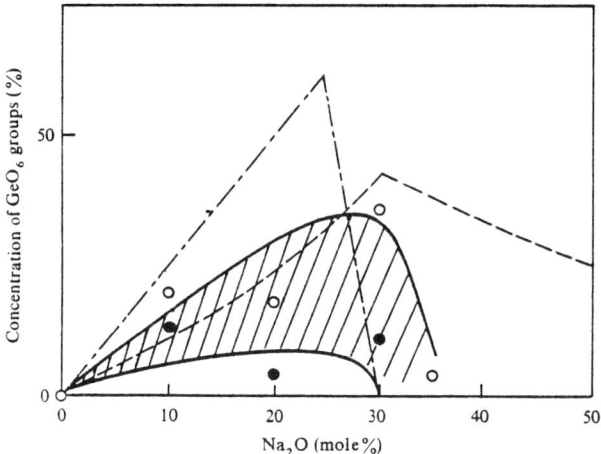

FIG. 32. Concentration of GeO_6 groups, N_6 lower (○) and upper (●) limits of Kamiya et al. (hatched area); (---) calculated by Yoshimura et al. (1971); (–·–·–) estimated by Takahashi and Yoshio (1977). (After Kamiya and Sakka, 1979.)

(1) For $0 < x < 0.18$: GeO_4 and GeO_6 groups. A structure similar to enneagermanate crystals with $2 - x$ bridging oxygens.

(2) $0.18 < x < 0.33$: Structures with $(2.8 - 5.4x)$ bridging, and $(4.4x - 0.8)$ nonbridging, oxygens. Behaves like a mixture of 18:82 and 33:67 structures, the latter similar to digermanate crystal structure.

(3) $0.33 < x < 0.5$: digermanate and metagermanate structures; with two nonbridging oxygens per Ge, $2 - 3x$ bridging oxygens.

Alkali germanate glasses also show phase separation (Milyukov et al., 1970), with transition and rare-earth elements expected to be enriched in one phase. Li_2O–GeO_2 glasses show high conductivity (Nassau et al., 1979). Glasses with high OH concentration were described by Weeks et al. (1980). Imaoka (1962) also documented wide fields of glass formation for binaries with Ba, Sr, Ca, and Zn oxides.

As in the case of SiO_2 from $SiCl_4$, GeO_2 can be doped with high-charge ions: Sb, P, Ti, and Si by co-oxidation of halides. This is of interest in the production of gradient index cross sections in light guides (Schultz and Smyth, 1974).

As with SiO_2, large amounts of PbO can combine with GeO_2 to form stable glasses, but in complex structures differing from those of the equivalent silica glasses. They appear to be related to those of the compounds (Morozov and Sharonova, 1969; Evstropiev, 1969) in the binary, which, according to Phillips and Schroger (1965), contains five compounds: P_4G, P_3G, PG, PG_2, PG_4. Raman and IR spectra indicate that with increase PbO content, GeO_6 groups are built up, saturating near PbO · GeO_2 (Oyamada and Hagiwaka, 1978; Kolesova, 1979). The introduction of V_2O_5 leads to conduction by polaron hopping and memory switching (Gahlmann and Brückner, 1977).

In contrast to earlier work, Topping and Murthy (1972) find immiscibility in this binary, even more in the ternary with silica (Topping et al., 1974a,b). Refractive indices are, of course, higher than in silica optical flint glasses (much above 2.0) but not enough to encourage wide use. Because of the higher mass of Ge combined with the low network concentration, lead germanate glasses have good IR transmission of 5.5 μm. AlF_3 additions extend IR transmission (Kroger and Res, 1977). Lead germanate (Eysel et al., 1973) and germanosilicate crystals are ferroelectric, and ferroelectric glasses or glass ceramics appear possible. Adding $PbSO_4$ or $(NH_4)_2SO_4$, one can obtain $PbO-GeO_2-SO_3$ glasses containing up to 12% ortho $(SO_4)^{2-}$ revealed by IR bands at 1200–950 and 600 cm^{-1} (Kolesova, 1980). In Tl_2O-GeO_2 glasses covalency increases with increasing Tl_2O content, as in $TP_2O-B_2O_3$ glasses where at 20–35 Tl_2O complex groups as in $PbO-B_2O_3$ glasses appear (Panek and Bray, 1977).

Bismuth germanate crystals and glasses have been investigated as laser hosts (Reisfeld, 1974) because of the larger number of phonons needed to match the energy gap for relaxation. Borogermanate and aluminogermanate glasses, somewhat analogous to borosilicate and aluminosilicate glasses exist over wide ranges (e.g., Yoshimura et al., 1971; Riebling, 1964; Murthy and Westman, 1966; Blinov, 1971; Estrop'ev et al., 1970; Urnes, 1971). In borogermanate glasses, GeO_6 octahedra extend into a wider range than in corresponding aluminogermanate glasses, probably because in the presence of BO_3 groups, fewer BO_4 than AlO_4 groups are available to compete with GeO_6 groups for alkali neighbors. In these glasses BO_4 never joins a BO_4 : GeO_6 never joins a GeO_6: and no nonbridging oxygens associate with either group (rather SiO_4 or BO_3) (Yoshimura et al., 1971). (See also Sakka, 1977.)

Alkali-free borogermanate glasses also show phase separation (Caslavska et al., 1968; Evstrop'ev et al., 1970). $CdO-B_2O_3-GeO_2$ glasses are photoconductors (Caslavska and Strickler, 1969). The binary $GeO_2-B_2O_3$ system was investigated by Baugher et al. (1972). Lanthanum borogermanate glasses attain a high index with low dispersion.

Other complex germanate glasses include titanogermanates (Cleek and Hamilton, 1964), alkali indium germanates (Fairweather and Murthy, 1973), antimony germanates (Riebling, 1974), silicogermanates (Topping et al., 1974a,b), arsenic germanates (Mochida and Takahashi, 1974; Kolesova and Kalinina, 1974), and tin germanates (Silver et al., 1977) with high refractive indices. Alkali germanosilicate glasses are used to obtain index gradients in waveguides because of the greater index difference obtainable by ion exchange than in alkali silicate glasses (Van Ass et al.,

1976; Galant, 1980). The influence of SiO_2 on the conversion of GeO_4 to GeO_6 groups was studied by Verweij et al. (1979).

B. Aluminate (Gallate and Beryllate) Glasses

Alumina itself does not appear to form a glass, although amorphous alumina has been obtained from the vapor phase under radical conditions. Rapid quenching in a hammer–anvil device was claimed to lead to a glassy $Nd_2O_3-Al_2O_3$ material by Coutures et al. (1975). But by and large only the binary $CaO-Al_2O_3$ system is the base for large areas of melted glass formation (Rawson, 1967, p. 199).

Glass formation in this system was explained by the simulation of the cristobalite structure in a "stuffed" structure where Si_2O_4 is analogous to $(CaAl_2) \cdot O_4$, but the structure has not been verified experimentally, nor does present theory require 4-coordination for glass formation. In crystalline $CaAl_2O_4$, aluminum is 6-coordinated.

Lime–alumina glasses were discovered first by Shepherd et al. (1909) and first studied with the aim of obtaining materials transmitting in the IR to above 5 μm (Stanworth, 1948; Sun, 1949). The best glasses are found near, but not at, 50 CaO, and it was necessary to add other constituents to open a new field of usable glasses, which ultimately led to formulations for large-scale fabrication (Florence et al., 1955; Hafner et al., 1958) using BaO, La_2O_3, Fe_2O_3, etc., in small percentages. A systematic study by Rawson (1956) describes the binary $CaO-Al_2O_3$ system in detail, associating glass formation on fast-cooled 20-mg samples with low liquidus ranges in contrast to the systems with Be, Mg, Sr, and Ba.

Calcium aluminate glasses resist alkali vapor (Burggraaf and Van Velzen, 1969) below T_g, the more so when crystallized to glass ceramics, which are applicable to thermoionic converters. They have high elastic moduli, among the highest (16 × 10^6 psi) in oxide glasses (Onoda and Brown, 1968), particularly in complex formulations with ZnO or MgO, and when filled with crystalline Al_2O_3.

$Gd_2O-Al_2O_3$ glasses were described by Shishido (1979). Complex aluminate glasses on a practically useful scale include those from the systems $K_2O(Cs_2O)-Nb_2O_5(Ta_2O_5)-Al_2O_3$; $Na_2O(K_2O,BaO)-TiO_2-Al_2O_3$; $PbO-Bi_2O_3-CdO-M_2O-MO-Al_2O_3$ (in which Al is 4-coordinated; BiO_4, CdO_4, and PbO_4 are network-forming groups; and Cd^{2+}, Pb^{2+}, and Bi^{3+} are modifiers). They have high IR transmission (to 6 μm) and high refractive indices (to 2.0) (Kokubo et al., 1974, 1976, 1979).

New combinations seem to become available through extreme quenching in laser spin techniques (Topol et al., 1973). Related glass systems contain Ga_2O_3 and Fe_2O_3 or Gd_2O_3 and Fe_2O_3 (Yajima et al., 1974) in large quantities in the absence of conventional glass formers. Other gallate glasses are also the subject of laser spin experimentation. $PbO-Ga_2O_3$

glasses were obtained in ultrafast quenching procedures by Kantor *et al.* (1973). Coutures *et al.* (1975) quenched La_2O_3 in combination with Ga_2O_3, as did Yajima *et al.* (1974). Lydina *et al.* (1975) prepared CaO and SrO glasses with down to 40 wt % Ga_2O_3. Fe_2O_3–BaO(CaO,PbO) glasses were reported produced by ultrafast quenching (Kantor *et al.*, 1973). Li_2O–Ga_2O_3 glasses show high electrical conductivity (Nassau *et al.*, 1979). Beryllate glasses have been obtained by quenching between copper plates (Mueller and Wood, 1972), e.g., 20CaO · 30BaO · 50BeO.

C. ANTIMONATE AND ARSENATE GLASSES

Glass formation decreases with increasing cation size from As_2O_3 via Sb_2O_3 to Bi_2O_3 (Rawson, 1967, p. 203). Vitreous As_2O_3 is obtained from its vapor by condensation on a cold surface (Rawson, 1967). Crystalline As_2O_3 exists in a cubic form, arsenolite, resembling P_4O_6, and a monoclinic form, claudetite, the sluggish transformation to which may be considered indicative of the vitrifiability of As_2O_3 (Rawson, 1967). The structure of vitreous As_2O_3 was described by a quasi-crystalline model with some interlayer expansion and some intralayer disorder (Plieth *et al.*, 1969).

It appears, from IR and Raman data (Lucovsky and Galener, 1980), to have a dimodal angle distribution of the $AsX_{3/2}$ pyramidal structure (but this does *not* mean two phases). Pair function analysis of x-ray radial distribution appears to exclude similarity to the structures of crystalline arsenolite and claudelite, rather suggesting layers formed from rigid three-member rings of AsO_3 pyramids with flexible connections (Imaoka and Hasegawa, 1980b). Binary As_2O_3 glasses were reported by Imaoka (1962) to exist in the ranges given in Table XXIX. $Ag_7I_4AsO_4$ glass is a highly conductive solid electrolyte (Grant *et al.*, 1978; Lazzari *et al.*,

TABLE XXIX

RANGES OF GLASS FORMATION IN BINARY As_2O_3 SYSTEMS[a]

Component	Range (mole%)
K_2O	40
Na_2O	54
Li_2O	48
BaO	52
SrO	48
CaO	—

[a] From Imaoka (1962).

1978). Pure Sb_2O_3 glass is hard to obtain. Its optical characteristics were compared to those of crystalline Sb_2O_3 by Wood et al. (1972). Considerable amounts of Sb_2O_3 can be introduced into complex oxide glasses containing BaO or PbO (Bishay, 1965). In these glasses Sb has coordination 3, subject to change in structure at high concentration. Antimonate glasses containing iodine were found by Turyanitsa and Kutsenko (1976). Sb_2O_3 glass (containing 5% B_2O_3) has a double-chain structure of four-membered SbO_3 pyramid rings resembling orthorhombic valentinite. A few SbO_3 pyramids turned over are one disorder feature (Hasegawa et al., 1978) (Sb–O = 1.99, in valentinite 2.00 Å). Sb_2O_3 glasses containing I, Br, and Cl were described by Turyanitsa and Kutsenko (1979).

D. Vanadate Glasses

Vanadium pentoxide forms a fluid melt that crystallizes easily and cannot be formed as a glass except by splat cooling (Rivoalen et al., 1976) or thermal decomposition (Sekiya and Matushita, 1977). However, small additions suffice to stabilize glasses high in V_2O_5 content. Roscoe (1867) obtained a glass 95 V_2O_5–5 P_2O_5. Glasses with over 60 V_2O_5 are obtained in the binary BaO–V_2O_5 (Denton et al., 1954). In these systems, indeed, substantial decreases in the liquidus occur (Rawson, 1967).

The structure of V_2O_5 differs greatly from that of P_2O_5, containing VO_5 groups in trigonal pyramids (Wells, 1962). Large areas of glass formation exist in polynary combinations containing V_2O_5, TeO_2, and P_2O_5. Although in such complex glasses V may be capable of substituting in a P position, there remains the important difference that V may assume lower valence states and bind fewer O atoms (Rawson, 1967).

Among complex systems are V_2O_5–TeO_2 (Chase et al., 1964), BaO–V_2O_5–P_2O_5 (Mackenzie, 1960a,b), (PbO)BaO–TeO_2–V_2O_5 (Mackenzie, 1960a, b), BaO–Ce_2O_3–V_2O_5 (Yakhkind et al., 1968), GeO_2–V_2O_5 (Yakhkind et al., 1968; Wells, 1962; Janakirama-Rao, 1965), CdO–TeO_2–V_2O_5 (Dimitrev et al., 1970), and borovanadate glasses (Matveev et al., 1970). In borovanadate glasses, the combination BVO_4 does not imitate the structure of the formally analogous Si_2O_4 and $AlPO_4$. Instead, as in $BAsO_4$, B remains in coordination 3 (Beekenkamp, 1964) and the VO_4 groups have three bridging and one nonbridging oxygen. If $AlVO_4$ is added, the structure with more bridging oxygen comes closer to that of the SiO_4 network.

Because of the facility with which V can be obtained in different valence states (particularly V^{4+} with a 3d configuration and V^{5+} (Munakata, 1960; Nester and Kingery, 1965; Rivoalen et al., 1976), vanadate glasses have the properties of semiconductors (Mackenzie, 1960; Rawson, 1967; Grechanik et al., 1961; Ioffe and Patrina, 1970; Miroshnichenko and

Klimashevskii, 1970; Denton et al., 1954; Baynton et al., 1957) of low electrical resistance. Pure amorphous V_2O_5 is a semiconductor with a hopping mechanism, having one order of magnitude higher resistance than the crystal, decreasing with increasing long range order on annealing (Rivoalen et al., 1976). V_2O_5 in combination with suitable components provides promising switching systems (Gaman, 1972; Gahlmann and Brückner, 1973/74; Higgins et al., 1975). The conductivity may change abruptly at high temperature (Schmid, 1968) and by orders of magnitude on crystallization (Blair et al., 1960). Thus switching can be achieved with stable switching voltage V_T (see, e.g., Regan et al., 1972; Higgins et al., 1975). Antiferromagnetic coupling between V^{4+} ions with a transition temperature of $-70°C$ was observed by Friebele et al. (1972). Glasses separated in two phases, especially under reducing conditions, and Neel temperatures of $-70°C$ and $-120°C$ for V_2O_4 and V_2O_3 became apparent. GeO_2 enhances V^{4+} formation, suppressing semiconduction (Janakirama-Rao, 1965). Conductance in $GeO_2-P_2O_5-V_2O_5$ glasses appears to go along chains, planes, or sheets in a structure containing VO_6 groups. [For details on electrical properties consult Adler (1979).]

E. TITANATE GLASSES

Glasses high in TiO_2 were described by many workers, e.g., Hamilton and Cleek (1958), Hirayama and Berg (1961), Herczog (1967), Trapp and Stevels (1960), Rao (1963, 1964). Rao has reported pure TiO_2 glass, which may (White, 1965) or may not (Rawson, 1967) have been aided by the presence of impurity. Alkali titanate glasses containing 50 or more TiO_2 were made (on a 20-g scale) by Baynton (1957) and Rao (1963, 1964) (1–5-g scale). Alkali titanate glasses can be stabilized to allow larger-sized preparation when small amounts of network formers (SiO_2, B_2O_3) are added. Glasses in the binary $BaO-TiO_2$ were prepared in an attempt to breed $BaTiO_3$ crystals from melts as close as possible to their composition (Herczog and Layton, 1967). In these formulations B_2O_3 and SiO_2 again promote, but Al_2O_3 hinders, glass formation. It appears that Al competes for the nonbridging oxygens. The addition of such glasses to polycrystalline $BaTiO_3$ may be used to modify its T characteristics via the influence of microstructure (Maki, 1968). Very high TiO_2 glasses are used also in the production of high-refracting glass beads for road signs (Searight and Alexander, 1965a,b,c,d); quenching of spheroids falling through flames allows TiO_2 content to 75% (Searight et al., 1971). In optical glass TiO_2 is used to replace PbO for lower-weight lenses of similar refraction (Schott, 1976). TiO_2-SiO_2 glasses prepared from ethyl silicate and titanium isopropoxide by hydrolysis may possibly be used for alkali resistant and other coatings (Sakurai et al., 1978).

Lead titanate glasses are made to obtain highly refractive beads (Searight et al., 1971) and matrices for crystallizing lead titanate (Grossman and Isard, 1970; Bergeron and Russell, 1965a,b; Brown and Ginell, 1962; Takahashi et al., 1978).

At high TiO_2 contents, some Ti is present in 4-coordination (Zorin and Zorina, 1966; Rao, 1963; Iwamoto et al., 1975) instead of 6-coordination as in the rutile polymorph of TiO_2. As in vanadate glasses, sharp breaks in log conductivity curves are observed (Schmid, 1968).

$TiCl_4$ added to vitreous SiO_2 in the hydrolysis of $SiCl_4$ to SiO_2 results in the low-expansion glass containing about 7 TiO_2 (Nordberg, 1939; Schultz and Smyth, 1972). Radio-frequency sputtered 70 $TiO_2 \cdot 30 SiO_2$ (0–7.8% Ti^{3+}) glass is semiconducting (Koffyberg, 1978a,b).

Under irradiation, two hole centers corresponding to one or two non-bridging oxygens can be identified by ESR (Kim et al., 1970). In barium silicate glasses, after irradiation, group organization like that in borate glasses can be followed when 10–16 TiO_2 is added (Bishay and Goma, 1969). In phosphate glasses, over 60 TiO_2 can be added, showing that TiO_6 groups participate in the PO_4 group polymer (Harrison and Hummel, 1959).

Glass formation in alkali titanate systems increases toward Cs_2O. Crystalline compounds 1:2 with Cs_2O and to a greater extent with Rb_2O show much less TiO_6 grouping than TiO_4 (Marfels, 1969), as do glasses with both Cs and Rb. Thus $Cs_2O–TiO_2$ glasses differ from $Cs_2O \cdot 2TiO_2$ crystals. Infrared bands previously assigned to definite alkali locations are in fact CO_2 bands (Marfels, 1969; White, 1965). Alkali titanate glasses devitrify to dititanates. The glasses are frequently colored dark brown to black, probably because of Ti^{3+}. Blue glasses are obtained in a hydrogen atmosphere. Lanthanide titanate glasses are obtained by laser-beam quenching (Shisido et al., 1978).

F. TELLURITE GLASSES

In crystalline TeO_2 and other crystalline tellurites, it is found that Te^{4+} forms trigonal pyramids with O^{2-} (coordination 3), trigonal dipyramids with one unoccupied equatorial position (coordination 4), or transitional arrangements (coordination 3+1) (Folger, 1973; Zeman, 1968, 1971). Presumably (Vogel et al., 1974), the same coordination is to be found in tellurite glasses. Earlier assignments of 6-coordination (Brady, 1956, 1957) have been based on older ideas about the structure of β TeO_2 (Ito and Sawada, 1940).* The following crystalline structures are known: (1)

---

* But in $V_2O_5–TeO_2$ glasses, Wright et al. (1977) find distorted TeO_6 octahedra (T–O = 1.98 for 40, 2.75 Å for 20).

three-dimensional networks, α TeO_2 (Vogel et al., 1974; Zeman, 1968); (2) layers, β TeO_2, Cd (Te_3O_8) (Zeman, 1971); (3) chains, 1:2, 1:3 zinc tellurites (Vogel et al., 1974; Zeman, 1968); (4) islands connected in loose chains, 1:1 zinc tellurite (Vogel et al., 1974; Zeman, 1968), $BaTeO_3$ (Folger, 1973), and (5) islands, $LiTeO_3$ (Folger, 1973).

Compounds in the binary R_2O-TeO_2 are listed in Table XXX. Glass formation is difficult near the open TeO_2 structures; no glass formation is found for the pure island structures. At high TeO_2 context, the binaries contain glasses apparently containing rather intact TeO_4 networks. Binaries ZnO(BaO)–TeO_2 have large second glass formation areas at high ZnO(BaO) contents where different structures may prevail, perhaps founded on the asymmetrical coordinations of Zn:5 and Ba:(7+2).

The glass formation range in R_2O-TeO_2 systems is very large, in RO–TeO_2 systems more restricted, with no glass found in the system CaO–TeO_2 (Table XXXI) (Vogel et al., 1974). The glass-forming regions in these systems are bordered by crystallization fields of TeO_2 and ditellurites. Other binaries have been investigated by Vogel et al. (1974), Sorrell (1968), Mochida et al., (1978), and Imaoka (1962), including those of MgO, Al_2O_3, CuO, MnO_2, WO_3, and Bi_2O_3. Figure 33 is a sketch of one example ($Al_2O_3-TeO_2$).

In all these systems (RO, R_2O, etc.), phase segregation tendencies are not pronounced; nor is the tendency to crystallization, except perhaps in the systems with Ti, Th, Nb, and La (Vogel et al., 1974). Tellurite glasses, with and without P_2O_5, containing large amounts of Fe_2O_3, were subjected to electron spin resonance analysis by Camara et al. (1980).

Tellurite glasses are characterized by the following:

(a) High refraction (Stanworth, 1952, 1954; Greco et al., 1972; Troitski et al., 1967), and they are used in recent optical and acousto-optical (Izumitani and Masuda, 1974) devices. The figure of merit for acousto-

TABLE XXX

Equilibrium Diagram Na_2O-TeO_2 (N–T) Data

	Na_2O (mole %)	Temperature (°C)
T	0	732
Eutectic	16.7	458
NT_4	20	470 (octohedral)
Eutectic	28	413
NT_2	$33\frac{1}{3}$	435
Eutectic	38	420
NT	50	710

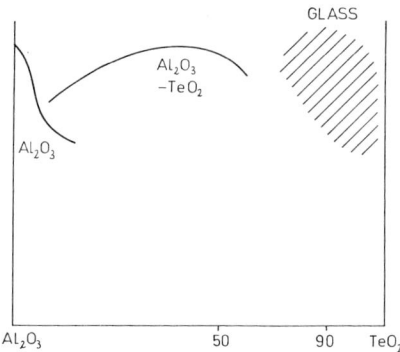

FIG. 33. Sketch of glass formation in the binary Al_2O_3–TeO_2 (after Imaoka, 1962).

optical materials contains a high power of the refractive index. Data are listed in Table XXXI.

(b) Infrared transmission to 6–7 μm, except for water bands at 3.2 and 4.4 μm (Vogel et al., 1974).

TABLE XXXI

GLASS FORMATION IN TELLURITE SYSTEMS[a,b]

Metal oxide	Range (mole %)	n_D
Li_2O	12.2–34.9	2.108–1.931
Na_2O	5.5–37.8	2.148–1.801
K_2O	6.5–19.5	2.122–1.955
Rb_2O	5.6–21.0	2.114–1.893
BeO	15 –27	2.109–2.075
MgO	11 –35	2.107–2.148
CaO	—	—
SrO	9.2–13.1	2.128–2.117
BaO	8.0–35.7	2.156–2.052
ZnO	17.3–37.2	2.124–2.038
Al_2O_3	7.6–16.8	2.082–1.946
Tl_2O	13.0–38.4	2.190–2.280
PbO	12.8–22.6	2.202–2.206
Nb_2O_5	2.2–24.0	2.192–2.191
Ta_2O_5	1.4–15.3	2.190–2.160
WO_3	8.5–44	2.192–2.191
La_2O_3	4.5–9.5	2.160–2.108
TiO_2	6.2–18.9	2.203–2.220
ThO_2	5.2–11.0	2.165–2.133

[a] From Vogel et al. (1974a).
[b] 20–100 g, cooled at 8–10°C/sec through transformation.

(c) High dielectric constant and expansion (Stanworth, 1952, 1954), especially in the softening range.
(d) Chemical stability (Stanworth, 1952, 1957).

The dispersion is characterized by V values from 18 to 15 for indices n_D 2.15–2.30 (Troitski *et al.*, 1967).

The uses so indicated have encouraged the study and development of numerous polynary tellurite glasses such as those based on PbO(BaO)–TeO_2 (Evstrop'ev and Yakhkind, 1965) (e.g., 55–65 TeO_2, 4–8 Ta_2O_5, 3–6 Bi_2O_3, 0–5 BaO, Tl_2O, 0–6 TiO_2, 1–3 PbO, 1–6 La_2O_3), V_2O_5–TeO_2 (Denton *et al.*, 1954; Baynton *et al.*, 1957; Wright *et al.*, 1977), Bi_2O_3–TeO_2 (Ulrich, 1964), WoO_3–TeO_2–T_2O_5–La_2O_3 (65:5:8:5) (Ovcharenko and Yakhkind, 1971), and also with As_2O_3 (Stanworth, 1954). Up to 15 mole % Ag_2O was added to improve optical properties of complex optical (75%) TeO_2 glasses also containing BaO, Nb_2O_5, PbO, and ZnO (Dimitriev *et al.*, 1980). With GeO_2 one obtains interesting glasses (Greco *et al.*, 1972), e.g. (wt %), TeO_2 83.2, La_2O_3 8.4, B_2O_3 0.6, K_2O 0.3, Ta_2O_5 3.2, GeO_2 3.4. ZnO–TeO_2 glasses soften around 350°C, recrystallize at 400°C, and are suitable as laser bases (Redman and Chen, 1967).

Tellurite glass ceramics have been investigated by IR spectroscopy (Morozova and Yakhkind, 1977).

1. Tellurite Glasses Containing Halides and Sulfates

Zinc fluoride was accommodated in complex tellurite glasses (Stanworth, 1952). Vogel *et al.* (1974a) developed, on the basis of introducing halides and sulfates into tellurite bases, new relations between dispersion and refraction (V–n relations). These workers synthesized glasses in the following systems, essentially free of water bands: $LiCl(Br)$–TeO_2; $NaF(Cl,Br)$–TeO_2; $KF(Cl,Br)$–TeO_2; K_2SO_4–TeO_2; $RbF(Cl,Br,SO_4)$–TeO_2; $BeSO_4$–TeO_2; $MgCl_2(Br_2,SO_4)$–TeO_2; $CaCl_2(Br_2)$–TeO_2; $SrCl_2(Br_2)$–TeO_2; $BaF_2(Cl_2,Br_2,SO_4)$–TeO_2; $LaCl_3(Br_3,SO_4)$–TeO_2; $BiOCL(Br_2)$–TeO_2; $CdCl_2(Br_2)$–TeO_2; $PbF_2(Cl_2,Br_2,SO_4)$–TeO_2; $ZnF_2(Cl_2,Br_2)$–TeO_2.

2. TeO_2–SiO_2, B_2O_3, GeO_2, P_2O_5 Glasses

These glasses with refractive indices n_D from 1.95 to 2.15 were developed by Vogel *et al.* (1974).

G. NIOBATE AND TANTALATE GLASSES[*]

Crystalline alkali niobates, applicable to transducers and conventionally obtained in the form of solid-reaction, poled, hot-pressed ceramics,

---

[*] Glasses high in ZrO_2, HfO_2, La_2O_3, and lanthanide oxides are also considered here.

can be obtained from glasses high in alkali niobate content (Herzog and Layton, 1964, 1967, 1969), e.g., Na_2O 40; Nb_2O_5 40; SiO_2 20, in the shape of loose dendritic 200 to 10,000-Å crystals. As little as 5% network formers may suffice. The more SiO_2 is added, the larger are the crystalline precipitates. Crystallized niobate glasses show electroptical effects (Borelli, 1967). The highest Nb_2O_5 and Ta_2O_5 (also ZrO_2, HfO_2, La_2O_3, and lanthanide oxide) contents have been obtained by laser spin techniques (Topol et al., 1973; Shisido et al., 1978). Niobate–tantalate–aluminate and other crystallizable glasses were prepared by Kokubo et al. (1974; Kokubo and Tashiro, 1978). Almost 50 mole % Nb_2O_5 are accepted in complex niobate–phosphate glasses (Shtin et al., 1977). Amorphous Nb_2O_5 with excess Nb is a semiconductor. It is obtained by anodizing Nb (Fuschillo et al., 1976). Alkali tantalate glasses were developed by Nassau et al. (1979). Ta_2O_5 films were obtained in amorphous form by Mead (1963). High electrical conductivity (10^5 Ω/cm) based on Li^+ transport is found in the glasses $Li(NbO_3)_{1-x-y}(SiO_2)_x(Fe_2O_3)_y$ with $32 < x < 39, 0 < y < 5$ (Prasad et al., 1979).

H. Tungstate and Molybdate Glasses

In a wide range of alkali tungstate glasses, a WO_4 tetrahedral chain structure is found. Minimum critical cooling rates are observed where WO_6 octahedra—the most frequent coordination in crystalline tungstates—appear (Gelsing et al., 1966; Gossink and Stevels, 1971; Rawson, 1967). Similar arrangements are found in alkali molybdate glasses (Gossink et al., 1971). In the system $AgI-Ag_2O-MoO_3$, glasses form below $Ag_2O/MoO_3 = 1$, having electrical conductivities of 10^{-5} to 10^{-2} Ω/cm (Minami et al., 1977). The coordination compares with that in nitrate or acetate rather than other network glasses, as indicated by the optical absorption of transition elements (Duffy, 1977). In WO_3–phosphate and MoO_3–phosphate glasses, O/P ratios as high as 7/1 can be obtained, indicating that W and Mo are not modifiers but contribute to the polymer structure. Optical glasses based on these compositions were developed by Rothermel et al. (1949) and studied by Franck (1955). Boynton et al. (1975a,b) find them semiconducting. Switching effects were observed in silicate and borate glasses containing 75 Nb_2O_5 (Higgins, 1977). The covalency in the tetragonal arrangement of molybdenyl $(MoO_2)^{2+}$ groups was demonstrated by Baugher and Park (1970). Alkaline earth tungstate glasses exhibit phototropy in the ultraviolet when doped with Bi (Sakka, 1969).

I. Other Oxide Glasses

ThO_2-H_2O glass acts like a solid gel—a metastable array of Th–O tetrahedra is connected by H bridges between OH groups (Guymont,

1977). A BaO–Fe_2O_3 glass was described by Takahashi *et al.* (1980), other Fe_2O_3 glasses by Messier and Roy (1978), a Li_2O–Bi_2O_3 glass (of high electrical conductivity) by Nassau *et al.* (1979), and a PbO–Bi_2O_3 glass by Nassau (also containing Tl_2O, CdO, BaO, and ZnO; transmitting to 9 μm; and having a density of 7 and n_D ~2.5).

X. Ionic Salt and Solution Glasses*

A. GENERAL CONSIDERATIONS

In all the inorganic systems considered up to this point, the glass-forming ranges may be regarded as binary or pseudobinary mixtures of acidic and basic oxides (or fluorides) in which the acid–base ratio is variable (though only on the acid side of the equivalence point). For instance, $Na_2O + SiO_2$ mixtures are glass-forming for small samples in the range 43–100 mole % SiO_2, with the equivalence point lying at 33% SiO_2 (Na_4SiO_4), $KF + BeF_2$ solutions with an equivalence point at 33% BeF_2 (K_2BeF_4) are glass-forming in the range 45–100% BeF_2. (See Section XI.) We come now to the consideration of systems of different character in which the components are all of saltlike, or equivalence point, compositions.

In many cases, these components may still be thought of as formed from acid and basic oxide combinations, but the acid–base origin of salt-type substances is not of particular relevance to the glass-forming properties. For instance, calcium nitrate, a key component of the popular mixed nitrate glasses, may be thought of as the product of reaction of equivalent proportions of the acidic oxide N_2O_5 and basic CaO. However, in glass-forming systems containing calcium nitrate, only the NO_3^- anion is identified, and whatever polymerizing possibilities N_2O_5 may have are of no relevance to the glass-forming properties of mixed nitrate solutions. Likewise, the glass-forming sulfate systems, e.g., $4K_2SO_4 \cdot 6ZnSO_4$ can be viewed as pseudobinary mixtures of the acidic oxide SO_3 and appropriate mixtures of the basic oxides ZnO and K_2O, but again this is not a relevant or useful approach, since only equivalent acid–base ratios (yielding SO_4^{2-} anions) are involved. In this respect the sulfate glasses are to be clearly distinguished from those involving the larger Group VI atoms, Se and particularly Te, in which oxygen bridging and consequent polymeric anion formation can occur. It is, indeed, the occurrence of glass-forming properties in the *absence* of polymeric oxyanions (or fluoroanions) or extended networks that lends scientific interest to this class of inorganic glasses; it is therefore preferable to place them in a separate category—*ionic salt* as distinct from *oxide* glasses—and to seek different interpretations of their glass-forming abilities. Such interpretations focus attention

---

* This section was written by C. A. Angell, Purdue University, West Layfayette, Indiana.

on the geometrical asymmetry of the anion and the packing problems that follow, particularly in unbalanced cation force fields. This class properly includes salts with simple organic anions such as the formates (HCO_2^-) and acetates ($CH_3CO_2^-$), which are best regarded as modified carbonate ions.

There is one class of system dealt with here to which the above considerations do not apply, and which should properly be included in the earlier section on standard Lewis acid–base polymeric anion-forming systems. This is the class of $ZnCl_2$-based halide glasses. However, because $ZnCl_2$ is usually thought of as the only common inorganic "salt" that can be vitrified in the pure state (rather than as a weakened analog of the network family SiO_2, GeO_2, BeF_2), we shall include it in this section. Because it is distinct, this class of system will be discussed separately from the others in a concluding section.

Many of the salts involved in the glass-forming mixtures are themselves rather soluble in ionizing solvents such as water and DMSO, and the range of glass-forming compositions can be extended almost indefinitely if such additional components are admitted. Only aqueous cases will be considered here. Intermediate in character, and often more interesting and relevant in practice, are systems containing a fixed ratio of solvent molecules to cations of one type (e.g., $Ca/H_2O = 1/4$), often referred to as molten hydrate glass.

Finally, a further and distinct class of salt glasses can be distinguished in which the anions are simple (e.g., Cl^-, Br^-) but the cations are complex (usually of organic character, e.g.,

ethylammonium (CH_3—CH_2—$\overset{+}{N}H_3$), α-picolinium $\left(\underset{CH_3}{\left\langle\bigcirc\right\rangle}\overset{+}{N}H\right)$

In the following sections we shall consider the various types of salt glasses so far identified and studied, under the subheadings nitrate glasses, carbonate glasses, formate and acetate glasses, thiocyanate glasses, sulfate and bisulfate glasses, dichromate glasses, aqueous salt glasses, organic salt glasses, and $ZnCl_2$-based halide glasses. An overview of these types of glass-forming systems and the link they provide between oxidic glasses and molecular and organic polymer glasses has been given elsewhere (Angell and Tucker, 1973).

B. Nitrate and Nitrite Glasses

The supercooling propensity of molten nitrate mixtures containing both divalent and monovalent cations has been known at least since the work of Rotskowsky (1930), but their study as glass-forming systems was not taken up seriously until the classical work of Dietzel and Poegel (1954) on the $Ca(NO_3)_2-KNO_3$ system. As usually found in nonnetwork systems, the glass-forming composition regions lie in the vicinity of the eutectic composition where the thermal energy at the liquidus is a minimum relative to the cohesive energy (Turnbull and Cohen, 1958).

Certain multicomponent mixtures of monovalent cation salts (e.g., $LiNO_3-AgNO_3-NH_4NO_3$) with low-lying eutectics can be vitrified at $\sim -50°C$ (Angell and Helphrey, 1971), but most detailed attention has been given to the mixed valence cation systems $Ca(NO_3)_2-KNO_3$, $Cd(NO_3)_2-NaNO_3$, and $Cd(NO_3)_2-TlNO_3$, which have wider glass-forming ranges, and glass temperatures near or above room temperature (e.g., 60°C for the widely studied, kinetically stable, $4Ca(NO_3)_2 \cdot 6KNO_3$ composition).

Systematic surveys of glass-forming composition ranges, initially of $Ca(NO_3)_2$ + alkali nitrates (Thilo et al., 1964) and more recently (but in less detail) of a wide range of mixed cation systems ranging up to $Th(NO_3)_4$ (van Uitert et al., 1971), have provided a general picture of the conditions needed for glass formation in these systems. The glass-forming ranges in relation to the phase equilibrium diagram from the work of Thilo et al. are shown in Fig. 34. Both Thilo and van Uitert concur that glass formation is most probable in those systems in which the electrical field

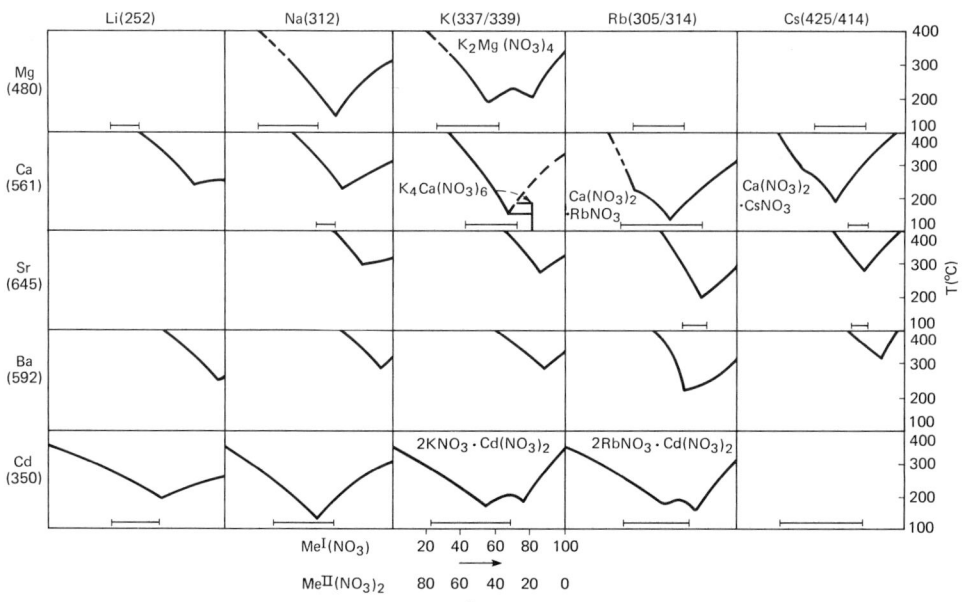

FIG. 34. Regions of glass formation (indicated by horizontal bars) in various binary nitrate systems (melting points of nitrates (°C) given in parentheses) (Thilo et al., 1964).

strengths z/r^2 of the two cations differ by 0.7 or more. The interpretation offered is that the high-field-strength cations form clusters with the anions that inhibit the nucleation of crystalline phases with different cation nitrate groupings. There is some doubt in the case of Van Uitert's glasses containing Mg^{2+} and rare earth cations about the efficiency of the dehydration procedures. In the light of difficulties previously encountered with dehydration in the presence of high-field cations (Wong, 1970) the sample heating methods employed by Van Uitert seem inadequate, and, as no analyses were reported, many of the new glasses may contain one or two moles residual water per mole of salt. This in itself would have been instrumental in stabilizing the liquid mixtures against crystallization. Where they have been measured, glass transition temperatures in the nitrate mixtures vary systematically with composition, increasing with increasing mole fraction of higher-valent cation, (Rao et al., 1973). At equal divalent–monovalent cation fraction, T_g tends to be lower in mixtures containing heavy metal cations with greater covalent bonding propensities.

Very little is known about the overall structure of nitrate glasses; no detailed x-ray or neutron scattering studies have so far been performed. Vibrational spectroscopy studies have established that minor perturbations of the nitrate ion ($D_{3h} \to C_{2v}$) occur because of unsymmetrical interactions with the higher valent cations (Hester and Krishnan, 1968), and theoretical arguments by Furukawa et al. (1977) have suggested models for the multivalent cation–nitrate configuration. Laboratory-based information is not available on the monovalent cation positions, but useful insight can be gained from the pioneering computer simulation study of the $0.4Ca(NO_3)–0.6KNO_3$ system by van Wechem (1976). As in other glasses, the monovalent cations remain the mobile species, and they dominate the low-temperature conductance properties (Howell et al., 1974).

Glass formation is also possible in mixed nitrite systems, such as $KNO_2 + Ca(NO_2)_2$, but in this case decomposition of the melts, particularly those high in $Ca(NO_2)_2$, is more of a problem. Glasses in the range 30–45% $Ca(NO_2)_2$ can be prepared by careful dehydration of the aqueous melt, and subsequent cooling to room temperature (Angell and Ringoen, 1978). It is possible that chemical stabilities might be higher in $Sr(NO_2)_2$-based systems and that the greater asymmetry of the NO_2 anion may lead to useful glass-forming ranges. A variety of mixed nitrate–nitrate glass-forming systems will exist.

C. CARBONATE GLASSES

Mixed carbonate glasses were first reported by Eitel and Skalics (1929). It is necessary to perform the fusions under moderate pressure to prevent escape of CO_2, and, probably for this reason as much as any other, these systems have not been given the attention they deserve. Glass formation in carbonate systems has similar origins to that in the nitrate systems. The simplest system is K_2CO_3–$MgCO_3$, which is glass-forming in the ranges of 40 to 60% K_2CO_3. Because of the larger charge on the triangular CO_3^{2-} anion, the carbonate liquids are more compatible with the higher melting basic-metal oxides, fluorides, and sulfates, and glasses with mixed anions may be formed in the absence of alkali metal cations. An example is the complex high-$CaCO_3$ glass described by Datta et al. (1964) in mole %:

24.3 CaF_2; 32.1 $Ca(OH)_2$; 37.1 $CaCO_3$; 6.5 $BaSO_4$

The carbonate glasses are almost completely uncharacterized at this time. Glass transition temperatures appear to be in the range of 200 to 300°C.

D. FORMATE AND ACETATE GLASSES

Formate and acetate glasses may be regarded chemically as derived from carbonate glasses by replacing one of the oxygens (—O^-) with a hydrogen (—H) or methyl group (—CH_3). The remaining single negative charge resonates between the two remaining oxides. The increased anion asymmetry enhances the glass-forming tendency of systems containing these ions over that of pure carbonate systems; the decreased anionic charge leads to lower glass transition temperatures.

Little is known about formate glasses. Although calcium formate is unstable, meaning that there is no formate equivalent of the much-studied Ca–KNO_3 glasses, the alkali formates can be fused and glasses containing up to 70 mole % Na formate (with 30 mole % Li acetate) have been prepared (Angell and Ringoen, 1978).

The acetate (Ac) glasses have been much more widely studied (Bartholomew and Holland, 1969; Duffy and Ingram, 1969; van Uitert et al., 1971) and seem to have some unique features among low-melting glasses (Fig. 35). Lithium acetate, for instance, is the only pure alkali metal salt that can be obtained in the glassy state without resort to splat-quenching methods. $PbAc_2$ is also glass-forming alone. KAc–$CaAc_2$ glasses have been shown to be uniquely resistant to radiation damage, even to transient effects on the microsecond time scale (Barkatt and Angell, 1978).

The most systematic study of glass formation in acetate systems has been that of van Uitert et al. (1971). These workers explored at least equimolar compositions in all the combinations of alkali metal (Li, Na, and K) acetates with the divalent metal (Mg, Ca, Sr, Ba, Zn, Cd, and Pb) acetates, finding glass formation in all cases except the $Ba(Ac)_2$ systems and, surprisingly, the K–Mg system (cf. carbonate glasses). The alkaline earth acetates are not soluble in water, and glasses containing them are

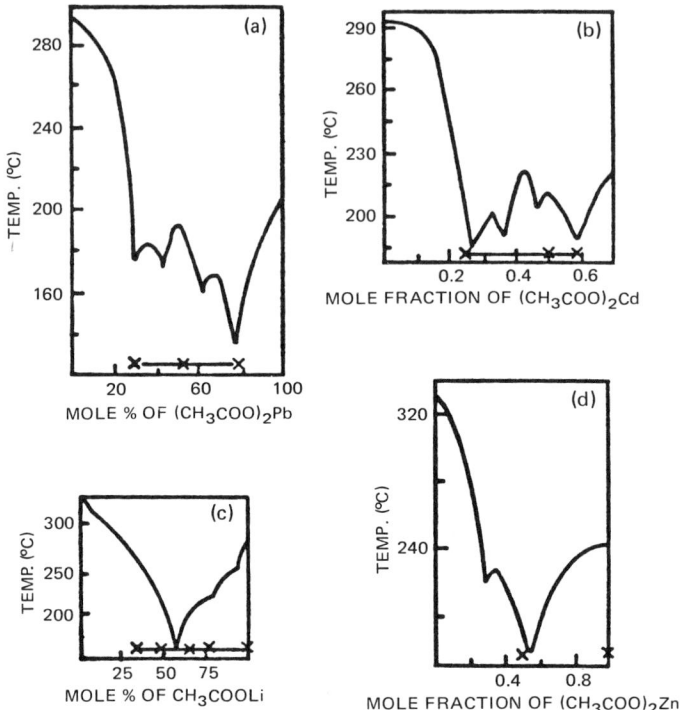

FIG. 35. Regions of glass formation in binary acetate systems: (a) $CH_3COOK–(CH_3COO)_2Pb$, (b) $CH_3COOK–(CH_3COO)_2Cd$, (c) $CH_3–COONa–CH_3COOLi$, and (d) $CH_3COONa–(CH_3COO)_2Zn$ (Bartholomew and Holland, 1969.)

more resistant to attack by atmospheric water than, for instance, the otherwise similar nitrate glasses. Glass transition temperatures for the mixed valence acetate glasses tend to be ~30° higher than for the corresponding nitrate glass (Bartholomew, 1970).

Van Uitert et al. (1971) also explored glass formation in rare earth acetates and trifluoroacetates ($CF_3CO_2^-$), finding that residual water was necessary to avoid thermal decomposition. A wide variety of combinations of alkali metal acetates and rare earth acetate are glass forming, the broadest ranges being found for systems with NaAc. The glasses formed probably contained residual water in most cases.

The structures of the formate and acetate glasses are totally uninvestigated. Alkali acetate mixtures (e.g., Li–Na) have been the subject of heat capacity (Angell Sichina, 1978), electrical conductivity (Bartholomew, 1970), radiation damage (Barkatt and Angell, 1978), and probe ion spectroscopy (Duffy and McDonald, 1970) studies.

E. Thiocyanate Glasses

It is clear that the contrast in glass-forming propensity between mixed valence nitrates and acetates and, for example, the corresponding unvitrifiable mixed valence chlorides, is a consequence of the presence of asymmetric anions, and the lower melting, higher viscosity characteristics that follow. It has been of interest to observe how far one can go in the direction of simplifying the particle geometrics before glass-forming ability is lost. Cyanides (CN^-) and thiocyanates (SCN^-) have been called pseudohalides because of chemical similarities with halides, yet their melting points tend to be much lower, particularly for the thiocyanates (e.g., KSCN, $T_m = 176°C$). It has been of interest therefore to explore the possibility of glass formation in these systems

Mixtures of thiocynates alone decompose very readily when heated, and great care and use of vacuum technique is needed to dehydrate them successfully (Barkatt and Angell, 1978). Glass formation has been observed in some $Ca(SCN)_2$ + KSCN mixtures, but there is difficulty in reproducing their properties. In the presence of sufficient residual water, glasses are easily formed on sufficient cooling, but glass temperatures are low. In mixtures with nitrates, on the other hand, decomposition is inhibited and $Ca(NO_3)_2$ + KSCN glasses containing up to 85 mole % KSCN have been prepared. Their glass transition temperatures are about the same as those for the corresponding all-nitrate mixtures.

F. Sulfate Glasses

The sulfate ion is quasi-spherical, hence less likely to lend itself to glass formation, than the ions discussed earlier. However, in a few mixed valence cation systems with very low liquidus temperatures (usually dependent on the presence of a transition metal cation with covalent bonding tendencies, e.g., Zn, Co, or Cu) glass formation in a limited composition range is encountered. The most useful example is $ZnSO_4$ + K_2SO_4, in which glass formation is obtained over a composition range of 40 to 70 mole % $ZnSO_4$ (Ishii and Akawa, 1965; Angell, 1965; Kolesova, 1975) (see Fig. 36). The physical properties of glasses in this system have been studied in some detail by Narasimham and Rao (1978). Glasses with the composition 60 mole % $ZnSO_4$ have also been used in visible and UV probe-ion studies of structure and basicity in ionic media (Angell, 1968a; Ingram and Duffy, 1968). A somewhat wider range of glass formation is found in the systems Tl_2SO_4–$ZnSO_4$ (Ishii and Akawa, 1965) and particularly $(NH_4)_2SO_4$ + $ZnSO_4$ (Wong, 1970). The bisulfate glass $KHSO_4$, which was the first case of sulfate glass formation reported (Förland and Weyl, 1950), owes its existence to the strong hydrogen bonding between

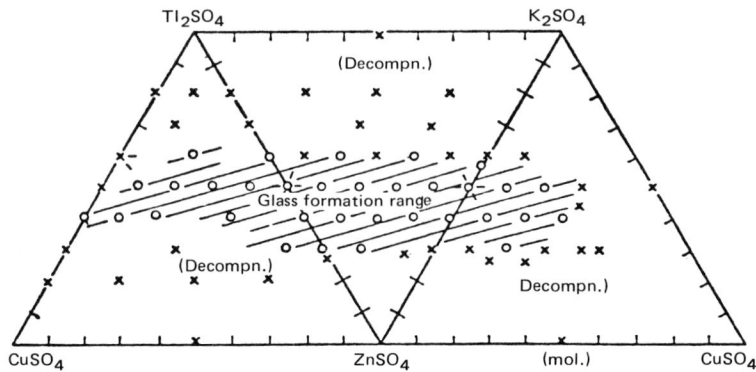

FIG. 36. Regions of glass formation in ternary systems containing $CuSO_4$, Tl_2SO_4, K_2SO_4, and $ZnSO_4$ (Ishii and Akawa, 1965).

HSO_4^- ions. An equimolar $NaHSO_4$–$KHSO_4$ composition, which is very stable against crystallization during cooling, has been useful in probe-ion basicity studies by Duffy and McDonald (1970). Glass formation with sulfates can be enhanced by addition of chlorides; e.g., K_2SO_4 + $CoCl_2$ solutions have also been observed to vitrify in small quantities (Angell, 1965a). If the glass former $ZnCl_2$ is introduced as a component, a wide range of glass-forming sulfate–chloride mixtures may be obtained (Schultz, 1957). Alkali sulfate melts do not mix with alkali silicate melts (Raask and Jessop, 1966), less for Na than K (Pearce and Beisler, 1965). This immiscibility accounts for the formation of so-called glass gall on top of soda-lime (window) glass melts in which salt cake (sodium sulfate) is conventionally used as an alkali source and fining agent.

G. Dichromate Glasses

Glass formation has been observed in some systems based on the alkali dichromates $M_2Cr_2O_7$. The mixtures $Li_2Cr_2O_7$ + $Na_2Cr_2O_7$ and $Li_2Cr_2O_7$ + $K_2Cr_2O_7$, for instance, are glass forming in the ranges 20–40% $Li_2Cr_2O_7$ and 40–50% $Li_2Cr_2O_7$, respectively, on quenching to 0°C or below ($T_g \sim 0°C$). Mixed systems such as $K_2Cr_2O_7$ + LiCl and $K_2Cr_2O_7$ + $LiNO_3$ form glasses in larger quantities, some of which are kinetically stable at room temperature (Angell and Parks, 1978).

H. Aqueous Hydrate Glasses

1. Molten Hydrate Glasses

Many simple ionic salts with multivalent cations form crystalline hydrates with sufficient water to complete or partially complete the cation hydration sphere. These substances usually have low melting points and in many ways behave like molten salts with large, hence weak field, cations. The first case studied from this point of view was calcium nitrate tetrahydrate (Angell, 1965), which is glass forming itself and was shown to yield glass forming solutions with anhydrous KNO_3 in which there was little composition dependence of the glass transition temperature (Angell and Helphrey, 1971). Glass-forming ranges and glass transition temperatures for this and some related $Ca(NO_3)_2 \cdot 4H_2O$-based systems are shown in Fig. 37. A great many molten hydrates of salts with asymmetric anions

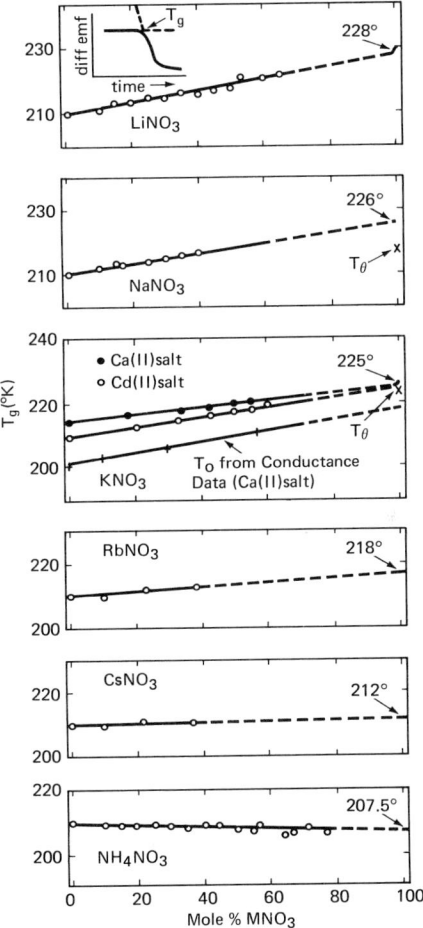

FIG. 37. Experimental glass transition temperatures for $Cd(NO_3)_2 \cdot 4H_2O$ + alkali metal nitrates and ammonium nitrate, showing extrapolations to obtain estimates for T_g of pure alkali nitrates. Also included are T_g and T_θ (from conductance measurements) values for $Ca(NO_3)_2 \cdot 4H_2O$ + KNO_3 solutions. Inset, differential emf–time trace showing definition of T_θ. (Angell and Helphrey, 1971.)

and multivalent cations prove to be glass-forming. Generally, these have glass transition temperatures below 0°C, though some of the rare earth acetates have higher values. There are also a range of low-T_g glass-forming hydrates in which the anions, rather than the cations, are hydrated (e.g., $K_2S_2O_3 \cdot 5H_2O$) (Moynihan, 1966).

Partly because of simplicity of experimentation, certain molten hydrates, particularly $Ca(NO_3)_2 \cdot 4H_2O$, have received very detailed attention in recent years. No x-ray or neutron scattering studies have been performed on these, but they have been probed structurally by Raman spectroscopy [for coordination equilibria (Hester et al., 1964)] and Raleigh light-scattering methods [for phonon velocities and dispersion (Ambrus et al., 1972a)]. Frequency-dependent electrical conductance in both liquid (Ambrus et al., 1972a) and glassy (Hodge et al., 1977) states has been studied, and a variety of thermodynamic (Angell and Tucker, 1974; Braunstein and Braunstein, 1971), NMR chemical shift (Sare et al., 1973), and proton relaxation (Braunstein et al., 1977) measurements have been made.

Mixtures of molten hydrates are usually glass forming, with simple linear variations in glass transition temperatures (Moynihan et al., 1969; Jain, 1978). In some cases where the salt itself can act as a Lewis acid (e.g., $ZnCl_2$), the cation will exchange H_2O ligands for excess anions to form, e.g., $ZnCl_4^{2-}$ anions, in which case pronounced deviations from simple mixing behavior result and very nonlinear T_g versus composition relations are found (Easteal et al., 1974). Glass-forming systems based on a salt hydrate as one component and an anhydrous Lewis acid salt [e.g., $ZnCl_2$ (Chin, 1971) or $CoCl_2$ and $NiCl_2$ (Islam and Ismail, 1976)] have also been investigated. Mixed anion systems of neutral character, such as $Ca(NO_3)_2 \cdot 4H_2O$ + KCNS also have wide glass-forming ranges (Islam and Ismail, 1975; Barkatt and Angell, 1978).

2. Ionic Solution Glasses

Departure from fixed salt–water ratios opens up a very broad range of glass-forming systems that are probably of more interest to solution chemists and radiation chemists than to glass scientists. Nevertheless, as these systems have been investigated more systematically than most other salt systems, their characteristics will be properly dealt with here. This is encouraged by the fact that the recent structural investigation of certain of these systems by sophisticated isotope-difference neutron-scattering techniques (Enderby et al., 1973) may serve as a model for future investigation of glass-forming systems of more practical interest.

Glass formation in aqueous salt solutions of high salt concentration was first studied by Vuillard (1957), who determined glass-forming composition regions in the systems KOH, LiCl, and Be(NO$_3$)$_2$, as well as several mineral acid systems. Glass-forming solution regions were more carefully defined in a more wide-ranging investigation by Angell and Sare (1970) in which some 30 inorganic chloride and nitrate salt + water systems were characterized. Glass-forming ranges for these systems are shown against their equilibrium phase diagrams in Figs. 38 and 39. Glass transition temperatures vary approximately linearly with composition except at the water-rich extremes, where some structural reorganization evidently occurs. Glass transition temperatures in these systems can be reduced to anion-dependent groups by normalizing the plots for uni-, di-, and trivalent cations to constant charge concentrations (N equiv/liter). The groups

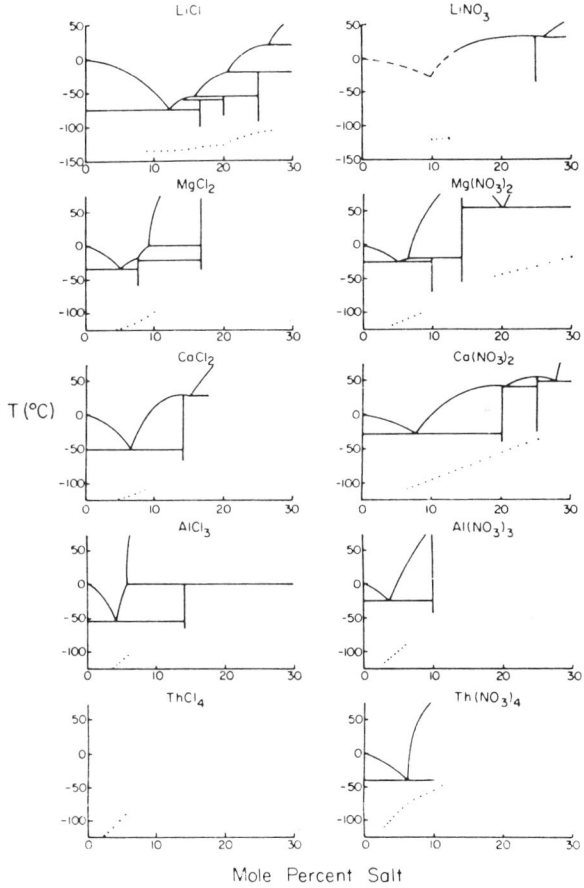

FIG. 38. Phase diagrams for salt–water systems showing glass-forming regions and glass transition temperatures: inert-gas-core-action chlorides and nitrates.

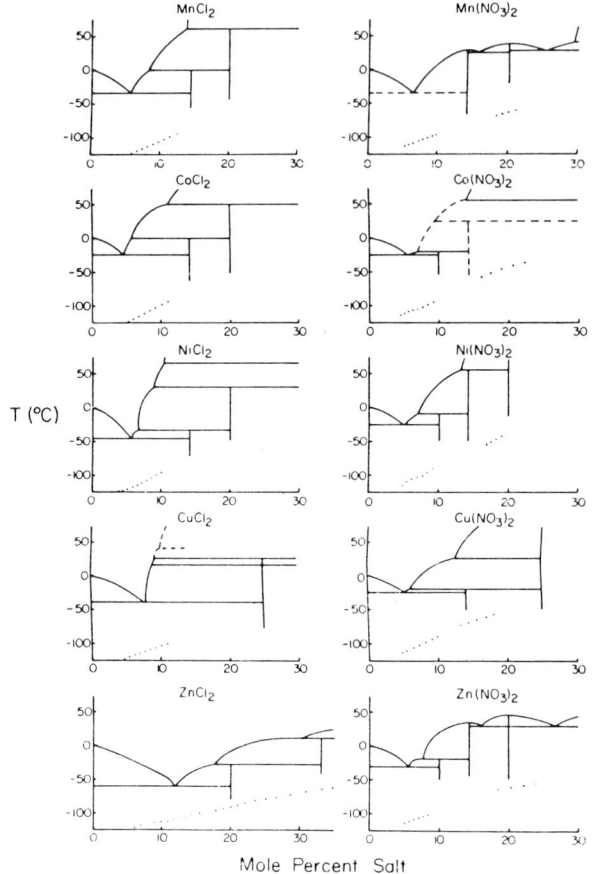

FIG. 39. Phase diagrams for salt–water systems showing glass-forming regions and glass transition temperatures: divalent-transition-metal-cation chlorides and nitrates.

for salts of different cations are differentiated according to the strength of anion–water hydrogen bonding interaction. Among salts so far studied the acetate ion yields the highest glass transition temperatures.

A number of structural studies of such solutions at temperatures well above T_g have been carried out by x-ray methods (Prins, 1935; Prins and Fonteyne, 1935; Beck, 1939) and have been used as the basis for the assertion that the solutions have a quasi-lattice structure. However, the amount of definitive information that can be obtained by this is very limited because of the large number of distinct pairs that contribute to the total structure factor. In normal neutron scattering studies similar problems are compounded by the problem of quantitative handling of the

Placzek corrections (correcting for neutron–atom recoil effects), which are particularly important in systems where the low-mass deuteron (which is always substituted for the otherwise dominant proton) is a primary scatter. Very recently the problem of sorting out the important ion–solvent structure from the superposition of many (typically 10) partial structure factors that combine to produce the experimentally recorded total structure factor, has been greatly simplified by the isotope replacement techniques pioneered for aqueous solutions by Enderby *et al.*, (1973; Howe *et al.*, 1974; Soper *et al.*, 1977). In these experiments neutron-scattering patterns are obtained for two solutions that are identical in every respect except for the abundance of one or other ionic isotope, e.g., ^{60}Ni and ^{62}Ni, or Cl(natural abundance) or ^{37}Cl. The difference between the two patterns can be shown, in favorable cases, to be dominated by the interesting structure factors, ion–oxygen and ion–deuterium. Because neutron-scattering techniques using high-flux sources (as at the Institute Laue–Langevin in France) are now yielding low-noise patterns, the differences, hence the important partial structure factors, can be isolated from the remainder in sufficiently well-defined form to yield definitive structural information. From the very limited studies performed so far using this approach, the coordination number of Ni^{2+} by H_2O, which is well known to be 6 from visible spectroscopy studies, has been confirmed, and the further information that the water protons lie inclined by 30° to the Ni–O axis has been provided. From the chlorine isotope substitution studies a coordination of Cl^- by 5.5 H_2O molecules (interacting through one of the protons for each molecule coordinated) has been established. The coordination number 6.0 has also been established for Ca^{2+} ions in concentrated solutions. With less certainty, a Ni^{2+}–Ni^{2+} distribution function was obtained that indicated a quasi-lattice arrangement within the solvent for these species—an apparent vindication of Prins' x-ray-based deductions and of the old lattice theory of concentrated ionic solutions based on the common $C^{1/3}$ dependence of the thermodynamic properties. It is clear that the oxide glass structure field could benefit greatly from the application of such techniques, and it will probably not be long before studies of key glass structure problems based on this approach are initiated. (A. C. Wright reported studies of this type at the Fifth International Congress on Physics of Non-Crystalline Solids, Montpellier, 1982.)

Other experimental studies that bear on the structure, and in particular on the dynamics of the structure, in aqueous glasses are exemplified by the light-scattering studies of Hsich *et al.* (1972) and the relaxation studies of Ambrus *et al.* (1972a), Johari and Goldstein (1970), and Hodge and Angell (1978), which are summarized by the following statements.

Hsich *et al.* studied the intensity of the Raleigh-scattered light line in a series of aqueous LiCl solutions as a function of temperature and concluded that LiCl solutions in which there are only ~7 H_2O per Li^+ split into two immiscible liquid phases at $-135°C$. This observation bears out an earlier speculation (Angell and Sare, 1968) that the network liquid, H_2O, and its solutions with salts might exhibit this behavioral analogy to the familiar phase-separating oxide network + ionic oxide binary solution systems.

Ambrus *et al.* (1972b), by a combination of frequency-dependent electrical susceptibility and viscoelasticity measurements, showed that the conductivity and mechanical relaxation times of $Ca(NO_3)_2$ + H_2O solutions, which are the same at high T/T_g, diverge increasingly as $T \to T_g$. Johari and Goldstein (1970) showed that some electrically active modes of motion are of very high frequency compared to the conductivity mode and are not frozen out until temperatures far below T_g. Hodge and Angell (1978), by studying the effect of changing water content, were able to relate this high-frequency mode to the even-higher-frequency mode observed in anhydrous Ca/KNO_3 (Hayler and Goldstein, 1977). This suggests that both involve some rocking or dentation exchange mode for NO_3^- in the first coordination sphere. Little is currently known about the structural origin of such low-energy-barrier motions in glasses.

3. Organic Salt Glasses

Glass-forming organic salts are part of everyday experience insofar as they provide the common noncorrosive flux cores of the familiar electrical solders. These, however, are complex cases and have been little studied from a fundamental viewpoint. Simpler cases based on derivatives of the hydrochloride of pyridine,

have received some attention recently because of interest in the chemical behavior of chloride ions in systems in which the cation field strength is very low. Pyridine hydrochloride (pyridinium chloride) is thermally a very stable salt suitable for mixing with the generally higher melting inorganic salts, but it is not glass forming in the pure state. Small asymmetry-producing complications of the molecular structure, such as attaching a methyl group to the pyridine ring in the α position, lowers the pure-substance melting point and results in glass-forming ability. The glass transition temperatures, however, are very low (Angell *et al.*, 1976).

Mixtures of salts with such organic cations are usually glass forming near their eutectic temperatures, and the glass transition temperatures of nonglass-forming end members may usually be estimated by short extrapolations of binary solution data (Abkemeier, 1972). There are a great many molten organic salt systems that are utilized by synthetic organic chemists (Gordon, 1969) that have not been researched for their glass-forming tendencies or characteristics.

Mixtures of glass-forming organic cation salts with inorganic salts are strongly glass forming over limited composition ranges. In most mixtures the chloride ion is "captured" by the inorganic cation to form specific complexes. The glass transition temperatures of these solutions show interesting variations that are dependent on the nature of the complex cation formed (Angell *et al.*, 1976). In the case of the system α-methyl pyridinium chloride + zinc chloride, solutions are glass forming over the entire binary composition range with the exception of a 5% range about the composition of the compound α-methyl pyridinium tetrachlorozincate. Reference to a detailed phase diagram and glass-formation study of the related but less extensively glass-forming system pyridinium chloride + $ZnCl_2$ is deferred to the next section on $ZnCl_2$-based systems.

I. Zinc-Chloride-Based Halide and Related Glasses

1. Chloride Systems

It has long been known that anhydrous zinc chloride, alone with BeF_2 among the pure inorganic halides, can be obtained in the glassy state by normal cooling of the liquid phase. The glass transition temperature is low (105°C at 10 deg/min by DTA determination), compared with that of BeF_2 ($\sim$350°C) Angell and Tucker, 1973), and, in view of the spectroscopic evidence for solidlike 4-coordination of Cl^- by Zn^{2+} in the liquid and glassy states, it has seemed reasonable to regard this substance as a weak analog of the BeF_2 and SiO_2 random network structures. Qualitative x-ray patterns have been published (Imaoka *et al.*, 1971) but no detailed analyses are in print at this time. This situation will soon be remedied as combined neutron scattering and x-ray scattering studies currently in progress become available (Wright, private communication; Narton, private communication). An attempt to computer-simulate liquid and vitreous states of $ZnCl_2$ by the methods of molecular dynamics has been made (Woodcock *et al.*, 1976) but the pair potentials utilized were of too simple a form for close reproduction of the laboratory material characteristics to be expected. Interestingly enough, however, the pair distribution functions obtained in this study yield the same average Zn-by-Cl coordination number, 4.7, suggested by the results of EXAFS investigations currently being reported (Wong and Lytle, 1977).

Zinc chloride is an interesting case for thermodynamic study, being intermediate between the strong network glasses, which show essentially no increase of heat capacity at T_g, and the molecular or salt glasses, which show large increases. It has recently been found to disobey the first

Davies–Jones equation connecting pressure dependence of T_g to the changes in expansion coefficient and heat capacity at T_g, the first well-documented case of such a failure (Angell *et al.*, 1976).

Like SiO_2 and BeF_2, $ZnCl_2$ is a strong Lewis acid and it is clearly of interest, in pursuit of the SiO_2 and BeF_2 network glass former analogy, to enquire about glass formation in $ZnCl_2$ + Lewis base chloride binary systems.

2. Zinc-Chloride-Based Binary Systems

a. Chlorides. Zinc chloride + alkali metal chloride mixtures are, unfortunately, not easily glass forming, although, if near-eutectic compositions in the $ZnCl_2$ + KCl system are poured dropwise into liquid nitrogen, a small fraction of the drops will vitrify. The glass transition temperature for the binary glass is above room temperature. The poor glass-forming ability compared with SiO_2- and BeF_2-based binary solutions is attributable to the improved competitive status of the alkali cation $[z/r\,(Zn^{2+}) < z/r\,(Be^{2+}, Si^{2+})]$ and the consequently greater degree of network disruption produced by the addition of a given amount of alkali salt.

If the alkali metal salt is replaced by the chloride of a weaker cation (e.g., of the benzenelike organic cation pyridinium, PyH^+, $C_5H_5NH^+$), then wide glass-forming composition ranges typical of the SiO_2- and BeF_2-based systems are restored. A rather detailed study of the system $ZnCl_2$ + PyHC has been made (Easteal and Angell, 1970), and it has been concluded that the system is a close analog of the system SiO_2 + Na_2O. A phase diagram for this system, locating the glass-forming region and glass transition temperatures (and also making a comparison with those for the SiO_2 + Na_2O binary glasses), is shown in Fig. 40. The T_g plateau region in the case of SiO_2 + Na_2O glasses corresponds with the composition range in which phase separation is observed in that system. It is presumed that in the corresponding $ZnCl_2$ + PyHCl plateau region an analogous phase separation occurs in the $ZnCl_2$-based glasses, though this is as yet unproven. The plateau region is absent from T_g versus composition plots in several other $ZnCl_2$ + organic cation chloride systems that have been studied (Hodge, 1974).

3. Mixed Halide Systems

Solutions of $ZnCl_2$ with KI rather than with KCl prove to be glass forming over wide composition regions (Schultz, 1957; Wong, 1970; Easteal and O'Rourke, 1974). This system has been studied in some detail by the latter authors, and a liquid–liquid immiscibility dome at $ZnCl_2$-rich compositions has been identified. The phase diagram is shown in Fig. 41. The top of the dome occurs just below the stable liquidus line.

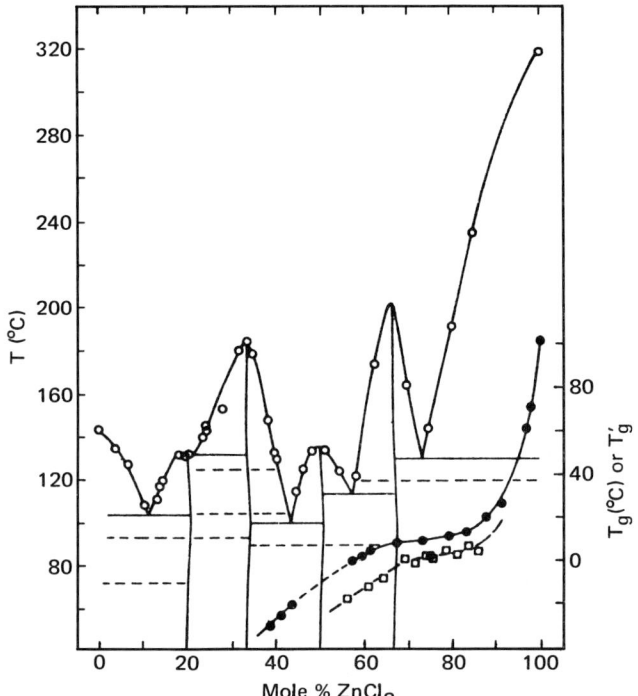

FIG. 40. Phase diagram: $ZnCl_2$ + PyHCl. (○), experimental liquidus temperatures; (--) solid-state transitions; (●) values of T_g for $ZnCl_2$ + PyHCl; (□) values of T_g' scaled down by 42% for SiO_2 + Na_2O (Easteal and Angell, 1970).

4. *Other Zinc-Chloride-Based Glasses*

Wide glass-forming composition ranges are found when nonhalide second components are added to $ZnCl_2$. Schultz (1957) discussed glasses in the system $ZnCl_2$ + K_2SO_4, and Angell *et al.* (1969) described the far-IR spectra of glasses in the system $ZnCl_2$ + KNO_3.

J. Bismuth-Chloride-Based Glasses

Although $ZnCl_2$ can itself be simply obtained in the glassy state, it does not generate a class of binary and ternary chloride glasses. In this respect it differs from its tetrahedrally coordinated halide relative BeF_2. Binary inorganic chloride glass can, however, be obtained using bismuth chloride ($BiCl_3$) as progenitor. While $BiCl_3$ itself is not glass forming, an extensive region of glass formation in the system $BiCl_3$ + KCl was identified some twenty years ago by Topol *et al.* (1960), but has not until recently been recognized and exploited. The glass-forming region, 25–45% KCl, can

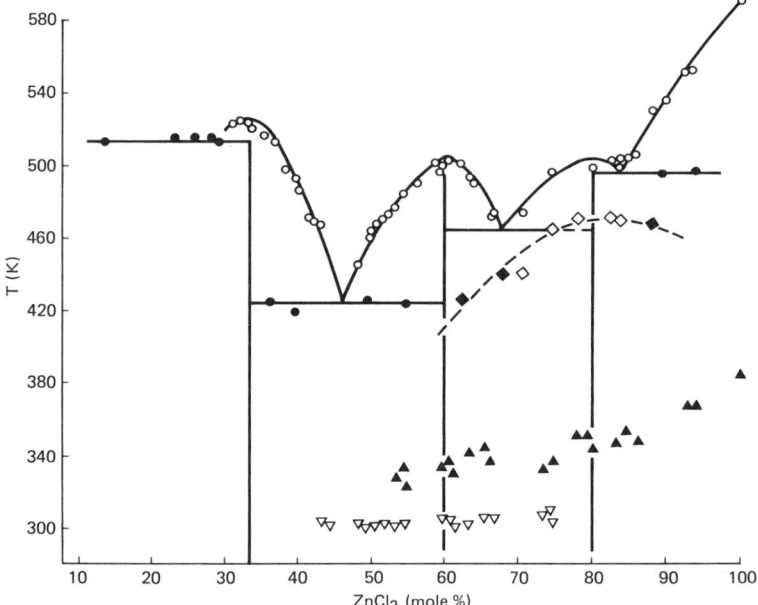

FIG. 41. Phase diagram for the $ZnCl_2$ + KI system: (○) liquidus temperatures; (●) eutectic temperatures; (--) postulated metastable immiscibility dome; (◇, ◆) two independent sets of observed phase separation temperatures; (▲, △) two apparently self-consistent sets of experimental glass transition temperatures (Easteal et al., 1974).

presumably be expanded by appropriate additions of other chlorides, such as $PbCl_2$. The glasses, which have T_g values in the range 20–40°C, are of considerable potential interest because of the low maximum phonon frequencies (~300 cm^{-1}) and the small competition for ligands offered to ions such as Nd^{3+}. Ions like the latter consequently enjoy well-defined coordination shells and consequently exhibit narrow absorption and emission lines. In fact, recent tests have shown (Weber et al., 1982) that the fluorescence cross section for Nd^{3+} in these glasses is considerably larger than that in any other glass studied to date. Detailed study of the physical properties of these glasses is now in progress.

XI. Fluoride and Oxyhalide Glasses

A. Fluoride Glasses

A large number of fluoride glasses exist in systems containing BeF_2, including BeF_2 itself (Goldschmidt, 1926; Mackenzie, 1976). BeF_2 has been conceived as a "weakened model" of SiO_2 (Goldschmidt, 1926), with ionic radii of Be and Si as well as F and O similar, with valencies halved (Si = +4, Be = +2, O = −2, F = −1). Indeed, crystalline BeF_2 shows a structure, polymorphism, and anomalous low-T heat capacity similar to that of SiO_2 (Roy et al., 1950, 1953; Leadbetter and Wright, 1970, 1972; Leadbetter and Wycherly, 1971; Zarzycki, 1971, Pelizzari, 1974) with lower density, refractive index, and melting point (540°C), and higher solubility. The Be–F distance is 1.43 Å in both glassy and crystalline BeF_2 (Batsanova et al., 1968). Yet there are considerable differences between SiO_2 and BeF_2 (Wright, 1969), as revealed by neutron scattering (Leadbetter and Wright, 1970). Ionic bonding is 80% in BeF_2 and 50% in SiO_2. BeF_2 may contain cristobalite and quartzlike regions that might separate (Gilev and Petrovski, 1968).

Complex fluoride glasses correspond to silicate glasses as indicated for their components: LiF–MgO; NaF–CaO; KF–SrO; MgF_2–TiO_2; CAF_2–ThO_2, ZrO_2; or for one complex sample: Na_2BeF_4–Ca_2SiO_4.

Fluoride glasses are particularly interesting to the designers of optical materials because of their low refractive index. Pure BeF_2 glass, under strict control of OH content, was prepared by Cline and Weber (1977), Baldwin and Mackenzie (1979), and Baldwin et al. (1981).*

Beryllium fluoride is toxic. Therefore, attempts were made early to find and expand fluoride systems free of BeF_2 (Enright and Marshall, 1957; Heyne, 1933). In these glasses AlF_3 is important. Neither AlF_3 nor alkali fluorides fit into the weakened model concept. AlF_3 uniquely may contribute (AlF_4) groups, allowing the novel combinations $AlLiF_4$ substituting for BeF_4. AlF_3 in this way contributes strongly to higher softening point and lower expansion in complex glasses (Krylova, 1967). An early example of a high-AlF_3 glass (Sun, 1949) is AlF_3 40; PbF_2 24; SrF_2 12; MgF_2 24. Phase separation has been observed in several fluoride glasses (Vogel et al., 1958).

Because a low refractive index and dispersion are the most important parameters for the desired low nonlinearity in laser glasses, fluoride glasses are the most promising candidate hosts (Weber, 1976) (Table XXXII) (Deutschbein, 1967; Deutschbein and Pautrat, 1967). Examples of base glasses for Nd^{3+} are (1) BeF_2 47; KF 24; CaF_2 14; AlF_3 10; ZrF_4 50; BaF_2 25 (Poulain et al., 1975); (2) ZrF_4 60; BaF_2 34; NdF_3 6 (Lucas et al., 1978); (3) ZrF_4–ThF_4–LaF_3–BaF_2–NaF–CaF_2 (Poulain and Lucas, 1978; Poulain et al., 1979; Matecki et al., 1978; Lecoq and Poulain, 1979; Leroy et al., 1978) (containing the important network-former ZrF_4; in these glasses electrical conduction is by F (Chandrashekhar and Shaper, 1980); (4) with Ag, Fe, Cr, V (Miranday et al., 1979); (5) with HfF_4 (transmitting farther out in the

---

* This is an excellent detailed review on halide glasses.

TABLE XXXII

Comparison of Merits for Laser Glasses[a]

Type of performance	Si–O	P–O	F–P–O	Be–F
Nonlinearity of index ($\times 10^{-13}$) (esu)	1.4	1.1	0.7	0.4
Cross section ($\times 10^{20}$) (cm²)	2.9	4.1	2.5	2.9
Relative gain coefficient	1	0.94	0.88	0.77
Relative figure of merit for rods	1	1.8	1.5	2.7
Relative figure of merit for disks	1	1.2	1.8	2.6

[a] After Weber (1976).

IR and suitable for fibers) (Drexhage et al., 1980); and (6) AlF_3–BaF_2, YF_3, YbF_3, SrF_2, MgF_2, NaF_2 (Videau et al., 1979a). A small amount of BeF_2 greatly increases resistance toward HF and gaseous F.

Cross sections in general vary from 1.6 to 4, lifetimes from 460 to 1030 μsec (Weber, 1979). The 8–9-coordination of Nd^{3+} applicable to silicate laser glasses is no longer valid in fluoride glasses. A suitable crystal field model was proposed by Brecher and Risenberg (1979). For pure BeF_2, a more complex model was required. Site-to-site differences may cause hole burning in laser glasses as well as reduced energy extraction. Laser-induced fluorescence spectroscopy (line narrowing as well as site selection) is used to assess such inhomogeneity and has become an invaluable and unique probe of fine structure (Weber, 1979; Brecher and Risenberg, 1979).

A complete bibliography on fluoroberyllate glasses was published by Cline and Kingman (1976).

BeF_2–$BeCl_2$ glasses were found by Petrovski et al. (1965).

B. Halides in Oxide Glasses

Fluoride–oxide glasses exist over wide ranges.

1. Phosphorous Pentoxide Systems

Systems of considerable practical importance contain P_2O_5 as the most important oxide and have been divided conveniently into three classes (Jahn, 1961):

(1) *Oxyfluoride glasses* containing BeF_2 with relatively little PO_4 (Jahn, 1961; Evstrop'ev et al., 1971).

(2) *Phosphate-bearing fluoride glasses free of* BeF_2, usually with high BaF_2 (Jahn, 1961).
(3) *Fluorophosphate glasses free of* BeF_2 (Khalilev *et al.*, 1975).

Much F can be introduced into stable $Al(PO_3)_3$ glasses if $R^{II}(PO_3)_2$, $R^{I}PO_3$, and AlF_3 are co-introduced (Jahn, 1961; Veksler *et al.*, 1976). F/P ratios from 4.8 to 26.1 were realized. Fluorophosphate glasses are promising laser host materials combining the high gain of phosphate glasses and the low dispersion and refractive index—advantageous for avoiding nonlinear effects—of fluoride glasses (Weber, 1976) (see Section XI.A). More recent studies of fluorophosphate glasses mostly aimed at laser host materials include those by Khalilev *et al.* (1975), Vrtanesyan and Khalilev (1977), Zatsepin *et al.* (1977), Veksler *et al.* (1978), Evstropiev and Nanor (1978), Vrtanesyan (1978), Videau *et al.* (1979b) (glasses containing Fe), Menil *et al.* (1979) (glasses containing Fe). P–F bonds were identified (Veksler *et al.*, 1978).

The fluorophosphate laser glasses such as those used at the NOVA laser fusion installation at Livermore represent a useful balance between (minimal) nonlinearity of refractive index and (maximal) cross section (Stokowski *et al.*, 1979). The ranges are 2.2 to about 4.5 for cross section, 310 to 570 μsec for lifetime (Weber, 1979). Glass is procured in large sizes from HOYA, Owens-Illinois, and Schott (Weber, 1980; Weber and Almeida, 1981), in part based on original experiments at the National Bureau of Standards. In these glasses, as in fluoride glasses, Nd^{3+} tends to have more than one site (Brecher and Risenberg, 1979). A typical composition is (mole %) 4 $Al(PO_3)_3$; 36 AlF_3; 10 MgF_2; 30 CaF_2 (LiF,NaF); 10 SrF_2; 10 BaF_2.

The cross section of about 3 compares to 4 for phosphate and 2.7 for silicate glass. The lifetime 495 μsec compares to 338 μsec for phosphate and 359 μsec for silicate glasses. In Russia, equivalent glasses are produced starting from systematic studies of $BaPO_3F$ unit structures (Vrtanesyan, 1978) with optimal properties reported to approach those of crystalline CaF_2 [$n = 1.43658$ (CaF_2, 1.4338), dispersion 95.7 (CaF_2, 95.3)] (Leidtorp and Petrovskii, 1980). For fiberization, HfF_4 is a preferred constituent. It transmits at longer wavelength; and a lower multiphonon loss compensates for higher scattering losses (Drexhage *et al.*, 1980). Fluorophosphate glasses were also applied to Faraday rotators (Myers, 1978).

2. Silicate Glasses

Infrared spectra suggest that $SiO_{4-x}F_x$ tetrahedra abound only in alkali silicate glasses containing K, Rb, and Cs, not Na or Li (Takusagawa, 1980).

a. Fluoroborosilicate Glasses. These glasses are used as low-melting sealers (Hoffmann, 1963).

b. Chlorine and Bromine in Borosilicate Glasses. While in some silicate glasses less than 3.5% Cl or Br is accepted, up to 15% is accepted in borosilicate glasses according to Polukhin *et al.* (1979).

c. Fluoroaluminosilicates. These are discussed in Section VII.E. They compromise compositions in the low-refraction "fluor crowns" of the optical glass industry.

3. Phosphate Glasses Containing Chlorine or Bromine

In metaphosphate glasses more than 30% $ZnCl_2$ can be accommodated (Polukhin *et al.*, 1979; Weber *et al.*, 1981). In such glasses doped with Nd^{3+} the highest cross section for laser emission (5.4×10^{-20} cm^2) in any glass has been achieved (Weber and Almeida, 1981). Compare 5.1 for potassium tellurite glass and 4.5 for fluorophosphate glass.

4. Fluorine in Sodium Calcium Silicate Glasses, Opal Glasses

While small amounts (less than 3%) of fluorine are used to lower the viscosity of sodium calcium silicate melts, higher concentrations cause the precipitation of crystalline fluorides, resulting in the formation of *opal glasses*. In the formulation of such glasses, opacifying phases are desired to have high solubility at high temperatures, and low solubility at low temperatures, to permit the control of light scattering properties by quenching and reheating cycles. Incompatibility of the cations in the network is thought to ease low temperature precipitation, such as in the case of Ca, Sr, Ba (Rothwell, 1956). The cubic forms of the most useful precipitates NaF_2, CaF_2, BaF_2, SrF_2 and some double fluorides are characteristic (Volf, 1961). The critical parameters of control are the refractive indexes of precipitate and glass, particle size, and separation. The critical parameters of the opal glass product are light transmission, scattering in angular dependence, and reflectivity, in different relations for different purposes.

The cations tending to have low oxygen coordination, on the other hand, such as Zn^{2+} (Stanworth, 1948), Pb^{2+}, Cd^{2+}, Mg^{2+}, and Be^{2+} (Callow, 1952), tend to remain in the vitreous phase. Increasing fluoride precipitation occurs in the series of increasing field strength: K^+, Ba^{2+}, Na^+, Ca^{2+}, etc. (Rothwell, 1956), except for cations capable of 4-coordination (Zn, Mg, and Be). Other opacifiers should be mentioned: phosphates (Das, 1967) (such as calcium fluoroapatite), ZrO_2, SnO_2, and TiO_2. In *enamels* part of the opacifier is added in powder form with clay to a glass frit for a brief (3 min) high-temperature reaction.

5. Fluoroborate Glasses

The system B_2O_3–NaF was investigated thoroughly by Maya (1977). Alkali borate glasses containing Cl are now being studied more extensively because of their relationship to the more complex photochromic glasses (Araujo, 1980). At the same time, an important finding is that more Li^+ can be introduced into borate glasses in the presence of Cl (and/or F). The high conductivity in high-Li^+ borate glasses is of interest in the search for solid electrolytes for advanced batteries. Considerable amounts of Cl are accepted, reducing significantly the activation energy of conduction (Button *et al.*, 1980).

6. Halogens in Lead Silicate Glasses

As much as almost 30 mole % halogen (F, Cl, Br, or I) can be accommodated in lead silicate glasses. In these glasses electrical resistance decreases appreciably with the addition of halogen and with decreasing halogen size (from I to F). The conductivity is anionic (Schultz and Mizzoni, 1973).

7. Silver Iodide–Oxide Glasses

Among solid electrolytes, a subgroup based on the introduction of AgI (and in part Ag_2SO_4) possesses interesting properties and can be obtained in the vitreous state by fast quenching (Lazzari *et al.*, 1978; Takuma and Minami, 1977; Minami, 1980). Glass formation and maximum conductivity are observed near the eutectics. Electronic conduction is usually less than ionic. For more detail on solid electrolytes containing Ag consult the appropriate oxide glass sections.

8. Other Glasses Containing Halides

Glasses were explored in the systems TeO_2–$ZnCl_2$(NaCl (Br,F), KCl(Br,F), CsCl)–WoO_3–BaO–Na_2O (Yakhkind, 1980).

XII. Chalcogenide Glasses

A. Introduction

1. General Considerations

The name chalcogen refers to the elements S, Se, and Te. These elements and many of their combinations with other elements form a vast class of nonoxide glasses (Rawson, 1967; Pearson, 1964; Dembovskii, 1969). Unlike oxygen, which is a gas at the temperature of synthesis, the chalcogens are solid or liquid and, therefore, can combine with other elements to glasses in wide ranges of proportion, while oxide glasses are essentially composed of stoichiometric oxides (Rawson, 1967). Unlike oxygen, which in the liquid state consists of O_2 molecules that crystallize on cooling, S, Se, and Te form elemental glasses containing disordered rings or chains. The structures of this large class of glasses are based on rings, chains, or three-dimensional networks as well as combinations of such structural groupings. The classical concept of network formers and network modifiers in oxide glasses thus can not be fruitfully transferred to them (Rawson, 1967); however, certain substitutes, e.g., Tl, Ag, Cl, and I can be considered depolymerizing, or ring, chain, and network breaking (Pearson, 1960, 1964, Flaschen et al., 1959) and present some analogy to network modifiers. As in oxide glasses, large areas of liquid–liquid phase separation have been observed.

The position of other elements participating in complex chalcogenide glass has been analyzed by Klemm and Niermann (1963) and by Krebs (1958), whose views are summarized by Rawson (1967). Often melts are more metallic and covalent than crystals, increasingly so at the higher temperatures from which they can be quenched to glasses, some of which are semiconductors. It is now believed that the anomaly of specific heat at low temperatures is a general characteristic of the vitreous state, so that chalcogenide glasses may resemble SiO_2 glass in this respect. For instance, the specific heat of 4 As–15 Ge–81 Te glass can be described by the general relation $\gamma T + AT^3$ at the lowest temperatures, with $\gamma = 0.027$ J/mole K (γ generally being 0.02–0.1), while at somewhat higher temperatures $(C_p - \gamma T)/T^3$ rises more rapidly (Jirmanus and Gerber, 1978).

2. Electronic States in Chalcogenide Glasses

The constitution of chalcogenide glasses has begun to be related quite successfully to electronic structure and electrical properties. Mott (1967) and Cohen et al. (1969) were among the first to modify the classical band model of crystalline solids for the complex, nonrepeating arrangements in noncrystalline solids (amorphous semiconductors). This early modification is generally referred to as to the MOTT-CFO model and has since then been the subject of various critiques and amendments (Emin et al., 1972; Leipold and Feuchtwang, 1976). It had become clear that there must exist in any transparent material whether crystalline or not a band gap that does *not* depend on order as in classical Bloch theory (Mott, 1967). Hundreds of papers on amorphous semiconductors have appeared since 1970. The general glass technologist finds an excellent introduction to amorphous semiconductors in Adler's (1971, 1980), and Cohen's (1971) reviews. Mott

(1978), who eventually received the Nobel prize, gave a fascinating, very personal account of this entire development.

The most importamt feature of these models for the generalist in glass technology is the substitution for the empty gap of states in the crystalline solid of a gap with nonzero density of states that are "localized" and have a *low mobility* in noncrystalline solids. In this connection one speaks of a "mobility gap" and a "mobility edge" (Cohen *et al.*, 1969) (Fig. 42). Localized states have a wave function extending over limited distances.

The electronic properties of the chalcogenide glasses were found to be qualitatively different (Emin *et al.*, 1972) from those of the high-mobility classical semiconductors such as silicon. The results of careful studies of such electronic properties as conductivity, thermoelectric power, and the Hall effect fit best models in which the predominant charge carriers are localized. The mobility is very low, much below 1 cm^2/V sec, and increases with temperature. Transport is between near neighbors assisted by phonons "hopping." Anomalies in the Hall effect sign are, in particular, experimental evidence for such transport characteristics, and high-mobility models may be considered discarded.

Anderson (1975) suggested that most localized states may be doubly occupied because electron–lattice interaction will outweigh columbic repulsion. His model was particularly appealing in the case of chalcogenide glasses and seemed to explain the absence of spin resonance signals.

The Mott–Davis–Street model (Mott *et al.*, 1975; Street and Mott, 1975), elaborating on Anderson, as found consistent with photoluminescence, photoconductivity, drift mobility, field effects, etc., in chalcogenide glasses. It was first considered for As_2Se_3 glass. The model is dependent on the acceptance of a rather large concentration of defects (10^{18}–$10^{19}/cm^3$) serving as charge traps.

The defects were described as dangling bonds imposed by constraints in the local disordered topography of the glass structure and can be occupied by 0, 1, and 2 electrons. The three variants were called D^+, D^0, and D^-,

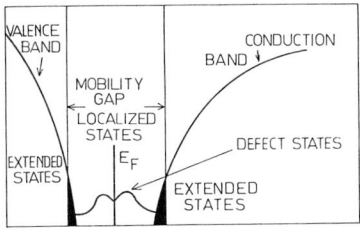

FIG. 42. The mobility gap (E_F = Fermi energy) (Tauc, 1976).

respectively, to indicate net charge. The exothermic character of the reaction $D^0 \rightarrow D^+ + D^-$ supposedly favors D^+ and D^-, relegating D^0 to excitation processes.

Dangling bonds (D centers) are illustrated schematically in Fig. 43 (Tauc, 1976). In chalcogenide glasses long-pair orbitals exist in the upper valence band; bonding states are deeper inside. Antibonding states are in the conducting band.

D^+ reacts strongly with the lone-pair bonding, distorting the environment. The antibonding orbital in the conduction band is formed by an extra electron with D^+. The two electrons of D^- cannot form a covalent bond with its neighbor but may occupy a valence-bandlike lone pair. D^0 is intermediate.

Effects of doping As_2Se_3 with Ag were explained by this model (Patel *et al.*, 1978).

A later model, designated the valence alternate-pair model (VAP) by Kastner *et al.* (1976), had more immediate appeal to the chemist working with chalcogenide glasses, because it assumed plausible structural errors in a typical chalcogenide chain or network. For instance, in As–Se glasses, Se may occur with 3 or 1 instead of 2 coordination, As with 2 or 4 instead of 3 coordination. The interaction between lone-pair electrons on chalcogenide atoms and their surroundings is described by the existence of C^- (singly coordinated chalcogen) and C_3^+ (triply coordinated chalcogen), due to a reaction $2C_2^0 \rightarrow C_3^+ + C_1^-$. A similar process is described for As. Fritzsche (1977; Fritzsche and Kastner, 1978; Fritzsche *et al.*, 1978) gave a lucid discussion of these models and their implication for doping. Ovshinsky (1977) discusses these effects under the name "deviant electronic configurations."

Kastner (1977, 1978) supplemented the VAP model by the concept of intimate valence alternate pairs (IVAP), the potential consequence of the approach (energetically favored) of C_3^+ and C^-. In terms of the glass chemist this would mean a preference for the close vicinity of a bridge Se–Se and a terminal Se in the case of a Se chain. Bishop *et al.* (1977)

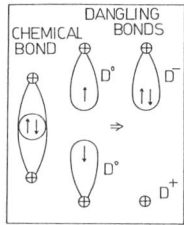

FIG. 43. Dangling bonds (Tauc, 1976). Two (neutral) dangling bonds (D^0) are generated from a chemical bond. Charged states D^+, D^- are formed if the reaction is exothermal.

interpreted photoluminescence on this basis. The localization of the defects was attempted by doping with Tl^+ and interpreting NMR signals (Jellison, 1977; Jellison and Bray, 1978a,b; Jellison *et al.*, 1977).

Yet the postulate of such high concentrations of defects suggests a different interpretation. If defects were the dominant cause of the low-mobility localized transport, methods of preparation should make much more difference than they do. On this basis Emin *et al.* (1972; Emin, 1977) prefer to attribute most of the transport to intrinsic sites. In their view, there is a pervasive strong interaction of electronic charges with atomic displacements, producing intrinsic localization ("small polarons"). In this respect chalcogenide glasses should be considered to contrast with amorphous silicon. One may say (Emin, 1980) that with their, on the average, lower coordination they are "softer" and less likely to form dangling bonds or show radical effects of doping. Defect concentration, however, might also be associated with microstructure (granularity). What is generally acknowledged is the important role of self-trapping in these noncrystalline solids.

3. Switching

At a specific voltage certain chalcogenide (and other) glasses can be "switched" to a high-conduction state generated by "channels" in which the voltage induces a structural change (Kolomiets, 1963; Ovshinsky, 1968). If the process is irreversible, one speaks of *memory switching*. In this case, crystallization may be observed. The memory can be destroyed by a short high-current pulse capable of remelting the structure. The crystallization in the filament was described in its complexity by Cohen *et al.* (1972), and, with some modification, in greater detail by Steventon (1976) and Esqueda and Henisch (1976). In some of these glasses, short-range order is quite closely related to that of corresponding crystals, e.g., in As_2S_3 (Se_3) or GeS_2 (Se_2). But this is not always the case, e.g., in As_2Te_3. Many compositions have, therefore, been proposed for different uses.

While relations to applications will be treated extensively in the volume on electrical properties, it may be of interest at this point to list names and glass composition fields of some candidate switching glasses under consideration in 1980: STAG, Si–Ge–As–Te; ECD, Ge–Sb–S–Te, As_2Te_3, $Ge_{15}Te_{81}Sb_2S_2$; SPAT, $(SiAsTe)_{95}P_5$; GAST, $Ge_{33}As_{17}S_{15}Te_{35}$, $Si_{12}Ge_{10}As_{30}Te_{48}$. Many theoretical and experimental studies are concerned with the electronic and/or thermal origin of the switching phenomenon. At least for thin layers, some evidence supports a purely electronic origin (see, e.g., Platakis, 1978). Even the crystalline nature of the *on* state of switching has not been ascertained in all cases.

Chalcogenide switching devices must compete fiercely (Tauc, 1976) with established, economical silicon technology. In the related data-processing field, silicon technology might be matched in special areas ("read-mostly" memories, e.g., in blood analyzers). But metal oxide semiconductors are competitive.

4. Other Applications

Selenium and, in specific or future applications, selenide glasses are the prototype materials for *photocopiers* (e.g., Xerox). More generally, *"dry" photography* is a promising application field with chalcogenide glasses excelling in contrast, resolution, and service stability (Tauc, 1976). Applications are specifically in the graphic arts, including microfilms, particularly when special wavelength sensitivity or contrast variability is required. Although the general level of sensitivity has so far been found low in chalcogenide glasses, active research continues.

Chalcogenide glasses are leading candidates as materials for *infrared optics*. Since larger ions with lower field strength participate in lower energy vibrations, good glasses transmitting to 15 μm have been obtained (Hilton, 1965, 1966; Hilton and Brau, 1963, 1968; Hilton and Jones, 1968; Hilton et al., 1964a,b, 1966a,b,c, 1974, 1975).

Because of a combination of high optical refractivity, low acoustical loss, low sound velocity, and high acoustic-optical effect, chalcogenide glasses are candidates for *sonic* uses (Krause et al., 1970).

B. SULFUR

Sulfur crystallizes in several allotropic forms, characterized by different stacking of S_8 rings (Prins, 1960; Rawson, 1967). The high-temperature form melts at 115°C to a liquid consisting of monomeric S_8 rings; at 160°C this liquid stiffens (10^3 P) to one consisting of polymeric chains (Myers and Felty, 1967) of 10^5–10^6 atoms. This viscous liquid can be quenched from above 180°C to a glass with $T_g = -27°C$ in which S_∞ chains persist with lengths to 10^5–10^6 atoms. Above 180°C the liquid contains a mixture of S_8 rings and S_∞ chains (Tobolsky et al., 1962a,b, 1964; Keezer and Balley, 1967; Massen et al., 1964; Lucovsky, 1969; Schenk, 1957; Myers and Felty, 1967). Above 300°C oriented fibers can be drawn that crystallize into patterns of winding chains with small stacks of S_8 rings tucked in. S_∞ chains can be broken by small concentrations of monovalent ions such as I^- (Rawson, 1967), resulting in a spectacular decrease in viscosity, or cross-linked by high-valency ions such as P^{5+}, increasing mechanical stability.

C. Selenium

1. Structural Principles

Like S, Se can form rings or chains, but Se_8 rings are not stable. The only crystalline form stable below the melting point (220°C) is the hexagonal metallic Se, whose atoms are arranged in spirals projecting normal to their axes as equilateral triangles (Bradley, 1924). In the melt—40% monomeric near the melting point—the average chain length is 10,000 (Eisenberg and Tobolsky, 1960) (for S, 60% monomeric, length is 368,000). Rings occur in the crystalline metastable monoclinic α and β forms, which resemble each other. In all 3 crystalline forms, the covalent bond lengths are about 2.35 Å, and the angle 103° is distorted by only +4° in the metastable forms. Van der Waals bond lengths between chains in trigonal and molecules in monoclinic crystals are about 3.5 Å. Placed between S and Te, Se resembles Te in its chains, S in its rings.

The Se melt may be quenched to a glass more easily than that of S (Rawson, 1967), doubtless because of the relative instability of Se_8 rings. The glass transition is at about 30°C (S − 27°C).

The glass structure contains rings as well as chains (Tobolsky, 1962a,b, 1964; and Keezer and Bally, 1967). Rings have been established by means of IR and Raman spectra (Zallen et al., 1971; Lucovsky, 1969; Kaplow et al., 1968), and spirals through viscosity studies (Keezer and Bally, 1967). Also, the insolubility of chains in CS_2 gives a clue (Keezer and Bally, 1967). Monte Carlo calculations lead to an estimate of 40% chains in sputtered films (Rechtin and Auerbach, 1973a,b). It is surprising to have Kaplow et al. (1968) in their thorough IR, optical, and x-ray study conclude that there are over 95% rings in the glass they studied. However, x-ray data might be interpreted as distorted spirals as well as rings (Henninger and Busher, 1967; Chang and Dove, 1974). Se–Se second peaks suggest that arrangements of rings are similar to the puckered rings in monoclinic (α) Se, and least like the chains in trigonal Se. Apparently, chains in vitreous Se are separated from each other like the rings in monoclinic (α) Se. Connell and Lucovsky (1978) stress that the comparison of spectra and models does not permit one to establish their connectivity. Not only does this leave the ratio of ring to spiral indeterminate, but this ratio depends on the process of fabrication (Rechtin and Auerbach, 1973a,b). More recently, Axmann et al. (1970) undertook neutron diffraction studies. Richter (1972) distinguishes three varieties (I, II, and III) of amorphous Se.

Large relaxation effects below T_g are due to the ring–chain equilibrium (Stephens, 1976). The activation energy corresponds to that of ring breaking in liquid Se. Large differences in the equilibrium observed for different methods of synthesis are probably often associated with small concentrations of impurity.

Chain length and chain–ring ratio affect properties that therefore depend strongly on melting history. Aging a Se glass, originally quenched from above T_g, at room temperature causes the moduli G, B, and E to increase and Poisson's ratio to decrease as the packing of Se atoms increases (Etienne et al., 1979). The process reverses as the temperature of treatment is raised toward T_g. The low absolute value of all moduli and the high B–G ratio suggest a large contribution of van der Waals forces. The dependence on melting temperature of chain length and chain–ring ratio is diminished by ultrasonic (22 kHz) treatment, which therefore allows one to control crystallization (as do applied fields) (Popov, 1978). For instance, the activation energy (in kilocalories per mole) of crystal growth varies with melting temperature and ultrasonic treatment as shown in the accompanying tabulation.

	Melt temperature (°C)			
	250	300	350	400
No ultrasonic treatment	16.2	13.5	12.5	8.3
Ultrasonic treatment	10.2	10.6	9.1	8.0

2. Doped Selenium

Monovalent donors and acceptors may significantly change the properties of vitreous Se (Ohsaka, 1975), e.g., reduce viscosity and enhance crystallization (Krebs, 1961) by reducing chain length (Krebs, 1958). Important examples are Na^+, Tl^+, and Cl^- (Rawson, 1967; Keezer and Bally, 1967). All chain ends may, for instance, be occupied by Tl^+ (Keezer and Bally, 1967). Increased doping, e.g., with Tl, may produce immiscibility (Hansen, 1958). The structure of vitreous TlSe (Zirke et al., 1977) resembles that of crystalline TlSe, exhibiting $(TlSe_2)_n^-$ chains. Oxygen in minute quantities (~1 ppm) introduces acceptor levels. If one eliminates all oxygen, an n-type thermoelectric effect is found instead of the p-effect in nominally pure Se. But those acceptor levels are easily compensated by other impurities. Conductivity increases 10^6-fold with 50 ppm O_2 (Lacourse et al., 1970); photoresponse is equally affected by O impurities and "getters" (McMillan and Shutov, 1977). Satisfactory predictions of T_g for binaries Se–Cl, S, Te, As, Sb, Bi, Si, Ge, and Sn can be based on average bond coordination (Berkes and White, 1977). Copolymers with organics (4,4'-diselenoazobenzine) were reported by Herrmann *et al.* (1978).

The ions As^{3+} and P^{3+} can substitute in the chain and triple its average length (Krebs, 1961), with the result that the viscosity increases and the glass becomes more stable. Eastman Kodak, around 1950, produced the first lenses for IR optics from As-stabilized Se glass.

Tellurium, with its lesser bond strength, as a dopant diminishes the viscosity of Se and its stability as a glass.

3. Copolymerization of Selenium

Selenium forms copolymers with S (Wood, 1968; Myers and Felty, 1967). A theory of the relation of monomer and polymer in such copolymers was developed by Tobolsky *et al.* (1962, 1964), explaining the dependence on temperature and S–Se ratio. With increasing S content the concentration of monomer increases and T_g shows the corresponding dependence on S content (plasticizing).

In the continuous series (Grison, 1951) of mixed-chain solid solutions of Se–Te, as many Se–Te bonds as possible form; they are stronger than the average of Se–Se and Te–Te bonds (Boolchand and Sukanyi, 1973). Changes of bonding with decreasing temperature or liquid Se–Te alloys have been investigated by Thurn and Ruska (1976).

In Se–Te amorphous alloys, microhardness and T_g increase with annealing time as inhomogeneities decrease, and they increase with Te content because molecular weight and/or chain concentration increases (Das *et al.*, 1972). Conductivity (at room temperature) increases; activation energy of conductivity decreases with Te content (Hulls and McMillan, 1974). For S–Se–Te copolymers see Suvarova and Skhol'nikov, 1976; Ohsaka, 1976. In these copolymers at high Se concentration, Se_3S_5 and Se_5Te_3 mixed rings were identified by IR spectra (Ohsaka, 1975).

4. Electrical Properties of Selenium Glass

The dc conductivity of purest Se is small ($\sim 10^{-17}$ Ω/cm) and thus difficult to measure. It is quite insensitive against most impurities, except oxygen. Doping with 50 ppm oxygen leads to an increase to 10^{-11} Ω/cm.

Conduction is mostly via the valence band (delocalization). The mobility of free holes dominates and is of the order of 20 cm²/V/sec (Adler, 1971). The mobility gap is of the order of 2 eV, equal to the gap of states. This must be connected with the similarity in short-range order of spirals and rings (no large positional fluctuations) (Adler, 1971). The concentration of localized states (10^{26}/cm³ eV) near the edges is hardly attributable to disorder; rather, it is attributable entirely to impurities and defects.

In ac conduction, one observes hopping, most likely via acceptors (density of states $10^{18}/cm^3$ eV).

5. Optical Properties of Selenium

The optical absorption edge is at 1.8 eV, according to numerous reproducible measurements. Photoconductivity is small, however, surpassing dark conductivity by so many orders of magnitude that it is of great technological importance (e.g., in photocopying). The large (0.6 eV) difference in the edges has therefore been well known for a long time. An extensive interpretation is by Lucovsky (1970). It is based on the postulate of Frenkel excitons, which cannot contribute to conduction. For the structure of the glass, it is significant that these excitons are excited states in the *ring*. Crystalline forms having the rings also exhibit this non-photoconductive edge (Adler, 1971).

Panchromatic response in xerography is improved by doping with (e.g., 3 at. %) Bi at the expense of absolute sensitivity. Se(3% Bi) glasses have a higher T_g (55°C versus 42°C). They are made by evaporating 100–1000-μm powder from a cooled sheet in a quartz crucible at 500–600°C on a 52°C Al substrate (Schottmiller *et al.*, 1969).

D. TELLURIUM

Unlike S and Se, Te occurs only in one crystalline form, trigonal Te, which consists of infinite spiral chains in parallel orientation (Bragg and Claringbull, 1965) and melts at 453°C. At and 10°C above the melting point, the chain structure persists, resulting in high viscosity; but at higher temperatures the chains break and the low-viscosity, metallic melt cannot be quenched to a glass (Moffat *et al.*, 1964). Presumably, quenching from near the melting point could result in glass formation. Amorphous layers have been made from vapor (Keller *et al.*, 1965; El Mouly, 1966), but recrystallize around 25°C (Rawson, 1967; Keller and Stoke, 1965). Structural investigations by neutron-spectroscopy were made by Axmann *et al.* (1970). In thin films Ichikawa and Ogawa (1972) determined chain structure. The behavior of Te is indicative of stronger interchain bonding than in either S or Se.

At T_g the electrical conductivity increases 10^3-fold (Keller and Stoke, 1965), then shows no strong T dependence (Sakurai and Munesul, 1951) (hole conduction). The mobility is 10^{-2} cm²/V sec (Spear, 1957, 1960). But Keller and Stoke (1965) find strong T dependence in the dark. The band gap of the amorphous phase is 0.87 eV (that of the crystalline phase is 0.32 eV). The absorption edge is higher by 0.4 eV than in the crystal. Photoconductivity is weak, at 1.2 eV (Keller and Stoke, 1965). It is likely that the amorphous phase is more cross-linked. At low T, conduction is impurity controlled. Details are found in Keller and Stoke (1965). (See also Suhrmann and Berndt, 1940.)

The effect of the addition of Se and S was reported by Suvarova *et al.* (1974; Suvarova and Skhol'nikov, 1976), that of Se and P by Kasatkin *et al.* (1974), that of Br and I by Shevchik and Kniep (1974). With Al, a three-dimensional network forms, similar to that in crystalline $AlTe_3$, with excess Te forming chains that may phase-separate (d'Anjou and Sanz, 1978).

E. Arsenic and Antimony Chalcogenide Glasses

1. Glass Formation and Structure

The compounds As_2S_3 and As_2Se_3 form glasses easily with structures closely related to the structure of crystalline As_2S_3 (orpiment) (Fig. 44) (As_2S_3: Schultz-Sellak, 1870; Frerichs, 1950; Petz *et al.*, 1961, Hopkins *et al.*, 1967, Bishop *et al.*, 1977; Leadbetter, 1974; Leadbetter and Apling, 1974; As_2Se_3: Vaipolin and Porai-Koshits, 1963; Vaipolin, 1966; Bishop *et al.*, 1977). X-ray data (Tsuchihashi and Kawamoto, 1969; Apling, 1973; Leadbetter, 1974; Leadbetter and Apling, 1974) suggest that the orpiment layer structure of the As_2S_3 glass retains short-range order but is heavily distorted. The layers are more widely separated and persist over a greater radial distance than intralayer correlations, with a greater range of bond length and angles. The signal from the 12-member ring structure in the crystal is absent. Bonds are predominantly covalent (Leadbetter, 1974; Leadbetter and Apling, 1974).

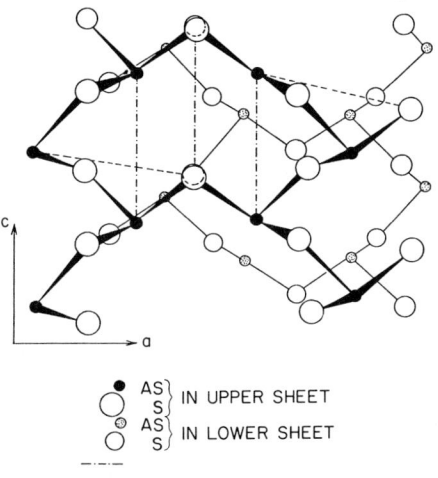

Fig. 44. Structure of orpiment (As_2S_3) (after Apling, 1973).

As_2Se_3 glass forms somewhat more easily than As_2S_3 glass. Bonding is somewhat more covalent (Onomichi et al., 1971). X-ray studies (Poltavtsev et al., 1972, Vaipolin and Porai-Koshits, 1963; Tsuchihashi and Kawamoto, 1969; Rechtin and Auerbach, 1973a,b, 1974) have established its orpiment structure; however, it is more complex because of a greater individuality of the As positions (Vaipolin and Porai-Koshits, 1963; Vaipolin, 1966). The difference is not great enough to affect coordination numbers, but is significant for explaining the greater ease of vitrification.

Noncrystalline As_2Te_3 glass is hard to obtain (Hruby and Stovrac, 1971; Brasen, 1972) and usually only as an evaporated film (Fitzpatrick and Maghrabi, 1971) or by doping. The structure of noncrystalline As_2Te_3 has no relation to that of crystalline As_2Te_3, rather it has the distorted orpiment structure of As_2S_3 and As_2Se_3 glasses (Fitzpatrick and Maghrabi, 1971). In contrast, some As-As "locks" may occur in the covalent $AsTe_{3/2}$ network (Cornet and Schneider, 1977).

In the binary As-S, the range of glass formation is large (Flaschen et al., 1959), with a sharp maximum of T_g (~200°C) at As_2S_3 (Soklakov and Zhanov, 1963; Myers and Felty, 1967). At very low As content, depolymerization of S chains leads to a low T_g (<20°C) (Pearson, 1964; Soklakov and Zhanov, 1963; Myers and Felty, 1967); and S_8 rings may be present up to about As/S = 1/10 (Tsuchihashi and Kawamoto, 1969, 1971). Then at As/S = 1/10, a linear increase of T_g is associated with the establishment of $AsS_{3/2}$ groupings (Myers and Felty, 1967). Thus between S and As_2S_3, glasses are copolymers consisting chiefly of S and $AsS_{3/2}$ units. Between $As_2S_{2.6}$ and $As_2S_{2.5}$, a plateau in T_g (176°C) suggests a miscibility gap (Maruno and Noda, 1972). As the first S is added to stoichiometric As_2S_3 glass, the orpiment layer structure is maintained but wrinkled as S-S bonds alter intralayer coordination (Vaipolin and Porai-Koshits, 1963). For As/S *larger* than As_2S_3, the network is broken by As_4S_4 groups similar to those in crystalline realgar (As_4S_4): T_g decreases once more and glass formation becomes difficult (Tsuchihashi and Kawamoto, 1969; Myers and Berkes, 1972). A separate, high-As region of glass formation reaching to As_2S was discovered by Hruby (1978). There is a eutectic near 56% As. Below 0°C, the glasses are of violet color and are hard, brittle, and stable. The melts from which these glasses quench must be held ~400°C above the liquidus for hours (as in the case of As-Te, As-Se glasses). β-As_4S_4 decomposes below a certain value of S (Bertoluzza et al., 1978; de Neufville et al., 1973/1974). A thorough, informative discussion of the topology in the system As-S (and also As-Se) was presented by Phillips (1979). An atomic model that is free of adjustable parameters predicts the chances of glass formation, T_g, and the phase diagram. X-ray studies of thin films show variation in grouping, depending on substrate temperature and other preparation parameters (Apling and Leadbetter, 1974; Apling et al., 1977; Nemanich et al., 1979; Daniel et al., 1979, 1980). The film structure contains a large portion of As_4S_4 (realgar) type groupings, with the amount depending on the conditions of film preparation. This is caused by the fact that the vapor above As_2S_3 in a sealed container contains mostly As_4S_4 molecules. On annealing, the film relaxes toward the bulk (orpiment) structure (Daniel et al., 1979, 1980). Extended x-ray absorption fine structure (EXAFS) evidence shows a decrease of As–As bonds on annealing (Nemanich et al., 1979; Street et al., 1979; de Neufville et al., 1973/1974). Films also convert to the bulk glass structure when exposed to band-gap (1.5 eV) illumination (photostructural transformation) (de Neufville et al., 1973/1974), with mostly irreversible property changes. Photostructural images can be developed by the diffusion of silver (Berkes et al., 1971; de Neufville et al., 1973/1974). Basically, the structures do not differ, but the assembly of different cross-linkings and molecules cannot be described by a simple model.

This is only one example of variations between bulk and film structure and variations of structures for films of variable origin in glass systems. The difference between the electronic structure of as-deposited and annealed (bulk structure) films can be described, in good agreement with experiment, by calculations based on the Hueckel theory (Shimizu and Ishi, 1978).

Since elemental Se glass contains a considerable fraction of Se_8 rings, the presence of Se_8 groupings in *As–Se binary* glasses makes their structures somewhat different from corresponding As–S glasses, as suggested by Raman and IR data (Schottmiller et al., 1970) and the nonlinear variation of T_g with composition. The increase in Se–Se bonds on adding Se to stoichiometric As_2Se_3 glass introduces a higher variability in the relative orientation of $AsSe_{3/2}$ units; while with excess As, the As–As bonds considerably increase steric hindrance between lone-pair electrons in the filled sp^3 orbitals (Apling, 1973) (Fig. 45) (Patel and Kreidl, 1975, 1978), favoring sheet formation.

In the binary system *As–Se* (Myers and Berkes, 1972; Rechtin and Auerbach, 1973; Nemilov and Petrovskii, 1963; Sholnikov et al., 1977), T_g again has a maximum at As_2Se_3, but glass formation continues above 55 As, probably in connection with the incongruent melting point (264°C) of As_4Se_4 (Myers and Berkes, 1972)—the equivalent of the congruently melting realgar, As_4S_4—which leads to the glass-forming melt structure of As_2Se_3. With excess As, As–Se glasses seem to contain more As–As bonds than As–S glasses, so that there is some controversy concerning the degree to which the orpiment structure is maintained (Renninger and Averbach, 1973; Sayers et al., 1974; Wright and Leadbetter, 1976).

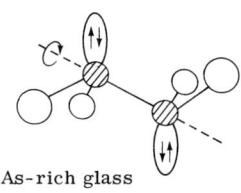

FIG. 45. Effect of excess of Se or As over As_2Se_3 on structure of As–Se glasses (Apling, 1973).

Immiscibility is observed for low As contents, as in the system As–S, and reaches into the ternaries with Te.

The structural relations as summarized by Kreidl and Ratzenboeck (1973) are reproduced in Table XXXIII.

TABLE XXXIII

STRUCTURAL RELATIONS IN As–X GLASSES[a]

	X		
	S	Se	Te
Crystalline form	X_8	$X_\infty(X_8)$	X_∞
Glassy form	$X_8 + X_\infty$	$X_8 + X_\infty$	—
X to As_2X_3 (gl)	$X_8 + X_\infty + o$	$X_8 + X_\infty + o$	$X_8 + X_\infty + o$
As_2X_3 (gl)	o	o	o
(cr)	o	o	m
$As_2X_3 + As$ (gl)	r + o, imm	o cross-linked	—
As_4X_4 (cr)	r	At T_m, $r \to As + o$	—

[a] X = S, Se, Te; cr = crystalline; gl = glassy; X_8 = rings; X_∞ = long chains; o = orpiment (As_2S_3) structure; r = realgar (As_4S_4) structure; m = monoclinic As_2Te_3 structure; imm = immiscibility, ( ) = unstable; T_m = melting point.

As in As_2S_3, As_4Se_6 "hard-sphere" dense random packing occurs in films of As_2Se_3 and is converted to the orpiment structure by 1.5-eV photons or annealing at 180°C for 1 h. When the annealed ("thermostructure transformation") film is illuminated, the absorption edge shifts ("photodarkening") (Deneufville et al., 1973, 1974; Leadbetter, 1976; Chang and Dove, 1974; Chang and Bestul, 1974). As in As_2S_3 films, As–As bonds are found (by EXAFS) and revert to the bulk structure on annealing (Nemanich et al., 1979; Street et al., 1979).

A high-pressure phase is observed to form above 8 kbar (Ota and Anderson, 1977).

Antimony glasses generally resemble equivalent arsenic glasses and are applied similarly, although somewhat less frequently. Among the systems studied are

Sb–S (Moss and Deneufville, 1972; Rawson, 1967; Hilton and Brau, 1963; 3M, 1968; Pauletti and Giamanio, 1967; Koperles et al., 1973);
Sb–Se (Mueller and Wood, 1972; Koperles et al., 1973; Brasen, 1974);
Sb–Te (Andrievski et al., 1963; Koperles et al., 1973).

These glasses have a surprisingly large band gap. The glass structure strongly deviates (Andrievski et al., 1963) from that of crystalline Sb_2Te_3 (Garner et al., 1974).

2. Properties

Arsenic sulfide glass replaced Se glass as an *IR optical material* after its rediscovery by Frerichs (1950). It was prepared by distillation in N_2 at 735°C, with S being added in excess, and cast in Al-covered molds (Fraser and Jerger, 1953; Glaze et al., 1957). As_2Se_3 glass exhibits *memory effects*.

In the binary As–Se system, electrical conductivity (at ambient temperature) varies from 10^{-16} Ω/cm at 5 As to 10^{-5} at 70 As (Hulls and McMillan, 1974). Conduction is by holes (Kolomiets, 1968; Adler, 1971; Scharfe, 1970; Rechtin, 1973). At As_2Se_3, one finds a maximum, at 5% As a minimum, lower than that of pure Se. Energy levels were proposed in detail by Kolomiets (1968). As–Se glasses, like many other chalcogenide glasses, tend to behave like crystalline semiconductors with conductivity inversely proportional to temperature, the gap being half the optical gap (Taylor et al., 1973). Most As–Se glasses show conduction into extended states (Hulls and McMillan, 1974). Seager and Quinn (1975) discuss the polaron hopping mechanism for As–Se glasses. The most recent interpretation was discussed in Section A.1. Gamma irradiation induces increased conduction (Minami et al., 1972).

In the As–Se system the average bond susceptibility—as calculated from dielectric constant measurements at optical frequencies—shows a maximum at the composition As_2Se_3, in contrast to the minimum at $GeSe_2$. The difference is attributed (Lucovsky et al., 1977) to the role of nonbonding electrons. In the As–Se system the lowest interband transitions are from nonbonding and antibonding states, while in Ge–Se, these states become important only at high Se contents. The slightly different situation in As–S is attributed to hopping conductivity (Feltz et al., 1974).

The thermally activated Hall mobility of As–Se films is one order of magnitude higher than that of bulk material (Klaffke and Wood, 1977). Such films might become applied as sensitive photoconductors in electrophotography. Fatigue might be recovered by high-frequency light (Bayer et al., 1977).

As–Se glasses are secondary electron emitters (Dunn et al., 1973); and As–Te glasses behave similarly (Hulls and McMillan, 1974; Brasen, 1972, 1974; Minami et al., 1972).

As_2Se_3 glass has been described as a photoconducting material (Adler, 1971). Its performance strongly depends on doping. Ag, Cu, Tl, In, and Ga increase the photoconductivity edge to a characteristic minimum concentration (Kolomiets et al., 1971). The nonphotoconductive region characteristic of Se is missing. As_2Se_3 is a candidate material for vidicon tubes and electrophotographic imaging (Adler, 1971). *Photoluminescence* indicates normal trap levels (Kolomiets et al., 1970). *Magnetic* properties were investigated by Bagley et al. (1975).

The *optical* spectra of liquid, glassy, and crystalline As–Se glasses are remarkably similar (Arai et al., 1972). The optical gap is 1.8 eV, almost twice the activation energy of electrical conduction; the Fermi energy is in the center of the mobility gap. Contamination by very small amounts of oxygen causes undesirable absorption bands, particularly at 10.6 μm. The interaction of O-induced absorption and silicon from melting containers is a critical factor (Hilton et al., 1975; Moynihan et al., 1975a,b). As–Se glasses darken reversibly under laser light, perhaps because of photon-induced As clustering (Asahara and Izumitani, 1975).

Acousto-optical merit factors for As_2Se_3-based glasses are high (Ohmachi and Uchida, 1972; Demboskii and Chernov, 1972).

3. Ternary Systems

For systems with Ge and Se see Sections XII.F.3 and XII.F.5. For systems with Ag, Cu, Tl, see Section XII.G.

When S is added to As–Se glasses, some of the discontinuities associated with definite groupings weaken (Myers and Felty, 1967), as more adaptable groupings form. The ternary field is very large (Fraser and Jerger, 1953; Kolomiets and Goryunova, 1956; Flaschen et al., 1959, 1960;

Tsugane et al., 1965; Ohsaka, 1976). Devitrification in the binary S–Se system is suppressed by small additions of As. Using this system and some polynary variants, Schnaus et al. (1970) first demonstrated the important independence of T_g and the Debye temperature θ. A correlation, once feared, would have precluded the combination of mechanical resistance and IR transmittance in new glasses. In As–S–Se(Te,O) glasses, θ varies from 300 to 650°C for compositions with T_g varying only 20°C up or down from 430°C.

Other *ternary arsenic chalcogenide* systems include the following:

As–Se–Te (Borisova et al., 1971; Kolomiets, 1964; Chang and Dove, 1974) includes (at 180°C/sec cooling rates) very large glass-forming areas (Cornet and Rossier, 1973; Cornet and Schneider, 1977). T_g generally increases with As, except for the maximum at $As_{40}Se_{60-x}Te_x$ near small ratios Te/(Se + Te). At constant As, T_g generally increases with Te/(Se + Te). At low (<20%) As, age hardening is observed as chains break to form rings. Ring structures induce phase separation. At high T_g, expansivity decreases.

Below 10 As, the structure consists of Se–Te rings and chains with As interconnections. With a relative increase in Te, at first T_g increases because of the metallic character of Te, but then it decreases, since As–Te bonds are weaker than As–Se bonds. For 20 to 40 As, tetrahedral units increase T_g and expansivity decreases. In $As_{40}Se_{60-x}Te_x$ glasses, T_g is insensitive to x in a three-dimensional network. At high As, layers are more interconnected for high Te. Immiscibility is suppressed by flash evaporation.

The resulting films are of interest, because of their photoconductivity, for page-reading (Cassanhiol et al., 1973) and pick-up tubes (Goto et al., 1974).

As–S–Se(Te) (Fraser and Jerger, 1953; Fraser and Upton, 1944; Tsugane et al., 1965; Ohsaka, 1976).

As–S(Se, Te)–Tl glasses containing up to 58% Tl and showing more than a 250°C reduction of the temperature at which the viscosity is 30 P. The field with Se is larger (Kolomiets et al., 1958). In the Te-rich region, semiconduction is more pronounced. In these systems Gubanov (1960, 1962) first demonstrated the validity of band models for noncrystalline solids, and Kolomiets and Goryunova (1956) as well as others (Pernot, 1968) confirmed the first concepts by photoconductivity data. Switching in a short period was observed in complex glasses in this system (Dorr and Kannewure, 1971; Uphoff and Healy, 1961; Kolomiets et al., 1963; Orlova, 1973; Koperles et al., 1973; Quinn and Johnson, 1972).

As–S–Cl (Pearson, 1964; Deeg, 1962); *As–S–Br* (Pearson, 1964;

Mel'nichenko et al., 1971); and As–S–I (Pearson, 1964) have low viscosities. $As_{17}S_{34}Br_{41}$ is liquid at room temperature ($T_g = -60°C$) (Pearson, 1964). Rather than chain termination by Br, $AsBr_3$ shows copolymerization with AsS_3 groups (Koudelka et al., 1979).

As(Sb, In)–S–I comprises large areas of glass formation (Turyanitsa et al., 1972, 1974; Dembovskii, 1969; Dembovskii and Chernov, 1972; Gerasimenko and Khiminets, 1977); with Se instead of S (Munir et al., 1973; Dembovskii et al., 1974; Koperles et al., 1973; Chernov et al., 1977).

As–S–Ni(Cu,Fe) (Timofeeva et al., 1972); As–S–Hg (Kirilenko and Polyakov, 1976).

As–Se–Ni(Cu,Fe) with up to 25 Cu (Savan et al., 1966; Timofeeva et al., 1972).

As–Sb–Se, a system that contains reported photoconductive, resistant (10^9 Ω cm) glasses (Li and Chaudhari, 1973; Orlova et al., 1974; Das et al., 1974).

As–Se–Pb (Borisova et al., 1973; Aggarwal et al., 1968; Moynihan et al., 1971).

As–Se–Be(Mg) (Liang and Bienenstock, 1975). Oxygen impurities are "gettered" by Be, Mg (Liang and Bienenstock, 1975), and Al.

As–Se–Hg. In this system a density minimum at 0.01Hg has been associated with decoupling of $AsSe_3$ groups without change in average coordination. Above 0.2Hg, the coordination changes as $HgSe_4$ zinc blende configurations occur (Cervinka et al., 1977).

As–S–Hg–I (Khiminets et al., 1980) has an As–S structure with Hg and I terminals and is applied to laser scanning.

Sb–Te–I (Koperles et al., 1973).

As–Sb–Se (Orlova et al., 1974; Das et al., 1974).

As–Te–I (Johnson and Quinn, 1978).

As–S–Pb (Bhat and Bhatia, 1978).

As–Se–Te–Pb (Owen, 1979) is applicable to ion selective electrodes.

As–Te–Al (Dunalev et al., 1980) with up to 10% Al, shows high-p conductivity).

4. Other Systems

Among other polynary chalcogenide glasses investigated more recently were those based on P–Se or P–S (Chernov et al., 1977; Tsarev et al., 1977; Ignatyuk and Stavnistyi, 1977; Rykova and Borisova, 1977; Ribes et al., 1979), P_2S_5–Na_2S (Orlova et al., 1979), and P–Se–Te. An interesting system is As–Se–I–biphenyl with a chain structure including As–C connective bonds (Herrmann and Schell, 1978).

Other compound studied are GaTe (which exhibits memory effects) (Romeo et al., 1977) and SnTe (Brown et al., 1970).

F. CHALCOGENIDE GLASSES CONTAINING GERMANIUM AND SILICON

1. General Considerations

Chalcogenide glasses of increased mechanical and thermal stability are generally obtained by the incorporation of cations of small ionic radii (Borisova and Timofeev, 1965), particularly Ge (Borisova and Timofeev, 1965; Kolomiets *et al.*, 1960, 1963, 1964) and Si (Skhol'nikov and Borisova, 1965a,b; Hilton, 1965, 1966; Hilton and Brau, 1963, 1968; Hilton and Jones, 1968a,b; Hilton *et al.*, 1964a,b, 1966a,b,c, 1974, 1975). Complex chalcogenide glasses thus have found numerous uses. The introduction of Si or Ge copolymerizes the three-dimensional orpiment network $AsSe_{3/2}$, forming $SiSe_{4/2}$ ($GeSe_{4/2}$), $AsAs_{3/3}$, and $SeSe_{2/2}$ groupings leading to density minima (Myuller *et al.*, 1966; Myuller and Tsai, 1967). Hilton's (Hilton, 1965, 1966; Hilton and Brau, 1963, 1968; Hilton and Jones, 1968a,b; Hilton *et al.*, 1964,a,b, 1966a,b,c, 1974, 1975) extensive studies led to the development of many polynary Ge–Se and Si–Te glasses, from which current IR optical materials were selected. Ovshinsky (1966, 1968) has stimulated in these systems the extensive exploration of amorphous semiconduction for threshold and memory switching.

Table XXXIV summarizes important crystalline compounds in the system Ge(Si)–As(Sb)–Se(Te), to which some of the glasses are structurally related and which may crystallize from them. As for the As–S and As–Se systems, atomic models without adjustable parameters were successfully based on topological considerations for the systems Ge–Se and Ge–Te (Phillips, 1979).

2. Germanium–Sulfur Glasses

a. Binary Glasses. In the binary Ge–S system (Nielsen, 1962; Hilton *et al.*, 1964; Bailey *et al.*, 1964; Myuller *et al.*, 1966; Inaoka and Yamazaki, 1967; Kawamoto and Tsuchihachi, 1971; Hruby and Štovrač, 1973; Nemanich *et al.*, 1979; Street *et al.*, 1979), two distinct regions may be distinguished:

(a) GeS_9 to GeS_2 with cross-linked S chains, and
(b) $GeS_{1.5}$ to $GeS_{1.3}$ with GeS_4 tetrahedra similar to those in crystalline GeS_2, and GeS_6 octahedra as in crystalline GeS.

In region (a), S rings form between GeS_4 and GeS_5, as indicated by linear property changes and solubility data. At lower Ge contents, a miscibility gap may exist. Infrared and Raman spectra (Lucovsky, 1974) are in agreement with nonrandom models. From the tetrahedral network at GeS_2, one proceeds to chain crossing, including the presence of S_8 rings,

TABLE XXXIV

CRYSTALLINE STRUCTURES AND COMPOUNDS IN THE SYSTEM Ge(Si)–As(Sb)–Se(Te)[a]

Compounds	System	Note	Ref.
GeSe	Orthorhombic		Okazaki (1958)
GeSe$_2$	Distorted fcc	Derives from GeSe$_2$	Liu (1962)
As$_2$Se$_3$	Orpiment (As$_2$S$_3$), parallel molecular layers		Vaipolin (1960)
As$_4$Se$_4$	realgar (As$_4$S$_4$), 8 at/molecule, bonded, van der Waals dec. at 264°C to As$_2$Se$_3$ + As	M–F's phase studies do not confirm Dembovski's	Myers and Felty (1967) Dembovski et al. (1964)
Sb$_2$Se$_3$	Orthorhombic Pb n m		Tideswell (1957)
GeAsSe (pd. with more As)		Powder diagram	Vinogradova (1908)
GeTe	NaCl trigonal (distorted NaCl)	High T Low T	Schubert (1951) Panson (1964)
Ge$_{10}$As$_{55}$Te$_{35}$	Rhombohedral (distorted NaCl)	Electron diffraction	Agaev (1966)
As$_2$Te$_3$	Centered monoclinic		Carron (1963)
Ge$_{10}$As$_{55}$Te$_{35}$		Not necessarily a compound	Uttretcht (1970)
(SiSe)$_n$	Polymeric isotypic with SiS		Emons et al. (1968)
SiSe$_2$	Isotypic with SiS$_2$		Weiss (1952)
SiTe	Complex cubic		Weiss (1953)
SiTe$_2$	Hexagonal dI$_2$ (6)	Perhaps rather Si$_2$Te$_3$?	Bebrick (1968)
Si$_2$Te$_3$	Hexagonal		Klein/Haneveld (1960)

[a] No structures or compounds are documented for the systems SiAsSe, SiSbSe, SiAsTe, SiSbTe. The binaries Ge(Si)–Sb(As) are not considered.

which sharply increase above GeS$_4$ (Lucovsky, 1974; Tronc et al., 1973; Ball and Chamberlain, 1978).

A very small excess of S over GeS$_2$ (e.g., GeS$_{2.01}$) aids glass formation considerably (Voigt et al., 1978). Defect centers were modeled and verified, in part, in ESR spectra by Cerny and Frumar (1979) for GeS$_2$: DCIV (GeS$_3$S*), DCV (Ge*S$_3$) for excess S; DCVI (GeS$_2$S$_n$S$_m^*$), and DCIII (S$_3$GeGe*S$_2$) for excess Ge. The centers DCIV and DCVI show ESR signals. (* stands for "not bonded.")

For Ge/S larger than 1/2, Ge–Ge bonds occur in addition to Ge–S bonds in a restricted network excluding S–S bonds (Lucovsky et al., 1977; Feltz *et al.*, 1974). This may relate to the sharp dielectric constant minimum at GeS$_2$ (Lucovsky *et al.*, 1977). The system forms the base for the development of acoustic materials (Dembovskii and Chernov, 1972). Unlike the situation, for Ge–Se glasses, spin resonance signals were found in the system Ge–S (Voigt *et al.*, 1978).

As in As$_2$S$_3$(Se$_3$), and for similar reasons, film and bulk structures of Ge–S glasses differ considerably. The film contains Ge–Ge and S–S bonds, many defects (ESR centers) and has a four orders of magnitude higher conductivity that is insensitive to Ag doping. The bulk glass consists mostly of ordered GeS$_4$ groups with a few Ge–Ge bonds, its lower conductivity increases largely by Ag doping (Watanabe *et al.*, 1980).

b. Complex Ge–S Glasses. The structure of ternary *Ge–As–S* glasses (Nielsen, 1962; Savage and Nielsen, 1965a,b; Myuller, *et al.*, 1966; Imaoka and Yamazaki, 1967; Schnaus *et al.*, 1972) might be characterized (Andreichin *et al.*, 1976) by eight groupings, among which the most important are those contributing to a three-dimensional network: GeS$_{4/2}$, AsS$_{3/2}$, AsS$_{4/2}$, and GeS$_{2/2}$ (Myuller *et al.*, 1966). At high S, chains of sulfur appear, and at high As, As + As$_{3/3}$ groups appear (Lucovsky, 1974). X-ray diffraction confirms the absence of Ge–As bonds (Andreichin *et al.*, 1976). The equivalent Ge–As–S glasses have become candidate materials for acousto-optical uses (Krause *et al.*, 1970) because the refractive index critical for the merit factor is as high as 2.5–3 while sound velocity and loss are reasonably low.

Among other polynary Ge–S systems studied extensively are those containing S (Feltz and Pfaff, 1979) (also prepared from solutions), S, Te (Makovskaya and Zhuko, 1980), P (Hilton and Jones, 1968), B, Si, Sn (Suvarova *et al.*, 1974), Pb (Feltz and Pfaff, 1978, 1979; Feltz and Schlenzig, 1974; Feltz and Voigt, 1973; Feltz *et al.*, 1973), Ga (Loireau–Lozach and Guittard, 1975), Tl (Artamanova, 1971; Imaoka and Yamazaki, 1967), Li, Na, K, Ag, Ba, Cd, Zn, Pb (Imaoka and Yamazaki, 1967; Ohsaka, 1976; Barrau *et al.*, 1980; Ribes *et al.*, 1979), Sn (Feltz *et al.*, 1974), Ga, Eu (Barnier *et al.*, 1980), organic-chalcogenide glasses, e.g., Na$_6$Ge$_2$S$_6$(Se$_6$)-4CH$_3$H (Feltz and Pfaff, 1978), Bi (Toghe *et al.*, 1979, 1980a,b), and Hg^{2+} (Feltz and Burkhardt, 1980). In complex X–Ge–S glasses with less than the stoichiometric quantity of S, Feltz *et al.*, 1974; Feltz and Pfaff, 1979) suggest divalent GeII and Ge–Ge bonding.

3. Germanium–Selenium Glasses

a. Binary Glasses. Binary Ge–Se containing up to 30 Ge are obtained easily from mixtures melted in sealed SiO$_2$ tubing at 1–10 Torr and quenched (Savage and Nielsen, 1965a,b), and compositions above 40 Ge have been reported (Azoulay *et al.*, 1975; Bienenstock *et al.*, 1976). When Ge is added to Se its chains and rings become cross-linked by single (minimum volume at GeSe$_3$), then netted GeSe$_{4/2}$ tetrahedra (Kolomiets *et al.*, 1962, 1963; Avetkyan *et al.*, 1969; Fawcett *et al.*, 1972; Nemilov and Pelrovskii, 1963; Wood, 1968; Uemura *et al.*, 1974; Ruska and Thurn, 1976). A second, lower-viscosity region exists between GeS$_2$ and GeSe. Structures now contain Ge–Ge bonds in Ge$_2$Se$_{6/2}$ groups and/or Ge^{2+} (Feltz *et al.*, 1973). Films can be deposited from higher-Ge mixtures (GeSe$_{0.7}$) (Molnar and Dove, 1974; Krebs and Ruska, 1974). As in As–Se and other binaries, the film structure differs greatly from the bulk structure, which it approaches on annealing or illumination (Nang *et al.*, 1979).

b. Complex Ge–Se Glasses. The extension of glass formation in ternary systems with Group V elements (As, Sb, P) is closely associated (Dietzel's rule) with the existence of binary compounds. The region thus is large in the system Ge–As–Se (Fig. 46) where compounds GeAs and GeAs$_2$ exist.

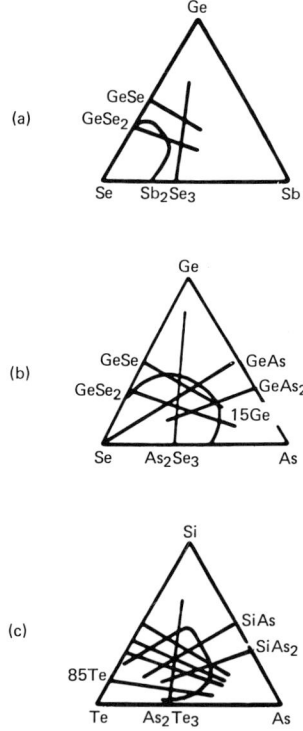

FIG. 46. Glass formation ranges in the systems (a) Ge–Sb–Se, (b) Ge–As–Se, and (c) Si–As–Te (Kreidl and Ratzenboeck, 1973).

It is more limited in the ternary Ge–Sb–Se system in which no binary compound Ge–Sb exists. The regions of glass formation is polynary Ge–As(Sb)–S(Se,Te) systems decrease generally from S to Se to Te.

Optimization between transmission (moving to higher wavelength from S to Te) and softening range (decreasing in that order) led to the selection of the Ge–As–Se system in the range 25–35 Ge, 50–60 Se, 10–20 As for the development of IR optical materials. A conventional glass has the composition 33 Ge · 12 As · Se; this has a softening temperature of 450°C as standardized by Jones and Hafner (1968).

The increase in softening temperature is associated with cross-linking by $GeSe_{4/2}$ as well as $AsSe_{4/2}$, and by $AsSe_{3/2}$ groupings (Hilton, 1965, 1966; Hilton and Brau, 1963, 1968; Hilton and Jones, 1968; Hilton et al., 1964a,b, 1966a,b,c, 1974, 1975; Krebs and Fischer, 1970). A study of IR spectra by Ohsaka (1976) supports, and gives details of, this structural principle and compares findings with those on other complex Se and Te glasses. Among Ge–Sb–Se glasses (Egorova and Kokorina, 1960, 1967, 1968a,b,c) the castable composition 28 Ge 12 Sb 60 Se represents another conventional IR material (Patterson and Brau, 1966; Jones and Hafner, 1968), which, however, has a much lower softening temperature (375°C) (Hilton, 1966).

In ternary Ge chalcogenide glasses, electrical conduction is generally by holes, yet the Hall mobility is generally n-type (Adler, 1971), explicable by microstructure (but see also Emin, 1980). The absence of photoconductivity in many of these glasses might be attributed to the absence of transitions characteristic of ring structures in Se.

Switching and memory effects are found in regions substantially lower in Ge and higher in chalcogen than in the high-softening-temperature optical glasses. The Te equivalents are more important in this application. Igo and Toyoshima (1973) formulated erasable light-based memories (photochromics) in the area 30 Ge · 25 As · 45 Se, in which Ge–As bonds appear.

In chalcogenide windows, laser damage can be minimized by using Ge and Si compositions of equal T_g and expansivity in a composite window, since the absorption for 10.6-μm laser radiation differs from layer to layer (Moynihan et al., 1975).

Conduction and structure were investigated in various sections of the ternary *Ge–Se–Tl* by Borisova et al. (1976).

Ge–As–Se–S glasses have been described by Ohsaka (1976), and *Ge–Ga–Se* glasses by Fichet-Ollitraut et al. (1977). They have good IR transmission and a T_g of 350° over wide ranges. Many other polynary Ge–Se glasses—including Hg, Pb, Sn–Ge–$Se(Te)$ systems—have been described. For such glasses containing less than the stoichiometric amount of Se, Feltz and Schlenzig (1974) and Feltz and Pfaff (1979) suggest the occurrence of divalent Ge^{II} and conceive such systems as containing regions resembling $GeSe_2$ and GeSe.

c. *Preparation of Complex Ge–Se Glasses.* Complex $Ge(Si)$–$As(Sb)$–$Se(Te)$ glasses are usually prepared in SiO_2 or Vycor containers. The system is evacuated to 10 Torr, heated in a rocking or rotating furnace, then cooled appropriately (see Hilton *et al.*, 1975). The ingot formed in the container or its fragments may be reheated in an O_2-free atmosphere (e.g., in forming gas) and cast or molded. For most (particularly IR-optics) uses, exceedingly careful control of impurities (<5 ppm O_2) is required. This is effected by distillation of the primary materials (Hilton *et al.*, 1975), at times in H_2 (Savage and Nielsen, 1965a,b; Ford and Savage, 1976). H_2 treatment, while suppressing O bands between 8 and 15 μm, may introduce weak H–Se bands at 4.55 and 4.95 μm. For smaller (1 kg) melts, an improved process uses a multiple chamber system (Hilton *et al.*, 1975).

4. Germanium–Tellurium Glasses

a. *Binary Glasses.* There is only one compound, GeTe, in the system Ge–Te, and both constituent atoms are 6-coordinated. The telluride glasses are more difficult to obtain than the corresponding S and Se glasses (Krebs and Fischer, 1970; Kokorina, 1971), and where glasses do form, their structures differ greatly from that of crystalline GeTe (Betts *et al.*, 1970, 1972a; Deneufville and Turnbull, 1972). Bonds are more covalent; below 33 Ge, the coordination of Ge is most likely 4; that of Te is 2 to 3; and the Ge–Te distance is about 2.7 Å. T_g has a maximum at $GeTe_2$, where the bonding and coordination are somewhat analogous to those in SiO_2. Above 33 Ge, the coordination may be in part three-fold, and the structure may resemble that of crystalline GeTe. Structural details have been worked out carefully by x-ray and neutron diffraction (Pickart *et al.*, 1979), particularly for the glass 17 Ge·85 Te, which is located near the Ge–Te eutectic (Bienenstock *et al.*, 1970; Betts *et al.*, 1972a). A structural model was based on Mössbauer spectroscopy by Aggarwal (1977). At the eutectic composition (80Te), T_g is 133°C, the crystallization temperature $T_x = 225$°C, and the liquidus temperature $T_m = 380$°C (Quinn and Johnson, 1973). Liquid–liquid phase separation is possible, with one phase close to pure Te (Takamori *et al.*, 1970; Feltz *et al.*, 1972), extending into ternaries, where it appears to be an important precursor in the memory-switching process. Yet Deneufville and Turnbull's (1972) T_g data do not show the characteristic plateau, so that phase separation could not be determined with certainty.

Crystalline (orthorhombic) GeTe is a *p*-type semiconductor with a small temperature dependence of conductivity (Kolomiets *et al.*, 1964a) and a carrier concentration of 10^{20}–10^{21}/cm^3. Glassy GeTe is also a p-type semiconductor (hopping mechanism). The room temperature resistance is 10 times that of the crystal, and the carrier concentration is 10^{18}/cm^3. The activation energy is 0.35 eV and the hole mobility 1 cm^2 v/sec. There are two traps: (a) near the Fermi level, and (b) about 0.2 eV below the mobility gap. Their nature is not known (Howard and Tsu, 1970). Switching effects in binary Ge–Te glasses were described by Evans *et al.* (1970) and Lewis (1974). Annealing studies are useful to study the crystallization of Te, which is related to these effects (Deneufville *et al.*, 1976; Brown, 1977). The optical gap (0.7 eV) is about twice the electrical gap.

b. Ge–As(Sb)–Te and Related Glasses. The area of glass formation (Savage *et al.*, 1966, 1971) (Fig. 45) extends from the narrow regions between As and Te toward the eutectic Ge–Te near 81 Te (Sample *et al.*, 1972). Near As_2Te_3, crystallization temperatures are much above T_g and many crystalline phases coexist. Near the Ge–Te eutectic, glass formation is less easy and the only well-defined crystalline phases are GeTe and Te. It is in this region that attractive switching compositions are found. In a $Ge_{15}Te_{18}S_2Sb_2$ threshold switching glass, Owen (1979) established an electrothermal switching mechanism. Softening ranges are much below those for the equivalent Se glasses (Hilton, 1966), so that infrared windows have not been made from stable ternary Ge–As–Te glasses. At elevated temperatures, Ge–As–Te glasses change in electronic structure (from overlapping sp^x hybrids to pσ), and are characterized by low viscosity and metal-like conduction (Krebs and Ruska, 1974). Related complex glasses such as $Ge_3As_3Se_6Te_8$ (Hilton and Jones, 1968) and $Ge_{15}Sb_2Te_{81}S_2$ (Moss and Deneufville, 1972; Chaudhari and Herd, 1972) have been described in connection with threshold switching experiments.

Ge–P–Te glasses have higher softening points (Hilton, 1966). Among other germanium telluride glasses are those with additional Si (Feltz *et al.*, 1972; Barth, 1973; Kirilenko and Dembrovski, 1974); Se (Noda and Maruno, 1974; Sarrach *et al.*, 1976); S (Maneglier-LaCordaire *et al.*, 1975); S and Se (Ohsaka, 1976); I (Feltz *et al.*, 1972); and Pb and Se (Feltz and Pfaff, 1979).

5. *Silicon–Tellurium Glasses and Other Chalcogenide Glasses Containing Much Silicon*

In his search for IR-transmitting glasses of high-temperature performance, Hilton (1966) verified the expected increase in softening point in replacing Ge by Si. But Si–As–Se glasses are unstable (Sholnikov, 1965; Hilton, 1966) so that Ge–As–Se glasses remained the most serviceable IR windows in the desired range.

The binary glasses Si_2Te_3 are probably biphasic (Seregin et al., 1979). Si–As–Te glasses are stable, surpass Ge–As–Te glasses and at least equal Ge–As–Se glasses in softening range. The region of glass formation is large, once more correlating with a large number of compounds in the system (Fig. 46c). The maximum softening temperatures are found near 40 Si 20 As 40 Te on the pseudobinary line SiTe–As, e.g., 35 Si 25 As 40 Te (Anthonis et al., 1973). According to Deneufville and Turnbull (1972), the structure contains characteristic 3-coordinated groups and a significant amount of Si–As bonds. Perhaps the isoelectronic make-up of As and GeTe is an important factor (Krebs, 1953; Suchet, 1971). In this connection the deposition of P and As to a vitreous material is noted in relation to their isoelectronic structure with GeS and GeSe (Rawson, 1967). The structures of Si–As–Te glasses do, however, differ from those of Ge–As–Te glasses (Anthonis et al., 1973). In the Ge–As–Te system, maxima suggest $GeTe_2$ grouping; in the Si–As–Te glasses, maxima suggest SiTe–As structures (Krebs, 1953; Deneufville and Turnbull, 1972).

The use of Si–As–Te glasses is handicapped by the affinity of Si for O, introducing oxide absorption bands. Hilton (1966) and Moynihan et al. (1975) proposed methods for their suppression in other chalcogenide glasses, which have not been applied to the Si–As–Te system.

Softening points can be raised by introducing Se as a fourth component in the presence of Ag (Anthonis et al., 1973; Amrhein et al., 1974), most likely producing a microcrystalline material. *Ge–Si–As–Te* glasses have been described as switching materials (Coward, 1971; Bagley and Blair, 1970; Bagley et al., 1975; Phillips et al., 1970; Bunton, 1971; Ormandroyd et al., 1975; Esqueda and Henisch, 1976).

Among many Si chalcogenide glass systems are $Si–S–I_2$ (Dembovski and Luzhnaya, 1971), Si–Sn–Ge–Te–Se–S (Kirilenko and Dembrovski, 1974), Ge–Si–S (Stepanek and Hruby, 1980), and Si–S–Na (Ribes et al., 1979). High-melting Si–S glasses are unstable as S evaporates. No glasses were found in the ternary Si–B–Te.

G. Silver, Copper, and Thallium in Chalcogenide Glasses

Electrical peculiarities have stimulated an interest in incorporating into the melt or by diffusion (Arai et al., 1972) the mobile ions of Ag, Cu, and Tl. Large glass-forming regions exist in the ternaries *Ag–As–S* (Maruno et al., 1973; Kolomiets et al., 1960; Kawamoto et al., 1973; Feichtner et al., 1974; Sherrell and Thompson, 1977), *Ag–As–Se* (Danilov and El Mosli, 1963; Patel and Kreidl, 1975), and *Ag–Ge–S* (Feltz and Schlenzig, 1974; Kazakova and Borisova, 1980a,b). When Ag diffuses into As_2S_3 or As_2Se_3 glass (at 175°C), As^{3+} may be reduced to As^{2+}, and As^+ and vitreous phases $Ag_2As_2S_3$ to $Ag_4As_2S_3$ may form (Holmquist, 1978).

Structures in these systems were investigated in detail. The conductivity rises sharply and becomes ionic with increasing Ag content. The drop in activation energy is attributed to groupings in which Ag (or Cu) is 4-coordinated carrying a charge of 3, compensated by dangling bonds D^+ (Mott, 1977).

Photoconductivity has been observed (Golovej et al., 1975; Kolomiets et al., 1970, 1971; Bonchev and Reichin, 1965; Borisova, 1970). Thermoelectric power was studied by Dong (1973) and Ninomiya et al. (1975) in Sb-Te and Bi-Te glasses. Other systems investigated include Ag-I-Sb-S (Turyanitsa et al., 1974), Ag-S-Te (Ida et al., 1974), Ag-Sb-Se (Tarasevich et al., 1971), Ag-Si(and Ge)-S (Imaoka and Yamazaki, 1967; Kostyshin et al., 1971; Kawamoto et al., 1973; Edmond, 1968; 3M Company (with Sb), 1968), and Ag-P-S (Kawamoto and Nishida, 1976). Photodoping was achieved by Deneufville et al. (1973/1974), Kokado et al. (1976a,b), Goldschmidt and Rudman (1976).

Photodissolution (Maruno, 1977; Goldschmidt and Rudman, 1977) (an expression preferred to photodoping because of the large amount (up to 35%) of Ag accepted) is a two-stage process: (a) radiation damage in the base glass, e.g., As_2S_3 and (b) photon-assisted dissolution across an Ag-chalcogenide interface activation barrier:

(a) $\quad As_2S_3 \xrightarrow{h\nu_1} As_2S_3^*$

(b) $\quad As_2S_2^* \xrightarrow{h\nu_2} As_2S_3(Ag)$

The phenomenon offers imaging possibilities based on the accompaniment of the visible by an electrical conductivity image. Ag is deposted under illumination with wavelengths shorter than those of the absorption edge (photosurface deposition, PSD) in 45 Ag 15 As 40 S glass (and other chalcogenide glasses). Optical information storage, printed circuits, and photosensitive images (Maruno, 1977) are potential applications.

The role of Cu in chalcogenide glasses was explored by Bonchev (1965), Borisova (1970), Kolomiets et al. (1971), Demovskii et al. (1971), Savan et al. (1966), Alimbarashvili and Baidakov (1973), Ashara and Izumitani (1974), Ninomiya et al. (1975), Tomofeeva et al. (1972) (by diffusion), and Liang (1975).

The addition of up to 40 Cu to an As-Se glass was claimed to lead to a highly transmittant optical material with a long wavelength edge at 20 μm and an increase in softening range and thermal resistance (Hoya, 1974). As Cu increases from 5 to 25%, the Se coordination seems to increase, and more As-As bonds are found (EXAFS technique) (Hayes, 1978). Cu may induce phase separation in Ge-S glasses, and eventually crystallization at Cu_2GeS_3 (Doupovec et al., 1979). A copper ion selective electrode can be obtained in As–Se(Te) glass bases containing up to 25 mole % Cu (Owen, 1979).

Thallium was introduced into chalcogenide glasses by many workers and studied in considerable detail by Majid et al. (1974), even in connection with incorporated mercury (Feichtner et al., 1974).

Conduction and structure were investigated in various sections of the ternary $Ge-Se-Tl$ by Borisova et al. (1976). The structure of TlSe glass resembles that of crystalline TlSe, which is characterized by long chains parallel to the c axis consisting of covalently bonded distorted tetrahedra, with Tl^{3+} being 4-coordinated and Tl^+ 8-coordinated ($\sim Tl^+(TlSe_2)^-$) as indicated by Raman spectra. Such spectra also exhibit differences due to vibrations of Tl^+ against the rigid chain typical of the crystal. In the structure of glasses with Se/Te above 1/1, both TlSe and amorphous Se groupings are recognized in a model of Tl_xSe_{1-x} ($x \leq 0.24$) (Zirke et al., 1977; Cervinka et al., 1979).

H. Chalcogens in Oxide Glasses

1. General Considerations

Many blends of oxides and sulfides can be obtained as glasses. It should be remembered that, unlike oxygen, chalcogens can be present in excess of formal stoichiometric relations. Sulfur and selenium are quite soluble in oxide glasses, first taking oxygen positions, but can (particularly on reheating) form chalcogen-rich groupings or crystallize out as chalcogenide compounds. The latter behavior has been used to obtain colored glasses containing crystals of specific absorption by controlled heating.

2. Soda–Silica–Sulfur

In the system Na_2S-SiO_2 (Handa et al., 1976), almost all S is retained. Glass formation extends from 100 to 45 SiO_2 and includes $Na_2S \cdot 2SiO_2$ and $Na_2S \cdot 3SiO_2$. At nonbridging sites, polysulfide groups may form and impart red coloration. The similarity in the glass-forming region suggests that the structures resemble those in the binary Na_2O-SiO_2 system.

3. Soda–Borate–Sulfur

In the system $Na_2O-B_2O_3$, retention of S is generally much lower than 100%, but increases as $(BO_4)^{5-}$ groups form on adding Na_2O. The maximum of S retention occurs where in the oxide system magnetic spin resonance indicates a maximum of $(BO_4)^{5-}$. The red polysulfide color appears where spin resonance indicates the beginning of nonbridging oxygen formation. Below this Na_2S concentration, the blue color of S_3^- appears (Ahmed et al., 1978a,b, 1979a,b). At very low Na_2O contents, S is hardly soluble. The following states of sulfur were recently identified from their absorption spectra in the visible and UV and their ranges of stability were determined: the S_2 molecule in glasses containing at least 15 mole % Na_2O, the S_3^- in glasses containing 20–30 mole % Na_2O, the S_2^- in glasses containing at least 20 mole % Na_2O, and the polysulfide ion in glasses of 30–35 mole % Na_2O (Ahmed et al., 1980).

4. Sulfur in Soda–Lime–Silica Glasses

In soda-lime glasses the replacement of O by S is supplemented by the addition of carbon to obtain sulfide colors by reduction. The color is intensified and modified by the presence of Fe. Such "carbon amber" glasses contain (Douglas and Zaman, 1969) S in various groupings, most importantly in $Fe^{3+}O_3S$ tetrahedra (Douglas and Zaman, 1969; Karlsson, 1969; Neumann and Dietzel, 1939). The O pressure must be low enough to dissolve enough S but not so low as to reduce most Fe^{3+} to Fe^{2+} (Brown and Douglas, 1965). Low nonbridging oxygens destroy the amber color (Brown and Douglas, 1965).

5. Cadmium Sulfide and Cadmium Selenide

Heat treatment of somewhat reduced silicate glasses containing Cd^{2+} and S leads to the precipitation of yellow cubic β CdS (Muller and Loeffler, 1933). In the presence of both S and Se, red mixed crystals of CdS and CdSe form, which permit the construction of sharp edge absorption materials over the entire visible spectrum (Handleman and Kauser, 1964). Zinc aids in the retention of S and Se, combining with these elements at high temperature and releasing them to Cd^{2+} at low temperature (Weyl and Marboe, 1964).

6. Selenium in Alkali Borate Glasses

Selenium is not stable in lithium borate glasses containing up to 20 mole % Li_2O. The polyselenides are stable in $K_2O-B_2O_3$ up to 20 mole % K_2O and in $Na_2O-B_2O_3$ up to 25 mole % Na_2O. The elemental Se is stable in the range of 25 to 30 mole % K_2O and 30 mole % Na_2O. Glasses containing higher contents of either Na_2O or K_2O may contain selenites or selenates (Ahmed et al., 1979b).

7. Tellurium in Alkali Borate Glasses

Glasses containing less than 15 mole % Na_2O or K_2O or up to 20 mole % Li_2O do not retain Te. The rose polytelluride color appears only in $K_2O-B_2O_3$ containing 15–20 mole % K_2O. Gray metallic Te is stable in glasses containing 15–30 mole % Na_2O or 25 mole % K_2O. Glasses containing higher contents of Na_2O or K_2O may contain tellurites and/or tellurates (Ahmed et al., 1978a,b).

References

Abarenkov, I., Amosov, A., Bratsev, V., and Yudin, D. (1970). *Phys. Status Solidi* **2**(4), 865.
Abe, K. (1970). *Eur. Conf. Opt. Fibre Commun., 2nd, Paris* (1970), p. 59.
Abe, T. (1952). *J. Am. Ceram. Soc.* **35**, 756.
Abkemeier, Rm. (1972). Ph.D. Thesis, Purdue Univ.
Abou-el-Leil, M., Heasley, J., and Omar, M. (1978). *Phys. Chem. Glasses* **19**(3), 37.
Adams, R., and Douglas, R. (1959). *J. Soc. Glass Technol.* **43**, 147T.
Adler, D. (1971). "Amorphous Semiconductors." CRC Press, Cleveland, Ohio.
Adler, D. (1979). Defects in Amorphous Semi-Conductors, MIT-10-36-79.
Adler, D. (1980). *Solar Cells* **2**, 199.
Aerojet (1968). *Glass Technol.* **9**(5), 139.
Aerojet (1969). British Patent 1 147 718.
Agaev, K. A., and Talybov, A. G. (1966). *Kristallografiya* (*Engl. transl.: Sov. Phys.-Crystallogr.*) **11**(3), 400.
Aggarwal, I., Moynihan, C., Macedo, P., Mecholsky, and J. Srinivasan, G. (1968). Tech. Rep. 21, N 00014-68-A-0506-0002, ONR 32-512/5-31-68/471.
Aggarwal, J. (1977). *J. Non-Cryst. Solids* **23**(3), 357.
Aggarwal, K., Suri, S., and Mendiratta, R. (1978). *J. Non-Cryst. Solids* **27**, 153.
Ahmed, A. (1978). Oral communication.
Ahmed, A., and Ashour, G. (1977). *Proc. Int. Cong. Glass, 10th, Kyoto,* p. 34.
Ahmed, A., and Ashour, G. (1981). *Glass Technol.* **22**, 24.
Ahmed, A., El-Shamy, T., and Sharaf, N. (1978a). *J. Non-Cryst. Solids* **30**, 225.
Ahmed, A., El-Shamy, T., and Sharaf, N. (1978b). *Arab Chem. Congr. 5th, Cairo.*
Ahmed, A., El-Shamy, T., and Sharaf, N. (1979). *J. Non-Cryst. Solids* **33**, 159.
Ahmed, A., El-Shamy, T., and Sharaf, N. (1980). *J. Am. Ceram. Soc.* **63**, 537.
Ainsworth, L. (1954). *J. Soc. Glass Technol.* **38**(185), 501.
Aleksandrov, V. *et al.* (1980). *Fiz. Khim. Stekla* **6**(2), 170.
Alekseeva, Z., Mazurin, O., Zver'yanov, V., and Glakhov, F. (1977). *Fiz. Khim. Stekla* **3**, 114.
Alexandre-Ferraris, V., Fernandez-Navarro, S., and Oteo-Mazo, J. (1977). *Proc. Int. Cong. Glass, 10th, Kyoto,* p. 7.
Alimbarashvili, N., and Baidakov, N. (1973). *Neorg. Mat.* **9**(12), 2108.
Ambrus, J., Dardy, H., and Moynihan, C. (1972a). *J. Phys. Chem.* **76**, 3495.
Ambrus, J., Moynihan, C., and Macedo, P. (1972b). *J. Phys. Chem.* **76**, 3287.
American Optical Company (1947). British Patent 585 287.
Amosov, A., Zakharov, V., Petrovskii, G., Prokhorova, T., and Yudin, D. (1969). *Kh. Prikl. Spektr.* **11**(4), 742.
Amrhein, E., (1963). *Glastech. Ber.* **36**(11), 425.
Amrhein, E., Day, D., and Kreidl, N. (1974). *Phys. Chem. Glasses,* **15**(16), 141.
Andersen, A., and Urnes, S. (1975). *Fiz. Khim. Stekla* **1**(5), 399.
Anderson, O. (1958). "Non-Crystalline Solids" (A. Frechette, ed.). Wiley, New York.
Anderson, O., and Bommel, H. (1955). *J. Am. Ceram. Soc.* **38**, 125.
Anderson, O., Halperin, B., and Varma, C. (1972). *Phil. Mag.* **27**, 1.
Anderson, P. (1975). *Phys. Rev. Lett.* **34**, 953.
Andreichin, R. *et al.* (1976). *J. Non-Cryst. Solids* **20**, 101.
Andreyev, N., Bokov, N., and Boiko, G. (1958). *Sov. Res. Glass Consult. B* **3**, 508.
Andreyev, N., Bokov, N., and Boiko, G. (1970). *J. Non-Cryst. Solids* **5**, 41.
Andreyev, N., Bokov, N., and Boiko, G. (1971). *Dolk. Akad. Nauk. SSSR* **201**(6), 1375.
Andrievski, A., Nabitovich, I., and Voloshchuk, Y. (1963). *Sov. Phys. Cryst.* **7**, 704.

3. INORGANIC GLASS-FORMING SYSTEMS

Angell, C. (1964). *J. Phys. Chem.* **68,** 218, 1917.
Angell, C. (1965a). *J. Am. Ceram. Soc.* **48**(10), 540.
Angell, C. (1965b). *J. Electrochem. Soc.* **112,** 1224.
Angell, C. (1968a). *J. Am. Ceram. Soc.* **51,** 124.
Angell, C. A. (1968b). *J. Am. Ceram. Soc.* **51**(3), 117–124.
Angell, C. (1971). *J. Phys. Chem.* **75,** 2360.
Angell, C. (1974). *J. Phys. Chem.* **78,** 278.
Angell, C., and Helphrey, D. (1971). *J. Phys. Chem.* **75,** 2306.
Angell, C., and Parks, J. (1978). Unpublished.
Angell, C., and Ringoen, D. (1978). Unpublished.
Angell, C., and Sare, E. (1968). *J. Chem. Phys.* **49,** 4713.
Angell, C., and Sare, E. (1970). *Chem. Phys.* **52,** 1058.
Angell, C., and Sare, E. (1971). *J. Chem. Phys.* **52,** 1058.
Angell, C., and Sichina, W. (1978). Unpublished.
Angell, C., and Tucker, J. (1974). "Glass Forming Molten Salt Systems" (*Chem. Proc. Metall., Richardson Conf.*). Imperial College Science, London.
Angell, C., Wong, J., and Edgell, W. (1969). *J. Chem. Phys.* **51,** 4519.
Angell, C., Hodge, I., and Cheeseman, P. (1976). "Molten Salts" (*Proc. Int. Conf.*) (J. Pemslerz, ed.), p. 138. Electrochemical Society.
Anthonis, H., and Kreidl, N. (1972). *J. Non-Cryst. Solids* **11**(3), 257.
Anthonis, H., Kreidl, N., and Ratzenböck, W. (1973/74). *J. Non-Cryst. Solids* **13,** 13.
Antonova, S., and Dyakova, V. (1979). *Fiz. Khim. Stekla* **5**(6), 679.
Apling, A. (1973). "Electronics and Structural Properties of Amorphous Semiconductors" (P. Comber and J. Mort, eds.), p. 234. Academic Press, New York.
Apling, A., and Leadbetter, A. (1974). "Amorphous Liquid Semiconductors" (J. Stuke and W. Brenig, eds.), p. 457. Taylor and Francis, London.
Apling, A., Leadbetter, A., and Wright, A. (1977). *J. Non-Cryst. Solids* **23**(3), 369.
Appen, A. (1959). *Int. Cong. Glass, 5th,* Session VI, p. 23. Glastechn. Ber., Munich.
Appen, A., and Fu-Si, J. (1959). *J. Appl. Chem. USSR* **32,** 1239.
Appen, A., and Hsi, F. (1959). *Sov. Phys.-Sol. State* **1,** 1400.
Appleman, D., and Clark, J. (1965). *Am. Min.* **50.**
Arai, K., Kuwahata, T., Namikawa, H., and Saito, S. (1972). *Jpn. J. Appl. Phys.* **11**(8), 1080.
Aramaki, S., and Roy, R. (1962). *J. Am. Ceram. Soc.* **45**(5), 229.
Araujo, R. (1966). *J. Chem. Phys.* **44**(3), 1299.
Araujo, R. (1979). *Phys. Chem. Glasses* **20**(5), 115.
Araujo, R. (1980). *J. Non-Cryst. Solids* **42**(1/3), 209.
Argyle, J., and Hummel, F. (1963). *Phys. Chem. Glasses* **4**(3), 103.
Armisted, W., (1946a). U.S. Patent 2 393 448.
Armisted, W. (1946b). U.S. Patent 2 393 449.
Armisted, W. (1946c). U.S. Patent 2 393 450.
Armstrong, D., Fortune, R., and Perkins, P. (1977). *J. Non-Cryst. Solids* **24,** 313.
Arndt, J. (1971). *Phys. Chem. Glasses* **12,** 1.
Arndt, J., and Stoeffler, J. (1969). *J. Phys. Chem. Glasses* **10**(3), 117.
Arndt, J., Hornemann, U., and Mueller, W. (1969). *J. Am. Ceram. Soc.* **52,** 285.
Arnold, G., and Borders, J. (1976). *Inst. Phys. Conf. Series 1975* (*Appl. Ion Beam Mat.*) **28,** 121.
Artamonova, G. (1971). *Steklo Tr. Nauch-Issl. Inst. Stekla* (1), 74.
Artamonova, M., Pavlvsukin, N., and Abdrashitova, E. (1978). *Tr. Mosk. Khim-Tekhnol. Inst. Im. D. I. Mendeleeva* **100,** 128. *CA* **91,** 293, 197706t (1980).
Artyushkina, N., Byrdina, V., Estropiev, K., and Khalilev, V. (1972). *Neorg. Mat.* **8**(5), 970.

Asahara, Y., and Izumitani, T. (1974). *Phys. Chem. Glasses* **15**(11), 583.
Asahara, Y., and Izumitani, T. (1975). *Phys. Chem. Glasses* **16**(2), 29.
Aslanova, M., and Kostareva, S. (1974). *Proc. Int. Cong. Glass, Kyoto, 10th* p. 813.
Assabghi, F., Arafa, S., Bishay, A., Boulos, E., and Kreidl, N. (1977). *J. Non-Cryst. Solids* **23**, 81.
Atalla, M., Tannenbaum, E., and Scheibner, E. (1959). *Bell Syst. Tech. J.* **38**(3), 749.
Averyanov, U., and Poraï-Koshits, E. (1966). "The Structure of Glass," p. 63. Consultants Bureau, New York.
Averyanov, V., Mazurin, Po., Poraï-Koshits, E., Reiss, H., Roskova, G., and Vogel, W. (1979). *Fiz. Shim Stekla* **5**(6), 637.
Avetkyan, G., Baidakov, L., and Strakhov, L. (1969). *Inorg. Mat.* **5**(10), 1411.
Axmann, A., Gissler, W., Kollmar, A., and Springer, T. (1970). *Proc. Faraday Conf. Vitr. State, London.*
Azoulay, R., Thiebierge, H., and Brenac, A. (1975). *J. Non-Cryst. Solids* **18**, 33.
Bacon, J. (1971a). Unitech. Aircraft Res. Lab., Final Rep. NASA Contracts, NASw 1301, 2013, January 31.
Bacon, J. (1971b). U.S. Patent 3 573 078.
Bagley, B., and Bair, H. (1970). *J. Non-Cryst. Solids* **2**, 155.
Bagley, B., Disalvo, F., and Waszczak, J. (1975). *J. Non-Cryst. Solids* **17**, 433.
Bailey, L., Brau, M., and Hilton, A. (1964). U.S. Patent 3 154 242, October 27.
Baldwin, C., and Mackenzie, J. (1979). *J. Am. Ceram. Soc.* **62** (9/10), 537.
Baldwin, C., Almeida, R., and Mackenzie, J. (1981). *J. Non-Cryst. Solids* **43**, 309.
Balk, P., and Eldrige, J. (1969). *Proc. IEEE* **57**(9), 1558.
Ball, G., and Chamberlain, J. (1978). *J. Non-Cryst. Solids* **29**, 239.
Ballard, C., and Pye, L. (1976). *J. Am. Ceram. Soc.* **59**(5/6), 266.
Barber, S., and Fajans, K. (1954). *J. Am. Chem. Soc.* **74**, 2761.
Barkatt, A., and Angell, C. (1978). Unpublished.
Barnier, S., Guittard, M., and Flahaut, J. (1980). *Mat. Res. Bull.* **15**(6), 689.
Barr, S., Mondy, J., and Rowe, A. (1970). *Int. Conf. Phys. Non-Cryst. Solids, 3rd, Sheffield, September.*
Barrau, B., Ribes, M., Maurin, M., Kone, A., and Souquet, J. (1980). *J. Non-Cryst. Solids* **37**, 1.
Bartenev, G., Brekhovskikh, M., Varisov, A., Landa, L., and Tsiganov, A. (1970). *Inorg. Mat.* **6**(9), 1371.
Barth, J. (1973). *Z. Anorg. All. Chem.* **396**(1), 103.
Bartholomew, R. (1970). *J. Phys. Chem.* **74**, 2507.
Bartholomew, R. (1972). *J. Non-Cryst. Solids* **7**, 221.
Bartholomew, R. (1973). *J. Non-Cryst. Solids* **12**, 321.
Bartholomew, R., and Holland, H. (1969). *J. Am. Ceram. Soc.* **52**, 402.
Bartholomew, R., Butler, B., Hoover, H., and Wu, C. (1980). *J. Am. Ceram. Soc.* **63**(9/10), 481.
Batsanova, L., Yur'ev, G., and Doronina, V. (1968). *Z Struct. Khim.* **9**(1), 79.
Baugher, J. (1972). *Phys. Chem. Glasses* **13**(7), 63.
Baugher, J., and Bray, P. (1969). *Phys. Chem. Glasses* **10**(3), 77.
Baugher, J., and Parke, S. (1970). *Int. Conf. Non-Cryst. Solids, 3rd, Sheffield, September.*
Bayer, E., Kempter, K., and Kiendla, (1977). *Phys. Non-Cryst. Solids Int. Conf., 4th*, p. 242.
Baynton, P., Rawson, H., and Stanworth, J. (1957a). *Nature (London)* **179**, 434.
Baynton, P., Rawson, H., and Stanworth, J. (1957b). *J. Electrochem. Soc.* **104**, 237.
Beall, G., and Reade, R. (1979). U.S. Patent 4 140 645.
Beall, G., Karstetter, B., and Rittler, H. (1967). *J. Am. Ceram. Soc.* **50**(4), 181.

Beattie, J. (1953). *Trans. Faraday Soc.* **49**, 1059.
Beck, N. (1939). *Phys. Z.* **40**, 479.
Beekenkamp, P. (1964). *Verres Refr.* **18**(1), 3.
Beekenkamp, P. (1965). *Phys. Non-Cryst. Solids Proc. Int. Conf.*, p. 512.
Beekenkamp, P. (1966). *Verres Refr.* **20**, 419.
Beeman, D., Lynds, R., and Anderson, M. (1980). *J. Non-Crystall. Solids* **42**(1/3), 61.
Bell, R. (1970). *Solid State Phys.* **3**, 2111.
Bell, R., and Dean, P. (1966). *Nature (London)* **212**, 1354.
Bell, R., and Dean, P. (1968). *Phys. Chem. Glasses* **9**(4), 125.
Bell, R., and Hibbins-Butler, D. (1975). *Solid State Phys.* **8**, 787.
Beltadze, P., Kovaleva, I., Kolobkov, V., Lunkin, S., and Mshvelidze, G. (1978). *Soobshch. Akad. Nauk. Gruz. SSR* **90**(2), 329.
Beltadze, P., Kovaleva, I., Kolobkov, V., Lunkin, S., and Mshvelidze, G. (1979). *CA* **90**, 42979.
Bennett, A., and Roth, C. (1971a). *J. Phys. Chem. Solids* **32**, 1251.
Bennett, A., and Roth, C. (1971b). *Phys. Rev. B* **4**, 2686.
Berezhnoi, A., Kumalagov, I., and Sark'sov, P. (1977). *Neorg. Mat.* **13**(8), 1509.
Bergeron, C., and Russell, C. (1965a). *J. Am. Ceram. Soc.* **48**(3), 115.
Bergeron, C., and Russell, C. (1965b). *J. Am. Ceram. Soc.* **48**(3), 162.
Berkes, J., and White, W. (1966). *Phys. Chem. Glasses* **7**, 191.
Berkes, J., and White, W. (1977). *Phys. Non-Cryst. Solids Int. Conf. 4th*, p. 405.
Berkes, J., White, W., Ing., B., and Hillegas, W. (1971). *J. Appl. Phys.* **42**, 4908.
Bertoluzza, A., Fagnano, C., Monti, P., and Semerand, G. (1978). *J. Non-Cryst. Solids* **29**, 49.
Betts, F., Bienenstock, A., and Ovshinsky, S. (1970). *J. Non-Cryst. Solids* **4**, 554.
Betts, F., Bienenstock, A., Keating, D., and Deneufville, J. (1972a). *J. Non-Cryst. Solids* **7**, 417.
Betts, F., Bienenstock, A., and Bates, C. (1972b). *J. Non-Cryst. Solids* **8–10**, 364.
Bhat, B., and Bhatia, K. (1978). *Phys. Chem. Glasses* **19**(4), 55.
Bienenstock, A., Betts, F., Keating, D., and Deneufville, J. (1970). *Bull. Am. Phys. Soc.* **15**, 1616.
Bienenstock, A., Mortyn, F., Narasimhan, S., and Rowland, S. (1976). *Monogr. Textbook Mat. Sci.* **8**, 1.
Billhardt, H. (1969). *Glastechn. Ber.* **42**(12), 498.
Biscoe, J., Pinius, A., Smith, C., Jr., and Warren, B. (1941). *J. Am. Ceram. Soc.* **24**, 116.
Biscoe, J., Robinson, C., and Warren, B. (1939). *J. Am. Ceram. Soc.* **22**, 6.
Bishay, A. (1965). *Int. Cong. Glass 7th Brussels* 24.
Bishay, A., and Goma, I. (1976). *J. Am. Ceram. Soc.* **50**(6), 302.
Bishay, A., and Goma, I. (1969). *Phys. Chem. Glasses* **9**(16), 193.
Bishay, A., and Hayek, M. (1977). *Phys. Non-Cryst. Solids Int. Conf. 4th*, p. 142.
Bishay, A., Boulos, E., Arafa, S., Assabghy, F., and Kreidl, N. (1978). "Borate Glasses" (L. Pye *et al.*, eds.), p. 239. Plenum Press, New York.
Bishop, S. (1964). Ph.D. Thesis, Brown Univ.
Bishop, S., Strom, U., and Taylor, P. (1977). "The Structure of Non-Crystalline Materials," p. 109. Trans. Tech. Publ., Aedermannsdorf, Switzerland.
Blair, H., and Milberg, M. (1974). *J. Am. Ceram. Soc.* **57**(6), 257.
Blair, G., and Urnes, S. (1961). *Glastech. Ber.* **34**(8), 391.
Blair, G., Hamblen, D., and Weidel, R. (1960). U.S. Patent 3 393 060.
Blau, H. (1951). *J. Soc. Tech.* **35**, 304.
Blinov, V. (1971a). *Steklobrazne Sist. Nov. Stekla IKh. OSN.* 91.

Blinov, V. (1971b). *Ref. Zh. Khim.* 208. *CA* **87**, 78830e.
Block, J. (1967). Oral communication.
Block, J., and Perloff, A. (1966). *Acta Cryst.* **19**(3), 297.
Block, J., and Piermarini, G. (1964). *Phys. Chem. Glasses* **5**(5), 138.
Bockries, J., and Lowe, C. (1954). *Proc. R. Soc. London Ser. A* **276**, 473.
Bockries, J., and Mackenzie, J. (1955). *Trans. Faraday Soc.* **51**, 1734.
Bockries, J., Kitchener, J., Ignatowicz, S., and Tomlinson, J. (1952). *Trans. Faraday Soc.* **48**, 75.
Bockries, J., Tomlinson, J., and White, J. (1956). *Trans. Faraday Soc.* **52**, 299.
Bockries, J., Tomlinson, J., and Heynes, M. (1958). *Trans. Faraday Soc.* **54**, 1872.
Boehm, H. (1972). *J. Non-Cryst. Solids* **7**, 192.
Boganov, A., Rudenko, V., and Baskina, G. (1966). *Inorg. Mat.* (2), 363.
Bonchev, L., and Reichin, R. (1965). *Dokl. Bolg. Akad. Nauk.* **18**, 805.
Bonchev, L., and Reichin, R. (1968). British Patent 3M 1 117 211 19.
Boolchand, P., and Sukanyi, P. (1973). *Phys. Rev.* **87**, 57.
Bordovski, G., and Izvochikov, V. (1967). *Kristalografia* **12**(5), 940.
Bordovski, G., and Izvochikov, V. (1968). *Crystallography* **12**(5), 818.
Borelli, N. (1963). *Phys. Chem. Glasses* **4**, 11.
Borelli, N. (1967). *J. Appl. Phys.* **38**(11), 4243.
Borelli, N. (1969). *Phys. Chem. Glasses* **10**(2), 43.
Borgardus, E. (1965). *J. Appl. Phys.* **36**(8), 2504.
Borisova, Z. (1970). *Stekloobraznoe Sost.* **5**(1), 89.
Borisova, Z., and Timofeev, V. (1965). *Khim. Tverd. Tela* **12d**, LGV 86.
Borisova, Z., Pazin, A., and Obraztsov, N. (1971). *Univ. Fiz. Khim.* (3), 121.
Borisova, Z., Dakulina, V., and Mironova, T. (1973). *Vestn. Leningr. Univ. Fiz. Khim.* (2) 107.
Borisova, Z., Kornienko, L., Obraztsov, A., and Durkina, E. (1976). *Neorg. Mat.* **12**(4), 592.
Boulos, E. (1971). Thesis, Univ. of Missouri, Rolla.
Boulos, E., and Kreidl, N. (1971a). *J. Am. Ceram. Soc.* **54**(6), 318.
Boulos, E., and Kreidl, N. (1971b). *J. Am. Ceram. Soc.* **54**(8), 364.
Boulos, E., and Kreidl, N. (1972). *J. Can. Ceram Soc.* **41**, 83.
Bradley, A. (1924). *Phil. Mag.* **48**, 477.
Brady, G. (1956). *J. Chem. Phys.* **24**, 477.
Brady, G. (1957). *J. Chem. Soc.* **27**, 300.
Brady, G. (1958). *J. Chem. Soc.* **28**, 48.
Bragg, L., and Claringbull, G. F. (1965). "Crystal Structures of Minerals." Cornell Univ. Press, Ithaca, New York.
Brandenberger, J. (1964). Master's Thesis, Brown Univ.
Brasen, D. (1972). *J. Non-Cryst. Solids* **11**, 113.
Brasen, D. (1974). *J. Non-Cryst. Solids* **15**(3), 395.
Braunstein, H., and Braunstein, J. (1971). *J. Chem. Thermo.* **3**, 419.
Braunstein, J., Bacarella, A., Benjamin, B., Brown L., and Girard, C. (1977). *J. Electrochem. Soc.* **124**, 844.
Brawer, S. (1975). *Phys. Chem. Glasses,* **16**, (1).
Brawer, S., and White, W. (1977). *J. Non-Cryst. Solids* **23**(2), 261.
Bray, P. (1958). *J. Chem. Phys.* **29**, 984.
Bray, P. (1964). *Phys. Chem. Glasses* **35**, 435.
Bray, P. (1966). "The Structure of Glass" (E. Porai-Koshits, ed.), Vol. 7. Consultant Bureau, New York.
Bray, P. (1978). "Borate Glasses" (L. Pye *et al.*, eds.), p. 321. Plenum Press, New York.

Bray, P., and O'Keefe, J. (1963). *Phys. Chem. Glasses* **4**(2), 37.
Bray, P., and Silver, A. (1962). *Mod. Asp. Vitr. State* 113.
Bray, P., Leventhal, N., and Hopper, H. (1963). *Phys. Chem. Glasses* **4**(2), 47.
Brebick, R. F. (1968). *J. Chem. Phys.* **49**(6), 2584.
Brecher, C., and Risenberg, L. (1979). *Univ. Conf. Glass Sci., 5th. Electr., Magn. and Opt. Prop Glasses* p. 469.
Brekhovskikh, S., Sidorov, T., and Grechishkin, V. (1971). *Zlatogorskii, Neorg. Mat.* **7**(9), 1596.
Brewster, G., Kreidl, N., and Pett, T. (1947). *Trans. Soc. Glass Technol.* **31**, 153.
Bridgeman, P. (1939). *Am. Sci.* **237**, 7.
Bridgeman, P., and Simon, I. (1953). *J. Appl. Phys.* **26**, 405.
Bril, T. (1976). Doctor's Thesis, Eindhoven.
Bronswijk, J. (1977). *Phys. Non-Cryst. Solids Int. Conf. 4th*, p. 101.
Bronswijk, J., and Strijks, E. (1977). *J. Non-Cryst. Solids* **24**(1), 145.
Brosset, L. (1958). *J. Soc. Electron. Technol.* **42**, 125.
Brown, D., and Douglas, R. (1965). *Glass Technol.* **6**(6), 190.
Brown, R. (1977). *J. Non-Cryst. Solids* **24**, 131.
Brown, R., Miilner, A., and Allgaier, R. (1970). *Thin Solid Films* **5**, 157.
Brown, S., and Ginell, R. (1962). *Symp. Nucl. Am. Ceram. Soc.* 109.
Brückner, R. (1970). *Glastechn. Ber.* **43**, 8.
Brückner, R. (1971). *J. Non-Cryst. Solids* **5**, 123, 177.
Brückner, R., Chun H.-U., and Goretzki, H. (1978). *Glastech. Ber.* **51**(1), 7.
Brückner, R., Chun, H.-U., Goretzki, H., and Sammet, M. (1980). *J. Non-Cryst. Solids* **42**(1/3), 49.
Brunauer, S., Emmet, P., and Teller, E. (1938). *J. Am. Chem. Soc.* **60**, 309.
Brungs, M., and Cartney, E. (1975). *Phys. Chem. Glasses* **16**(2), 48.
Buerger, M., Klein, G., and Donnay, G. (1954). *Am. Min.* **39**(9/10), 805.
Bunton, G. (1971). *J. Non-Cryst. Solids* **6**, 72.
Burckhardt, W. (1980). *Z. Anorg. Allg. Chem.* **461**, 35.
Burggraaf, A., and Van Velzen, J. (1969). *J. Am. Ceram. Soc.* **52**(5), 238.
Burggraaf, T., Gaiant, G., Dobychin, D., Relshakrit, A., and Tolstoi, M. (1971). *Inorg. Mat.* **7**(6), 909.
Burnett, D., and Douglas, R. (1970). *Phys. Chem. Glasses* **11**(5), 125.
Burnett, D., and Douglas, R. (1971). *Phys. Chem. Glasses* **12**(5), 117.
Burro, E., and Ardelean, I. (1979). *Mat. Res. Bull.* **14**(11), 1425.
Button, D., Tandon, R., Tuller, H., and Uhlmann, D. (1980). *J. Non-Cryst. Solids* **42**(1/3), 297.
Cable, M. (1978). "Borate Glasses" (L. Pye *et al.*, eds.), p. 399. Plenum Press, New York.
Cable, M., and Haroon, M. (1970). *Glass Technol.* **11**, 48.
Cable, M., and Martlew, D. (1971). *Glass Technol.* **12**, 142.
Cahn, J., and Charles, R. (1965). *Phys. Chem. Glasses* **6**, 18.
Cahn, J., and Hillard, J. (1959). *J. Chem. Phys.* **31**, 688.
Cahn, J., and Hillard, J. (1961). *Acta Met.* **9**, 195.
Callow, R. (1952). *J. Soc. Glass Technol.* **36**, 266, 270.
Calvert, P., and Shaw, R. (1970). *J. Am. Ceram. Soc.* **53**(6), 350.
Camara, B., Kozhukarov, V., and Oel, H. (1980). *Glastech. Ber.* **53**(1), 10.
Carlson, E. (1932). *J. Res. Nat. Bur. Std.* **9**, 830.
Carpenter, L. (1972). U.S. Patent 239 626.
Carpenter, L. (1973). French Patent 2 178 175.
Carron, G. J. (1963). *Acta Cryst.* **16**, 338.
Carwile, L., and Hage, H. (1966). U.S. Army Tech. Rep. 67-7-PR.

Caslavska, V., and Strickler, D. (1969). *J. Am. Ceram. Soc.* **52**(3), 154.
Caslavska, V., Strickler, D., Gibbon, D., and Roy, R. (1968). *J. Mat. Sci.* **3**, 440.
Cassanhiol, B., Cornet, J., and Rossier, D. (1973). *Int. Coll. Cathod. Sputtering, 1st, Montpellier.*
Cerny, V., and Frumar, M. (1979). *J. Non-Cryst. Solids* **33**(1), 23.
Cervelle, B., Jaulmes, S., Larvelle, P., and Loireau-Lazach, A. (1980). *Mat. Res. Bull.* **15**(2), 159.
Cervinka, L., and Hruby, A. (1979). *J. Non. Cryst. Solids* **34**, 275.
Cervinka, L., Trkal, V., and Lezal, T. (1977). *J. Non-Cryst. Solids,* **23**, 2.
Chance-Pilkington (1977). *Glass Ind.* **58**(11), 8.
Chandrashekhar, G., and Shaper, M. (1980). *Mat. Res. Bull.* **15**(2), 221.
Chang, J., and Dove, D. (1974). *J. Non-Cryst. Solids* **16**, 72.
Chang, S., and Bestul, A. (1974). *J. Chem. Thermod.* **6**(4), 325.
Charles, R. (1963). *J. Am. Ceram. Soc.* **46**(5), 235.
Charles, R. (1964). *J. Am. Ceram. Soc.* **47**(11), 539.
Charles, R. (1966). *J. Am. Ceram. Soc.* **49**(7), 55.
Charles, R. (1967). *J. Am. Ceram. Soc.* **50**(12), 631.
Charles, R. (1968). *J. Am. Ceram. Soc.* **51**(1), 16.
Charles, R. (1969). *Phys. Chem. Glasses* **10**, 169.
Chase, G., and Phillips, C. (1964). *J. Am. Ceram. Soc.* **47**(9), 467.
Chaudhari, P., and Herd, S. (1972). *J. Non-Cryst. Solids* **8–10**, 56.
Chernov, A., Babievskaya, I., Ol'khovskii, V., and Kalinnikov, V. (1977). *Fiz. Khim. Stekla* **3**(3), 208, 231.
Chin, D. (1971). Ph.D. Thesis, Purdue Univ.
Cleek, G., and Hamilton, E. (1964). U.S. Patent 3 119 703.
Cline, C., and Kingman, D. (1976). Fluoroberryllate Glasses and Crystals. Bibliography, ERDA Contract, #W-7405-Eng.-48.
Cline, C., and Weber, M. (1977). Rep. UCRL-81168, CONF 780720-1. Available NTIS, *Energy Abstr.* **3**(24), #57229; *CA* **91**, 222, #8628k (1979).
Cohen, A., and Smith, H. (1958). *J. Phys. Chem. Solids* **7**, 301.
Cohen, M. (1971). *Phys. Today* **26**, May.
Cohen, M., Fritzsche, H., and Ovshinsky, S. (1969). *Phys. Rev. Lett.* **22**, 1065.
Cohen, M., Neale, R., and Paskin, A. (1972). *J. Non-Cryst. Solids* **8/10**, 855.
Cojocaru, L. (1966). *Rev. Roum. Phys.* **11**(7), 593.
Coleman, M., Duncan, J., and Sithamarad, D. (1970). *Rev. Chem. Min.* **7**(6), 1129.
Collins, D., and Mulay, L. (1970). *J. Am. Ceram. Soc.* **53**(2), 74.
Connell, G., and Lucovsky, G., (1978). *J. Non-Cryst. Solids* **31**, 123.
Cooper, A. (1978). "Borate Glasses" (L. Pye, V. Frechette, and N. Kreidl, eds.), p. 167. Plenum Press, New York.
Cornet, T., and Rossier, D. (1973). *J. Non-Cryst. Solids* **12**, 85.
Cornet, T., and Schneider, J. (1977). *Phys. Non-Cryst. Solids Int. Conf. 4th,* p. 397.
Corning Glass (1928). British Patent 1 213 603.
Corning Glass (1977). French Patent 2349421-1977-11-25.
Corsaro, R. (1976). *Phys. Chem. Glasses* **17**(1), 13.
Couty, R., and Gabatier, G. (1978). *J. Chim. Phys.-Phys. Chim. Biol.* **75**(9), 843.
Coutures, J. *et al.* (1974). *Rev. Int. Hautes Temp. Ref.* **4**, 263.
Coutures, J., Benezech, G., and Aivtic, E. (1975). *Mat. Res. Bull.* **10**(6), 539.
Coward, L. (1971). *J. Non-Cryst. Solids* **6**(2), 107.
Dale, S., and Stanworth, J. (1949). *J. Soc. Glass Technol.* **33**, 167.
Dale, S., and Stanworth, J. (1951). *J. Soc. Glass Technol.* **35**, 185.

Daniel, M., Leadbetter, A., Wright, A., and Sinclair, N. (1979). *J. Non-Cryst. Solids* **32**, 271.
Daniel, M., Leadbetter, A., Wright, A., and Sinclair, N. (1980). *J. Non-Cryst. Solids* **41**, 127.
Daniels, W., and Moore, R. (1974). *J. Am. Ceram. Soc.* **57**(11), 480.
Danilov, A., and El Mosli, M. (1963). *Sov. Phys.-Solid State* **5**, 1472.
d'Anjou, A., and Sanz, F. (1978). *J. Non-Cryst. Solids* **28**(3), 319.
Dannheim, H., Del, H., and Tomandl, G. (1976). *Glastech. Ber.* **49**(7), 170.
Das, G. (1967). *Glass Ceram. Bull., India* **14**(3), 85.
Das, G., Bever, M., Uhlmann, D., and Moss, S. (1972). *J. Non-Cryst. Solids* **7**, 251.
Das, G., Patakis, N., and Bever, M. (1974). *J. Non-Cryst. Solids* **15**, 30.
Da Silva, J., Pinatti, D., Anderson, C., and Rudee, M. (1975). *Phil. Mag.* **31**, 713.
Datta, R., Roy, D., Faile, S., and Tuttle, D. (1964). *J. Am. Ceram. Soc.* **47**(3), 153.
Davis, E., and Greeson, R. (1979). U.S. Patent 4 146 655 (March 27).
Day, D. (1972). "Amorphous Materials" (R. Douglas and B. Ellis, eds.). Wiley, New York.
Day, D. (1976). *J. Non-Cryst. Solids* **21**, 343.
Day, D., and Rindone, G. (1962). *J. Am. Ceram. Soc.* **45**(12), 579.
Day, D., and Steinkamp, W. (1969). *J. Am. Ceram. Soc.* **52**(11), 571.
Day, D., and Stevels, J. (1974). *J. Non-Cryst. Solids* **14**, 165.
Day, R. K. (1953). "Glass Research Methods." Industrial Publ.
DeCarli, P., and Jamieson, J. (1959). *J. Chem. Phys.* **31**, 1675.
DeCarli, P., Jamieson, J., and Milton, P. (1964). *Science* **147**, 144.
Decuottignigs, M., Philippou, J., Zarzycki, J. (1977). *CR-C* **285**(8), 265. *CA* **88**, 234, 77781e.
Deeg, E. (1962). *Advan. Glass Tech.* 348.
Dembovskii, S. (1969). *Inorg. Mat.* **5**(3), 385.
Dembovskii, S., and Chernov, A. (1972). *Mekh. Tepl. Sov. Str. Neorg. Stekol* 380.
Dembovskii, S., and Luzhnaya, N. (1964). *Zh. Neorg. Khim.* **9**, 660.
Dembovskii, S., and Luzhnaya, N. (1971). *Neorg. Mat.* **7**, 328.
Dembovskii, S., Kirilenko, V., and Khvorostenko, A. (1971). *Inorg. Mat.* **7**(10), 1659.
Dembovskii, S., Chernov, A., Vinogradova, G., Kirilenko, V., Kirilenko, I., and Khvorostenko, A. (1974). *Stekloobr. Sost.* 279.
Deneufville, J., and Turnbull, D. (1970). *Faraday Soc. Conf. Vitreous State, September, 1970.*
Deneufville, J., and Turnbull, D. (1970/1972). Con. Pr. DAHC 15-70-C-0187, CARPA, 1570-ODIO 5/8/70-5/8/72.
Deneufville, J., and Turnbull, D. (1972). *J. Non-Cryst. Solids* **8–10**, 85.
Deneufville, J., Turnbull, D., Moss, S., and Ovshinsky, S. (1973/1974). *J. Non-Cryst. Solids* **13**, 191.
Deneufville, J., Turnbull, D., and Ockstad, H. (1974). "Amorphous and Liquid Semiconductors" (J. Stuke and W. Brenig, eds.), p. 419. Taylor and Francis, London.
Deneufville, J., Turnbull, D., Gerard, P., and Devenyi, J. (1976). *J. Non-Cryst. Solids* **22**, 77.
Denker, B., Kornienko, L., Maksimova, G., Osiko, Vo., Rybaltovski, A., and Tikhomirov, V. (1979). *Fiz. Khim. Stekla* **5**(6), 720.
Denton, E., Rawson, H., and Stanworth, J. (1954). *Nature (London)* **173** (4413), 1030.
De Paolis, P., and Mauer, P. (1947). U.S. Patent 3 250 721.
Deutschbein, O. (1967). French patent 502 709, 24, November.
Deutschbein, O., and Pautrat, C. (1967). *Rev. Phys. Appl.* **2**, 29.
DeVos, W., and Volger, J. (1967). *Physica* **34**, 272.
De Waal, H. (1969). *Phys. Chem. Glasses* **10**(3), 181.
Dgebuadze, T., and Averyanov, V. (1970). *Steklo. Sist.* **5**(1), 72.

Dietzel, A. (1941). *Z. Electrochem.* **48**, 9.
Dietzel, A., and Sheybany, H.-A. (1948). *Verres Refr.* **6**(2), 63.
Dietzel, A. (1966). *Naturwissenschaften* **53**(1), 16.
Dietzel, A., and Poegel, H. (1954). *Int. Congr. Glass, 3rd, Venice, 1953* Vol. 3, p. 219. Stabilimento Grafica di Roma, Rome.
Dimarcello, F., Treptow, W., and Baker, L. (1968). *Bull. Am. Ceram. Soc.* **47**, 511.
Dimitriev, Y., Marinov, Ivanova, I., and Popov, M. (1970). *Dokl. Bolg. Akad. Nauk* **23**(5), 507.
Dimitriev, Ya., Bankov, A., Nanova, I., Dimitrov, V., Petrakiev, A., Tomova, M. (1980). *Stroit. Mat. Sil. Prom. St.* **21**(1), 22; *CA* **93**, 366, 52442k.
Disalvo, N., Roy, D., and Mulay, L. (1972). *J. Am. Ceram. Soc.* **55**(10), 536.
Distefano, T., and Eastman, D. (1971). *Phys. Rev. Lett.* **27**, 1560.
Dobychin, D. (1962). *Zh. Prikl. Khim.* **35**(1), 51.
Domenici, M., and Pozza, F. (1970). *J. Mater. Sci.* **5**, 746.
Dong, N. (1973). *J. Non-Cryst. Solids* **12**, 161.
Dong, N. (1976). *Phys. Status Solidi (a)* **33**, 195.
Dong, N., Danh., T., Auguin, B., and Defrisne, A. (1977). *Phys. Non-Cryst. Solids Int. Conf. 4th*, p. 172.
Donnay, G., Schairer, J., and Donnay, J. (1959). *Min. Mag.* **32**(245), 93.
Doremus, R. (1962). "Mod. Asp. Vitr. State" (J. Mackenzie, ed.). Butterworth, London.
Doremus, R. (1969). "Ion Exchange," Vol. 2. Dekker, New York.
Doremus, R. (1970). *J. Non-Cryst. Solids* **3**, 369.
Doremus, R. (1973). "Glass Science." Wiley, New York.
Doremus, R. (1974). *J. Am. Ceram. Soc.* **57**(11), 478.
Dorr, R., and Kannewure, C. (1971). *J. Non-Cryst. Solids* **6**(2), 113.
Douglas, R., and Zaman, M. (1969). *Phys. Chem. Glasses* **10**(4), 125.
Daupovec, J., Thurzo, I., and Ulasak, G. (1979). *J. Am. Ceram. Soc.* **62**, 16.
Drake, C., and Scanlan, I. (1970). *German Patent* 2 027 941, p. 12.
Drexhage, M., Hoynihan, C., and Saleh, M. (1980). *Mat. Res. Bull.* **15**(2), 213.
Dubrovo, S., and Shmidt, Y. (1959). *Zh. P. Khim.* **32**(4), 742 (see also Mazurin, 1973).
Duering, O. (1966). *Acta Cryst. Suppl. A* **21**, Part 7, 232.
Duffy, J. (1970). *Phys. Chem. Glasses* **11**(1), 1.
Duffy, J. (1971). *Phys. Chem. Glasses* **12**(3), 87.
Duffy, J. (1972). *Phys. Chem. Glasses* **13**(3), 65.
Duffy, J. (1977). *J. Am. Ceram. Soc.* **60**(9), 440.
Duffy, J., and Ingram, M. (1969). *J. Am. Ceram. Soc.* **52**, 224.
Duffy, J., and McDonald, W. (1970). *J. Chem. Soc. A* 4411.
Duke, D., McDowell, J., and Karstetter, B. (1967). *J. Am. Ceram. Soc.* **50**(2), 67.
Dumbaugh, W., and Schultz, P. (1969). "Kirk-Othmer Encyclopedia of Chemical Technology," p. 18. Wiley, New York.
Dunalev, A., Borisova, Z., Mikhailov, M., and Bratov, A. (1980). *Fiz. Khim. Stekla* **6**(2), 174.
Dunicz, B. (1966). *Science* **153**, 737.
Dunlevy, F., and Cooper, A. (1972). *Bull. Am. Ceram. Soc.* **51**, 374.
Dunn, B., Ooka, K., and Mackenzie, J. (1973). *J. Am. Ceram. Soc.* **56**(9), 494.
Dzevuskaya, T., Khodskii, L., Vorotinskaya, D., Milevskaya, R. (1977). *Vst. Ak. Nauk. USSR, Ser. Khim. Nauk* (4) 53. *CA* **82**, 236, 171756b.
Easteal, A., and Angell, C. (1970). *J. Phys. Chem.* **74**, 2987.
Easteal, A., and O'Rourke, P. (1974). *Aust. J. Chem.* **27**, 35.
Easteal, A., Sare, E., Moynihan, C., and Angell, C. (1974). *J. Solids Chem.* **3**, 807.
Edmond, J. (1968). *J. Non-Cryst. Solids* **1**(1), 39.

Edwin, D., Seager, C., and Quinn, R. (1972). *Phys. Rev. Lett.* **28**, 813.
Edwin, D., Seager, C., and Quinn, R. (1977). *Phil. Mag.* **35**, 1188.
Elimov, A., and Mikhailor, B. (1979). *Fiz. Khim. Stekla* **51**(6), 692.
Egorova, E., and Kokorina, V. (1960). "Structure of Glass." Vol. 2, p. 430. Consultant Bureau, New York.
Egorova, E., and Kokorina, V. (1967). *Steklo. Tr. Gos. Nauch. ISS, Inst. Stekla* (1), 109.
Egorova, E., and Kokorina, V. (1968a). *Zh. Prikl. Khim.* **3**(4), 440.
Egorova, E., and Kokorina, V. (1968b). *Ref. Zh. Khim.* 138737; *CA* **86**, 81388c (1971).
Egorova, E. A., and Kokorina, V. F. (1968c). *Zh. Prikl. Khim.* (*Engl. Transl.: J. Appl. Chem. U.S.S.R.*) **41**(6), 1142.
Eguchi, K., Tarumi, Sh., and Hanaoka, H. (1979). *Yogyo Kyokaishi* **87**(12), 602.
Eguchi, K., Kurita, H., and Kato, T. (1980). *Yogyo Kyokaishi* **88**(1), 21.
Eisenberg, A., and Tobolsky, A. (1960). *J. Polym. Sci.* **46**, 19.
Eissa, N., Shaishta, E., and Hussivn, A. (1974). *J. Non-Cryst. Solids* **16**(2), 206.
Eitel, W., and Skalics, W. (1929). *Z. Anorg. Allg. Chem.* **183**(3), 263.
Ellis, J. (1971). U.S. Patent 3 564 587, 02-16-1971.
Elmer, T. (1965). *Int. Cong. Glass, 7th, Brussels*.
Elmer, T. (1976). *Ceram. Bull.* **55**(11), 999.
Elmer, T., and Meissner. (1976). *J. Amer. Ceram. Soc.* **59**, 206.
El Mouly, M. (1966). *Vestn. Leningr. V. Fiz. Khim.* (1), 152.
Emin, D. (1977). *Phil. Mag.* **35**, 1188.
Emin, D. (1980). Electrical and optical properties of amorphous thin films. *In* "Polycrystalline and Amorphous Thin Films and Devices" (L. Kazmerski, ed.), pp. 6, 17, 27–28. Academic Press, New York.
Emin, D., Seager, C., and Quinn, R. (1972). *Phys. Rev. Lett.* **28**, 813.
Emons, H. H., and Theisen, L. (1968). *Z. Anorg. Allg. Chem.* 321–327.
Endell, K. (1940). *Z. Angew. Chem. Beith.* 38.
Enderby, J., Howells, W., and Howe, R. (1973). *Chem. Phys. Lett.* **21**, 109.
Enright, D., and Marshall, P. (1957). ONR Tech. Rep., Penn. State Univ., April, p. 43.
Efimov, A., and Mikhailov, B. (1979). *Fiz. Khim. Stekla* **5**(6), 692.
Eppler, R. (1963). *J. Am. Ceram. Soc.* **46**(2), 97.
Ernsberger, F. (1974). U.S. Patent 3 756 789.
Esqueda, P., and Henisch, H. (1976). *J. Non-Cryst. Solids* **22**, 97.
Etienne, S., Guenin, G., and Perez, J. (1979). *J. Phys. D* **12**(12), 2189.
Evans, D., and King, S. (1966). *Nature* (*London*) **212**, 1352.
Evans, E., Helber, J., and Ovshinsky, S. (1970). *J. Non-Cryst. Solids* **2**, 334.
Everstein, F., Stevels, J., and Waterman, H. (1960). *J. Phys. Chem.* **1**, 123.
Evstrop'ev, K. (1960). "The Structure of Glass," Vol. II, p. 237. Consultant Bureau, New York.
Evstrop'ev, K. (1969). *Dokl. Khem. Tekh.* **188/89**, 197.
Evstrop'ev, K. (1970). *Stekl. Sost.* **5**(1), 139.
Evstrop'ev, K., and Davlovskii, V. (1967). *Neorg. Mat.* **3**(4), 673.
Evstrop'ev, K., and Nanov, I. (1978). *Zh. Prikl. Khim.* (*Leningr.*) **51**(J), 985.
Evstrop'ev, K., and Yakhkind, A. (1965). British Patent appl. 979, 193.
Evstrop'ev, K., and Yakhkind, A. (1966). U.S. Patent 3 291 620.
Evstrop'ev, K., Petrovski, G., and Khalilev, W. (1971). *Proc. Int. Cong. Glass, 9th. Versailles*, Vol. 1, p. 485.
Evstrop'ev, K., Krupkin, Y., Galimov, D., Tarlakov, Y., and Shevyakov, A. (1970a). *Zh. Prikl. Spektrosk.* **13**(4), 655.
Evstrop'ev, K., Krupkin, Y., and Milyukov, E. (1970b). *Neorg. Mat.* **6**(6), 1126.
Eysel, W., Wolfe, W., and Newham, R. (1973). *J. Am. Ceram. Soc.* **56**(4), 185.

Fagan, M. (1951). *Proc. Natl. Electron Conf.* **7**, 380.
Faile, S., and Roy, D. (1970). *Mat. Res. Bull.* **5**(6), 385.
Faile, S., Roy, D., and Schmidt, J. (1967). *Science* **156**, 1593.
Fairweather, M., and Murthy, K. (1973). *J. Am. Ceram. Soc.* **56**(7), 349.
Fajans, S., and Barber, S. (1954). *J. Am. Chem. Soc.* **74**, 2761.
Fajans, S., and Kreidl, N. (1948). *J. Am. Ceram. Soc.* **31**(4), 106.
Faulstich, M. (1961). *Glastech. Ber.* **34**, 102 (see also U.S. Patent 3 248 238).
Fawcett, R., Wagner, C., and Cargill, G. III (1972). *J. Non-Cryst. Solids* **8–10**, 369.
Feichtner, J., Isaacs, T., and Price, A. (1974). *Infrared Phys.* **14**(1), 3.
Feltz, A., and Burckhardt, W. (1980). *Z. Anorg. Allg. Chem.* **461**, 35.
Feltz, A., and Lippmann, F. (1973). *Z. Anorg. Ch.* **398**(2), 157.
Feltz, A., and Pfaff, G. (1978). *Z. Anorg. Allg. Chem.* **442**, 41.
Feltz, A., and Pfaff, G. (1979). *Rev. Chim. Min.* **16**(4), 381; *CA* (1980). **92**, 312, 133789k.
Feltz, A., and Schlenzig, E. (1974). *Proc. Conf. Amorph. Liqu. Semicond., 5th, Garmisch-Partenkirchen* (J. Stuke and W. Brenig, eds.), p. 261. Taylor and Francis, London.
Feltz, A., and Voigt, B. (1973). *Z. Anorg. Ch.* **403**(1), 61.
Feltz, A., Buettner, H., Lippmann, F., and Haul, W. (1972). *J. Non-Cryst. Solids* **8–10**, 64.
Feltz, A., Voigt, B., and Schlenzig, E. (1973). *Proc. Conf. Amorph. Liqu. Semicond., 5th, Garmisch-Partenkirchen* (J. Stuke and W. Brenig, eds.), p. 261. Taylor and Francis, London.
Feltz, A., Schlenzig, E., and Arnold, D. (1974). *Z. Anorg. Alleg. Chem.* **403**(3), 243.
Ferguson, G., and Hass, M. (1970). *J. Am. Ceram. Soc.* **53**(2), 109.
Fichet-Ollitraut, R., Cholie, R., and Rivet, J. (1977). *Ann. Chim.* **2**(1), 31.
Fitzpatrick, J., and Maghrabi, C. (1971). *Phys. Chem. Glasses* **12**(4), 105.
Flamenbaum, J., and Schultz, P. (1973). French Patent 2178177.
Flaschen, S., Pearson, A., and Northover, W. (1959). *J. Am. Ceram. Soc.* **42**, 450.
Flaschen, S., Pearson, A., and Northover, W. (1960). *J. Am. Ceram. Soc.* **43**, 274.
Fleming, J., and Shiever, J. (1979). *J. Am. Ceram. Soc.* **62**(9,10), 526.
Floerke, O. (1967). *Z. Krist.* **125**, 134.
Florence, J., Glaze, F., and Black, M. (1955). *J. Res. Nat. Bur. Std.* **55**, 231.
Förland, T., and Weyl, W. (1950). *J. Am. Ceram. Soc.* **33**, 186.
Förland, T., and Tashiro, M. (1951). *Glass Ind.* **37**, 38.
Folger, F. (1973). Thesis, Jena.
Fontanella, J., Johnston, R., Siegel, G., and Andeen, C. (1979). *J. Non-Cryst. Solids.*, **31**(3) 401.
Ford, E., and Savage, J. (1976). *J. Phys. E.* **9**(8), 622.
Franck, H. (1955). *Tag. Ber. Chem. Ges.* DDR, 119.
Francel, J., and Hagedorn, E. (1964). U.S. Patent 3 127 278.
Franz, H. (1978). "Borate Glasses" (L. Pye *et al.*, eds.), p. 567. Plenum Press, New York, 1978.
Franz, M., and Kelen, T. (1967). *Glast. Ber.* **40**, 141.
Fraser, W., and Jerger, J. (1953). *J. Opt. Soc. Am.* **43**, 1153.
Fraser, W., and Upton, L. (1944). *J. Am. Ceram. Soc.* **27**(4), 121.
French, W., Mac Chesney, J., and Pearson, D. (1975). *Ann. Rev. Mat. Sci.* **5**, 375.
Frerichs, R. (1950). *Phys. Rev.* **78**, 643.
Friebele, E., Wilson, L., and Kinser, D. (1972). *J. Am. Ceram. Soc.* **55**, 164.
Friebele, E., Griscom, D., and Sigel, G. (1977). *Phys. Non-Crys. Solids Int. Conf. 4th*, p. 154.
Frischat, G., and Schrimpf, C. (1980). *J. Am. Ceram. Soc.* **63**(11, 12), 714.
Fritz, I. (1978). *J. Appl. Phys.* **49**(8), 4623.

Fritzsche, H. (1971). *J. Non-Cryst. Solids* **6,** 49–71.
Fritzsche, H. (1973). "Electronic and Structure Properties Amorphous Semiconductors" (P. Le Comber and J. Mort, ed.), p. 55. Academic Press, New York.
Fritzsche, H. (1977). *Proc. Int. Conf. Amorph. Liqu. Semicond. 7th, Edinburgh* (W. Spear, ed.), p. 3.
Fritzsche, H., and Kastner, M. (1978). *Phil. Mag.* **37**(5), 285.
Fritzsche, H., Gaczi, P., and Kastner, M. (1978). *Phil. Mag. B* **37**(5), 593.
Fulrath, R., and Hollar, E. (1968). *Ceram. Bull.* **47**(5), 493.
Furakawa, T., and White, W. (1980). *Phys. Chem. Glasses* **21**(2), 85.
Furukawa, T., Brawer, S., and White, W. (1977). *J. Chem. Phys.* (in press).
Fuschillo, N., Lalevic, B., and Annamalai, N. (1976). *J. Non-Cryst. Solids* **21**(1), 85; **22,** 159.
Gahlmann, H., and Brückner, R. (1973/74). *J. Non-Cryst. Solids* **13,** 355.
Gahlamnn, H., and Brückner, R. (1977). *Phys. Non-Cryst. Solids, Int. Conf. 4th,* p. 260.
Gakurai, S., and Mochizuki, H. (1978). *Nagoya-shi Kogyo Kenkyusho Kenkyu Hokoku* **58,** 9; (1979). *CA* **91,** 250, 43431-n.
Galakhov, F., Aver'yanov, V., Arsehev, M., Vavilonova, V., Kolesova, V., and Makeeva, N. (1977). *Fiz. Khim. Stekla* **3**(4), 347.
Galant, E. (1980). *Fiz. Khim. Stekla* **6**(1), 121.
Galimov, D., Krupkin, Y., Burba, A., Sheryanov, A., Evstrop'ev, K., and Konitskii, I. (1969). *Dokl. Chem. Tech.* (188/89), 197; (1971). *CA* 485.
Galimov, D., Krupkin, Y., Burba, A., and Sheryanov, A. (1971). *Zh. Prikl. Spektrosk.* **14**(2), 337.
Gallagher, P., Kurkjian, C., and Bridenbaugh, P. (1965). *Phys. Chem. Glasses* **6**(3), 95.
Gaman, V. (1972). *Izv. VUZ Fiz.* (2) **57**(3), 45.
Ganguli, D., and Saha, P. (1965). *Centr. Glass Res. Inst., India* **12,** 24.
Gani, M., and McPherson, R. (1977). *J. Aust. Cer. Soc.* **12**(2), 21.
Gannon, J. (1980). *J. Non-Cryst. Solids* **43**(1/3), 239.
Garner, C., Gilbert, L., and Wood, C. (1974). *J. Non-Cryst. Solids* **15,** 63.
Gaskell, P., and Johnson, P. (1976). *J. Non-Cryst. Solids* **20**(2), 171.
Gaskell, P., and Ward, R. (1967). *Trans. Met. Soc. AIME* **239**(2), 269.
Gates, L., and Lent, W. (1967). *Bull. Am. Ceram. Soc.* **46**(7), 202.
Gdula, R., and Tompkins, R. (1970). *Glass. Tech.* **11**(6), 164.
Gehlhoff, G., and Thomas, M. (1926). *Z. Tech. Phys.* **7,** 105.
Gelsing, R., Stein, H., and Stevels, J. (1966). *Phys. Chem. Glasses* **7**(6), 185.
Gerasimenko, V., and Khiminets, O. (1977). *Fiz. Khim. Stekla* **3**(4), 343.
Gerhardt, U. (1973). *Rev. Sci. Instrum.* **44**(5), 657.
Gibbons, R., and Kleeman, J. (1970). *Trans. Am. Geophys. Un.* **51**(11), 770.
Gilev, I., and Petrovski, G. (1968). *Neorg. Mat.* **4**(8), 1259, 1264.
Gill, W. (1972). *J. Appl. Phys.* **43,** 5033.
Glaze, F., Blackburn, D., Osmalov, J., Hubbard, D., and Black, M. (1957). *J. Res. Nat. Bur. Std.* **59,** 83.
Godron, Y. (1966). Canadian Patent 779 708.
Goetz, J., and Vosahlova, E. (1968). *Glast. Ber.* **41**(2), 67.
Goetz, J., Hoebbel, D., and Wieker, W. (1976a). *J. Non- Cryst. Solids* **20**(3), 413.
Goetz, J., Hoebbel, D., and Wieker, W. (1976b). *J. Non-Cryst. Solids* **22**(2), 391.
Goldschmidt, D., and Rudman, P. (1976). *J. Non-Cryst. Solids,* **22**(2), 229.
Goldschmidt, D., and Rudman, P. (1977). *Phys. Non-Cryst. Solids Int. Conf. 4th,* p. 305.
Goldschmidt, V. (1926). *Skrifter Norske Videnskaps Akad.* (*Oslo*), *Mat.-natur.* **1**(8), 7.
Golovej, M. *et al.* (1975). *Neorg.* (*Uzhgorod Gos. Univ.*) **11**(4), 745.
Golubkov, V., Poraï-Koshits, E., and Titov, A. (1975). *Fiz. Khim. Stekla* **1**(5), 396.

Golubkov, V., Titov, A., and Poraï-Koshits, E. (1977). *Proc. Int. Cong. Glass, 11th, Prague* (discussed by Porai-Koshits, 1977).
Gonzales-Oliver, C., Johnson, P., and James, P. (1979). *J. Mat. Sci.* **14**(5), 1159.
Gorbachev, V., Bartenev, G., and Tsyganov, A. (1978). *Steklo* (2), 61.
Gordon, J. (1969). Applications of fused salts in organic chemistry. *In* "Techniques and Methods of Organic and Organometallic Chemistry" (D. B. Denney, ed.). Dekker, New York.
Gossink, R. (1977). *J. Non-Cryst. Solids* **26**(1–3), 112.
Gossink, R., and Stevels, J. (1971). *J. Non-Cryst. Solids* **5**(3), 217.
Goto, N., Isozaki, Y., Shidara, K., Maruyama, E., Hirai, T., and Fujita, T. (1974). *IEEE Trans. Electron. Dev.* **21**, 662.
Gough, E., Isard, J., and Topping, J. (1969). *Phys. Chem. Glasses* **10**(3), 89.
Grant, R., Ingram, M., Turner, L., and Vincent, C. (1978). *J. Phys. Chem.* **82**(26), 2838.
Greaves, G. S. (1979). *J. Non-Cryst. Solids* **32**(1,3), 295.
Grechanik, L., Petrovykh, N., and Karschenko, V. (1961a). *Sov. Phys. Solid State* **2**, 1908.
Grechanik, L., Petrovykh, N., and Karschenko, V. (1961b). *Fiz. Tverd Tela* **2**, 2131.
Greco, J., Blair, G., Rindone, G. (1972). German Patent 2 146 682.
Greene, C. (1943). U.S. Patent 234 824.
Greene, C., and Lee, H. (1965). *J. Am. Ceram. Soc.* **48**(10), 528.
Greene, C., and Platts, D. (1969). *J. Am. Ceram. Soc.* **52**(2), 106, 109.
Gresch, C., Müller-Warmuth, W., and Dutz, H. (1976). *J. Non-Cryst. Solids* **21**(1), 31.
Griscom, D. (1977). *J. Non-Cryst. Solids* **24**(2), 155.
Griscom, D. (1978). *In* "Borate Glasses" (L. Pye, V. Frechette, and N. Kreidl, eds.), pp. 11, 139. Plenum Press, New York.
Grison, E. (1951). *J. Chem. Phys.* **19**, 1109.
Grodkiewicz, W. (1971). *Mat. Res. Bull.* **6**(4), 283.
Grossman, D. (1972). *J. Am. Ceram. Soc.* **55**, 446.
Grossman, D., and Isard, J. (1970). *J. Phys. D* **3**(7), 1058.
Guaker, R., and Urnes, S. (1973). *Phys. Chem. Glasses* **14**(2), 21.
Gubanov, A. (1960). *Sov. Phys. Solid State*, **2**, 605.
Gubanov, A. (1962). *Sov. Phys.-Solid State* **4**, 2104.
Gupta, Y., and Mishra, U. (1969). *J. Phys. Chem. Solids* **30**, 1327.
Guymont, M. (1977). *C.R. Acad. Sci. Paris Ser. C* **285**(10), 345.
Haack, D. (1977). *Phys. Non-Cryst. Solids Int. Conf. 4th*, p. 284.
Hafner, H., Kreidl, N., and Weidel, R. (1958). *J. Am. Ceram. Soc.* **41**, 315.
Haisty, R. W., and Krebs, H. (1969). *J. Non-Cryst. Solids*, **1**, 399–426, 427–436.
Hakim, R., and Uhlmann, D. (1967). *Phys. Chem. Glasses* **8**(5), 174.
Haller, W. (1974). *J. Chem. Phys.* **42**, 686.
Haller, W., and Macedo, P. (1968). *Phys. Chem. Glasses* **9**(5), 153.
Haller, W., Blackburn, D., Wagstaff, F., and Charles, R. (1970). *J. Am. Ceram. Soc.* **53**(1), 34.
Hamilton, E., and Cleek, G. (1958). *J. Res. Nat. Bur. Std.* **60**, 593.
Hammel, J. (1965). *Int. Cong. Glass* (36, 1, 2, 3).
Hammel, J. (1967). *J. Chem. Phys.* **46**(6), 2234.
Hammel, J., and Ohlberg, S. (1965). *J. Appl. Phys.* **36**(4), 442.
Hammond, C. (1978). *Phys. Chem. Glasses* **19**(3), 41.
Hammond, C., and Norman, S. (1977). *Opt. Quantum Electron.* **9**(5), 399.
Hanada, T., Soga, N., and Kunugi, M. (1976). *J. Non-Cryst. Solids* **21**(1), 651.
Handleman, E., and Kauser, W. (1964). *J. Appl. Phys.* **35**(12), 3519.

Hansen, K. (1965). *J. Electrochem. Soc.* **112**(10), 994.
Hansen, M. (1958). "Constitution of Binary Alloys." McGraw-Hill, New York.
Hanst, P., Early, U., and Klemperer, W. (1965). *J. Chem. Phys.* **42**(3), 1097.
Harris, I., Jr., and Bray, P. (1980). *Phys. Chem. Glasses* **21**(4), 156.
Harrison, P., and Hummel, F. (1959). *J. Am. Ceram. Soc.* **42**, 487.
Hartleif, Z. (1938). *Z. Anorg. Allg. Chem.* **238**, 353.
Hartung, E., and Heide, K. (1978). *Kristall and Technik* **8–13**, K57.
Hasegawa, H., Sone, M., and Imaoka, M. (1978). *Phys. Chem. Glasses* **19**(2), 28.
Hayami, R., and Terai, R. (1972). *Phys. Chem. Glasses* **13**(4), 102.
Hayes, T. (1978). *J. Non-Cryst. Solids* **31**, 57.
Hayes, T., Allen, J., Tauc, J., Giessen, B., and Hauser, J. (1978). *Phys. Rev. Lett.* **40**, 1282.
Hayler, L., and Goldstein, M. (1977). *J. Chem. Phys.* **66**, 4736.
Hayward, P. (1977). *Phys. Chem. Glasses* **18**(6), 121.
Heidenreich, E., Ehrt, R., and Vogel, W. (1976). *Silikattechnik* **27**(12), 602.
Heidenreich, E., Ehrt, R., and Vogel, W. (1977a). *Proc. Int. Cong. Glass, 11th.*
Heidenreich, E., Ehrt, R., and Vogel, W. (1977b). *Silikattechnik* **28**(2), 45.
Heindorf, W., and Vogel, W. (1970). German Patent 1 302 538.
Hendrickson, J., and Bray, P. (1972a). *Phys. Chem. Glasses* (Part I) **13**(2), 43.
Hendrickson, J., and Bray, P. (1972b). *Phys. Chem. Glasses* (Part II) **13**(4), 107.
Henninger, E., and Busher, R. (1967). *J. Chem. Phys.* **46**, 586.
Herczog, A., and Layton, M. (1964). *J. Am. Ceram. Soc.* **47**(3), 107.
Herczog, A., and Layton, M. (1967). *J. Am. Ceram. Soc.* **50**(7), 360.
Herczog, A., and Layton, M. (1969). *Glass Technol.* **10**(2), 50.
Hermann, D., and Schlenzig, M. (1978). *Chemistry* **18**(3), 94.
Hermann, D., Lemnitzer, M., and Monheim, W. (1978). *Z. Anorg. Allg. Chem.* 438.83.
Hermann, D., and Schell, H. (1978). *Z. Anorg. Allg. Chem.* **442**, 119.
Hermann, D., Monheim, W., Gollmick, R., and Wagner, W. (1978). *Z. Anorg. Allg. Chem.* **442**, 125.
Hester, R., and Krishnan, K. (1955). *J. Chem. Soc. A.*
Hester, R., and Krishnan, K. (1968). *J. Chem. Phys.* **49**, 4356.
Hester, R., Krishnan, K., and Plane, R. (1964). *J. Chem. Phys.* **40**, 411.
Heyne, G. (1933). *Angew. Ch.* **46**, 473.
Hicks, J. (1967). *Science* **155**, 459.
Higashi, G., and Kastner, M. (1979). Charged Defect Pair Luminescence in a-As_2S_3, MIT 10-51-79.
Higgins, J. (1977). *J. Non-Cryst. Solids* **23**(3), 321.
Higgins, J., Buesch, L., Volterra, V., Moynihan, C., and Macedo, P. (1973). *J. Am. Ceram. Soc.* **56**(6), 334.
Higgins, J., Lewis, J., Lowe, M., and Roue, F. (1975). *J. Non-Cryst. Solids* **18**(1), 77.
Hill, R. (1966). *Science* **151**, 194.
Hillig, W. (1962b). *Proc. Symp. Mech. Strength Glasses, Charleroi, Belgium,* p. 206. Union Scientifique Continentale du Verre.
Hilton, A. (1965). Final Rep. N-ONR, 3810, (00), 9.
Hilton, A. (1966). *Appl. Opt.* **5**, 1877.
Hilton, A. (1968). *Phys. Chem. Glasses* **9**(5), 148.
Hilton, A., and Brau, M. (1963). *Infrared Phys.* **3**, 69.
Hilton, A., and Brau, M. (1968). U.S. Patent 3 371 211.
Hilton, A., and Jones, C. (1968). U.S. Patent 3 370 964, 3 370 965.
Hilton, A., Jones, C., and Brau, M. (1964a). *Infrared Phys.* **4**, 213.

Hilton, A., Balley, L., and Brau, M. (1964b). U.S. Patent 3 154 242.
Hilton, A. R., Jones, C. E., and Brau, M. (1966a). *Phys. Chem. Glasses* **7**(4), 105.
Hilton, A., Jones, C., and Brau, M. (1966b). *Phys. Chem. Glasses* **7**, 112.
Hilton, A. R., Jones, C. E., Dobrott, R. D., Klein, H. M., Bryant, A. M., and George, I. D. (1966c). *Phys. Chem. Glasses*, **7**(4), 116.
Hilton, A. R., Jones, C. E., and Brau, M. (1966). *Infrared Phys.* **6**, 183–194.
Hilton, A., Hayes, D., and Rechtin, M. (1974). Tech. Rep. 1, NOOO 14-73-C-0367.
Hilton, A., Hayes, D., and Rechtin, M. (1975). *J. Non-Cryst. Solids* **17**, 319, 339.
Hinz, W., and Mitsch, J. (1968). *Silikattech.* **19**(3), 90.
Hirayama, C., and Berg, D. (1961). *Phys. Chem. Glasses* **2**, 145.
Hirayama, C., and Subbard, E. (1962). *Phys. Chem. Glasses* **3**(4), 111.
Hiroshima, H., and Yoshida, T. (1977). *Yogyo Kyo Kai Shi* **85**(9), 434.
Hodge, I. (1974). Ph.D. Thesis, Purdue Univ.
Hodge, I., Angell, C., and Miller, J. (1977). *J. Chem. Phys.* **67**, 4.
Hoffman, L. (1963). U.S. Patent 3 115 415.
Hoffman, L., and Weyl, W. (1957). *Glass Ind.* **38**(2), 81.
Hoffmann, W., Fijcher, P., and Maier, G. (1966). *Naturwissenschaften* **53**, 16.
Hollabaugh, C., and Ernsberger, F. (1960). American Ceramic Meeting, extensively reproduced in Doremus (1973, p. 33).
Holland, A., and Segnit, E. (1966). *Aust. J. Chem.* **19**, 905.
Holmquist, G. (1978). Rep LBL-6922. Energy Res. Abstr. (NTIS) **8**(3), 19186; CA **89**, 459, 116631.
Hood, H. (1926). *Glass Ind.* **7**, 787.
Hood, H., and Nordberg, M. (1934a). U.S. Patent 2 106 744.
Hood, H., and Nordberg, M. (1934b). U.S. Patent 2 035 318.
Hood, H., and Nordberg, M. (1940). U.S. Patent 2 221 709.
Hopkins, T., Pasternak, R., Gould, E., and Herndon, J. (1967). *J. Phys. Chem.* **66**, 733.
Horozova, I., and Yakhkind, A. (1977). *Fiz. Khim. Stekla* **3**(3), 197.
Hoshikawa, T., and Nambu, T. (1978). Japanese Patent 78108111 (1978-09-20).
Howard, W., and Tsu, R. (1970). *Phys. Rev. B* **1**, 4709.
Howe, R., Howells, W., and Endersy, J. (1974). *J. Phys. C. Solid State Phys.* **7**, 111.
Howell, F., Bose, R., Macedo, P., and Moynihan, C. (1974). *J. Phys. Chem.* **78**, 639.
Hoya Glass Works (1974). U.S. Patent 3 841 739.
Hruby, A. (1978). *J. Non-Cryst. Solids* **28**(1), 139.
Hruby, A., and Štovrač, L. (1971). *Mat. Res. Bull. J. Phys. B* **6**, 465.
Hruby, A., and Štovrač, L. (1973). *Mat. Res. Bull. J. Phys. B* **23**(11), 1263.
Hsich, S., Gammon, R., Macedo, P., and Montrose, C. (1972). *J. Chem. Phys.* **56**, 1666.
Huang, Y., Sakar, A., and Schultz, P. (1978). *J. Non-Cryst. Solids* **27**(1), 29.
Hubbard, D., and Cleek, G. (1952). *J. Res. Nat. Bur. Std.* **49**(4), 267.
Huettenlocher, H. (1935). *Z. Krist.* **90**, 508.
Huggins, M., and Abe, T. (1957). *J. Am. Ceram. Soc.* **40**, 287.
Hughes, R. (1975). *Appl. Phys. Lett.* **26**, 436, 476.
Hulls, K., and McMillan, P. (1974). *J. Non-Cryst. Solids* **15**(3), 357.
Hummel, F. (1965). U.S. Patent 3 169 072.
Hummel, F., and Harrison. E. (1964). *Electrochem. Soc.* **103**, 491.
Hurt, J. (1967). Rutgers Thesis, American Ceramic Society.
Hurt, J., and Philips, C. (1970). *J. Am. Ceram. Soc.* **53**(5), 269.
Ichnikawa, T., and Ogawa, S. (1972). *Acta Cryst. A.* **28** (Pt. 4 Suppl.) S129.
Ida, K., Honma, K., and Okazaki, H. (1974). *J. Jpn. Inst. Mat.* **38**(8), 682.
Ignatyuk, V., and Stavnistyi, N. (1977). *Neorg. Mat.* **13**(7), 1303.

Igo, T., and Toyoshima, Y. (1973). *J. Non-Cryst. Solids* **11**, 304.
Imaoka, M. (1954). *J. Ceram. Assoc. Jpn.* **62**, 24.
Imaoka, M. (1962). "Advances in Glass Technology," Part I, p. 149. Plenum Press, New York.
Imaoka, M., and Hasegawa, H. (1980a). *Yogyo Kyokaishi* **88**(3), 141.
Imaoka, M., and Hasegawa, H. (1980b). *Phys. Chem. Glasses* **21**(2), 67.
Imaoka, M., and Mazurin, O. (1966). "Advances in Glass Technology" p. 14g. Plenum Press, New York.
Imaoka, M., and Sakamura, H. (1974). *Glass Technol.* **15**(4), 105.
Imaoka, M., and Yamazaki, T. (1967). *Seisan Kenkyu* **19**(9), 261.
Imaoka, M., Konagaya, Y., and Hasegawa, H. (1971). *J. Ceram. Soc. Jpn.* **79**, 97.
Ingram, M., and Duffy, J. (1968). *J. Chem. Soc. A* 2575.
Inoue, E., Kokado, H., and Shimizu, I. (1974). *Oyo Buturi Suppl.* **43**, 101.
Inoue, E., Kokado, H., and Shimizu, I. (1976). *Abstr. Phys. Chem. Glasses* **17**(1).
Ioffe, V. (1961). *Sov. Phys. Solid State* **3**, 1387.
Ioffe, V., and Patrina, I. (1970). *Phys. Status Solidi* **40**, 389.
Irany, E. (1943). *Ind. Eng. Ch.* **35**, 1290.
Isard, J. (1969). *J. Non-Cryst. Solids* **1**, 235.
Isard, J. (1972). *Extended Abstr. Symp. Mass Transport Am. Solids* Electrochem. Soc. Glass Technology.
Isard, J., and Lacharme, A. (1978). *J. Non-Cryst. Solids* **27**(3), 381.
Ishii, A., and Akawa, K. (1965). *Rep. Res. Lab. Asaki Glass Co.* **15**, 1.
Islam, N., and Ismail, K. (1975). *J. Phys. Chem.* **79**, 2180.
Islam, N., and Ismail, K. (1976). *J. Phys. Chem.* **80**, 1929.
Ito, T., and Sawada, H. (1940). *Z. Krist.* **102**, 13.
Ito, T., Morimoto, N., and Sadanaga, R. (1951). *Acta Cryst.* **4**, 310.
Itoh, H., and Mori, Sh. (1977). *Fukuoka Daigaku-Shuho* **19**, 13. *CA* **80**, 47810s (1978).
Iwamoto, N., Tsunawaki, Y., Fuji, M., and Hatfori, T. (1975). *J. Non-Cryst. Solids* **18**(2), 303.
Izumitani, T., and Masuda, I. (1974). *Int. Cong. Glass, 10th, Kyoto, July* Sect. 5, p. 74.
Jabra, R., Phalippou, J., and Zarzycki, J. (1979). *Rev. Chim. Miner.* **16**(4), 245; **92**, 269, 11511h (1980).
Jack, K. (1972). *Nature (London) Phys. Sci.* **238**(80), 28.
Jack, K. (1976). *J. Mat. Sci.* **11**(6), 1135.
Jack, K. (1977). "Nitrogen Ceramics" (E. R. Ley, ed.), p. 257. Noordoof.
Jack, K. (1978). *J. Mat. Sci.* **13**(12), 1327.
Jack, K., and Hardie, D. (1957). *Nature* **180**, 332.
Jaeger, R. (1968). *J. Am. Ceram. Soc.* **51**, 57.
Jaeger, R., and Smyth, H. (1971). American Ceramic Society, Chicago Convention (unpublished).
Jagdt, R. (1960). *Glastech. Ber.* **33**, 10.
Jahn, W. (1961). *Glastech. Ber.* **34**(3), 107.
Jahn, W. (1961b). German Patent 1 088 674, C1 326.
Jain, S. (1978). *J. Phys. Chem.* **82**, 1272.
James, P., and McMillan, P. (1970). *Phys. Chem. Glasses* **11**, 59, 64.
Janakirama-Rao, V. (1965). *J. Am. Ceram. Soc.* **48**(6), 311.
Jankowski, P., and Risbud, S. (1980). *J. Am. Ceram. Soc.* **63**(5,6), 350.
Jellison, G. (1977). *Phys. Non-Cryst. Solids Int. Conf. 4th.*
Jellison, G., and Bray, P. (1978a). "Borate Glasses" (L. Pye *et al.*, eds.), p. 353. Plenum Press, New York.

Jellison, G., and Bray, P. (1978b). *Non-Cryst. Solids* **29**, 187.
Jellison, G., Paner, L., Bray, P., and Rouse, G. (1977). *J. Chem. Phys.* **66**(2), 802.
Jenaer Glasswerke, (1965). French Patent 1 492 750.
Jirmanus, M., Gerber, V., Sample, H., and Neuringer, L. (1978). *J. Non-Cryst. Solids* **27**, 1.
Johari, G., and Goldstein, M. (1970). *J. Chem. Phys.* **53**, 2372.
Johnstone, J. (1967). US AE Comm., IST-T-213, available from GFSTI, abstr. *Nucl. Sci. Abstr.* **22**(8), 15564 (1968).
Johnson, J. (1951). *J. Am. Ceram. Soc.* **34**, 165.
Johnson, R., Jr., and Quinn, R. (1978). *J. Non-Cryst. Solids* **28**(3), 369.
Joiner, B., and Thompson, J. (1976). *J. Non-Cryst. Solids* **21**, 215.
Jones, C. (1967). T Rep. AFAL-TR-67, November, Contr. AF 33 (615)-3963.
Jones, C., and Hafner, H. (1968). AFAL-TR-68-348, Final Rep. 03-68-68, Contr. AF 33 (615)-3963. Washington, DC.
Jost, K., and Wodtke, F. (1962). *Makromol. Chem.* **53**, 1.
Kaes, H. (1972). *Glastech. Ber.* **45**(6), 234.
Kaes, H. (1973). German Patent 2216064.
Kamiya, K., and Sakka, S. (1979). *Phys. Chem. Glasses* **20**(3), 60.
Kamiya, K., Sakka, S., and Yamanaka, I. (1974). *Int. Cong. Glass, 10th, Kyoto, July,* Sect. 13, p. 44.
Kamiya, K., Sakka, S., and Ito, S. (1977). *Yogyo Kyokai Shi* **85**(2), 8999.
Kanbara, T. (1969). *Yogyo Kyokai Shi* **76**(879), 385.
Kantor, P., Revcolevschi, A., and Collongues, R. (1973). *J. Mat. Sci.* **8**(9), 1359.
Kaplow, R., Rowe, T., and Auerbach, B. (1968). *Phys. Rev.* **168**(3), 1068.
Karlsson, K. (1969). *Glasteck. Tidskr.* **24**(1), 13.
Karlsson, K. (1970). *Suom. Kem. B* **43**, 489.
Karlsson, K. (1973). *Abstr. Phys. Chem. Glasses* **14**(2), 21A.
Kasatkin, B., Borisova, Z., and Orlova, G. (1974). *Inorg. Mat.* **10**(9), 1387.
Kasimova, S., and Milyukov, E. (1977). *Izv. Akad. Nauk. Vzb. SSR. Fiz.-Mat. Nauk* (3), 72; *CA* **88**, 26603 (1978).
Kastner, M. (1977). *Proc. Int. Conf. Amorphous and Liquid Semanicond., 7th* (W. Spear, ed.), p. 504. Univ. of Edinburgh.
Kastner, M. (1978). *J. Non-Cryst. Solids* **31**, 223.
Kastner, M., and Hudgens, S. (1978). *Phil. Mag.* **37**, 665.
Kastner, M., Adler, D., and Fritzche, H. (1976). *Phys. Rev. Lett.* **37**, 1504.
Kato, A., Ono, Y., Kawazoe, S., and Machipa, I. (1972). *Yogyo-Kyokai Shi* **80** (3), 114.
Katsuda, T., and Minami, T. (1979). *J. Phys. Chem.* **83**, 1306.
Katzschmann, R. (1965). *Glass Technol.* **6**(3), 156.
Kawai, N., Mochizuki, S., and Fujiza, H. (1971). *Phys. Lett.* **34A**(2), 107.
Kawamoto, Y. (1974). *J. Am. Ceram. Soc.* **57**(11), 489.
Kawamoto, Y., and Nishida, M. (1976). *J. Non-Cryst. Solids* **20**, 393.
Kawamoto, Y., and Tsuchihacki, S. (1971). *J. Am. Ceram. Soc.* **54**(3), 131.
Kawamoto, Y., Nagura, N., and Tsuchihachi, S. (1973). *J. Am. Ceram. Soc.* **56**(5), 289.
Kawamoto, Y., Gata, M., and Tsuchihachi, S. (1974). *Yogyo Kyokai Shi* **82**(9), 502.
Kawazoe, H., Hosono, H., and Kanazawa, T. (1978a). *J. Non-Cryst. Solids* **29**, 159.
Kawazoe, H., Hosono, H., and Kanazawa, T. (1978b). *J. Non-Cryst. Solids* **29**, 173.
Kawazoe, H., Hosono, H., and Kanazawa, T. (1978c). *J. Non-Cryst. Solids* **29**, 249.
Kawazoe, H., Hosono, H., Kokumai, H., and Kanazawa, T. (1979). *Yogyo Kyokaishi* **87**(5), 237.
Kazakova, E., and Borisova, Z. (1980a). *Fiz. Khim. Stekla* **6**(2), 143.
Kazakova, E., and Borisova, Z. (1980b). *Fiz. Khim. Stekla* **6**(2), 148.

Keck, D., Maurer, R., and Schultz, P. (1973). *Appl. Phys. Lett.* **22**, 307.
Keezer, R., and Balley, M. (1967). *Mat. Res. Bull.* **2**, 185.
Keller, H., and Stoke, J. (1965). *Phys. Status Solidi* **8**, 831.
Kennedy, G. (1960). Oral communication to R. Sosman (see R. Sosman, 1965, p. 152).
Kerr, J., and Jorgensen, L. (1971). *J. Non-Cryst. Solids* **5**, 306.
Khalilev, V., Petrovskaya, M., and Nikolina, G. (1975). *Fiz. Khim. Stekla* **1**(6), 508.
Khalilev, V., Pronkin, A., Vakhrameev, V., and Vasylyak, Ya. (1979). *Fiz. Khim. Stekla* **5**(2), 188.
Khiminets, V., Puga, P., Rosola, I., Puga, G., and Chepur, D. (1980). *Zh. Struct. Khim.* **21**(1), 88; *CA* **93**, 393 78105c.
Kilroy, W. (1978). *J. Am. Ceram. Soc.* **61**(9/10), 457.
Kim, H.-H., and Yamane, M. (1980). *Yogyo Kyokaishi* **88**(4), 191.
Kim, Y., and Bray, P. (1970). *J. Chem. Phys.* **53**(2), 716.
King, S. (1967). *Nature (London)* **213**(5018), 1112.
Kinser, D. (1970). *J. Electrochem. Soc.* **11**(4), 540.
Kirilenko, V., and Dembrovski, S. (1974). *Neorg. Mat.* **10**(3), 542.
Kirilenko, V., and Polyakov, Y. (1976). *Neorg. Mat.* **12**(2), 336.
Kislitskaya, E. A., and Kokorina, V. F. (1971). *Zh. Prikl. Khim.*, **44**(3) 646.
Kittel, C. (1967). "Introduction to Solid State Physics," p. 194. Wiley, New York.
Klaffke, G., and Wood, C. (1977). *Phys. Non-Cryst. Solids, Int. Conf. 4th*, p. 236.
Klemm, W., and Niermann, H. (1963). *Angew. Ch. Int. Ed.* **2**, 523.
Kley, G., and Linke, I. (1979). *J. Non-Cryst. Solids* **33**, 299.
KleinHaneveld, A. J., van der Veer, W., and Jellinek, F. (1968). *Rev. Trav. Chim. PaysBas* **87**(3), 255.
Kobayashi, K. (1979). *J. Am. Ceram. Soc.* **62**(9/10), 440.
Kobayashi, K., and Okuma, H. (1976). *J. Am. Ceram. Soc.* **59**(7), 354.
Koenines, J., Kueppers, D., Lydtin, H., and Wilson, H. (1975). *Proc. Conf. Chem. Vap. Depl. Int. Conf.* **5**, 270. *CA* **84**, 21343 (1976).
Koffyberg, F. (1978a). *J. Non-Cryst. Solids* **28**, 231.
Koffyberg, F. (1978b). *J. Non-Cryst. Solids* **28**, 241.
Kokado, H., Shimizu, I., and Inoue, E. (1976a). *J. Non-Cryst. Solids* **20**, 131.
Kokado, H., Shimizu, J., and Inoue, E. (1976b). *J. Non-Cryst. Solids* **21**, 225.
Kokorina, B. (1971). "Stekloobraznoe Sost." (*Proc. All-Union Cong., 5th, Leningrad, 1969*), p. 95. Nauka.
Kokubo, T., and Tashiro, M. (1978). *J. Mat. Sci.* **13**(5), 930.
Kokubo, T., Nishimura, M., and Tashiro, M. (1974). *J. Non-Cryst. Solids* **15**, 329.
Kokubo, T., Nishimura, M., and Tashino, M. (1976). *J. Non-Cryst. Solids* **22**, 125.
Kokubo, T., Naito, S., and Tashiro, M. (1979). *Yogyo Kyokaishi* **87**(91), 653.
Kolb, K., and Hansen, K. (1965). *J. Am. Ceram. Soc.* **48**(8), 438.
Kolb, K., and Hansen, K. (1966). *J. Am. Ceram. Soc.* **49**(2), 105.
Kolesova, V. (1971). *Neorg. Mat.* **7**(2), 348.
Kolesova, V. (1975). *Fiz. Khim. Stekla* **1**, 290.
Kolesova, V. (1979). *Fiz. Khim Stekla* **5**(3), 367.
Kolesova, V. (1980). *Fiz. Khim. Stekla* **6**(2), 247.
Kolesova, V., and Kalinina, M. (1974). *Int. Cong. Glass, 10th, Kyoto, July,* Sect. 13, p. 53.
Kolomiets, B. (1963). "Vitreous Semiconductors." LDNTP, Leningrad.
Kolomiets, B. (1968). *Proc. Int. Conf. Phys. Semicond., 9th, Moscow.*
Kolomiets, B., and Goryunova, N. (1956). *Bull. Acad. Sci. USSR Phys. Ser.* **20**, 1372.
Kolomiets, B., and Lebedev, G. (1963). *Radio Tekh. Elktron.* **8**, 2097.

Kolomiets, B., Goryunova, N., and Shilo, V. (1958). *Sov. Phys.-Tech. Phys.* **3**, 912.
Kolomiets, B. T., Goryunova, N. A., and Shilo, V. P. (1959). *Proc. All-Un. Conf. Glassy State, 3rd, Leningrad* Vol. II, p. 410. Consultants Bureau, New York.
Kolomiets, V., Goryunova, N., and Shilo, V. (1960). *Steklobraznoe Sost.* **2**, 410.
Kolomiets, V., Nazarova, T., and Shilo, V. (1962). *Rep. Int. Conf. Phys. Semicond., Exeter* p. 259.
Kolomiets, B., Aio, G., and Kokorina, V. (1963). *Opt.-Mekh. Promyshl.*
Kolomiets, B., Lev, E., and Sysoeva, L. (1964a). *Sov. Phys.-Solid State* **5**, 2101.
Kolomiets, B., Lev, E., and Sysoeva, L. (1964b). *Sov. Phys.-Solid State* **6**, 551.
Kolomiets, B., Mamantova, G., and Babaev, A. (1970). *J. Non-Cryst. Solids* **4**, 289.
Kolomiets, B., Rukhlyadev, Yu., and Shild, U. (1971). *J. Non-Cryst. Solids* **5**(5), 389–402.
Kondratyev, Y., and Smirnova, L. (1970). *Neorg. Mat.* **6**(2), 340.
Konijnendijk, W., and Buster, J. (1976). *J. Non-Cryst. Solids* **22**, 379.
Konijnendijk, W., and Buster, J. (1977). *J. Non-Cryst. Solids* **23**, 401.
Konijnendijk, W., and Stevels, J. (1975). *J. Non-Cryst. Solids* **18**(3), 307.
Konijnendijk, W., and Stevels, J. (1976). *J. Non-Cryst. Solids* **20**, 193.
Konijnendijk, W., and Stevels, J. (1978). "Borate Glasses" (L. Pye *et al.*, eds.), p. 259. Plenum Press, New York.
Konijnendijk, W., and Verweij, H. (1976). *J. Am. Ceram. Soc.* **59**(9, 10), 459.
Kornilova, E., and Petrovskii, G. (1980). *Dokl. Adad. Nauk SSSR* **251**(2), 409.
Konnert, J., and Karle, J. (1972). *Nature (London) Phys. Sci.* **236**, 92.
Konnert, J., and Karle, J. (1973). *Acta Cryst.* **A29**, 702.
Konnert, J., Karle, J., and Ferguson, G. (1973). *Science* **179**, 177.
Konnert, J., Karle, J., and Ferguson, D. (1974). *Science* **184**(4132), 93.
Koperles, B., Puga, P., Turyanitsa, I., Borets, A., and Chepur, D. (1973). *Vyssh. Vcheb. Zaved. Fiz. (Uzhgorod)* **16**(10), 99. CA **80**, 8974, 123965c.
Kornilova, E., and Petrovskii, G. (1980). *Dokl. Ak. Nauk SSSR* (Chem. Tech.) **25**(2), 409.
Kostyshin, M., Romanenko, P., Dembovskii, S., and Vinogradova, G. (1971). *Inorg. Mat.* **7**(2), 187.
Koudelka, L., Horak, J., Psarcik, M., and Sakal, L. (1979). *J. Non-Cryst. Solids* **31**, 339.
Kozlova, L., Tarasova, B., and Chepizhnyi, K. (1968). *Steklo.* (3), 6.
Kozlova, L., Tarasova, B., and Chepizhnyi, K. (1969). *Proizv. Issled. Stekla Silik. Mat.* **88**. CA 1075304 (1971).
Kracek, F. (1930a). *J. Am. Chem. Soc.* **52**, 1436.
Kracek, F. (1930b). *J. Phys. Chem.* **34**(12), 2641.
Krause, J., and Kurkjian, C. (1978). "Borate Glasses" (L. Pye *et al.*, eds.), p. 577. Plenum Press, New York.
Krause, J., Kurkjian, C., Pinnow, D., and Sigets, E. (1970). *J. Appl. Phys.* **17**(9), 367.
Krause, J., Testardi, L., and Thurston, R. (1979). *Phys. Chem. Glasses* **20**(6), 135.
Krebs, H. (1953). *Angew. Chem.* **65**, 293.
Krebs, H. (1958). *Angew. Chem.* **71**, 615.
Krebs, H. (1961). *Z. Anorg. Allg. Chem.* **268**, 156.
Krebs, H., and Fischer, P. (1970). *Faraday Soc. Conf. Vitr. State,* paper No. 4.
Krebs, H., and Ruska, J. (1974). *J. Non-Cryst. Solids* **16**, 329.
Kreidl, N. (1974). "Handbook of Glass Manufacturing," p. 967. Books for Ind., New York.
Kreidl, N., and Blair, G. (1956). *Nucleonics* **14**, 82.
Kreidl, N., and Hensler, J. (1958). "Modern Materials" (H. Hausner, ed.), Vol. I, p. 217. Academic Press, New York.
Kreidl, N., and Maklad, M. (1969). *J. Am. Ceram. Soc.* **52**(9), 508.
Kreidl, N., and Ratzenboeck, W. (1973). "Recent Advances in Science" (A. Bishay, ed.). Plenum Press, New York.

Kreidl, N., and Rood, J. (1965). "Applied Optics and Optical Engineering" (R. Kingslake, ed.), Vol. I. Academic Press, New York.
Kreidl, N., and Weyl, W. (1941). *J. Am. Ceram. Soc.* **20**(11), 372.
Krishna-Murthy, M., and Hill, H. (1962). *J. Am. Ceram. Soc.* **45**(12), 616.
Krishna-Murthy, M., and Scroogie, B. (1965). *Phys. Chem. Glasses* **6**, 162.
Kriz, H., Park, M., and Bray, P. (1971). *Phys. Chem. Glasses* **12**(2), 45.
Kroger, K., and Res, M. (1977). *Glass Technol.* **18**(1), 5.
Krogh-Moe, J. (1960). *Phys. Chem. Glasses* **1**, 26.
Krogh-Moe, J. (1962). *Phys. Chem. Glasses* **31**, 101.
Krogh-Moe, J. (1965). *Phys. Chem. Glasses* **6**, 30, 46.
Krylova, L. (1967). *Steklo* (1), 101 (Ref. (1968). *Zh. Khim.* Abstr. 9 B-616).
Kuehne, K. (1955). *Silikattech.* **6**, 190.
Kueppers, D., and Koenings, J. (1976). *Eur. Conf. Opt. Fibre Commun., 2nd, Paris* p. 49.
Kueppers, D., Koenings, J., and Wilson, H. (1977). *Proc. Eur. Conf. Opt. Commun., 3rd* p. 12. VDE, Berlin.
Kumar, S. (1963). *Phys. Chem. Glasses* **4**, 1061.
Kumar, B., and Rindone, G. (1979a). *Phys. Chem. Glasses* **20**(6), 119.
Kumar, B., and Rindone, G. (1979b). *Phys. Chem. Glasses* **20**(6), 148.
Kumato, K., Yamamoto, K., and Namikawa, H. (1977). *Yogyo Kyokai Shi* **85**(7), 359.
Kurkjian, C. (1965). *Phys. Chem. Glasses* **6**, 95.
Kurkjian, C., and Peterson, G. (1974). *Phys. Chem. Glasses* **15**(1), 12.
Kurosaki, S., and Usui, Y. (1978). *Jpn. Kokai Takkyo Koho* **78**, 134, 810. *CA* **90**, 285, 191386 (1979).
Kutateladze, K., Sarukhanishvili, A., and Kutateladze, N. (1974). *Int. Cong. Glasses, 10th, Kyoto, July* Sect. 13, p. 49.
Kuznetsova, M., and Estrop'ev, K. (1972). *Neorg. Mat.* **8**(2), 347.
Labino, D. (1958). U.S. Patent 2 823 117.
Lacharme, J., and Isard, J. (1978). *J. Non-Cryst. Solids* **27**, 381.
Lachman, I., and Armistead, W. (1978). U.S. Patent Appl. 899 369, (1978-04-24).
Lacourse, W., Twadell, A., and Mackenzie, J. (1970). *J. Non-Cryst. Solids* **3**, 234.
Lacourse, W., Twadell, A., and Mackenzie, J. (1976). *J. Non-Cryst. Solids* **21**, 431.
Lacy, E. (1963). *Phys. Chem. Glasses* **4**(6), 234.
Lagzdon, Yu., Shults, I., Vitina, I., Sedmalis, V., and Buka, J. (1980). *Fiz. Khim. Stekla* **6**(1), 90.
Lam, D., Palikas, A., and Veal, B. (1980). *J. Non-Cryst. Solids* **42**(1/3), 41.
Laubengeyer, A., and Morton, D. (1932). *J. Am. Chem. Soc.* **54**, 7303.
Lazzari, M., Sckojati, B., and Vincent, C. (1978a). *J. Am. Ceram. Soc.* **61**(9/10), 451.
Lazzari, M., Scrojati, B., and Vincent, C. (1978b). *J. Am. Ceram. Soc.* **61**(9/10), 651.
Leadbetter, A. (1974). *J. Non-Cryst. Solids* **15**, 250.
Leadbetter, A. (1976). *J. Non-Cryst. Solids* **21**(1), 47.
Leadbetter, A., and Apling, A. (1974). *J. Non-Cryst. Solids* **15**, 250.
Leadbetter, A., and Wright, A. (1970). *J. Non-Cryst. Solids* **3**, 239.
Leadbetter, A., and Wright, A. (1972a). *J. Non-Cryst. Solids* **7**, 37.
Leadbetter, A., and Wright, A. (1972b). *J. Non-Cryst. Solids* **7**, 156.
Leadbetter, A., and Wycherley, K. (1971). *Phys. Chem. Glasses* **12**(2), 4.
Lebedev, N. (1921). *Proc. State Opt. Inst., Leningr.* **2**(10).
Lecoq, A., and Poulain, M. (1979). *J. Non-Crystal Solids* **34**, 101.
Lee, D. H., and McPherson, R. (1980). *J. Mat. Sci.* **15**(1), 25.
Leedecke, C., and Baca, H. (1978). Corning, British Patent 1,520,228.
Leedecke, C., and Baca, H. (1979). Rep. Sand-79-0248 (NTIS), *Energy Res. Abstr.* **4**(13), 37049.

Leedecke, C., and Baca, H. (1980). *J. Am. Ceram. Soc.* **63**(7,8), 479.
Leedecke, C., and Loehman, R. (1980). *J. Am. Ceram. Soc.* **63**(3,4), 190.
Leidtorp, R., and Petrovskii, G. (1980). *Dokl. Akad. Nauk SSR* **251**(2), 343.
Leipold, W., and Feuchtwang, T. (1976). *J. Non-Cryst. Solids* **21**(2), 181.
Leko, V., Gusakova, N., Meshcheryakova, E., and Prokhorova, T. (1977). *Fiz. Khim. Stekla* **3**(3), 219.
Lengyel, B., and Boksay, Z. (1954). *Z. Phys. Chem.* **203**, 93.
Lengyel, B., and Boksay, Z. (1955). *Z. Phys. Chem.* **204**, 157.
Lentz, C. (1964). *Inorg. Chem.* **3**, 574, 579.
LeRoy, D., Lucas, J., Poulain, M., and Ravaine, D. (1978). *Mater. Res. Bull.* **13**, 1125.
Levand, V., Thomasson, G., and Holcomb, R. (1971). U.S. Patent 3 588 315.
Levand, V., Thomasson, G., and Holcomb, R. (1972). U.S. Patent 3 652 302.
Levenberg, V. A., and Lunter, S. G. (1979). *Fiz. Khim. Stekla* **5**(6), 140.
Levin, E. (1965). *J. Am. Ceram. Soc.* **48**(9), 491.
Levin, E. (1967). *J. Am. Ceram. Soc.* **50**(1), 29–38.
Levin, E., and Block, S. (1957). *J. Am. Ceram. Soc.* **40**, 95, 113.
Levin, E., and Block, S. (1958). *J. Am. Ceram. Soc.* **41**(2), 49.
Levin, E., and Cleek, G. (1958). *J. Am. Ceram. Soc.* **41**, 175.
Levin, E., and McMurdie, H. (1975). "Phase Diagrams." Am. Ceram. Soc. Columbus, Ohio.
Levy, R., Lupis, C., and Flinn, P. (1976). *Phys. Chem. Glasses* **17**(4), 94.
Lewis, J. (1974). *J. Non-Cryst. Solids* **15**, 351.
Li, Ch. (1968). *Z. Krist.* **127**, 327.
Li, J., and Uhlmann, D. (1970). *J. Non-Cryst. Solids* **3**(2), 205.
Li, L. (1962). *J. Am. Ceram. Soc.* **45**(7), 83, 89.
Li, S., and Bray, P. (1962). *Phys. Chem. Glasses* **3**(2), 37.
Li, S., and Chaudhari, P. (1973). *J. Non-Cryst. Solids* **11**, 285.
Liang, K. (1974). *Bull. Am. Phys. Soc.* **19**, 213.
Liang, K. (1975). *J. Non-Cryst. Solids* **18**, 197.
Liang, K., and Bienenstock, A. (1975). *J. Non-Cryst. Solids* **17**, 289.
Liang, K., Bienenstock, A., and Bates, C. (1974). *Phys. Rev. B* **13**(4), 1528.
Liedberg, D., Ruderer, C., and Bergeron, D. (1965). *J. Am. Ceram. Soc.* **48**(8), 440.
Lisenkov, A., and Vasil'ev, A. (1979). *Fiz. Khim. Stekla* **5**(5), 537.
Litvin, B., Chudinova, N., Bebikh, L., and Tanasov, I. (1979). *Dokl. Adad. Nauk SSSR* **249**(5), 1124.
Litvinov, P., and Zhura'leva, R. (1965). *Steklo Tr. Inst. Stekl* **2**, 38.
Liu, Ch., Pashinkin, A. S., and Novoselova, A. V. (1962a). *Zh. Neorg. Khim.* **7**, 2159–2161.
Liu, Ch., Pashinkin, A. S., and Novoselova, A. V. (1962b). *Russ. J. Inorg. Chem.* **7**, 1117.
Loehman, R. (1979). *J. Am. Ceram. Soc.* **62**(9,10), 491.
Logan, J., and Yung, M. (1976). *J. Non-Cryst. Solids* **21**(1), 151.
Loh, E. (1964). *Solid State Commun.* **2**, 269.
Loireau-Lozach, A., and Guittard, M. (1975). *Ann. Chim.* **10**(2), 101.
Lonsdale, T., and Whitaker, A. (1978). *J. Mat. Sci.* **13**(7), 1503.
Love, W. (1973). *Phys. Rev. Lett.* **31**(13), 822.
Low, M., Ramasubramanian, N., and Ramamurthy, P. (1969). *J. Am. Ceram. Soc.* **52**(3), 124.
Lucas, J., Chanthanasinh, M., Poulain, Ma., Poulain, Mi., Brun, P., and Weber, M. (1978). *J. Non-Cryst. Solids* **27**(2), 273.
Lucovsky, G. (1969). *Mat. Res. Bull.* **4**(8), 505.

Lucovsky, G. (1970). *Proc. Int. Conf. Phys. Semicond.*, 10th, Cambridge, Massachusetts, p. 799. USAEC Div. Tech. Inf., Oak Ridge, Tennessee.
Lucovsky, G. (1974). *Phys. Rev. Bull.* **10**(12), 5134.
Lucovsky, G., and Galener, F. (1980). *J. Non-Cryst. Solids* **37**, 53.
Lucovsky, G., Geils, R., and Keezer, R. (1977). *Phys. Non-Cryst. Solids. Int. Conf. 4th,* p. 299.
Lukas, R. (1950). U.S. Patent 2 522 523/4.
Lydina, L., Sidorov, T., Sheinina, T., Milova, L., Ealaktionova, E., and Lebedova, E. (1975). *Neorg. Mat.* **11**(3), 513, 578.
Lyon, J., Fox, T., and Lyons, J. (1966). *Ceram. Bull.* **45**(12), 1078.
Macedo, P. B. (1975). Oral Communication.
Macedo, P., and Litovitz, T. (1976). U.S. Patent 3 938 976 (1976-02-17).
Macedo, P., Gupta, P., Droxhage, M., and Litovitz, T. (1977). *Phys. Non-Cryst. Solids, Int. Conf. 4th,* p. 471.
Macedo, P., Gupta, P., and Drexhage, M. (1978). *Japanese Patent* 118 412 (1978-10-16).
Mackenzie, J. (1956). *Trans. Faraday Soc.* **52**, 1564.
Mackenzie, J. (1959a). *J. Phys. Chem.* **63**(11), 1875.
Mackenzie, J. (1959b). *J. Am. Ceram. Soc.* **42**(6), 310.
Mackenzie, J. (1960a). *J. Am. Ceram. Soc.* **43**, 615.
Mackenzie, J. (1960b). *Mod. Asp. Vitr. State* **I**, 189.
Mackenzie, J. (1975/1976). Interim Prog. Rep., Lawrence Livermore Lab., May 1–April 30.
Mader, K.-H., and Lorentz, T. (1978). "Borate Glasses" (L. Pye *et al.*, eds.), p. 549. Plenum Press, New York.
Maeda, K., and Sato, J. (1977). *Denki Kagaku Oyobi Kogyo Butsuri Kagaku* **45**, (10), 654. *CA* **88**, 1094012 (1978).
Mahle, S., and McCammon, R. (1969). *Phys. Chem. Glasses* **10**, 222.
Majid, C., Prager, P., Fletcher, N., and Bretell, J. (1974). *J. Non-Cryst. Solids* **16**, 365.
Maki, T. (1968). *J. Ceram. Assoc. Jpn.*, **76**, 320.
Makishima, A., Tamura, Y., and Sakaino, T. (1978). *J. Am. Ceram. Soc.* **61**(5/6), 247.
Maklad, M., and Kreidl, N. (1969). *J. Am. Ceram. Soc.* **52**(9), 508.
Maklad, M., and Kreidl, N. (1971). *Int. Congr. Glasses 9th Paris.*
Makovskaya, Z., and Zhuko, E. (1980). *Neorg. Mat.* **16**(2), 25.
Malugani, J.-P., Wasniewski, A., Doreau, M., and Robert, G. (1978). *C. R. Acad. Sci. Paris Ser. C* **287**(11), 455–457.
Maneglier-La Cordaire, S., Besancon, P., Rivet, J., and Flahaut, J. (1975). *J. Non-Cryst. Solids* **18**, 439.
Marfels, H. (1969). *Glastech. Ber.* **42**(5), 161.
Margaryan, A., and Arutyunan, D. (1972). *Inorg. Mater.* **8**(5), 852.
Margaryan, A., Manbeiyan, M., and Akopyan, R. (1971). *Arm. Khim. Zh.* **24**(11), 1022.
Marinov, M., and Dimitriev, I. (1964). *Dokl. Bolg. Akad. Nauk.* **17**(8), 717.
Marinov, M., and Radenkova-Jancva, M. (1967). *C. R. Acad. Bulg. Sci.* **20**(12), 1305.
Maruno, S. (1977). *J. Non-Cryst. Solids* **24**, 301.
Maruno, S., and Noda, M. (1972). *J. Non-Cryst. Solids* **7**, 1.
Maruno, S., Noda, M., and Yamada, T. (1973). *J. Ceram. Assoc. Jpn.* **81**(10), 445.
Mashkovich, M., and Udenko, N. (1965). *Sov. Phys. Solid State* **7**(2), 417.
Massen, C., Wents, A., and Poulis, J. (1964). *Trans. Faraday Soc.* **60**, 317.
Masson, C., Smith, I., and Whiteway, S. (1970). *Can. J. Chem.* **48**(1), 201.
Matecki, M., Poulain, M., and Lucas, J. (1978). *Mater. Res. Bull.* **13**, 1039.
Matusita, K., Sakka, S., and Hiyanishi, K. (1975). *J. Non-Cryst. Solids* **17**(3), 436.

Matveev, M., Melnik, M., Glasova, M., and Kotel'niskova, G. (1970). *Stekloobr. Sost.* **5**(1), 189, (1971). *CA* 66974.
Matveev, M., Rzhevushkaya, T., and Rachkovjkaya, G. (1971b). *Stekloobr. Sost. Nov. Stekla I. Kh. Osn.* 147; (1972). *CA* 78848s.
Maurer, R. (1962). *J. Appl. Phys.* **33**(6), 2132.
Maya, L. (1977). *J. Am. Ceram. Soc.* **60**(7-8), 323.
Mazurin, O., and Borisovski, E. (1957). *Zh. Tekh. Fiz.* **27**(2), 275.
Mazurin, D., and Potselueva, L. (1978). *Fiz. Khim. Stekla* **4**(5), 570.
Mazurin, O., and Strel'tsina, M. (1972). *J. Non-Cryst. Solids* **11**(3), 199.
Mazurin, O., Tret'yakova, N. (1970). *Neorg. Mat.* **6**(11), 2022.
Mazurin, O., Roskova, G., and Koyev, V. (1970). *Faraday Conf. Vitr. State*.
Mazurin, O., Strel'tsina, M., and Shvaiko-Shvaiskovskaya, T. (1973). "Prop. Glasses and Glass-forming Melts I Vitr. SiO_2 and Binary Silicate Systems." Nauka, Leningrad. "Svoistva Stekolistekl. Obrazuyushtshi Kh." Rasplavo, Nauka, Leningrad.
McChesney, J., O'Connor, P., and Presby, H. (1974). *Proc. IEEE* **62**, 1280.
McChesney, J., O'Connor, P., Simpson, P., and Lazay, P. (1974). *Int. Congr. Glass, 10th, Kyoto* **6**, 40.
McDowell, J. (1966). *Ind. Eng. Cherm.* **58**(3), 38.
McDowell, J., and Beal, G. (1969). *J. Am. Ceram. Soc.* **52**(1), 17.
McMarlin, R. (1971). U.S. Patent 3 620 787.
McMillan, P. British Patent 1 022 681.
McMillan, P. British Patent 1 020 573.
McMillan, P. (1962). "Advanced Glass Technology." Plenum Press, New York.
McMillan, P. (1964). "Glass Ceramics." Academic Press, New York.
McMillan, P., and Matthews, C. (1976). *J. Mater. Sci.* **11**(7), 1187.
McMillan, P., and Shutov, S. (1977). *J. Non-Cryst. Solids* **24**, 307.
McMillan, P., and Tummala, R. (1978). U.S. Patent application 875-703, February 6.
McMillan, P., and Tummala, R. (1979). British Patent Appl. 2013650, August 15.
McQueen, R. (1963). *J. Geophys. Res.* **68**, 2319.
McVay, G., and Day, D. (1970). *J. Am. Ceram. Soc.* **53**(9), 508.
McVay, G., and Farnum, E. (1972). *J. Am. Ceram. Soc.* **55**(5), 275.
Mead, C. (1963). *Phys. Rev.* **128**, 2088.
Meiling, G., and Uhlmann, D. (1967). *Phys. Chem. Glasses* **8**(2), 62.
Melling, P., and Duncan, J. (1980). *J. Am. Ceram. Soc.* **63**(5/6), 264.
Mel'nichenko, Mikhal'ko, I., Semak, D., Turyanitsa, I., and Chepur, D. (1971). *Neorg. Mat.* **7**(6), 1065.
Menil, F., Fournes, L., Dance, J.-M., and Cideau, J. J. (1979). *J. Non-Cryst. Solids* **34**, 209.
Messier, R., and Roy, R. (1978). *J. Non-Cryst. Solids* **28**(3), 107.
Milberg, M., and Peters, C. (1969). *Phys. Chem. Glasses* **10**, 46.
Milberg, M., O'Keefe, J., Verhelst, R., and Hooper, H. (1972). *Phys. Chem. Glasses* **13**, 79.
Miller, R., and Mercer, R. (1960). *Mineral Mag.* **35**, 250.
Milyukov, E., Rieshakit, A., and Tolstoi, M. (1970). *Fiz. Tverd. Tela.* **12**(7), 525.
Minami, T. (1980). *Kenkyu Hokuku Asahi Garusa Kogyo Gijufsu Shareikai* **37**, 195.
Minami, T., and Tanaka, M. (1979). *Rev. de Chim. Min.* **16**, 283.
Minami, T., and Tanaka, M. (1980). *J. Solid State Chem.* **32**(1), 51.
Minami, T., Yoshida, A., and Tanaka, M. (1972). *J. Non-Cryst. Solids* **7**(4), 328.
Minami, T., Hibino, M., and Tanaka, M. (1974). *J. Non-Cryst. Solids* **15**, 141.
Minami, T., Nambu, H., and Tanaka, M. (1977). *J. Am. Ceram. Soc.* **60**(5/6), 283, (9/10), 467.
Minami, T., Imai, K., and Tanaka, M. (1979). *Kaen Yoshishu-Kotai Lonikusu Toronkai* **53**; (1980). *CA* **92**, 262, 219688 p.

Min'ko, N. (1973). *Neorg. Mat.* **9**(10), 1816.
Miranday, J. P., Jacoboni, Co., and De Pape, R. (1979). *Rev. Chim. Min.* **16**(4), 277; (1980). *CA* **92**, 312, 133788j.
Miroshnichenko, O., and Klimashevskii, L. (1970). *Neorg. Mat.* **6**(10), 1893.
Misawa, M., Price, D., and Suzuki, K. (1980). *J. Non-Cryst. Solids* **37**, 85.
Mita, Y. (1979). *CA* **92**, 264, 98390z.
Mitra, N., Mukherjee, M., Vasitsha, S., Das Podder, P. (1971). *Indian Ceram.* **15**(4), 133.
Mochida, N., and Takahashi, K. (1974). *Int. Cong. Glasses 10th, Kyoto,* Sect. 13, 29.
Mochida, N., Takahashi, K., and Nakata, K. (1978). *Yogyo Kyokai Shi* **86**(7), 316.
Moffat, T., Pearsall, G., and Wulff, J. (1964). "Structure." Wiley, New York.
Molchanova, O. (1957). *Steklo. I Ker.* **14**(3), 5.
Molchanova, O. (1958). *Structure Glasses* **1**, 109.
Molnar, B., and Dove, D. (1974). *J. Non-Cryst. Solids* **16**(2), 149.
Momii, R., and Nachtrieb, N. (1968). *J. Phys. Chem.* **72**(10), 3416.
Moon, D., Aitken, J., MacCrone, R., and Cieloszyk, G. (1975). *Phys. Chem. Glasses* **16**, 91.
Moore, H., and Carey, M. (1957). *J. Soc. Glass Technol.* **35**(162), 43.
Moore, H., and Lyle, A. (1947). *Glass* **28**, 563.
Moori, T., Morikawa, M., Iwai, S., and Tagai, H. (1970). *J. Ceram. Soc. Jpn.* **78** (904/12), 396.
Morey, G. (1931). *J. Am. Ceram. Soc.* **14**, 529.
Morey, G. (1938). "Properties of Glass," p. 74. Van Nostrand Reinhold, New York.
Morey, G. (1939). U.S. Patent 2 150 694.
Morey, G. (1954). "Properties of Glass," p. 319. Van Nostrand Reinhold, New York.
Morey, G., and Bowen, N. (1925). *J. Soc. Glass Technol.* **9**, 226.
Morey, G., and Bowen, N. (1931). *Glass Ind.* **12**, 133.
Morey, G., and Merwin, H. (1932). *J. Opt. Soc. Am.* **22**(11), 632.
Morgan, D. (1947). U.S. Patent 3 746 556.
Moriya, Y. (1970a). *Osaka Kogyo Gijutsu Shikensho Kiho* **21**(1), 1.
Moriya, Y. (1970b). *Yogyo Kyokai Shi* **78**, 898, 196.
Moriya, Y. (1970c). *Osaka Kogyo Gijutsu Shikensho Kiho* **21**(4), 221.
Moriya, T., Tanaka, Y., and Nogami, M. (1968). *Osaka Kogyo Gijutsu Skikensho Kiho* **29**(4), 436. *CA* **91**, 272, 983902.
Morozov, V., and Sharonova, N. (1969). *Opt. Spektr.* **26**(3), 256.
Morozova, I., and Yakhkind, A. (1977). *Fiz. Khim. Stekla* **3**(3), 108.
Mosel, B., Mueller-Warmuth, W., and Dutz, H. (1974). *Phys. Chem. Glasses* **15**(6), 154.
Mosesman, M., and Pitzer, K. (1941). *J. Am. Chem. Soc.* **63**, 2348.
Moss, S., and Deneufville, J. (1972). *J. Non-Cryst. Solids* **8–10**, 45.
Mott, N. (1967). *Adv. Phys.* **16**, 49.
Mott, N. F. (1969). *Phil. Mag.* **19**, 835.
Mott, N. (1977). *Phys. Non-Cryst. Solids, Int. Conf. 4th,* p. 3.
Mott, N. (1978). *J. Non-Cryst. Solids* **28**, 147.
Mott, N., and Davis, E. (1971). *Electr. Proc. Non-Cryst. Mat.,* Oxford.
Mott, N., Davis, E., and Street, R. (1975). *Phil. Mag.* **32**, 961.
Moynihan, C. (1966). *J. Phys. Chem.* **70**, 3399.
Moynihan, C., Smalley, C., Angell, C., and Sare, E. (1969). *J. Phys. Chem.* **73**, 2287.
Moynihan, C., Macedo, P., Aggarwal, I., and Schnaus, U. (1971a). *J. Non-Cryst. Solids* **6**, 322.
Moynihan, C. T., *et al.* (1971b). Presentation 6-G-71, Annual Meeting, American Ceramic Society, Chicago, Illinois, April 26.

Moynihan, C., Macedo, P. Panielson, P., and Elterman, P. (1975a). Tech. Rep. No. 032551, p. 33. Catholic Univ., Washington, D.C.
Moynihan, C., Macedo, P., Maklad, M., and Mohr, R. (1975b). *J. Non-Cryst. Solids* **17**, 369.
Moynihan, C., Casteal, A., Tran, D., Wilder, J., and Donovan, E. (1976). *J. Am. Ceram. Soc.* **59**(3/4), 137.
Moynihan, C., Saad, N., Tran, D., and Lesikar, A. (1980). *J. Am. Ceram. Soc.* **63**(7/8), 458.
Mozzi, R., and Warren, B. (1969). *J. Appl. Cryst.* **2**, 164.
Mozzi, R., and Warren, B. (1970). *J. Appl. Cryst.* **3**, 251.
Mueller, G. (1972). *J. Non-Cryst. Solids* **7**(4), 433.
Mueller, J., and Loeffler, G. (1933). *Zs. Anorg. Chem.* **46**, 538.
Mueller, R., and Wood, C. (1972). *J. Non-Cryst. Solids* **7**(4), 301.
Muir, G., and Glasser, F. (1974). *Phys. Chem. Glasses* **15**(1), 6.
Muir, G., and Glasser, F. (1976). *Phys. Chem. Glasses* **17**(3), 47.
Mukherjie, S., Zarzycki, J., Badie, J., Traverse, J. (1976). *J. Non-Cryst. Solids* **20**(3), 455.
Mulfinger, H. (1966). *J. Am. Ceram. Soc.* **49**, 462.
Munakata, M. (1960). *Solid State Electron.* **1**, 159.
Munir, Z., Fuke, L., and Kay, E. (1973). *J. Non-Cryst. Solids* **12**(3), 435.
Murthy, M., and Westman, A. (1966). *J. Am. Ceram. Soc.* **49**(6), 310.
Mydlar, M., Kreidl, N., Hendren, S., Clayton, G. (1970). *Phys. Chem. Glasses* **11**(6), 196.
Myers, J. (1978). NVO 1552-1 (from NTIS) *Energy Res. Abstr.* **4**(9), 24034.
Myers, M., and Berkes, J. (1972). *J. Non-Cryst. Solids* **8–10**, 804.
Myers, M., and Felty, E. (1967). *Mat. Res. Bull.* **2**(7), 715.
Myuller, R., and Tsai, T. (1967). *Zh. Prikhl. Khim.* **35**(4), 714.
Myuller, R., Orlova, G., Timofeevna, V., and Ternova, G. (1966). "Solid State Chemistry." Consultant's Bureau, New York.
Nagel, D. (1970). "Advances in X-Ray Analysis," p. 182. Plenum Press, New York.
Nagel, S., and Bergeron, C. (1974). *J. Am. Ceram. Soc.* **57**(3), 29.
Nakagawa, K., and Izumitani, T. (1972). *J. Non-Cryst. Solids* **7**(2), 168.
Nang, T., Okuda, M., and Matsushita, T. (1979). *Phys. Rev. B: Condens. Matter* **19**(2), 947.
Nambu, H., and Minami, T. (1977). *J. Am. Ceram. Soc.* **60**(5/6), 293, 467.
Narasimhan, P., and Rao, K. J. (1978). *J. Non-Cryst. Solids* **27**, 225.
Nassau, K., Wang, C., and Grasso, M. (1979). *J. Am. Ceram. Soc.* **62**(9/10), 503.
Naudin, F., and Zarzycki, J. (1968). *CR* 266c, (11), 729.
Neilson, G. (1969). *Phys. Chem. Glasses* **10**(2), 54–62.
Neilson, G. (1972). *Phys. Chem. Glasses.* **13**, 70.
Neilson, G., and Weinberg, M. (1977). *J. Non-Cryst. Solids* **23**, 43.
Neilson, G., and Weinberg, M. (1978). *J. Non-Cryst. Solids* **28**, 209.
Nemanich, R., Connell, G., Hayes, T., and Street, R. (1979). *Phys. Rev.* (to be published).
Nemec, L. (1977). *J. Am. Ceram. Soc.* **60**(9–10), 436.
Nemilov, S. (1964). *Zv. An. SSRR. Ser.* **28**, 1783.
Nemilov, S. (1968). *Inorg. Mat.* **4**(6), 835.
Nemilov, S. (1969). *Zh. P. Khim.* **42**(1), 55.
Nemilov, S., and Petrovskii, G. (1963). *Zh. P. Khim.* **36**, 977.
Neov, S., Gerasimova, I., Kozhukarov, V., and Marinov, M. (1980). *J. Mat. Sci.* **15**(5), 1153.
Nester, H., and Kingery, J. (1965). *Int. Cong. Glasses, 7th Brussels* A, p. 106.
Neumann, C., and Dietzel, A. (1939). *Glastech. Ber.* **17**, 286.
Neumann, C., and Dietzel, A. (1940). *Glastech. Ber.* **18**, 267.
Nielsen, S. (1962). *Infrared Phys.* **2**, 117.
Nielsen, S. (1965a). *Phys. Chem. Glasses* **6**(3), 90.

Nielsen, S. (1965b). *Int. Cong. Glass, 7th.* Glastech. Ber., Frankfurt.
Nielsen, S. (1965c). *Infrared Phys.* **5**, 195.
Nielsen, S. (1966). *Phys. Chem. Glasses* **7**, 56.
Ninomiya, Y., Nakamura, Y., and Shimoji, M. (1975). *J. Non-Cryst. Solids.* **17**, 231.
Nishida, T., and Takashima, Y. (1980). *J. Non-Cryst. Solids* **37**(1), 37.
Noda, M., and Maruno, S. (1974). *Yogyo Kyokai Shi* **82**(4), 234.
Nogami, M., and Moriya, Y. (1980). *J. Non-Cryst. Solids* **37**(2), 191.
Noll, W. (1963). *Angew. Chem.* **75**, 123.
Nordberg, M. (1939). Oral communication.
Nordberg, M. (1950a). U.S. Patent 2 494 259.
Nordberg, M. (1950b). U.S. Patent 2 505 001.
Nowak, W. (1977). *Phys. Status Solidi (a)*, **44**, 265.
Nukui, A., Tagai, H., and Morikawa, H. (1978). *J. Am. Ceram. Soc.* **61**(3/4), 174.
Oancea, C., Teodorescu, J., and Cristea, P. (1967). *Kristalografia* **7**, 560.
Oberlies, T. (1964). *Glast. Ber.* **37**, 122.
O'Connor, P., McChesney, J., Presby, H., and Cohen, L. (1976). *Am. Ceram. Soc. Bull.* **55**(5), 513.
Ohlberg, S., and Golob, H. (1973). *J. Am. Ceram. Soc.* **56**(16), 300.
Ohlberg, S., and Parsons, J. (1964). *Conf. Non-Cryst. Solids, Delft.*
Ohlberg, S., Golob, H., and Strickler, D. (1962). *Symp. Nucl. Cryst. Glasses Melts* p. 55. American Ceramic Society, Columbus, Ohio.
Ohlberg, S., Hammel, J., and Golob, H. (1965). *J. Am. Ceram. Soc.* **48**(4), 178.
Ohmachi, Y., and Uchida, N. (1972). *J. Appl. Phys.* **43**(4), 1709.
Ohno, M. (1965). *Int. Cong. Glass 7th, Brussels,* Paper 8.
Ohsaka, T. (1975). *J. Non-Cryst. Solids* **17**, 121.
Ohsaka, T. (1976a). *J. Non-Cryst. Solids* **21**(1), 23.
Ohsaka, T. (1976b). *J. Non-Cryst. Solids* **22**, 89, 359.
Okazaki, A. (1958). *J. Phys. Soc. Japan* **13**, 1151–55.
Olshanskii, Ya. (1951). *Dokl. Akad Nauk SSSR* **76**, 93.
Omar, M., and Stevels, J. (1978). *J. Non-Cryst. Solids.* **27**, 51.
Onoda, G., and Brown, S. (1968). Final Rep. (R 7363) on Contr. N 00019-67-C-301, Naval Air Systems Command, February.
Onoda, G., and Brown, S. (1970). *J. Am. Ceram. Soc.* **53**(6), 311.
Onomichi, M., Arai, T., and Kudo, J. (1971). *J. Non-Cryst. Solids* **6**, 362.
Ordway, F. (1964). *Science* **143**, 800.
Ordway, F. (1969). *Acta Cryst.* **25A**(3), 524.
Orlova, G. (1973). *Vestn. Leningr.* **4**, 94.
Orlova, G., Rasina, O., and Krivenkova, N. (1974). *Zh. Prikl. Khim.* **47**(3), 510.
Orlova, G., Kolomeitseva, S., Timonov, A., and Kuznetsova, O. (1979). *Fiz. Khim. Stekla* **5**(5), 546.
Ormandroyd, R., Thompson, M., and Allison, J. (1975). *J. Non-Cryst. Solids* **18**, 375.
Ota, R., and Anderson, O. (1977). *J. Non-Cryst. Solids* **24**(2), 235.
Otto, K., and Milberg, M. (1967). *J. Am. Ceram. Soc.* **50**(10), 513.
Otto, K., Milberg, M., and Peters, C. (1969). *Phys. Chem. Glasses* **10**(2), 46.
Ovcharenko, N., and Yakhkind, A. (1971). *Opt. Mekh. Prom.* **38**(3), 37.
Ovshinsky, S. (1966). U.S. Patent 3 271 591.
Ovshinsky, S. (1968). *Phys. Rev. Lett.* **21**, 1450.
Ovshinsky, S. (1977). *Int. Conf. Amorphous Liqu. Semicond, 7th, Edinburgh* p. 519.
Owen, A. (1961). *Phys. Chem. Glasses* **2**, 87, 152.
Owen, A. (1979). *U. Conf. Glass Sci. 5th*.

Oyamada, R., and Hagiwaka, H. (1978). *Yogyo Kyokai Shi* **8b**(4), 151.
Panek, L., and Bray, P. (1977). *J. Chem. Phys.* **66**(8), 3822.
Panson, A. J. (1964). *Inorg. Chem.* **3**, 940–943.
Pant, A., and Cruikshank, D. (1968). *Acta Cryst.* **24B**, I, 13.
Parfenov, A., Klimov, A., and Mazurin, O. (1959). *Vestn. LGO* **10**, *Ser. Fiz. Khim.* (2), 129.
Park, M., and Bray, P. (1972). *Phys. Chem. Glasses* **13**(2), 50.
Partridge, G. (1979). *Glass Technol.* **20**(6), 246.
Patel, P., and Kreidl, N. (1975). *J. Am. Ceram. Soc.* **58**(5,6), 263.
Patel, P. (1978). Oral communication.
Patterson, R., and Brau, M. (1966). unpublished talks at Electrochem. Society Meeting Infrared, Symp., Fort Monmouth, New Jersey.
Paul, A., and Youseff, A. (1978). *J. Mat. Sci.* **13**(1), 97.
Paul, A., Donaldson, J., Donoghue, M., and Thomas, M. (1977). *Phys. Chem. Glasses* **18**(6), 125.
Paul, A., Parker, J., and Ward, A. (1979). *Phys. Chem. Glasses* **20**(5), 97.
Pauletti, G., and Giamanio, F. (1967). *Vetr. Sil.* **11**(64), 12.
Pauling, L. (1952). *J. Phys. Chem.* **56**, 361.
Pauling, L. (1960). "The Nature of the Chemical Bond," p. 98. Cornell Univ. Press, Ithaca, New York.
Pavlovskii, V. (1970). *Stekloobraznoe Sost.* **5**(1), 148; (1971). *CA* 66981.
Pavlushkin, N., and Zhuravlev, A. (1969). *Neorg. Mat.* **5**(3), 595.
Pavlushkin, N., Nurbekov, T., and Egorova, L. (1968). *Neorg. Mat.* **4**(8), 1390.
Pavlushkin, N., Agakkov, A., and Lomakina, O. (1975). *Material* **1**, 93; (1976). *CA* **84**, 78399y.
Pearce, M., and Beisler, J. (1965). *J. Am. Ceram. Soc.* **48**, 40.
Pearson, A. (1960). *J. Appl. Phys.* **31**, 219.
Pearson, A. (1964). "Modern Apsects of the Vitreous State" (J. MacKenzie, ed.), p. 29. Butterworths, London.
Peddle, C. (1920). *J. Soc. Glass Technol.* **4**, 320.
Pelizzari, C. (1974). Thesis. Univ. of Michigan.
Perloff, A., and Block, S. (1966). *Acta Crystallog.* **20**(2), 274.
Pernot, F. (1968). *Verres Refr.* **22**(6), 595.
Peters, F. (1966). *Ceram. Bull.* **45**(11), 1017.
Peterson, G., Kurkjian, C., and Carnevale, A. (1974). *Phys. Chem. Glasses* **15**, 59.
Peterson, G., Kurkjian, C., and Carnevale, A. (1975). *Phys. Chem. Glasses* **16**(3), 63.
Peterson, G., Carnevale, A., and Kurkjian, C. (1977). *J. Non-Cryst. Solids* **23**(2), 245.
Petrovskii, G. (1959). *Silikaty* **3**(4), 336.
Petrovskii, G., Leko, E., and Mazurin, O. (1965). "Structure of Glass" (O. Mazurin, ed.), Vol. 4, p. 198. Consultants Bureau, New York.
Petz, J., Kruh, R., and Amstutz, G. C. (1961). *J. Chem. Phys.* **34**(5), 76.
Phillips, B., and Roy, R. (1964). *Phys. Chem. Glasses* **5**(6), 172.
Phillips, B., and Schroger, M. (1965). *J. Am. Ceram. Soc.* **48**(8), 398.
Phillips, J. (1979). *J. Non-Cryst. Solids* **34**, 153.
Phillips, S. V., Booth, R. E., and McMillan, P. W. (1970). *J. Non-Cryst. Solids* **4**, 510.
Phillips, W. (1972). *Low Temp. Phys.* **7**, 351.
Phipps, K., and Sullenger, D. (1964). *Science* **145**, 1049.
Phipps, K., Sullenger, D., Jones, L., Tucker, P., and Wittenberg, L. J. (1964). *Ceram. Bull.* **43**(2), 31.
Pickart, S., Sharma, Y., and De Neufville, J. (1979). *J. Non-Cryst. Solids* **34**, 183.
Pierre, A., Uhlmann, D., and Molea, F. (1972). *J. Appl. Cryst.* **5**(3), 216.

Pincus, A., and Warren, B. (1941). *J. Am. Ceram. Soc.* **23**, 301.
Pinnow, D., Van Uitert, L., Rich, T., Ostermayer, F., and Grodkiewicz, W. (1975). *Mat. Res. Bull.* **10**, 133.
Pirooz, P. (1963). U.S. Patents 3 088 833, 3 088 835, 1963.
Pivinskii, Y. (1970). *Ogneupory* **35**(8), 54.
Platakis, N. (1978). *J. Non-Cryst. Solids* **27**, 331.
Plieth, K., Reuber, E., and Zschoerper, K. (1969). *Glastech. Ber.* **42**(9), 359.
Plumat, E. R. (1968). *J. Am. Ceram. Soc.* **51**(9), 499.
Pockels, F. (1902). *Ann. Phys.* 745.
Polk, D. (1971). *J. Non-Cryst. Solids* **5**, 365.
Poltavtsev, Y., Zaicharov, V., and Remizovich, T. (1972). *Inorg. Mat.* **9**, 813.
Polukhin, V., Zimina, M., Smirnova, R., and Sololev, L. (1968). *Zh. Prikl. Khim. Leningr.* **41**(7), 1447.
Polukhin, V., Urusovskaya, L., Smirnova, T., and Barbova, T. (1979). *Fiz. Khim. Stekla* **5**(3), 303.
Poole, J. (1949). *J. Am. Ceram. Soc.* **32**(7), 230.
Popov, A. (1978). *Phys. Chem. Glasses* **19**(3), 43.
Poraï-Koshits, E. (1936). *Z. Krist.* **95**, 195.
Poraï-Koshits, E. (1955). *Struct. Glass* 145.
Poraï-Koshits, E. (1965). *Struct. Glass* **5**, 1.
Poraï-Koshits, E. (1977). *J. Non-Cryst. Solids* **25**(1-3), 87.
Poraï-Koshits, E., and Averyanov, U. (1969). *Proc. Russ. Symp. Phase Sep. Glass, 1st, Leningrad,* p. 26.
Poraï-Koshits, E., Averyanov, U., Andreev, N., and Golubkov, V. (1971). *Sci. Tech. Commun. Int. Congr. Glass 9th, Paris,* pp. 341, 391.
Poraï-Koshits, E., Golubkov, V., and Titov, A. (1978). *In* "Borate Glasses" (L. Pye *et al.,* eds.), p. 183. Plenum Press, New York.
Poulain, M. I., and Lucas, J. (1978). *Verres Refr.* **32**, 505.
Poulain, M., Lucas, J., and Brun, P. (1975). *Mat. Res. Bull.* **10**(4), 243.
Poulain, M. I., Poulain, M. A., and Lucas, J. (1979). *Rev. Chim. Min. Min.* **16**(9), 267; (1980j). *CA* **92**, 312, 133787h.
Powell, J., and Frieser, R. (1979). *Glass Technol.* **20**(3), 96.
Powers, D. (1978). *J. Am. Ceram. Soc.* **61**(7/8), 295.
Prasad, R., and Isard, J. (1967). *Phys. Chem. Glasses* **8**(6), 218.
Prasad, E., Sayer, M., and Vyas, H. (1980). *J. Non-Cryst. Solids* **40**(1/3), 119.
Prebus, A., and Michener, J. (1952). *Phys. Rev.* **87**, 201.
Prebus, A., and Michener, J. (1954). *Ind. Eng. Chem.* **16**, 147.
Preisinger, A. (1962). *Naturwissenschaften* **49**, 345.
Preston, E., and Turner, W. (1936). *J. Soc. Glass Technol.* **20**, 144.
Primak, W. (1962). *Phys. Rev.* **128**(6), 2580.
Primak, W., Fuchs, L., and Day, P. (1953). *Phys. Rev.* 1064.
Primak, W., Fuchs, L., and Day, P. (1955). *J. Am. Ceram.Soc.* **38**(4), 135.
Prins, J., (1935). *J. Chem. Phys.* **3**, 72.
Prins, J. (1960). *In* "Non-Crystalline Solids" (V. Frechette, ed.), p. 322. Wiley, New York.
Prins, J. (1965). "Non-Crystalline Solids," p. 39. North-Holland Publ., Amsterdam.
Prins, J., and Fonteyne, E. (1935). *Physica* **2**, 570.
Prjanishnikov, V., Bartenev, G., Tsiganov, A., and Gorbachev, V. (1971). *Int. Cong. Glass, 9th Paris* **1**, 119.
Protsenko, P., and Belova, Z. (1957). *Russ. J. Inorg. Chem.* **2**, 220.
Protsenko, P., and Bergman, A. (1950). *J. Gen. Chem. USSR* **20**, 1421.

Protsenko, P., and Popovskaya, V. (1953). *J. Gen. Chem. USSR* **23**, 1246.
Protsenko, A., and Shevchenko, N. (1952). *J. Gen. Chem. USSR* **22**, 1357.
Provenzano, V., Macedo, P., and Volterra, V. (1971). *Electr. Prop. of Sodium Trisilicate Gl., Cath. U. Am., Washington, D.C.* **20**, 17 (quoted by Mazurin, 1973).
Quincke, G. (1880). *Ann. Phys.* **10**, 160, 374, 512.
Quinn, R., and Johnson, R. (1972). *J. Non-Cryst. Solids* **7**, 53.
Quinn, R., and Johnson, R. (1973). *J. Non-Cryst. Solids* **12**, 213.
Raask, E., and Jessop, R. (1966). *Phys. Chem. Glasses* **7**(16), 200.
Rao, B. (1963). *Phys. Chem. Glasses* **4**, 22.
Rao, B. (1963). *Phys. Chem. Glasses* **46**(3), 107.
Rao, B. (1964). *J. Am. Ceram. Soc.* **47**(9), 455.
Rao, B., Helphrey, D., and Angell, C. (1973). *Phys. Chem. Glasses* **14**, 26.
Rathmann, C., Mann, C., and Nordberg, M. (1968). *Appl. Opt.* **7**(5), 819.
Rawal, B., and McCrone, R. (1977). *Phys. Non-Cryst. Solids, Int. Conf. 4th,* p. 248.
Rawson, H. (1956). *Proc. Int. Congr. Glass, 4th* p. 62.
Rawson, H. (1960). *Phys. Chem. Glasses* **1**, 170.
Rawson, H. (1967). "Inorganic Glass-Forming System," p. 249. Academic Press, New York.
Ray, N., Lewis, C., Laycock, J., and Robinson, W. (1973a). *Glass Technol.* **14**(2), 50.
Ray, N., Laycock, J., and Robinson, W. (1973b). *Glass Technol.* **14**(2), 55.
Rechtin, M., and Auerbach, P. (1973a). *J. Non-Cryst. Solids* **12**, 391.
Rechtin, M., and Auerbach, P. (1973b). *Solid State Commun.* **13**, 491.
Rechtin, M., and Auerbach, P. (1974). *J. Non-Cryst. Solids* **16**, 1.
Redman, M., and Chen, J. (1967). *J. Am. Ceram. Soc.* **50**(10), 523.
Redwine, R., and Field, M. (1968). *Mat. Sci.* **3**, 380.
Redwine, R., and Field, M. (1969). *Mat. Sci.* **4**(8), 713.
Regan, M., and Drake, C. (1972). *Mat. Res. Bull.* **7**(12), 1559.
Reggiani, J-C., Makugani, J-P., and Bernard, J. (1978). *J. Chim. Phys. Phys.-Chim. Biol.* **75**(9), 970.
Reilly, M. (1970). *J. Phys. Chem. Solids* **31**, 1041.
Reisfeld, R., and Boehm, I. (1974). *J. Non-Cryst. Solids* **16**, 83.
Reisfeld, R., and Eckstein, Y. (1973a). *J. Non-Cryst. Solids* **12**, 357.
Reisfeld, R., and Eckstein, Y. (1973b). *Solid State Commun.* **13**, 741.
Reisfeld, R., and Lieblich, N. (1973). *J. Phys. Chem. Solids,* **34**, 1467.
Renninger, A., and Averbach, B. (1973). *Phys. Rev.* **8**, 1507.
Revesz, A. (1972). *J. Non-Cryst. Solids* **7**, 77.
Rhee, C., and Bray, P. (1971). *Phys. Chem. Glasses* **12**(6), 156, 165.
Ribes, M., Ravaine, D., Souquet, J., and Maurin, M. (1979). *Rev. Chim. Min.* **16**(4), 339.
Richter, H. (1972). *J. Non-Cryst. Solids* **8–10**, 388.
Riebling, E. (1964). *J. Am. Ceram. Soc.* **47**, 478.
Riebling, E. (1968). *J. Am. Ceram. Soc.* **51**(7), 406.
Riebling, E. (1971). *J. Chem. Phys.* **55**(2), 804.
Riebling, E. (1972a). *J. Chem. Phys.* **56**(4), 1811.
Riebling, E. (1972b). *J. Mat. Sci.* **7**(1), 40.
Riebling, E. (1973). *J. Am. Ceram. Soc.* **56**(1), 25.
Riebling, E. (1974). *J. Am. Ceram. Soc.* **57**(9), 373.
Riebling, E., and Dalton, D. (1970). U.S. Patent 3 542 572.
Riebling, E., and Puke, P. (1966). *Mat. Sci.* **1**(1), 33.
Ring, P. (1962). Thesis, Brown Univ.

Risbud, S., and Pask, J. (1977). *J. Am. Ceram. Soc.* **60**(9,10), 418.
Risbud, S., and Pask, J. (1978). *J. Am. Ceram. Soc.* **61**, 63.
Risbud, S., and Pask, J. (1979). *J. Am. Ceram. Soc.* **62**, 214.
Rivoalen, L., Revcolevischi, A., Livage, J., and Collongues, R. (1976). *J. Non-Cryst. Solids* **21**, 171.
Robinson, J. (1965). *Phys. Chem. Solids* **26**, 209.
Robinson, M., Pastor, R., Turk, R., Devor, D., Braunstein, M., and Braunstein, R. (1980). *Mat. Res. Bull.* **15**(6), 735.
Rockett, T., and Foster, W. (1965). *J. Am. Ceram. Soc.* **48**(2), 75.
Rockett, T., Foster, W., and Ferguson, R. (1965). *J. Am. Ceram. Soc.* **48**(6), 329.
Roetger, H. (1958). *Glastech. Ber.* **31**(2), 54.
Rogozhin, Y., Zaionts, L., Yantseva, O. (1971a). *Stekloobr. Sist. Nov. Stekla Ikh. Osn., 101 Ref. Zh.,* 20B500. *CA* 78833h (1972).
Rogozhin, Y., Zaionts, L., Yantseva, O., and Chekharin, U. (1971b). *Stekloobr. Sist. Nov. Stekla Ikh. Osn. Ref. Zh. Khim.* 20B516. *CA* 78844n (1972).
Rogozhin, Y., Zaionts, L., Yantseva, O., and Rodina, O. (1977). *Sist. Nov. Stekla Ikh. Osn., Ref. Zh. Khim.* 20B505. *CA* 78838 (1972).
Romeo, N., Sberveglieri, G., and Tarricone, L. (1977). *J. Non-Cryst. Solids* **23**, 7.
Roscoe, H. (1867). *Phil. Trans. R. Soc.* **158**, 1.
Rossini, F. (1952). Selected Values of Chemical Thermodyn. Properties, Circular, 500, p. 629. National Bureau Standards, Washington, D.C.
Roth, M., and Zarzycki, J. (1974). *J. Non-Cryst. Solids* **16**(1), 93.
Rothermel, J., Sun, K., and Silverman, A. (1949). *J. Am. Ceram. Soc.* **32**, 153.
Rothwell, G. (1956). *J. Am. Ceram. Soc.* **39**(12), 407.
Rotskowsky, A. (1930). *Zh. Russ. Fiz. Khim. Obsch.* **61**, 2055.
Rouse, G. Miller, P., and Risen, W. (1978). *J. Non-Cryst. Solids* **28**, 193.
Rowe, J. (1974). *Appl. Phys. Lett.* **25**(10), 576.
Roy, D., Roy, R., and Osborn, E. (1950). *J. Am. Ceram. Soc.* **33**, 85.
Roy, D., Roy, R., and Osborn, E. (1953). *J. Am. Ceram. Soc.* **36**, 185.
Roy, R., and Cohen, H. (1961). *Nature (London)* **190**, 798.
Roy, R., and Cohen, H. (1974). *Science* **184**(4132), 91.
Roy, R., and Osborn, E. (1949). *J. Am. Chem. Soc.* **71**, 2056.
Rudenberg, H. (1962). *Proc. Electr. Components Conf.* 90.
Ruffa, A. (1973). *J. Non-Cryst. Solids* **13**, 37.
Rusert, E., Drennan, D., and Biggs, M. (1979). *Sci. Technol. Aerospace Rep.* **17**, 20, Abstr. N79-29333.
Rusetskaya, E., and Ermolenko, N. (1970). *Stekloobr. Sost.* **5**(1), 178.
Ruska, J., and Thurn, H. (1976). *J. Non-Cryst. Solids* **22**, 277.
Russak, M., and McLaren, M. (1973). *Ceram. Bull.* **52**(3), 271.
Rykova, T., and Borisova, Z. (1977). *Fiz. Khim. Vestn., Leningrd. Univ.* (2), 153. *CA* **87**, 234, 156011m (1977).
Rza-Zade, P., Abdulaev, G., Eyabova, N., and Jamedov, F. (1971). *Inorg. Mat.* **7**(11), 1872.
Sakka, S. (1969). *J. Am. Ceram. Soc.* **52**(2), 69.
Sakka, S. (1977). *Yogyo Kyokai Shi* **85**(6), 299.
Sakka, S., and Kamiya, K. (1976). *Proc. U.S.-Jpn. Seminar Basic Sci. Ceram., 1975,* p. 135.
Sakka, S., and Matsuita, K. (1976). *J. Non-Cryst. Solids* **22**, 57.
Sakka, S., and Senga, A. (1978). *J. Mat. Sci.* **13**(3), 505.
Sakurai, S., and Mochizuki, H. (1978). *Nagoya-shi Kogyo Kenkyusho Kenkyu Hokoku* **58**(9). *CA* **91**, 250, 43431 (1979).

Sakurai, T., and Munesue, S. (1951). *Phys. Rev.* **85,** 921.
Sample, H., Neuringer, L., Gerber, J., and Deneufville, J. (1972). *J. Non-Cryst. Solids* **8–10,** 50.
Santt, R. (1978). *Mat. Technol. (Paris)* (7,8), 266; (1979). *CA* **90,** 274, 191193f.
Sare, E., Moynihan, C., and Angell, C. (1973). *J. Phys. Chem.* **77,** 1869.
Sarrach, D., Deneufville, J., and Haworth, W. (1976). *J. Non-Cryst. Solids* **22,** 245.
Sastry, B., and Hummel, F. (1959). *J. Am. Ceram. Soc.* **42,** 81.
Saunders, J. (1942). *J. Res. Nat. Bur. Std.* **28,** 51.
Savage, J. (1971). *J. Mat. Sci.* **6**(7), 964.
Savage, J., and Nielsen, S. (1965a). *Infrared Phys.* **5,** 195.
Savage, J., and Nielsen, S. (1965b). *Phys. Chem. Glasses* **6**(3), 90.
Savage, J., and Nielsen, S. (1966). *Phys. Chem. Glasses* **7**(2), 56.
Savan, J., Kozhina, I., and Borisova, Z. (1966). *Vest. Leningr. Gos., Univ. Ser. Fiz. Khim* **22**(7), 141.
Savva, M., and Newns, G. (1971). *Int. Cong. Glass, 9th,* **1** (*Sci. Tech. Commun.*), Paris, p. 419.
Sayer, D., Lytle, E., and Stern, E. (1974). "Amorphous and Liquid Semiconductors" (J. Stuke and W. Brenig, eds.), Vol. 1, p. 403. Taylor and Francis, London, 1974.
Scharfe, M. (1970). *Phys. Rev.* **2,** 5025.
Scheerer, J., Muller-Warmuth, W., and Dutz, H. (1973). *Glastech. Ber.* **46,** 109.
Schenk, J. (1957). *Physics* **23,** 325.
Schlichting, J. (1978). *Glastech. Ber.* **51**(7), 21.
Schlichting, J., and Schubert, P., (1978). *J. Non-Equilib. Thermodyn.* **3**(1), 245.
Schmid, A. (1968). *J. Appl. Phys.* **39**(7), 3140.
Schmitt, J. (1962). *Verr. Ref.* **16,** 344.
Schnaus, V., Moynihan, C., Gammon, R., and Macedo, P. (1970). *Phys. Chem. Glasses* **11**(6), 213.
Schnaus, V., Marshall, A., and Moynihan, C. (1972). *J. Am. Ceram. Soc.* **55**(4), 180.
Schnaus, V., Schroeder, J., and Haus, J. (1976). *Phys. Lett.* **57A**(1), 92.
Scholes, S. (1973). *J. Non-Cryst. Solids* **12**(2), 266.
Scholze, H. (1959). *Glastech. Ber.* **32,** 81, 142, 178, 374, 381.
Scholze, H. (1977). "Glas." 2nd ed. Vieweg and Sons, Braunschweig.
Scholze, H. (1966). *Glass Ind.* **546**(10,11,12).
Scholze, H. (1969). *Int. Cong. Glass, 8th Soc. Glass Technol., Sheffield,* p. 69.
Schott (Faulstich) (1976). German Patent 22 59 183.9-45.
Schottmiller, J., Taylor, T., and Ryan, F. (1969). *Appl. Opt. Suppl.* **3,** 55.
Schottmiller, J., Tabak, M., Kukovsky, G., and Ward, A. (1970). *J. Non-Cryst. Solids* **2**(4), 80.
Schroeder, J. (1977). "Treatise on Materials Science and Technology," Vol. 12, Glass I, Academic Press, New York. pp. 157–222.
Schubert K., and Fricke, H. (1951). *Z. Naturforsch.,* **6a,** 781–782.
Schulmann, J., Ginther, R., and Klick, C. (1951). *J. Appl. Phys.* **22,** 1479.
Schulmann, J. and Etzel, H. (1953). *Science* **118,** 184.
Schultz, I. (1957). *Naturwissenschaften* **44,** 536.
Schultz, I., and Hinz, E. (1956). *Glastech. Ber.* **29,** 319.
Schultz, P. (1976). *J. Am. Ceram. Soc.* **59**(5/6), 214.
Schultz, P., and Mizzoni, M. (1973). *J. Am. Ceram. Soc.* **56**(2), 65.
Schultz, P., and Smyth, H. (1972). "Amorphous Materials" (R. W. Douglas and B. Ellis, eds.). Wiley, New York.
Schultz, P., and Smyth, H. (1974). *CA* **81,** 667.

Schultz-Sellak, C. (1870). *Ann. Phys.* **139**, 162.
Schulze, G. (1913). *Ann. Phys. Leipzig* **40**, 327.
Seager, C., and Quinn, R. (1975). *J. Non-Cryst. Solids* **17**, 386.
Searight, C., and Alexander, M. (1965a). U.S. Patent 3 198 641.
Searight, C., and Alexander, M. (1965b). U.S. Patent 3 894 051.
Searight, C., and Alexander, M. (1965c). U.S. Patent 3 294 558.
Searight, C., and Alexander, M. (1965d). U.S. Patent 3 547 517.
Searight, C., Alexander, M., Ryan, J., and Brasfield, S. (1971). U.S. Patent 3 560 074.
Segnit, E. (1954). *J. Am. Ceram. Soc.* **37**(6), 273.
Segnit, E., and Holland, A. (1965). *J. Am. Ceram. Soc.* **48**(8), 409.
Seifert, F., Virgo, D., and Mysen, B. (1978). Yearbook, Carnegie Inst., Washington, D.C. *CA* **92**, 264, 219705s (1980).
Sekiya, T., and Matushita, T. (1977). *Yogyo Kyokai Shi* **85**(10), 481.
Sella, C., Trant, C., Navez, M., and Trillat, J. (1964). *J. Silic. Ind.* **29**, 15.
Semin, E., and Kentov, V. (1970). *Neorg. Mat.* **6**(12), 2213.
Sen, I., and Tooley, F. (1950). *J. Am. Ceram. Soc.* **33**, 178.
Seregin, P., Turaev, E., Andreev, A., and Melekh, B. (1979). *Fiz. Khim. Stekla* **5**(6), 659.
Settarova, Z., Sergeev, O., and Nikalaeva, Z. (1973). *Phys. Chem. Glasses* **10**(3), 577.
Seward, T., Uhlmann, D., and Turnbull, D. (1968a). *J. Am. Ceram. Soc.* **51**(5), 278.
Seward, T., Uhlmann, D., and Turnbull, D. (1968b). *J. Am. Ceram. Soc.* **51**(11), 634.
Seward, T., Uhlmann, D., and Turnbull, D. (1978). "Borate Glasses" (L. Pye *et al.*, eds.), p. 427. Plenum Press, New York.
Shabanova, E. (1967). *Tr. Gor'k. Politekh. Inst.* **23**(4), 38.
Shackelford, J. (1978). "Borate Glasses" (L. Pye *et al.*, eds.), p. 377. Plenum Press, New York.
Shadid, K., and Glasser, F. (1971). *Phys. Chem. Glasses* **12**(2), 50.
Shand, E. (1958). "Glass Engineering Handbook," p. 30. McGraw-Hill, New York.
Sharp, D. (1933). *Ind. Chem. Eng.* **25**, 755.
Sharp, D. (1940). *Glass Ind.* **21**, 158.
Shartsis, L., and Shermer, H. (1953). *J. Am. Ceram. Soc.* **36**, 35.
Shartsis, L., and Shermer, H. (1954). *J. Am. Ceram. Soc.* **37**, 544.
Shartsis, L., and Spinner, S. (1951). *J. Res. Nat. Bur. Sta.* **46**(5), 385.
Shartsis, L., Spinner, S., and Capps, W. (1952). *J. Am. Ceram. Soc.* **35**, 155.
Shaw, R., and Breedis, J. (1972). *J. Am. Ceram. Soc.* **55**(8), 422.
Shaw, R., and Heasley, J. (1967). *J. Am. Ceram. Soc.* **50**(6), 297.
Shaw, R., and Uhlmann, D. (1969). *J. Non-Cryst. Solids* **1**(6), 474.
Shaw, R., and Uhlmann, D. (1971a). *J. Non-Cryst. Solids* **5**(3), 237.
Shaw, R., and Uhlmann, D. (1971b). *J. Am. Ceram. Soc.* **54**(3), 170.
Shcheglova, Z. (1971). *Stekloobr. Sist. Nov. Stekla Kh. Osn.* 107; (1972) *CA* 78835k.
Shelby, J. (1979). *J. Appl. Phys.* **50**(12), 801.
Shelby, J., and Day, D. (1969). *J. Am. Ceram. Soc.* **52**(4), 169.
Shelby, J., Mattern, P., and Ottesen, D. (1979). *J. Appl. Phys.* **50**(12), 801.
Sheperd, E., Rankin, G., and Wright, F. (1909). *Am. J. Sci.* **28**, 293.
Sherrel, P., and Thompson, J. (1977). *J. Non-Cryst. Solids* **24**, 69.
Shevchik, N., and Kniep, R. (1974). *J. Chem. Phys.* **60**(8), 3011.
Shimizu, T., and Ishi, N. (1978). *J. Non-Cryst. Solids* **27**, 109.
Shiren, N., Arnold, W., and Kazyaka, T. (1977). *Phys. Rev. Lett.* **39**(4), 239.
Shishido, T. (1979). *J. Mat. Sci.* **14**(4), 823.
Shishido, T., Okamura, K., and Yajima, S. (1978). *J. Mat. Sci.* **13**(5), 1006.
Shi-Tuan, Y., and Regel, A. (1962). *Sov. Phys.-Solid State* **3**, 2627.

Shmidt, Y., and Alekseeva, Z. (1964). *Zh. Pr. Kh.* **37**(10), 2299.
Skhol'nikov, E. (1965). *Vestnik Leningrad Univ.* **4**, 115.
Skhol'nikov, E. (1970). *Stekloobr. Sost.* **5**(1), 94.
Skhol'nikov, E., and Borisova, Z. (1965a). *Vestnik Leningrad. Univ.* **4**, 120.
Skhol'nikov, E., and Borisova, Z. (1965b). *Vestnik Leningrad. Univ.* **4**, 199.
Skhol'nikov, E., Gerasimenko, V., and Borisova, Z. (1977). *Fiz. Khim. Stekla* **3**(4), 338.
Shtin, A., Galaktionov, A., Makarov, V., and Mamoshin, V. (1977). *Fiz. Khim. Stekla* **3**(3), 201.
Sidorov, T., and Tyul'kin, U. (1968). *Neorg. Mat.* **4**(4), 578.
Sigel, G. (1973/74). *J. Non-Cryst. Solids* **13**, 372.
Sigel, G. (1977). "Treatise on Materials Science and Technology," Vol. 12, Glass I, p. 5. Academic Press, New York.
Silver, A. (1960). *J. Chem. Phys.* **32**, 959.
Silver, A., and Bray, P. (1958). *J. Chem. Phys.* **29**, 984.
Silver, J., White, R., and Donaldsen, J. (1977). *J. Mat. Sci.* **12**(4), 827.
Simmons, J. (1973). *J. Am. Ceram. Soc.* **56**(5), 284.
Simon, I. (1953a). *J. Chem. Phys.* **21**, 23.
Simon, I. (1953b). *J. Am. Ceram. Soc.* **36**, 160.
Skatula, W., and Kuehne, K. (1959). *Silicattech.* **10**, 105.
Skatula, W., Vogel, W., and Wessel, H. (1958). *Silicattech.* **9**, 51, 323.
Smakota, N., Barinov, Y., Kozlov, A., Belozub, G., and Es'kov, A. (1974). *Steklo. Keram.* (1), 14; (1976). *CA* 99402u.
Smart, K., and Glasser, F. (1974). *J. Am. Ceram. Soc.* **57**(9), 378.
Smirnova, E. (1966). "Structure of Glass," Vol. 7, p. 25. Consultant's Bureau, New York.
Smith, C. (1978). "Borate Glasses" (L. Pye *et al.*, eds.), p. 307. Plenum Press, New York.
Smith, G. (1963). *Acta Cryst.* **16**(7), 594.
Smith, G., and Isaacs, P. (1964). *Acta Cryst.* **17**, 842.
Smith, J. (1954). *Acta Cryst. Cambridge* **7**, 479.
Smith, P., Leadbetter, A., and Apling, A. (1975). *Phil. Mag.* **31**(1), 57.
Snyder, L. (1978). "Borate Glasses" (L. Pye *et al.*, eds.). p. 151. Plenum Press, New York.
Soklakov, A., and Zhanov, G. (1963). *Sov. Phys. Cryst.* **7**, 447.
Soper, A. K., Neilson, G. W., Enderby, J. E., and Howe, R. A. (1977). *J. Phys. C* **10**, 1793.
Sorrel, C. (1968). Oral communication.
Sosman, R. (1964). *J. Am. Ceram. Soc.* **43**(3), 213.
Sosman, R. (1965). "The Phases of Silica." Rutgers.
Spear, W. (1957). *Proc. Phys. Soc.* **70-B**, 669.
Spear, W. (1960). *Proc. Phys. Soc.* **76**, 827.
Spear, W. (1974). *Proc. Conf. Amorph. Liquid Semicond, 5th* (J. Stuke and W. Brenig, eds.), p. 1. Taylor and Francis, London.
Spear, W., Loveland, R., and Al-Sharbaty, A. (1974). *J. Non-Cryst. Solids* **15**, 410.
Speranskaya, E., Skorikov, V., Safrandov, G., and Mitkina, G. (1968). *Izv. Akad. Nauk. SSR Neorg. Mat.* **4**(8), 1374.
Spinner, S. (1962). *J. Am. Ceram. Soc.* **45**, 394.
Srinivasan, G., Mahoney, R., Tweer, I., Macedo, P., and Haller, W. (1970). *J. Am. Ceram. Soc., Spring Meeting* Paper # 12-6-70.
Srinivasan, G., Mahoney, R., Tweer, I., Macedo, P., and Haller, W. (1971). *J. Non-Cryst. Solids* **6**, 221.
Stadler, L., and Cronin, D. (1977). *Glass Ind.* **58**(12), 10.
Stadler, L., and Ladue, A. (1978). *Glass Ind.* **59**(1), 10.
Stanworth, J. (1948). *J. Soc. Glass Technol.* **32**, 154.
Stanworth, J. (1952). *J. Soc. Glass Technol.* **36**, 217.

Stanworth, J. (1954). *J. Soc. Glass Technol.* **38**, 425.
Stanworth, J. (1957). *J. Electrochem. Soc.* **104**, 237.
Steinkamp, W., Shelbey, J., and Day, D. (1967). *J. Am. Ceram. Soc.* **50**(5), 271.
Stepanek, B., and Hruby, A. (1980). *J. Non-Cryst. Solids* **37**(3), 343.
Stepanov, A., and Novikov, S. (1972). *Neorg. Mat.* **8**(6), 1120.
Stepanov, A., and Novikov, S. (1976). *Fiz. Khim. Stekla* **2**(3), 238.
Stephens, R. (1976). *J. Non-Cryst. Solids.* **20**, 75.
Stevels, J.(1946). *J. Soc. Glass Technol.* **30**, 54.
Steventon, A. (1976). *J. Non-Cryst. Solids* **21**, 319.
Stewart, D., Rindone, G., and Dachille, F. (1967). *J. Am. Ceram. Soc.* **50**(9), 467.
Stokowski, S., Martin, W., and Yarema, S. (1950). *Electr., Magn. Opt. Prop. Glasses, 5th Univ. Conf. Glass Sci. 1979* (M. Tomazawa *et al.*, eds.), p. 469. North-Holland Publ., Amsterdam.
Stookey, D. (1950). U.S. Patent 2 290 971.
Stookey, D. (1970). U.S. Patent 3 498 803.
Street, R., and Mott, N. (1975). *Phys. Rev. Lett.* **35**(19), 1293.
Street, R., Nemanich, R., and Connell, G. (1979). *Phys. Rev.* (to be published).
Strickler, D. (1968). Thesis, Penn State.
Strong, S., and Kaplow, R. (1968). *Acta Cryst.* **B24**(8), 1032.
Stroud, J., and Lell, E. (1971). *J. Am. Ceram. Soc.* **54**(11), 554.
Stroud, J., Schrever, J., and Tacher, R. (1965). *Int. Cong. Glass, 9th Paris* p. 42.
Strugach, L., Soklakov, A., Kuzmenkov, and Pechkovskii, V. (1979). *Fiz. Khim. Stekla* **5**(4), 492.
Suchet, J. (1971). *Mat. Res. Bull.* **6**, 491.
Suhrmann, R., and Berndt, W. (1940). *Z. Phys.* **115**, 70.
Sullivan, E., and Taylor, W. (1915). "Invention of Pyrex Glass." U.S. Patent 36 136.
Sun, K. U.S. Patent 2 466 409.
Sun, K. (1946). *Glass Ind.* **27**, 552, 580.
Sun, K. (1947). *J. Am. Ceram. Soc.* **30**, 277.
Sun, K. (1949). *Glass Ind.* **30**, 199, 232.
Sunners, B., and Narken, B. (1965). *Ceram. Bull.* **44**(8), 619.
Suvarova, L., and Skhol'nikov, E. (1976). *Neorg. Mat.* **12**(4), 610.
Suvarova, L., Borisova, Z., and Orlova, G. (1974). *Neorg. Mat.* **10**(3), 441.
Suzuki, Y., and Muranaka, K. (1964). *Asahi Glass Co. Res. Rep.* (1), 1. Summarized in (1965). *Phys. Chem. Glasses* **6**(7), 32A.
Svanson, S., Forslind, E., and Krogh-Moe, J. (1962). *J. Phys. Chem.* **66**, 174.
Swift, H. (1947). *J. Am. Ceram. Soc.* **30**, 171.
Syritskaya, Z. (1971). *Stekloobr. Sist. Nov. Stekla I Kh. Osn.* **60**; (1972). *CA* 78835k.
Syritskaya, Z., and Kutukova, E. (1971). *Proc. Int. Cong. Glass Sci. Technol., 9th, Versailles* **1**, 557.
Syritskaya, Z., Rogozhin, Y., and Shakova, P. (1971). *Stekloobr. Sist. Nov. Stekla I Kh. Osn* **55**; (1972). *CA* 78840h.
Takahashi, K. (1962). *Adv. Glass Technol., Washington, D.C.* p. 366.
Takahashi, K., and Goto, Y. (1971). *Mem. School Eng. Okayama Univ.* **6**(1), 47.
Takahashi, K., and Yoshio, T. (1977). *Yogyo Kyokai Shi* **85**, 65.
Takahashi, K., Shirasaki, S., Yamamura, H., and Kakegawa, K. (1978). *Proc. Meeting Ferro-electric Mat. and Their Appl., 1st* F-4, p. 281.
Takahashi, K., Yoshio, T., and Kanamura, F. (1980). *J. Non-Cryst. Solids* **42**(1/3), 157.
Takamori, T., and Tomozawa, M. (1976). *J. Am. Ceram. Soc.* **59**(9/10), 377.
Takamori, T., Roy, R., and McCarthy, G. (1970). *A. Ceram. Soc. Meeting.* Paper 10 G, p. 70.

Takamori, T., Reismann, A., and Berkenblit, M. (1976). *J. Am. Ceram. Soc.* **59**(7/8), 312.
Takuma, Y., and Minami, T. (1977). *J. Electrochem. Soc.* **124**(11), 1659.
Takusagawa, N. (1980). *J. Non-Cryst. Solids* **42**(1/3), 35.
Tammann, G., and Elbraechter, A. (1932). *Z. Anorg. Allg. Chem.* **207**(3), 268.
Tanaka, H. (1977). *Osaka Kogyo Gijutsu Sh. Kensho Kiho* **28**(2), 107; (1978). *CA* **88**, 26608w.
Tanaka, M., and Minami, T. (1965). *Jpn. J. Appl. Phys.* **4**, 1023.
Tarasevich, S., Kovaleva, I., Medvedeva, Z., and Antonova, Z. (1971). *Neorg. Khim.* **16**(10), 2862; (12), 3341.
Tauc, J. (1975). Optical Properties of Highly Transported Solids" (S. Mitra and B. Bemdow, eds.), pp. 245–260. Plenum Press, New York.
Tauc, J. (1976). *Physics Today*, October, p. 23.
Taylor, I., and Owen, D. (1980). *J. Non-Cryst. Solids* **42**(1/3), 143.
Taylor, M., and Brown, G. (1979a). *Geochim. Cosmoch. Acta* **43**, 61.
Taylor, M., and Brown, G. (1979b). *Geochim. Cosmoch. Acta* **43**, 1467.
Taylor, M., Brown, G., and Fenn, PH. (1980). *Geochim. Cosmoch. Acta* **44**(1), 109; *CA* **92**, 258, 167916g.
Taylor, P., and Friebele, E. (1974). *J. Non-Cryst. Solids* **16**, 375.
Taylor, P., Biship, S., and Mitchell, D. (1973). *Rep. NRL Prog.* March.
Terai, R. (1971). *J. Non-Cryst. Solids* **6**(2), 121.
Terai, R., and Sugita, A. (1978). *Osaka Kogyo Gijutsu Shikensho Kiho* **29**(6), 353. *CA* **91**, 272, 127712t.
Terai, R., Kitoaka, T., and Ueno, T. (1969). *Yogyo Kyokai Shi* **77**(883), 88.
Terai, R., Sawashita, E., and Okawa, E. (1973). *Osaka Kogyo Gijutsu Shikensho Kiho* **24**, 280.
Tewhey, J., and Hess, P. (1979). *Phys. Chem. Glasses* **20**(3), 41.
Thilo, E. (1965). *Angew. Chem. Int. Ed.* **4**, 1061.
Thilo, E., Wieker, C., and Wieker, W. (1964). *Silikat. Technol.* **15**, 109.
3M (1968). British Patent 1 117 211, June 19.
Thurn, H., and Ruska, J. (1976). *J. Non-Cryst. Solids* **22**, 331.
Tideswell, N. W., Kruse, F. H., and McCullough, J. D. (1957). *Acta Cryst.* **10**, 99.
Tiede, R. (1964). U.S. Patent 3 127 277.
Tiede, R., Machlan, G., and McKinnis, C. (1971). IR 37709 (II), April, USAF F 3361 5-71-C-1024.
Tilton, L. (1957). *J. Res. Nat. Bur. Std.* **59**, 139.
Timofeeva, N. *et al.* (1972). *Probl. Fiz. Khim.* **6**, 234. *CA* **79**, 262 (1973).
Tindyala, M., and Ott, W. (1978). *Ceram. Bull.* **57**(4), 432.
Tobolsky, A., Owen, G., and Eisenberg, A. (1962a). *J. Polym. Sci.* **59**, 324.
Tobolsky, A., Owen, G., and Eisenberg, A. (196b). *J. Colloid Sci.* **17**, 717.
Tobolsky, A., Owen, G., and Eisenberg, A. (1964). *Am. Sci.* **52**, 385.
Tohge, N., Yamamoto, Y., Minami, T., and Tanaka, M. (1979). *Appl. Phys. Lett.* **34**, 640.
Tohge, N., Minami, T. and Tanaka, M. (1980a). *J. Non-Cryst. Solids* **37**, 23.
Tohge, N., Minami, T., Yamamoto, Y., and Tanaka, (1980b). *J. Appl. Phys.* **51**, 1048.
Tomazawa, M., and Obara, R. (1973). *J. Am. Ceram. Soc.* **56**, (7), 378.
Tomazawa, M., and Takamori, T. (1980). *J. Am. Ceram. Soc.* **63**(5/6), 276.
Tomazawa, M., Herman, H., and McCrone, R. (1968). *Int. Symp. Phase Transf.*, Manchester.
Tomazawa, M., Herman, H., and McCrone, R. (1972). *Phys. Chem. Glasses* **13**(6), 6.
Topol, L., Hengstenberg, D., Blander, M., Happe, R., Richardson, N., and Nelson, I. (1973). *J. Non-Cryst. Solids* **12**, 377.
Topping, J., and Murthy, M. (1973). *J. Am. Ceram. Soc.* **56**, 270.

Topping, J., and Isard, J. (1971). *J. Phys. Chem. Glasses* **12**(6), 145.
Topping, J., and Murthy, M. (1977). *J. Can. Ceram. Soc.* **46**, 19.
Topping, J., Fuchs, P., and Murthy, M. (1974a). *J. Am. Ceram. Soc.* **57**(5), 205.
Topping, J., Harrower, I., and Murthy, M. (1974b). *J. Am. Ceram. Soc.* **5**(5), 209.
Toropov, N., and Khotimchenko, V. (1969). *Neorg. Mat.* **5**(2), 402.
Toropov, N., Galakhov, S., and Bondar', I. (1956). *Izv. Akad. Nauk. Khim. Nauk.* **6**, 641.
Toropov, N., Golakhov, F., and Bondar', L. (1956). *Bull. Acad. USSR Div. Chem. Sci.* 651.
Toyuki, H. (1980). *Yogyo Kyokaishi Shi* **88**(1016), 168.
Tran, T. (1965). *Glass Technol.* **6**(5), 161.
Trapp, H. (1978). *Verres Refr.* **32**(1), 17–23.
Trapp, H., and Stevels, J. (1960). *Phys. Chem. Glasses* **1**, 107, 181.
Troitski, B., Yakhkind, O., and Martishchenko, N. (1967). *Inorg. Mat.* **3**(4), 661.
Tronc, B., Bensoussan, M., Brenac, A., and Sebenne, C. (1973). *Phys. Rev. B* **8**, 5947.
Tsai, S., and Nemanich, P. (1980). *J. Non-Cryst. Solids* **35**(2), 1203.
Tsarev, V., Lyamtsev, O., and Baidokov, L. (1977). *Fiz. Khim. Stekla* **3**(3), 231.
Tsekhomski, V., Mazurin, O., and Estrep'ov, K. (1963). *Phys. Solid State* **5**(7), 426.
Tsuchihashi, S., and Kawamoto, Y. (1969). *Yogyo Kyokai Shi* **77**(2), 35, 77.
Tsuchihashi, S., and Kawamoto, Y. (1971). *J. Non-Cryst. Solids* **5**, 286.
Tsuchiya, T., Horiuchi, T., and Moriya, T. (1979). *Yogyo Kyokai Shi* **87**(5), 223.
Tsugane, S., Haradame, M., and Hioki, R. (1965). *Jpn. J. Appl. Phys.* **4**.
Turnbull, D., and Cohen, M. (1958). *J. Chem. Phys.* **29**, 1049.
Turnbull, D., and Polk, D. (1972). *J. Non-Cryst. Solids* **8–10**, 20.
Turner, W., and Winks, F. (1926). *J. Glass Technol.* **10**, 102.
Turyanitsa, I., and Kutsenko, Y. (1976). *Fiz. Khim. Stekla* **2**(2), 188.
Turyanitsa, I., and Kutsenko, Ya. (1979). *Fiz. Khim. Stekla* **5**(2), 247.
Turyanitsa, I., Koperless, B., and Mel'nichenio, T. (1972). *CA* 51862e.
Turyanitsa, I., Golevei, V., Golovei, M., Bogdanova, A., Khodolii, V., and Voroshilov, Y. (1974). *Izv. Vyssh. Ucheb. Saved. Khim. Technol.* **17**(4), 613; *CA* **81**, 357, 125851.
Tyulkin, V., and Shalunenko, N. (1971). *Inorg. Mat.* **7**(12), 1959.
Uemura, O., Sagara, Y., and Satow, T. (1974). *Phys. Status Solidi A* **26**(1), 99.
Uhlmann, D., and Shaw, R. (1969). *J. Non-Cryst. Solids* **1**, 347.
Uhlmann, D., and Wicks, G. (1979). *Wiss. Z. Friedrich Schiller Univ. Jena, Math Naturwiss Reihe* **28**(2/3), 231.
Ukyo, Y., Goto, K., and Inomata, Y. (1979). *J. Am. Ceram. Soc.* **62**(7/8), 410.
Ulrich, D. (1964). *J. Am. Ceram. Soc.* **47**(11), 595.
Uphoff, H., and Healy, J. (1961). Contract NONR 2965 (00) A 1 (NR 017 446).
Urnes, S. (1958). *Glastech. Ber.* **31**, 337.
Urnes, S. (1959). *Glass Ind.* **40**(5), 237.
Urnes, S. (1960). "Modern Aspects of the Vitreous State," Vol. I, p. 10. Butterworth, London.
Urnes, S. (1961). *Glastech. Ber.* **34**, 213.
Urnes, S. (1969). *Phys. Chem. Glasses* **10**(2), 69.
Urnes, S. (1971). *Phys. Chem. Glasses* **12**(3), 82.
Urnes, S. (1972). *Phys. Chem. Glasses* **13**(3), 77.
Uttrecht, R., Stevenson, H., Sie, C. H., Griener, J. D., and Raghavan, K. S. (1970). *J. Non-Cryst. Solids* **2**, 358–370.
Vaipolin, A. (1960). *Sov. Phys. Cryst.* **10**, 509.
Vaipolin, A. (1966). *Sov. Phys.-Krist.* **10**, 509.
Vaipolin, A., and Porai-Koshits, E. (1963). *Sov. Phys.-Solid State* **5**, 178.
Valenkov, A., and Porai-Koshits, E. (1936). *Z. Krist.* **95**, 195.

Valters, A., and Lunter, S. (1979). *Fiz. Khim. Stekla* **5**(6), 711.
Van Ass, H., Geittner, P., Gossink, R., Kueppers, D., and Seuerin, P. (1976). *Philips Tech. Rev.* **36**(7), 182.
Van der Plas, H., and Bube, R. (1977). *J. Non-Cryst. Solids* **24**, 377.
Van der Steen, G. (1977). *Phys. Non.-Cryst. Solids, Int. Conf., 4th*, p. 486.
Van der Steen, G., and Van den Boom, H. (1977). *J. Non-Cryst. Solids* **23**(2), 279.
Van Uitert, L., Bonner, W., and Grodkiewicz, W. (1971). *Mat. Res. Bull.* **6**(6), 283, 513.
Van Wazer, J. (1958). "Phosphorous and Its Compounds," Vol. I. Wiley (Interscience), New York.
Van Wechem, H. (1976). Thesis, Univ. of Amsterdam.
Vargin, V., and Stepanov, A. (1964). *Zh. Prikl. Khim.* **37**(6), 1366.
Vargin, V., and Milyukov, E. (1968). *Zh. Prikl. Khim.* **41**, 194.
Vasilevskay, T., Golubkov, V., and Poraï-Koshits, E. (1980). *Fiz. Khim. Stekla* **6**(1), 51.
Vasilos, T. (1960). *J. Am. Ceram. Soc.* **43**, 517.
Vaughn, J., and Kinser, D. (1975). *J. Am. Ceram. Soc.* **58**(7/8), 326.
Veksler, A. (1975). Deposited Doc., Viniti 3479-75, 50-3. *CA* **88**, L53, 1255227c (1978).
Veksler, A., Estro Pev, K., and Pronkin, A. (1976). *Zh. Prikl. Khim.* **49**(6), 1427.
Verebeichik, N., and Odeleskii, V. (1960). "Structure of Glass," Vol. 2, p. 248. Consultant's Bureau, New York.
Verweij, J. (1979). *J. Am. Ceram. Soc.* **62**(9/10), 450.
Verweij, H., and Buster, J. (1979). *J. Non-Cryst. Solids* **34**, 81.
Verweij, H., and Konijnendijk, W. (1976). *J. Am. Ceram. Soc.* **59**(11/12), 517.
Verweij, H., Buster, J., and Remmers, G. (1979). *J. Mat. Sci.* **14**(4), 931.
Videau, J. J., Portier, J., and Piriou, B. (1979a). *Rev. Chim. Min.* **16**(4), 393.
Videau, J. J., Portier, J., and Piriou, B. (1979b). *J. Non-Cryst. Solids* **34**, 203.
Vinogradova, G. Z., Dembovskii, S. A., Luzhnaya, S. A., and Luzhnaya, N. P. (1968). *Zh. Neorg. Khim.* **13** (*Russ. J. Inorg. Chem.* **13**(5), 758–761).
Virgo, D., Seifert, F., and Mysen, B. (1978). Yearbook, Carnegie Institute, Washington, D. C.; *CA* **92**, 264, 219704.
Vogel, W. (1959). *Silikat.* **10**(5), 241.
Vogel, W. (1960). *Struct. Glass* **2**, 17.
Vogel, W. (1966). *Phys. Chem. Glasses* **7**, 15.
Vogel, W., and Byhan, H. (1964). *Silicattech.* **15**, 212, 239.
Vogel, W., and Gerth, K. (1958). *Glastech. Ber.* **31**, 15.
Vogel, W., Skatulla, W., and Wessel, H. (1958). *Silicattech.* **9**, 51.
Vogel, W., Schmidt, W., and Horn, L. (1969). *Z. Chem.* **9**(7), 276.
Vogel, W., Buerger, H., Mueller, B., Zerge, G., Mueller, W., and Forkel, K. (1974a). *Silicattech.* **26**(6), 205.
Vogel, W., *et al.* (1974b). *Silicattech.* **25**(6), 206.
Vogel, W., *et al.* (1974c). *Silicattech.* **25**(6), 207.
Vogel, W., Buerger, H., Winterstein, G., Ludwig, C. H., and Jackel, W. (1974d). *Silicattech.* **25**(6), 209.
Voigt, B., Feltz, A., and Schorder, B. (1978). *Z. Chem.* **18**(2), 77.
Voigt, B., Senf, G., and Dresler, G. (1979). *Wiss. Z. Friedrich-Schiller-Univ. Jena Math. Naturwiss. Reihe* **28**(L/3), 327. *CA* **92**, 316, 27125f (1980).
Volf, P. (1961). "Technical Glasses," Pitman and Sons, London.
Voronko, Y., Denker, B., Neustruev, V., Osiko, V., and Prokhorov, A. (1976). *Opt. Commun.* **18**(1), 88.
Vrtanesyan, G. (1978). *Fiz. Khim. Stekla* **4**(5), 581.

Vrtanesyan, G., and Khalilev, U. (1977). *Fiz. Khim. Stekla* **3**(3), 265, 282.
Vuillard, G. (1957). *Ann. Chim. (Paris)* **2**, 233.
Vukcevich, M. (1972). *J. Non-Cryst. Solids* **11**, 25.
Wachtel, A. (1970). *J. Electrochem. Soc.* **117**(5), 708.
Wagstaff, F., and Charles, R. (1968). *J. Am. Ceram. Soc.* **51**(8), 449.
Warren, B. (1941). *J. Am. Ceram. Soc.* **24**, 256.
Warren, B., and Pincus, A. (1940). *J. Am. Ceram. Soc.* **23**, 301.
Warren, B., Krutter, H., and Morningstar, O. (1936). *J. Am. Ceram. Soc.* **19**, 202.
Warren, B., Krutter, H., and Morningstar, O. (1938). *J. Am. Ceram. Soc.* **21**, 259.
Warren, B., Biscoe, J., and Robinson, H. (1939). *J. Am. Ceram. Soc.* **22**, 180.
Watanabe, K., Sumiyoship, Y., and Anbo, L. (1970). *J. Ceram. Soc. Jpn.* **78**, (897/5), 165.
Watanabe, K., Maeda, T., and Shimizu, T. (1980). *J. Non-Cryst. Solids* **37**(3), 335.
Waxler, R., and Cleek, G. (1971). *J. Res. Nat. Bur. Std.* **75A**(4), 279.
Weber, M. (1976). *In* "Critical Materials Problems in Energy Production" (C. Stein, ed.), p. 261. Academic Press, New York [details in the less accessible "Study of Fluoride Laser Glasses" by M. Deutschbein *et al.*, #873 PEC, Centre National d'Etudes des Telecommunication, Issy-Les-Moulineaux, France (1968).].
Weber, M. (1979). *U. Conf. Glass Sci., 5th.*, p. 622.
Weber, M. (1980). *J. Non-Cryst. Solids* **42**(1/3), 189.
Weber, M., and Almeido, R. (1981). *J. Non-Cryst. Solids* **43**(1), 99.
Weber, M., Layne, C., Saruyan, R., and Milam, D. (1976). *Opt. Commun.* **18**, 171.
Weber, M., Hagarty, J., and Blackburn, J. (1978). "Borate Glasses" (L. Pye *et al.*, eds.), p. 215. Plenum Press, New York.
Weber, M., Ziegler, D. C., and Angell, C. A. (1982). *J. Appl. Phys.* **53**, 4344.
Wedgwood, F., and Wright, A. (1976). *J. Non-Cryst. Solids* **21**(1), 95.
Weeks, R., Magruder, R., III, and Kinser, D. (1980). *J. Non-Cryst. Solids* **42**(1/3), 307.
Weiss, A. (1954). *Z. Anorg. Allg. Chem.* **276**, 95.
Weiss, A., and Weiss, A. (1952). *Z. Naturforsch.* **7b**, 483.
Weiss, A., and Weiss, A. (1953). *Z. Anorg. Chem.* **273**, 124.
Wells, A. (1950). "Structural Inorganic Chemistry," 2nd ed., p. 485. Oxford Univ. Press (Clarendon), London and New York.
Wells, A. (1962). "Structural Inorganic Chemistry" 3rd ed. Oxford Univ. Press (Clarendon), London and New York.
Wemple, S. (1973). *Solid State Commun.* **12**, 701.
Westbrook, J. (1960). *Phys. Chem. Glasses* **1**, 32.
Westman, A. (1960). *Mod. Asp. Vitr. State* **1**, 63.
Westman, A., and Beatty, R. (1966). *J. Am. Ceram. Soc.* **49**(2), 63.
Westman, A., Smith, M., and Cartaganis, P. (1959). *Can. J. Chem.* **37**, 1764.
Westrum, E. (1959). Thermodyn. and Transport Prop. ASME.
Weyl, W. (1945). *J. Soc. Technol.* **29**(35), 291.
Weyl, W. (1953). *J. Phys.* **57**, 753.
Weyl, W., and Barber, S. (1952). *J. Am. Chem. Soc.* **74**, 2761.
Weyl, W., and Marboe, E. (1964). "The Constituion of Glasses." Wiley, New York.
White, E., and Gibbs, G. (1958). Oral communication.
White, E., McKinstry, H., and Bates, T. (1958). *Proc. Conf. Ind. Appl. X-Ray Anal., 7th, Denver, Colorado* p. 239.
White, P. (1971). *Phys. Chem. Glasses* **12**(1), 11.
White, W. (1965). *J. Am. Ceram. Soc.* **48**(2), 108.

White, W., Brawer, S., Furukawa, T., and McCarthy, G. (1978). "Borate Glasses" (L. Pye *et al.*, eds.), p. 281. Plenum Press, New York.
Wieker, W., Hoebbel, E., and Goetz, J. (1979). *Wiss. Z. Friedrich-Schiller-Univ. Jena Math. Naturwiss. Reihe* **28**(2/3), 277; *CA* **92**, 258, 46086k.
Wignall, G., Rothon, R., Longmann, G., and Woodward, G. (1977). *J. Mat. Sci.* **12**(5), 1039.
Williamson, J., and Glasser, F. (1965). *Science* **148**, 1589.
Wills, J. (1969). *Encycl. Chem. Tech.* **18**, 134.
Winchell, P. (1970). *Phys. Chem. Glasses* **11**(4), 115.
Wittmann, A., and Beulich, W. (1965). *Naturwissenschaften* **52**(7), 157.
Wong, J. (1970). Ph.D. Thesis, Purdue Univ.
Wong, J. (1978). "Borate Glasses" (L. Pye *et al.*, eds.), p. 297. Plenum Press, New York.
Wong, J., and Lytle, F. (1977).
Wood, C. (1968). Dept. of Physics, Northern Illinois, Private communication.
Wood, C., Van Pelt, B., and Wight, A. (1972). *Phys. Status Solidi B* **54**(2), 701.
Woodcock, L., Angell, C., and Cheeseman, P. (1976). *J. Chem. Phys.* **65**, 1565.
Wright, A. (1969). Ph.D. Thesis, Bristol.
Wright, A. (1974). *Adv. Struct. Res. Dittr.* **5**, 1.
Wright A. (1977a). *Phys. Non-Cryst. Solids, Int. Conf., 4th*, p. 598.
Wright, A. (1977b). *J. Non-Cryst. Solids* **23**, 369.
Wright, A. Private communication.
Wright, A., and Leadbetter, A. (1976). *Phys. Chem. Glasses* **17**(5), 122.
Wright, A., Yarker, C., Johnson, P., and Wedgewood, F. (1977). *Phys. Non-Cryst. Solids, Int. Conf., 4th*, p. 118.
Wu, Ch-K. (1980a). *J. Am. Ceram. Soc.* **63**(7/8), 453.
Wu, Ch-K. (1980b). *J. Non-Cryst. Solids* **41**(3), 381.
Wuellner, A., and Wien, H. (1902). *Ann. Phys.* **9**, 1217.
Wuellner, A., and Wien, H. (1903). *Ann. Phys.* **11**, 619.
Yajima, S., Kokamura, T., and Shishido, T. (1974a). *Chem. Lett.* (6), 545.
Yajima, S., Kokamura, T., and Shishido, T. (1974b). *Chem. Lett.* (12), 1531.
Yakhkind, A. (1980). *Fiz. Khim. Stekla* **6**(2), 164.
Yakhkind, A., Ovcharenko, N., and Semenov, D. (1968). *Opt. Mekh. Prom.* **35**(5), 35.
Yamane, M., and Okand, S. (1979). *Yogyo Kyokai Shi* **87**(8), 434.
Yasui, I., Hasegawa, H., and Imaoka, M. (1979). *Yogyo Kyokai Shi* **87**(5), 242.
Yip, K., and Fowler, W. (1974). *Phys. Rev. B.* **10**(4), 1400.
Yoldas, B. (1971). *Phys. Chem. Glasses* **12**, 28.
Yoshikagawa, M., Kaite, Y., Ikuma, T., and Kishimoto, T. (1980). *Electr. Magn. Opt. Prop. Glasses, 5th Univ. Conf. Glass Sci. 1979* (M. Tomazawa *et al.*, eds.). North-Holland Publ., Amsterdam.
Yoshimura, T., Fukunagn, J., and Ihara, M. (1971). *Yogyo Kyokai Shi* **79**(915), 428.
Young, J., Glaze, F., Faick, C., and Finn, A. (1939). *J. Res. Nat. Bur. Std.* **22**(4), 453.
Yun, Y., and Bray, P. (1978). *J. Non-Cryst. Solids* **27**(3), 363.
Zachariazen, W. (1932). *J. Am. Chem. Soc.* **54**, 3841.
Zachariazen, W. (1965). *Acta Cryst.* **18**(4), 705, 710, 714.
Zagar, L. (1971). *Glastech. Ber.* **44**(9), 345.
Zallen, R., and Lucovsky, G. (1971). "Selenium." Van Nostrand-Reinhold, New York.
Zallen, R., Lucovsky, G., Slade, M., and Ward, A. (1971). *Phys. Rev. B* **3**, 4257.
Zarzycki, J. (1957). *Verres Refr.* **11**, 3.
Zarzycki, J. (1971). *Phys. Chem. Glasses* **12**(4), 97.
Zarzycki, J. (1978). "Borate Glasses" (L. Pye *et al.*, eds.), p. 201. Plenum Press, New York.
Zarzycki, J., and Mezard, R. (1962). *Transm. Elm. Phys. Ch. Glasses*, p. 163.

3. INORGANIC GLASS-FORMING SYSTEMS

Zarzycki, J., and Naudin, F. (1967). *C. R. Acad. Sci. Paris* **265,** 1456.
Zatsepin, A., Dimitriev, I., Startsev, V., and Bryuna, L. (1977). *Fiz. Khim. Stekla* **3**(3), 276.
Zeman, J. (1968). *Z. Krist.* **127,** 319.
Zeman, J. (1971). *Monatsh. Chem.* 209.
Zhabrev, V., Moisseev, V., Nekrasov, A., and Sviridov, S. (1978). *Fiz. Khim. Stekla* **4**(5), 597.
Zhdanets, N., and Kheifeis, V. (1967). *Zh. Prikl. Khim.* **40**(11), 2418.
Zhdanov, S. (1974). *Proc. Int. Cong. Glasses, 10th, Kyoto, Japan,* Sect 13, p. 58. Ceramic Society of Japan.
Zhdanov, S., Koromaldi, E., Smirnova, L., Gavrilova, T., and Bryzgalova, N. (1973). *Neorg. Mat.* **9**(10), 1852.
Zhilova, A., Ratobyl'skaya, V., and Bozhko, Y. (1972). *Tr. Mosk. Khim-Tekhn. Inst.* **73,** 13.
Ziemba, B. (1963). *Ker.* **13,** 321.
Ziemba, B. (1964). *Ker.* **15**(2), 36.
Zirke, J., Tausend, A., Winter, H., and Wobig, D. (1977a). *Phys. Non-Cryst. Solids, Int. Conf., 4th,* p. 178.
Zirke, J., Dromer, C., Tausend, A., Wobig, D. (1977b). *J. Non-Cryst. Solids* **24,** 283.
Zorin, A., and Zorina, M. (1966). *Acta Cryst. Suppl. A* **21,** Part 7, 239.
Zorin, A., and Zorina, M. (1967). *Inorg. Mat.* **3**(12), 1991.
Zorina, L., and Vakhrameev, V. (1969). *Neorg. Mat.* **5**(10), 1834.

CHAPTER 4

Glazes and Enamels

Richard A. Eppler

PEMCO PRODUCTS
MOBAY CHEMICAL CORPORATION
BALTIMORE, MARYLAND

I. Introduction		301
II. Ceramic Glazes		303
	A. Role of the Various Oxides in a Glaze Formulation	304
	B. Raw Materials and Frits	309
	C. Leadless Gloss Glazes	311
	D. Lead-Containing Gloss Glazes	317
	E. Opaque Glazes	322
	F. Satin and Matte Glazes	326
III. Porcelain Enamels		328
	A. Adherence of Porcelain Enamels to Their Substrates	328
	B. Ground-Coat Porcelain Enamels	330
	C. Cover-Coat Enamels	332
IV. Summary		336
	References	336

I. Introduction

Glazes and enamels are the names given to two types of vitreous coatings. When the substrate is a ceramic, and usually a whiteware, the coating is referred to as a glaze. When the substrate is a metal, the coating is referred to as a porcelain enamel. In both cases, the coatings are thin layers of glass that have been fused onto the surface of the substrates.

Ceramic coatings are applied to substrates for many reasons. In any given example a ceramic coating may serve to render the substrate impervious, mechanically stronger, more resistant to abrasion and scratching, chemically more inert, more readily cleanable, and last, but by no means least, aesthetically more pleasing to the touch and eye.

The first requirements for a ceramic coating material derive from the fact that it must be applied to and bond with the substrate (Singer and

Singer, 1963). It must be of such composition that it will fuse to a homogeneous, viscous glass at an appropriate temperature. That temperature either will be coincident with that at which the substrate matures or will be sufficiently low that distortion of the substrate during firing does not occur.

During and after the fusion of the coating, it must react with the surface of the substrate to form an intermediate bonding layer. In the case of porcelain enamels, particular components of the coating called adherence oxides may be included in order to produce this reaction. In any case, the formulation of the coating must be such that just the right amount of interaction with the body occurs. If the reaction is insufficient, the coating will subsequently fall off the substrate. On the other hand, if the reaction is excessive, the composition of the body or the coating may be affected adversely.

On cooling the fired ware, the whole coated substrate contracts. If the coefficients of expansion of the coating and the substrate are not sufficiently close together, stresses and strains will be set up, resulting in defects such as spalling or crazing. Although the expansions of the coating and the substrate should be close, they should not be identical (Rado, 1969). Ceramic coatings and ceramic bodies, unlike metals, are much stronger in compression than in tension. Therefore, the weaker of the coating–substrate combination, namely, the coating, is put in compression. To do this a coating is selected with a coefficient of thermal expansion somewhat lower than that of the substrate. Thus during the cooling the coating shrinks less than the substrate and is therefore compressed. Finally, the coating must have a low surface tension so that it will spread uniformly over the substrate and not crawl away from edges and holes.

Beyond the coating requirements necessary for its application and bonding are those related to its use. In almost all cases, the coating is expected to be homogeneous, smooth, and hard. A smooth, hard surface is required in order to resist abrasion and scratching. The sole exception is a textured coating, when an aesthetically pleasing pattern is imposed.

A smooth surface is not only visually more appealing and more resistant to abrasion and scratching, but also more apt to be impervious to liquids and gasses and as a result more readily cleanable. Many of the applications involving the use of ceramic coatings are for contact with food and drink. In such applications, sanitary requirements impose high standards of cleanability. Moreover, for many if not most ceramic coatings, chemical durability in service is a prime concern and a prime reason for selection of the ceramic coating material. Ceramic coatings are formulated to be resistant to many reagents, including hot water, acids, alkalies, and most if not all organic media. About the only reagent for which these materials cannot be considered is hydrofluoric acid.

In any surface coating material the optical and appearance properties will be of prime concern. Various possibilities can be called for in a given requirement. The coating may be transparent or it may be opaque. It may have a high-gloss finish or it may be satin or even a complete matte. Finally, various surface finishes may be provided for, such as textures, crystals, and colors. The latter subject, color, is of such vast extent that it will not be possible to cover it in this chapter. Interested readers are referred to appropriate texts on this subject (Eppler, 1977; Shaw, 1971; Singer and Singer, 1963).

II. Ceramic Glazes

Ceramic glazes are the vitreous coatings applied to various ceramic ware. A wide variety of glaze formulations are used. Variety is so great that it is not possible to classify glazes in a simple manner.

One might first ask whether it is really necessary to have such a wide variety of glaze compositions. In fact, there are several reasons why there is a real need for a variety of glaze compositions (Rhodes, 1973). In the first place, ceramic ware is fired over a very wide range of temperature. The lowest fired ware is heated to about 800°C and the highest is fired at about 1400°C. Obviously, the same glaze composition would not be satisfactory at all temperatures. A glaze that will melt at a low temperature will volatilize and run off ware at a higher temperature. In practice, any given composition is useful for a temperature range of only about 30°C. Thus various recipes are required for each range of temperature.

Another reason for a variety of glaze compositions is the need for a variety of surface qualities. Glazes may be bright or dull, opaque or transparent, glossy or matte, thick or thin, as well as all the gradations in between. All these surface qualities result from varying the composition of the glaze.

Although it is not possible easily to systematize the discussion of glazes, some classification must be adopted in order to treat so large a body of information. Therefore, the sections that follow will begin by discussing in a general sort of way the role of the various oxides in the composition of glazes. Then the various raw materials that can be used and the role of glass frits in the preparation of glazes will be covered. Section II.C is concerned with leadless gloss glazes. Because of its unique character, and because of the many problems resulting from the poisonous nature of lead oxide, glazes containing lead will be treated in Section II.D. Section II.E introduces the subject of opaque glazes and the techniques used for opacification. In Section II.F the development of satin and matte glazes is treated.

Before proceeding, however, it is necessary to describe the Seger (1902) method of describing the formulas of glazes. Although related to conventional mole percent and readily calculatable therefrom, it is sufficiently different to require explanation. This method is the one commonly used in glaze composition work. Moreover, it has certain advantages in understanding the relationship of one glaze to another. The oxides present are arranged in three columns in the form

Basic oxides : amphoteric oxides : acidic oxides

The basic oxides are those monovalent and divalent oxides that occupy the modifier positions in the glass structure according to the Zachariasen (1932) concept of glass structure. The amphoteric oxides are those trivalent oxides such as alumina and boron oxide. In some glazes that melt at high temperatures the only acidic oxide listed is silica, but for glazes maturing at lower temperatures other tetravalent and/or pentavalent ions may be present. It is customary that the sum of the basic oxides be made equal to unity. Thus, for example, the following illustrates the use of the Seger molecular formula for a typical earthenware glaze (Singer and German, 1964).

Na_2O 0.16 Al_2O_3 0.34 SiO_2 2.70
K_2O 0.14 B_2O_3 0.32
CaO 0.46
PbO 0.24

For comparison, the mole percent and weight percent formulas of this same glaze are shown in the accompanying tabulation.

Mole %	Oxide	Weight %
3.7	Na_2O	3.1
3.2	K_2O	4.1
10.6	CaO	8.0
5.5	PbO	16.0
7.8	Al_2O_3	10.8
7.3	B_2O_3	6.9
61.9	SiO_2	50.5

A. Role of the Various Oxides in a Glaze Formulation

A significant portion of the periodic table has been used at one time or another in the formulation of glazes. Among the more commonly used oxides are Li_2O, Na_2O, K_2O, MgO, CaO, SrO, BaO, ZnO, PbO, B_2O_3,

Al_2O_3, SiO_2, and ZrO_2. Several others have been occasionally used for one purpose or another. In addition, fluorine is sometimes used as a partial replacement for oxygen. In a given formulation, each oxide has a particular contribution to make to the glaze.

Silica is by far the most important of the oxides (Rhodes, 1973). By itself it will form a glass, given sufficient temperature. Most formulations contain more silica than all other constituents together. Therefore, one way to look at glazes is to view them as a network of silica tetrahedra to which the other materials have been added as modifiers. Low-firing glazes, those that mature at 1050°C or less, contain from one to two parts of silica to each part of all other ingredients. Higher-firing glazes, such as those melting at 1250°C or higher, will have three to five times as much silica as all other ingredients combined.

About the only serious deficiency of silica, as a glaze component, is its very high melting point (about 1700°C), which makes it unsuitable for use alone as a glaze. Therefore, the first and probably foremost reason for the addition of other oxides is to provide fluxes to reduce the melting point of the glaze. The chart in Fig. 1 gives the approximate temperatures at which the most commonly used fluxes are effective. Each of these oxides has a

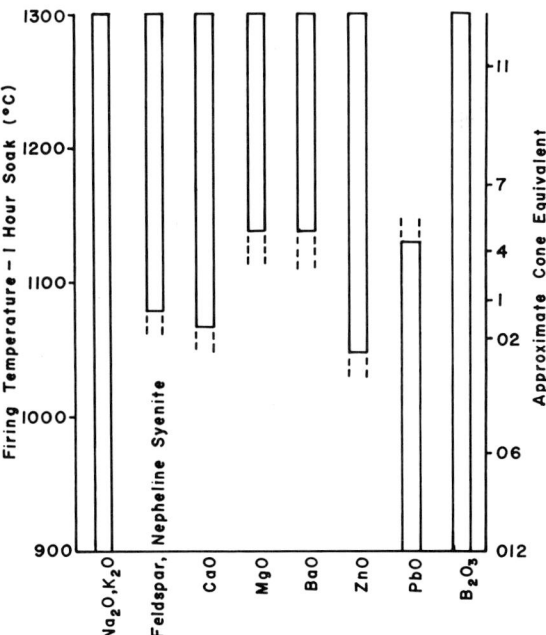

FIG. 1. Use temperature(s) for the various fluxes.

specific melting point, and it will be seen that these melting points vary widely—from 2800°C for magnesium oxide to 886°C for lead oxide. It must be remembered, however, that when used in combination with other oxides, even a material that by itself has a very high melting point may function as a flux.

This chart shows that potash and soda are useful as fluxes throughout the temperature range (Rhodes, 1973). Moreover, both of these oxides are very potent fluxes. The alkaline earths, calcia, magnesia, and baria are active fluxes only at high temperatures. Below 1100°C they may even inhibit rather than promote fusion. Zinc oxide is useful as a flux from about 1000°C up to the highest temperatures. Moreover, when used in small amounts it serves as a catalyst to promote the fusion of other oxides. Lead oxide is an active flux from the lowest temperatures up to about 1150°C. Above this temperature lead volatilization becomes excessive. Boric oxide is effective from the lowest temperatures up to the highest temperatures. Moreover, because it is itself a network former rather than a modifier, it can be used in conjunction with the other oxides to obtain a higher level of fluxing than is possible solely with oxides that destroy the silica network. The primary limitation on the use of boric oxide as a flux is that above a concentration of about 12–15 wt % the durability of the resulting glaze deteriorates drastically.

Considering the network-modifying ions as a whole, their fluxing power at moderate temperatures (1100°C, for instance) is as follows, in decreasing order (Rado, 1969):

Li_2O PbO Na_2O K_2O BaO CaO SrO MgO ZnO.

Sodium oxide is very active chemically and functions in glazes as one of the strongest fluxes (Rhodes, 1973). It is useful from the lowest temperature range to the highest. Moreover, glazes that contain a high concentration of soda may be brilliantly colored by the additions of appropriate coloring oxides. The primary disadvantage to the use of soda is the very high coefficient of thermal expansion that soda imparts to a glaze. Therefore, glazes high in soda tend to craze on most bodies. Another disadvantage of high-soda glazes is their tendency to be soft and easily abraded. Moreover, they are soluble in acids and tend to weather and deteriorate. On the other hand, when used in moderate amounts and in combination with other fluxes, soda is a very useful oxide in glazes over a wide range of temperature.

Lithium oxide is the most active of the fluxing oxides. Its behavior is otherwise quite similar to sodium. Because it is more expensive than the other alkali oxides, its use is normally restricted to those glazes where its powerful fluxing action is truly needed.

Potassium oxide is very similar to sodium oxide in its action in glazes.

Their behavior is so much alike that they are frequently used interchangeably in order to specify the use of naturally occurring minerals as raw material sources. There are, however, a couple of minor differences. In the first place, potash improves the gloss of the glaze, relative to soda. Second, in aluminosilicate formulations the viscosity at a given temperature of a potash system is higher than that of an equivalent soda system.

Most glazes contain calcium oxide. It is a common and inexpensive material and contributes desirable properties to glazes. Although it has a very high melting point, its principal function in a glaze is that of a flux. Its action does not derive so much from its melting point as from its contribution of a very low viscosity in the molten glaze once formed. Thus, although it may be the principal flux in high-fired glazes, in lower temperature glazes other fluxes, such as lead oxide, zinc oxide, or sodium oxide, must be used along with calcium to produce the melting. If too high a percentage of calcium oxide is used in a glaze, it will produce a matte surface. This can be caused by inadequate firing and also by crystallization of wollastonite. This phenomenon will be discussed in more detail in the section on matte glazes.

The action of the other alkaline earth oxides is similar to that of calcium. Strontium oxide is somewhat more fusible than calcium and may be substituted when a more active melt is desired. Conversely, barium oxide is more refractory than calcium and cannot be used in very high concentrations. Magnesium oxide has a use primarily as a high-temperature flux. It is far too refractory to be of much use in low-fired glazes, except that it may contribute to a lower coefficient of thermal expansion. On the other hand, because magnesium oxide has the steepest temperature coefficient of viscosity, it is a powerful flux at high temperatures. Moreover, it is doubly useful because, among modifier oxides, its expansion is low and most high-fired bodies tend to be low in expansion.

Zinc oxide is a very useful flux in the middle and higher firing temperatures. In small amounts it may be a very active flux, although when added in larger quantities it may produce a matte. It is little used below 950°C because at the lower temperatures it does not have much fluxing power. In conjunction with other fluxes, such as lead, alkalies, and boric acid, zinc may be a very valuable material, contributing to the creation of a smooth, trouble-free glaze surface. Conversely, when used in large quantities, zinc oxide may lead to crawling and to pitting and to pinholing. Finally, the presence of zinc has a profound effect on the colors obtained from the various pigmenting materials. Therefore, in some glazes zinc cannot be used.

There are several reasons why lead oxide is used in glazes (Eppler, 1974b; Singer and German, 1964). In the first place, the strong fluxing action of lead oxide allows the formulation of glazes that mature at temperatures relatively low in comparison with their lead-free counterparts. Moreover, this fluxing action leads to greater flexibility in the formulation of the glaze to obtain other desired properties such as low expansion and smooth surface. Second, lead oxide in the glaze allows for satisfactory maturing of the glaze over a wider firing range. Thus such glazes are more adaptable to the varying conditions that might occur in production-scale equipment. Moreover, lead oxide imparts low surface tension, leading to a smooth surface with a high gloss. Lead-containing glazes also more readily heal over blisters, drying cracks, and other defects of the glaze surface. Further, lead oxide imparts to a glaze high index of refraction, which results in a brilliant appearance. Finally, lead oxide in the glaze reduces any tendencies toward surface crystallization or devitrifcation of the glaze. This particular combination of desirable properties cannot at this time be reproduced on a production scale in leadless glazes.

On the other hand, there are several disadvantages to the use of lead oxide in glazes. In the first place, lead-glazed ware must be fired in an oxidizing atmosphere, since lead is very readily reduced. Another limitation of lead oxide is that above about 1200°C it becomes highly volatile. For this reason lead glazes are seldom used above cone 6. By far the most serious disadvantage, however, is the fact that lead oxide is poisonous and therefore appropriate precautions must be taken in using it, to avoid the possibilities of lead poisoning. Lead poisoning is a serious matter, and the possibility of contracting it while working with glazes must not be discounted. It is caused by the ingestion of soluble lead compounds into the system, usually by mouth, although it can also result from breathing vapors or dust. The disease is very difficult to diagnose since its various symptoms are similar to many other ailments. Therefore, every possible precaution must be taken when preparing lead glazes to avoid poisoning.

Moreover, if glazes are not properly formulated, they are subject to attack, primarily by acidic media, which results in release of lead into the solution. If such glazes are used in contact with food or drink, lead poisoning may result. This subject will be treated in detail in the section on lead glazes.

Boric oxide can be used as the primary flux in a glaze or as an auxilliary flux. It is an oxide of strong fluxing power, comparable to that of lead oxide or sodium oxide. It can be used from the lowest to the highest temperatures. Because it is a network-forming oxide, it generally contributes to a lowering of the expansion of the glaze, which makes it useful in prevention of crazing. The primary limitation on the use of boric oxide is its effect on the durability of the glaze. In sodium borosilicate glazes, additions of boric oxide up to 12 wt % improve the durability (Mellor, 1934). Above that, however, the durability rapidly deteriorates. In lead-containing glazes, any additions of boric oxide reduce the durability.

Alumina contributes importantly to the working properties of a glaze, and the only glazes that are normally made up without any alumina are those that are intended to develop crystals during cooling (Rhodes, 1963). In the presence of alkali and/or alkaline earth oxides, alumina takes over a function similar to boric oxide, forming tetrahedra coordinated by oxygen ions taken from the alkalies and alkaline earths (Rado, 1969). The presence of alumina in a glaze increases the viscosity so that it is less apt to run off vertical surfaces. In glazes containing boric oxide or alkaline earth oxides, the presence of alumina serves to retard phase separation and/or crystallization. Alumina also adds to the hardness, durability, and tensile strength of glazes. If used to excess, it can result in a matte.

Zirconium oxide is primarily used in glazes as an opacifying agent. This aspect will be treated later in the section on opaque glazes. However, small amounts of zirconium, less than its solubility in the glaze, may often be added. The primary purpose of small additions of zirconium oxide is to improve the alkali resistance.

The fluoride ion has an ionic size very close to that of oxygen. Hence, it can replace oxygen in the lattice of a glaze. However, because it is monovalent it serves to terminate the lattice structure of the glaze. Therefore, the amount that can be used is limited. Also, the effect of fluorine is to reduce drastically the melting temperature and the stability of the glaze.

B. Raw Materials and Frits

Primarily for economic reasons, glazes are not made up by mixing the separate oxides. Rather, as far as possible, naturally occurring minerals are used (Rado, 1969). Many of these minerals are the source of more than one of the constituent oxides. In Table I will be found a compilation of the more common raw materials used in the formulation of glazes. For reasons that will be discussed no materials that are highly soluble in water are included. These materials are used only after fritting.

Although no water-soluble materials will be found in Table I, there are materials, such as baria, litharge, and lead monosilicate, that are toxic and acid soluble. Care must be taken with these materials to avoid ingestion.

Glazes for high temperatures (greater than 1200°C) as used, for example, in hard porcelain, are prepared completely from crystalline raw materials as given in Table I (Rado, 1969). Lower-temperature glazes, on the other hand, require the use of a frit, or premelted glass, for essentially all or part of the glaze formulation.

The primary reason for the use of frits is that some of the ingredients needed for the lower temperature glazes are soluble in water (Rado, 1969). If such materials were used, the normal methods of glaze preparation by water grinding and the application of glaze as an aqueous suspension (slip) would not be possible. It is therefore necessary that the water-soluble

TABLE I

GLAZE RAW MATERIALS

Oxide desired	Possible raw materials	Other oxides introduced
Silica (SiO_2)	Quartz sand	—
	Clay	Al_2O_3
	Feldspars	Al_2O_3, Na_2O, K_2O
	Wollastonite	CaO
	Nepheline syenite	Al_2O_3, Na_2O, K_2O
	Talc	MgO
	Zircon	ZrO_2
Zirconia (ZrO_2)	Zircon	SiO_2
Alumina (Al_2O_3)	Corundum	—
	Clay	SiO_2
	Feldspars	SiO_2, Na_2O, K_2O
	Nepheline syenite	SiO_2, Na_2O, K_2O
	Alumina hydrate	—
Magnesia (MgO)	Talc	SiO_2
	Heavy magnesium oxide	—
	Dolomite	CaO
Lime (CaO)	Calcium carbonate	—
	Dolomite	MgO
	Wollastonite	SiO_2
Strontia (SrO)	Strontium carbonate	—
Baria (BaO)	Barium carbonate	—
Zinc oxide (ZnO)	Zinc oxide	—
Lead oxide (PbO)	Litharge	—
	Lead monosilicate[a]	SiO_2
	Lead bisilicate[a]	SiO_2
Lithia (Li_2O)	Spodumene	SiO_2, Al_2O_3
Soda (Na_2O)	Feldspar	SiO_2, Al_2O_3, K_2O
	Nepheline syenite	SiO_2, Al_2O_3, K_2O
Potash (K_2O)	Feldspar	SiO_2, Al_2O_3, Na_2O
	Nepheline syenite	SiO_2, Al_2O_3, Na_2O

[a] Lead monosilicate and lead bisilicate, as sold in commerce, are, in fact, frits rather than crystalline raw materials.

ingredients be made insoluble. This is done by mixing them dry with all or some of the other glaze materials and premelting them to a glasslike substance that is called a frit.

The solubility problem is of particular importance with respect to boric oxide containing frits. As can be seen from its absence from Table I, there are no common sources of boric oxide that are insoluble. Other difficulties arise in the case of the various alkalies, where the amount required is often well in excess of that which can be supplied by the mineral raw materials given in Table I.

There are several other reasons for using a frit. One of these is the poisonous nature of lead oxide. When lead oxide is fritted with sufficient quantities of appropriate oxides, it becomes insoluble and hence much less dangerous to handle (Hinton and Williams, 1976; Singer and German, 1964). Second, some glaze raw materials are substantially different in density from the others. As a result, during slip preparation, these materials may settle out and layered sedimentation may occur. This problem is avoided by fritting these raw materials (Singer and Singer, 1963). When the glaze raw materials have been fritted, the reactions between them have already largely been carried to completion. Therefore, less heat work is required in the eventual firing of the glaze coating. As a result, the surface of a high-frit glaze is superior to that of a raw glaze of identical composition fired at the same temperature (Rado, 1969). Alternatively, the time required to fire the glaze coating onto the substrate is reduced (Singer and Singer, 1963). Finally, certain raw materials such as magnesium carbonate, if used, would cause difficulty because of their fine particle size and great bulk (Parmalee and Harman, 1973).

A ground frit does, of course, tend to sediment, just like a raw material, and under some circumstances it will form a very hard deposit in doing so. This is prevented by retaining some of the insoluble parts of the glaze formulation from the frit and adding them in the mill. Thus a very common formulation for a high frit glaze is 90% of frit and 10% of clay (Singer and Singer, 1963). The 10% clay serves to suspend the frit in the glaze slip, preventing sedimentation.

Frits are normally prepared in a high-volume continuous process. The finely ground materials are mixed dry and fed to a simple glass-melting furnace at a controlled rate. The melt is then run into open tanks filled with water and subsequently dried, or it is quenched by pouring it through water-cooled rotating rollers, which produces very thin flakes (Rado, 1969).

Because frit making is a high-volume continuous process, this work is normally done by firms specializing in this process (Pemco, 1975). It is inefficient for a glaze manufacturer to make small batches of frit. More important, it is difficult to maintain quality control when small batches are prepared. Quality frit must be made in large batches using continuous smelting under stable conditions.

C. Leadless Gloss Glazes

When lead oxide is not used as a flux, one must rely on basic oxides such as calcium oxide, magnesium oxide, and the alkali oxides for fluxing, together with boric oxide (Singer and German, 1964). Nevertheless, a number of leadless glaze systems have been known for many years (Marquis and Eppler, 1974). In the first place, glazes that are fired to a temperature greater than 1150°C must be leadless (Singer and German, 1964). This is because glazes containing lead break down at about 1150°C with excessive volatilization of lead oxide. These high-temperature leadless glazes have empirical formulas in one of two families, the high-silica type and the high-alkali type (Marquis and Eppler, 1974). Some typical examples are given in Table II. These high-temperature glazes are those used on hard-paste porcelain and on sanitary ware.

The high-silica types are used on the green body of hard-paste porcelain (Marquis and Eppler, 1974). The origin of the glazes designed for the highest firing porcelains is studies of feldspar ($K_2O \cdot Al_2O_3 \cdot 6SiO_2$) (Rado, 1969) and of the system calcia–alumina–silica (Kerl, 1907; Parmalee and Lyon, 1934). Seger (1902) found that the best formula for a hard-paste porcelain is as given in A1 on Table II.

Descending to the porcelains fired at lower temperatures, the so-called soft porcelains or hard stoneware, formulas such as A2 through A5 are satisfactory (Marquis and Eppler, 1974; Rado, 1969). As can be seen by studying these formulas, the overall amount of alumina and silica used is adjusted for the firing temperature required (Parmalee and Harman, 1973).

TABLE II

HIGH-TEMPERATURE LEADLESS GLAZES

No.	Reference	Cone	K_2O	Na_2O	CaO	MgO	ZnO	Al_2O_3	B_2O_3	SiO_2
A. High-Silica Types											
A1	a	15	0.3	—	0.7	—	—	1.0	—	10.0
A2	b	11	0.2	—	0.7	0.1	—	0.9	0.2	8.0
A3	c,d	11	0.3	—	0.7	—	—	0.8	—	7.0
A4	e	10	0.2	—	0.7	0.1	—	0.56	—	4.66
A5	f	7	0.2	—	0.7	0.1	—	0.4	—	3.5
B. High-Alkali Types											
B6	c,d	11	0.25	0.25	0.5	—	—	0.9	—	8.0
B7	g	9	0.3	0.2	0.5	—	—	0.6	1.0	5.2
B8	e	4	0.25	0.25	0.5	—	—	0.4	0.5	3.2
B9	b	11	0.1	0.1	0.6	—	0.2	0.55	—	3.0

[a] Rado (1969).
[b] Singer and German (1964).
[c] Kitamura (1921).
[d] Parmalee and Harman (1973).
[e] Steger (1927).
[f] Pukall (1910).
[g] Granger (1908).

The usual range is the following:

For cones 11–16 the approximate range of compositions will be RO : 0.5–1.0 Al_2O_3 : 6.0–15 SiO_2

For cones 7–10 the approximate range of compositions will be RO : 0.4–0.8 Al_2O_3 : 3.0–5.0 SiO_2

The ratio of the silica to the alumina is held within the narrow range from 7 : 1 to 10 : 1.

The high-alkali type of high-temperature leadless glaze, which normally has 0.4–0.5 alkali, can also be used on porcelain bodies, if the expansion is large enough. Some examples of these glazes are given in B6 through B8 in Table II.

The primary use for this type of glaze is for sanitary ware. Sanitary ware glazes mature at a sufficiently high temperature that limitations imposed by fusibility do not apply (Singer and German, 1964). However, since tin oxide was a customary addition for opacity and it is an extremely expensive material, the composition must be considered in relation to the solubility of the opacifier in the glaze. To this end, boric oxide is normally omitted and a substantial concentration of barium oxide or zinc oxide is added. Thus a typical fire-clay sanitary ware glaze is as given in B9 in Table II.

The Bristol glaze is a variation of the soft porcelain glaze that has been developed to produce an opaque white coating on stoneware and similar colored clay bodies (Marquis and Eppler, 1974). The opacity arises from the high concentrations of zinc oxide that are used. A Bristol glaze can be formulated from porcelain glazes through substitutions of large quantities of zinc oxide for the alkalies and alkaline earths. Some typical Bristol glaze formulas are given in Table III. They find application on stoneware, terra cotta, and exterior building tiles.

In recent years this type of glaze has been further modified to produce glazes suitable for the glazing of wall tile at rapid firing rates (Orth, 1967). Some examples of this type of glaze are given in section B of Table III. The primary alteration has consisted of reducing the silica content in order to lower the firing temperature. In addition, boric oxide has been added in low concentrations to further enhance the melting rate. Glazes B13–B15 are suitable for firing in a kiln with a 1–4-h total cycle while glaze B16 is suitable for firing in a 10–30-min cycle. For an adequate melting to be achieved it is necessary to use a high frit, or even a totally fritted glaze in fast-fire application. The best results are often obtained by using two frits, a higher firing one and a lower firing one.

The development of glazes for semivitreous and vitreous dinnerware is a more difficult subject than has been discussed so far. Among the

TABLE III
Bristol Glazes and Wall Tile Glazes

No.	Reference	Cone	K_2O	Na_2O	CaO	MgO	BaO	ZnO	Al_2O_3	B_2O_3	SiO_2	ZrO_2	F
						A. Bristol Glazes							
A10	a	10	0.1	0.1	0.4	0.2	—	0.2	0.4	—	3.5	—	—
A11	b	7	0.2	0.15	0.35	—	—	0.3	0.55	—	3.5	—	—
A12	c	7	0.2	0.2	0.3	—	—	0.3	0.6	—	3.55	—	—
						B. Wall Tile Glazes							
B13	d	1	0.07	0.06	0.48	0.03	0.05	0.31	0.19	0.12	1.88	0.05	—
B14	d	1	0.05	0.23	0.35	0.05	—	0.32	0.28	0.26	2.88	—	—
B15	d	1	0.04	0.27	0.35	0.01	—	0.32	0.26	0.05	2.65	—	—
B16	d	1	0.04	0.14	0.33	—	0.04	0.45	0.28	0.10	2.12	0.06	0.05

[a] Rado (1969).
[b] Parmalee and Harman (1973).
[c] Watts (1916).
[d] Pemco (n.d.).

required properties are a firing temperature of approximately cone 4, compatibility with essentially all pigment systems stable at cone 4, and a coefficient of thermal expansion of no greater than $7.0 \times 10^{-6}/°C$ for semivitreous ware and $5.5 \times 10^{-6}/°C$ for vitreous hotel china.

Shortly after World War II an extensive investigation of leadless glazes suitable for semivitreous dinnerware was undertaken (Marquis, 1952; McCutchen, 1944; Orlowski and Marquis, 1945, 1946). Some of these formulas are tabulated in section A of Table IV. An important development in these formulas is the use of more than one alkaline earth. In several cases three or four such materials have been used. It was found that the properties of a glaze with several alkaline earths was superior in melting and in surface to any one of these materials alone and in larger concentrations. This work has continued and there are several leadless glazes that find some application in the semivitreous dinnerware industry (Pemco n.d.). Three of these glazes are listed in section B of Table IV.

The study of strontium-containing leadless glazes for semivitreous earthenware having higher coefficients of thermal expansion than is customary for American practice has been carried out by Shteinberg (1971, 1974; Gray, 1979). Several of her more important glazes are given in section C of Table IV. She finds that the most suitable glazes involve a combination of strontium oxide with magnesium oxide and calcium oxide. Both boric-oxide-containing (B25) and boric-oxide-free (B26) glazes have been used.

The development of leadless glazes for the lower expansion vitreous dinnerware imposes even more stringent requirements because of the lower coefficient of thermal expansion. As a result, it has only recently been attempted (Marquis and Eppler, 1974). Some examples of these glazes are given in section D of Table IV. It was found that suitable glazes could be prepared only within a fairly narrow range of compositions. The alkali concentrations were suitable between 0.1 and 0.19 and preferably from 0.1 to 0.15. Too little alkali raises the firing temperature; too much alkali raises the coefficient of thermal expansion. The strontium concentration varies from 0.3 to 0.65. Too little strontia raises the firing temperature unacceptably at these low alkali concentrations. Too high strontia results in too low calcia, which causes problems with color compatibility.

These latter glazes, however, do have severe limitations. In the first place, it is very difficult to eliminate blister-type defects from these glazes. A more severe limitation, however, arises from the fact that these glazes are stiff enough that they duplicate exactly the coating that is applied. Therefore, it is very difficult to obtain acceptable surface quality in these glazes.

When the high durability requirements of a food contact surface are not required, these defects can be eliminated through use of a more fluid glaze

TABLE IV
Dinnerware Glazes

No.	Reference	Cone	K_2O	Na_2O	Li_2O	CaO	MgO	SrO	BaO	ZnO	Al_2O_3	B_2O_3	SiO_2	ZrO_2	MoO_3	F
						A.	Glazes by Orlowski and Marquis									
A17	a	4	0.04	0.12	0.1	0.44	0.15	0.15	—	—	0.3	0.23	2.8	—	—	—
A18	a	3	0.12	0.06	—	0.43	—	—	0.26	0.13	0.33	0.31	2.86	—	—	—
A19	a	5	0.12	0.06	—	0.4	0.1	0.05	0.1	0.17	0.27	0.31	2.6	—	—	—
A20	a	5	0.144	0.124	—	0.425	0.154	0.154	—	—	0.349	0.236	3.159	—	—	—
A21	a	5	0.093	0.138	0.041	0.423	0.153	0.153	—	—	0.347	0.235	3.146	—	—	0.041
						B.	Modern Semivitreous Dinnerware Glazes									
B22	b	4	0.062	0.081	0.104	0.454	0.149	0.15	—	—	0.33	0.236	3.121	—	—	—
B23	b	4	0.09	0.091	—	0.581	—	0.238	—	—	0.339	0.357	3.08	—	—	—
B24	b	1	0.115	0.081	0.047	0.58	0.066	0.11	—	—	0.367	0.171	2.721	—	—	—
						C.	Russian Semivitreous Dinnerware Glazes									
C25	c	4	0.067	0.416	—	0.23	0.057	0.23	—	—	0.246	0.198	2.66	—	—	—
C26	c	4	0.08	0.32	—	—	0.20	0.40	—	—	0.18	—	2.47	—	—	—
C27	c	04	0.11	0.485	—	—	0.135	0.27	—	—	0.244	0.39	3.30	—	—	—
						D.	Vitreous Dinnerware Glazes									
D28	d	4	0.15	—	—	0.42	—	0.43	—	—	0.322	0.36	3.053	0.05	—	—
D29	d	4	0.12	0.03	—	0.34	—	0.51	—	—	0.322	0.36	3.053	0.05	—	—
D30	d	4	0.12	0.03	—	0.42	—	0.43	—	—	0.322	0.36	3.453	0.05	—	—
D31	d	4	0.12	0.03	—	0.42	—	0.43	—	—	0.393	0.36	3.531	0.056	0.01	—
D32	d	4	0.075	0.015	0.06	0.315	0.11	0.425	—	—	0.39	0.25	3.30	—	0.01	—
						E.	Glazes for Technical Ceramics									
E33	e	4	0.07	0.07	0.04	0.21	0.26	0.16	0.14	0.05	0.14	0.32	2.04	—	—	—
E34	f	4	0.02	0.30	0.03	0.17	—	0.02	0.35	0.10	0.35	1.12	3.52	—	—	—

[a] Orlowski and Marquis (1945).
[b] Pemco (n.d.).
[c] Shteinberg (1974).
[d] Marquis and Eppler (1974).
[e] Hinton (1978).
[f] Knapp (1978).

(Hinton, 1978; Knapp, 1978). Some examples of glazes for use on alumina bodies for spark plugs and similar technical applications are given in section E of Table IV.

One final type of leadless glaze remains to be considered, the low-expansion glazes suitable for zircon and cordierite bodies. The thermal expansion of zircon bodies is low, making for difficult glaze fit problems (Parmalee and Harmon, 1973). The coefficient of thermal expansion is between 4 and 5 × 10^{-6} mm/mm °C. Glazes high in magnesium oxide have been the usual solution (Luttrell, 1949; van Gordon, 1951), as shown in Table V.

Cordierite bodies are even lower in expansion, from 1.5 to 4 × 10^{-6} mm/mm °C. It is not possible to glaze these bodies with fully vitreous glazes. However, it is possible to glaze cordierite by inducing in an appropriately formulated glaze the precipitation of a low-expansion phase (Eppler, 1971a, b, 1972a, b, 1974a; Eppler and O'Conor, 1973, 1974, 1975). Some typical examples of these glazes are given in section B of Table V. These glazes are based on the crystallization of the low expansion stuffed quartz phase in a vitreous matrix. Glazes B38 and B39 are translucent, whereas glaze B40, which contains substantial amounts of zircon as well as stuffed quartz, is opaque.

D. LEAD-CONTAINING GLOSS GLAZES

As was mentioned earlier, when the firing temperature is reduced below cone 6, it is difficult to obtain a high-quality glaze, particularly at expansions below 8.0 × 10^{-6} mm/mm °C, without the use of lead oxide (Laurs, 1974). The benefits that accrue from the use of lead oxide make it worthwhile, in spite of the difficulties inherent in working with a toxic substance.

Some examples of lead-containing glazes for use at cone 4 are given in Table VI. It is instructive to compare these glazes (A41–A44) with glazes A17–B24 on Table IV. In general, the contents of silica, boric oxide, alumina, and calcia are similar. The use of lead oxide in glazes A41–A44 permits the use of generally higher concentrations of alkali oxide than are found for the leadless glazes. Said another way, lower concentrations of lead oxide appear in the lead-containing glazes than the concentrations of alkaline earths other than calcia that appear in the leadless glazes. The generally low thermal expansion behavior of lead oxide permits the use of the higher concentrations of alkali oxide, which results in greater fluidity and thus improved surface.

To further reduce the maturing temperature, alumina and silica are lowered and the lead oxide content is increased (Rado, 1969). Some examples of commercial clear glaze formulas suitable for cone 06 are also

TABLE V
Low-Expansion Glazes

No.	Reference	Cone	K_2O	Na_2O	Li_2O	CaO	MgO	ZnO	Al_2O_3	B_2O_3	SiO_2	ZrO_2
						A. Glazes for Zircon						
A35	a	12	0.13	0.07	—	0.15	0.65	—	0.58	—	5.7	—
A36	b	12	0.12	0.06	—	0.26	0.56	—	0.57	—	5.15	0.25
A37	c	10	0.14	0.09	—	0.17	0.49	0.12	0.42	0.18	4.21	0.69
						B. Glazes for Cordierite						
B38	d	2	0.15	—	0.85	—	—	—	0.74	0.20	2.51	—
B39	d	1	0.13	—	0.87	—	—	—	0.71	0.09	2.56	0.05
B40	e	4	0.13	—	0.83	—	—	0.04	0.68	0.09	2.72	0.33

[a] Luttrell (1949).
[b] van Gordon (1951).
[c] Luttrell (1949).
[d] Eppler and O'Conor (1973).
[e] Eppler and O'Conor (1974).

TABLE VI
Lead Glazes

No.	Reference	K_2O	Na_2O	CaO	PbO	ZnO	SrO	Al_2O_3	B_2O_3	SiO_2	ZrO_2
				A.	Cone 4 Dinnerware Glazes						
A41	a,b	0.066	0.179	0.494	0.261	—	—	0.340	0.314	3.369	—
A42	a,b	0.013	0.182	0.572	0.233	—	—	0.290	0.360	2.971	—
A43	a,b	0.090	0.090	0.580	0.240	—	—	0.320	0.360	3.064	—
A44	a	0.070	0.177	0.290	0.140	0.322	—	0.319	0.257	2.662	—
				B.	Cone 06 Clear Glazes						
B45	a,c	—	0.053	0.072	0.875	—	—	0.272	0.172	2.580	0.008
B46	a,c	—	0.157	0.218	0.625	—	—	0.273	0.507	2.792	0.023
B47	a,c	—	0.051	0.068	0.691	—	0.190	0.262	0.238	2.561	0.008
B48	a,c	—	0.155	0.215	0.565	—	0.065	0.272	0.525	2.783	0.023
				C.	Lead Bisilicate Glazes						
C49	a,d	—	—	—	1.000	—	—	0.268	—	2.460	—
C50	a	—	—	—	1.000	—	—	0.255	—	1.945	—

[a] Eppler (1975).
[b] Marquis (1971).
[c] Eppler (1974b).
[d] Pemco (n.d.).

given in Table VI. These glazes are typical of those suitable for use on artware and hobbyware bodies.

At the lowest maturing temperatures, glazes become simpler again, involving the use of only one basic oxide, lead oxide, and eliminating boric oxide. These glazes are based on lead bisilicate with small additions of alumina to retard phase separation. Some examples of lead bisilicate are also given in Table VI. These glazes can be fired as low as cone 010 for special effects in artware and hobbyware glazing.

The great limitation on the use of lead oxide in glazes is the potential for lead poisoning. Unfortunately, over the years, occasional episodes of lead toxication have resulted from the use of improperly fired and formulated lead-containing glazes on ceramic ware (Cole, 1971). In most cases, these episodes of lead toxication have resulted from the use of a poorly glazed vessel for acidic beverages consumed regularly in rather large quantities for a prolonged period of time. In most cases, the wares were manufactured by hobbyists or by artware manufacturers ignorant of the proper methods for assuring production of safe glazes.

Studies of this problem began as early as 1900, when the problem of lead poisoning among British pottery workers was addressed (Thorpe, 1899, 1901). This work mandated detailed procedures for the handling of lead compounds by workers and in addition banned in England the use of soluble sources of lead oxide such as white lead (Singer and German, 1964).

In the early 1970s the problem of lead released from already fired ceramic glazes while in service was addressed (ILZRO, 1970; Marquis, 1971; Merwin, 1971; Nordyke, 1971). A major accomplishment resulting from these studies was the development of a test for lead release that is suitable for the screening of formulations (ASTM, 1980). The test is applicable to ceramic glazes ranging from those with negligible lead release to those with high release (Marquis, 1971). It is the basis for federal regulations applicable to food contact surfaces, and promulgated by the FDA, which places a limit of 7.0 ppm for the lead release from any ceramic glaze when tested according to this method (Monteith et al., 1970; Steele, 1970). Generally similar tests for lead release from glazes have been developed in many countries (Laurs, 1974), and cooperative work is now in progress to develop a worldwide standard method of testing (WHO, 1976).

There are a number of factors that must be considered in the formulation of a glaze in order to achieve low lead release (Eppler, 1974b, 1975; Marquis, 1971). The lead release is influenced by the total glaze composition, including opacifiers and colorants when used; the thermal history of the glaze during firing; the thickness and uniformity of the glaze application; the extent of glaze–body solution at the interface; and the atmospheric conditions that exist during the firing process. When focusing on the glaze composition, it has been observed that there is a correlation between lead release and acid resistance of the glaze (Eppler, 1974b, 1975; Kitamura, 1921; Lehman et al., 1978; Marquis, 1971; Wood and Blachere, 1978a, b). In general, those glazes with excellent acid resistance also show low lead release.

It has been shown quite conclusively that the typical cone 4 glazes used on both vitreous and semivitreous dinnerware in the United States produce consistently low lead release values, in most cases less than 0.5 ppm and with low scatter (ILZRO, 1970; Marquis, 1971). For example, note the lead release from glazes A41–A44 (Table VI), which will be found in Table VII. Not only do these glazes have low lead release, but also they are not significantly affected by glaze–body interface reactions and show high acid resistance when applied in either a very light or very heavy coating. The acid resistance of these glazes is not significantly affected when they are fired in a static atmosphere compared with a moving atmosphere. Thus one can reasonably predict that these glazes will consistently produce low lead release under normal operating parameters.

Although most cone 4 glazes are found to be of a high acid resistance and a low lead release, a vastly different situation is found when attention is turned to artware and hobbyware glazes that are conventionally fired at cone 06 or lower. Some of these glazes, such as those given in Table VI, do have low lead release, as is indicated in Table VII (Eppler, 1974b). On the other hand, there are many other glazes that have produced high, erratic, and sometimes unpredictable lead release values. The compositions of these glazes vary widely, depending on the required end use properties. They are often glazes containing more than 20% lead oxide and they often contain other acid-soluble constituents as well. They generally have low acid resistance. They are sensitive to application thickness as well as to the atmosphere in the firing kiln. Finally, they often produce erratic lead release. Some examples of these glazes are given in Table VIII.

A procedure has been developed to predict the magnitude of the lead release of the glaze from its composition (Eppler, 1975). This procedure was developed from the observation that lead release is correlated with acid resistance. It was first noted that SiO_2, Al_2O_3, ZrO_2, and oxides of similar ions, such as TiO_2 and SnO_2, are effective in lowering the lead release of a glaze. Therefore, the concentrations of these ions, as expressed in the Seger molecular formula are summed:

$$\text{Good} = 2\,[Al_2O_3] + [SiO_2] + [ZrO_2] + [TiO_2] + [ZrO_2]. \tag{1}$$

The factor 2 arises because each unit of Al_2O_3 contains two Al^{3+}. On the other hand, the alkalies, alkaline earths, B_2O_3, F, P_2O_5, ZnO, CdO, and PbO are all more or less detrimental to the lead release in a glaze. Summing the molecular formula concentrations of these ions,

$$\begin{aligned}\text{Bad} = {}& 2\{[Li_2O] + [Na_2O] + [K_2O] + [B_2O_3] + [P_2O_5]\} \\ & + [MgO] + [CaO] + [SrO] + [BaO] + [F] \\ & + [ZnO] + [CdO] + [PbO]. \end{aligned} \tag{2}$$

In a study of 77 different glazes it was found that the factor

$$\text{FM} = \text{Good}/\sqrt{\text{Bad}} \tag{3}$$

correlated with the lead release of the glaze. When the value of this figure of merit exceeded a value of 2.05, the lead release was always less than 7 ppm. On the other hand, when the figure of merit was less than 1.80, some measurements of lead release always exceeded 7 ppm, although individual readings were often less. The figure of merit was not able to discriminate the lead release acceptably in the interval 1.80 to 2.05. Only 17% of the glazes studied fell into this intermediate region. Thus, in the vast majority of the cases studied, the figure of merit was able to predict whether the glaze would have a low lead release or not.

Only one qualification to this figure of merit has been noted (Eppler, 1974b). The figure of merit apparently does not apply to those cases where a soluble crystalline phase has precipitated. Thus this figure of merit is

TABLE VII

LEAD RELEASE FROM GLAZES

No.	Reference	No. samples	Lead release			Figure of merit
			Avg.	Min.	Max.	
A. Cone 4 Dinnerware Glazes						
A41	a	2	0.16	0.14	0.18	2.96
A42	a	9	0.07	0.03	0.10	2.57
A43	a	6	0.31	<0.10	0.60	2.69
A44	a	3	0.90	0.28	1.71	2.49
B. Clone 06 Clear Glazes						
B45	a	6	0.42	0.26	0.61	2.65
B46	a	21	0.47	<0.10	0.70	2.28
B47	a	6	0.24	0.09	0.50	2.50
B48	a	21	0.49	<0.10	1.39	2.26
C. Lead Bisilicate Glazes						
C49	a	6	0.79	0.69	0.85	3.00
C50	a	4	2.51	0.60	3.34	2.46

[a] Eppler (1975).

largely limited to gloss glazes. In opacified and matte glazes, crystalline phases are normally present. In such cases, each phase must be considered separately, and that phase with the lowest acid resistance governs.

The ability of the figure of merit to predict the lead release of glazes is illustrated in Tables VII and VIII. All of the glazes in Table VII show low lead release and a figure of merit in excess of 2.05. Conversely, all of the glazes in Table VIII show high lead release and a figure of merit below 1.80.

E. OPAQUE GLAZES

Opaque glazes are those sufficiently low in light transmitted that they effectively hide the body from view. They are usually white, although this is not a requirement.

Whiteness or opacity is introduced into ceramic coatings by the addition to the coating of a substance that will disperse into the coating as discrete particles, which will scatter and reflect some of the incident light (Singer and German, 1964). In order to do this, the dispersed substance must have a low solubility in the molten glaze and a refractive index that differs appreciably from that of the clear ceramic coating. The refractive index of most ceramics is 1.5–1.6 and therefore the refractive indices of opacifiers must be either greater or less than this. As a practical matter, opacifiers of high refractive index are used. Some possibilities for this

TABLE VIII

Some Glazes with High Lead Release[a]

No.	K$_2$O	Na$_2$O	CaO	PbO	B$_2$O$_3$	Al$_2$O$_3$	S:O$_2$	Lead release				Figure of merit
								No. samples	Avg.	Min.	Max.	
51	—	—	—	1.000	1.613	0.010	3.027	3	2347	561	3360	1.48
52	0.085	0.041	0.483	0.391	0.440	0.099	1.890	3	194	147	266	1.48
53	—	0.280	—	0.720	0.561	0.117	1.129	3	267	163	369	0.88

[a] From Eppler (1974b).

TABLE IX

Opacifiers

Material	Refractive index
SnO_2	2.04
ZrO_2	2.40
$ZrSiO_4$	1.85
TiO_2 (anatase)	2.5
TiO_2 (rutile)	2.7

purpose are given in Table IX. They include tin oxide with a refractive index of 2.04, zirconia with a refractive index of 2.40, zircon with a refractive index of 1.85, and titania with a refractive index of 2.5 for anatase or 2.7 for rutile.

In glazes fired at temperatures in excess of 1000°C, zircon is the opacifier of choice. It has a solubility in many ceramic glazes of about 5% at 1200°C and 2–3% at 700°C (Booth and Peel, 1959). A customary total addition would be 8–10% zircon. Consequently, most opacified glazes contain both zircon that was placed in the mill and went through the firing process unchanged and zircon that dissolved in the molten glaze during firing but recrystallized on cooling.

The more often the path of light is broken by the dispersed crystals, the greater is the opacity of the glaze. Therefore, the finer the particle size of the crystals, the greater is the surface area and hence the greater the reflection (Rado, 1969). The particle size for optimum opacity is about 0.4 μm. Therefore, among the various zircon opacifiers available on the market, presenting a range of average particle size, maximum opacity is given by that opacifier that has the finest particle size. Since this greater fineness is achieved by milling, the finer zircons are also the most expensive. On the other hand, zircon for smelting into a frit is optimum when intermediate in particle size. The effectiveness of the zircon opacifier can therefore be improved in partially or fully fritted glazes by smelting some of the zircon into the frit (Pemco, n.d.). Some examples of typical opacified glazes as used in the wall tile industry are given in Table X. Note that in all cases a portion of the zirconia is added in the frit, whereas the rest is added as a mill additive.

Zirconia is not much used as an opacifier because in the vast majority of ceramic glazes it reacts with the silica in the glaze to produce zircon as the opacifying agent (Jacobs, 1954). Therefore, since zircon is much less expensive than zirconia, it is the opacifying agent that is added. Tin oxide is

TABLE X
Opaque Glazes for Wall Tile[a]

No.	K_2O	Na_2O	CaO	BaO	ZnO	Al_2O_3	B_2O_3	SiO_2	ZrO_2 Added in frit	ZrO_2 Added in mill
54	0.118	0.159	0.288	0.034	0.401	0.480	0.170	2.252	0.150	0.141
55	0.083	0.111	0.407	0.028	0.371	0.406	0.143	2.019	0.126	0.122
56	0.088	0.200	0.304	—	0.409	0.543	0.175	2.310	0.101	0.196
57	0.055	0.139	0.409	0.031	0.366	0.388	0.142	2.002	0.082	0.167
58	0.084	0.267	0.340	0.048	0.261	0.504	0.352	2.985	0.125	0.193
59	0.060	0.137	0.389	0.033	0.380	0.416	0.145	2.229	0.103	0.160

[a] From Pemco (n.d.).

a more effective opacifying agent than any of the other possibilities because its solubility is lower, being much less than 1% in most glazes (Singer and German, 1964). However, the price of tin oxide is so high that it is no longer an economic solution. In coatings where the firing temperature is considerably less than 1000°C, titania in the anatase crystal phase is the opacifying agent of choice. Because of its very high index of refraction, titania is a very effective opacifying agent. However, at a temperature of 850°C, anatase inverts to rutile in silicate systems. Once inverted to rutile, titania crystals are able to grow rapidly to sizes that are no longer effective for opacification. Moreover, because the absorption edge of rutile is very close to visible, as the rutile particles grow, the absorption edge extends into the visible, leading to a pronounced cream color.

F. SATIN AND MATTE GLAZES

Opaque glazes can be regarded as a halfway stage toward satin and matte glazes (Singer and German, 1964). This effect is again due to the presence of small crystals dispersed in the glaze. It is the result of the devitrification produced when a completely fused glaze cools and part of the fused mass crystallizes. The crystals must be very small and evenly dispersed in order to give the glaze surface a smooth and velvet appearance. It should be possible to write on a matte or a satin glaze with ordinary pencil and then rub the mark off with the finger (Singer and Singer, 1963. Matte glazes are always more or less opaque because the crystals, as in normal opaque glazes, break up the rays of light. The crystals are of zinc silicate (willemite) as in the case of zinc mattes or calcium silicate (wollastonite) or calcium aluminum disilicate (anorthite) in the case of lime mattes. If barium oxide is added to the formulation, barium aluminum disilicate (celsian) may crystallize. Matte effects can also be obtained by undissolved material in the glaze, such as bone, feldspar, or talc. However, the quality of the matte is inferior to those produced by crystallization (Singer and German, 1964).

Since crystallization is required, the glaze should not be overfired and the proper cooling cycle must be maintained (Singer and German, 1964). Matte glaze compositions cooled too rapidly have glossy surfaces.

Most glazes can be converted to a matte glaze by the addition of a matting agent. Zinc mattes (see Table XI, part A) are produced by addition of a mixture of zinc oxide and clay in more or less equal proportions to the glaze formulation. The purpose of the clay is to introduce alumina and thus reduce the size of the zinc silicate crystals, which tend to be too large (Singer and German, 1964).

Additions of whiting or wollastonite to a glaze will give lime matte glazes (see Table XI, part B). Barium oxide, strontium carbonate, and

TABLE XI
SATIN AND MATTE GLAZES

No.	Cone	K_2O	Na_2O	Li_2O	CaO	MgO	SrO	BaO	PbO	ZnO	CdO	Al_2O_3	B_2O_3	SiO_2	ZrO_2	F	P_2O_5
							A. Zinc Matte Glazes[a]										
A60	02	—	0.083	—	0.062	—	—	—	0.713	0.138	—	0.085	0.214	0.696	0.241	—	—
A61	1	0.052	0.087	—	0.152	0.097	—	—	0.247	0.364	—	0.443	0.142	1.566	0.210	—	—
A62	1	0.045	0.131	—	0.216	—	—	—	—	0.608	—	0.314	0.155	1.291	0.186	—	—
A63	2	0.100	0.146	—	0.211	0.110	—	0.118	0.097	0.218	—	0.600	0.060	1.781	0.227	—	—
							B. Lime Matte Glazes[a]										
B64	06	0.032	0.009	—	0.301	—	—	0.215	0.444	—	—	0.124	0.177	0.942	0.148	—	—
B65	06	0.035	0.188	—	0.350	—	—	—	0.378	—	0.049	0.072	0.248	1.691	0.201	0.044	—
B66	1	0.059	0.040	—	0.524	—	—	—	0.377	—	—	0.262	0.176	1.746	0.211	—	—
B67	1	0.006	0.173	—	0.821	—	—	—	—	—	—	0.237	0.380	1.801	0.227	—	—
B68	4	0.078	0.052	—	0.645	—	0.225	—	—	—	—	0.324	0.228	2.194	0.249	—	—
B69	4	0.056	0.038	—	0.549	—	—	—	0.357	—	—	0.271	0.167	2.144	0.238	—	—
							C. Textured Tear Glazes[b]										
C70	1	—	0.230	0.294	0.073	—	—	—	0.291	0.113	—	0.101	0.709	0.758	0.264	—	—
C71	1	—	0.496	—	0.196	—	—	—	—	0.307	—	0.219	0.601	1.152	0.317	1.184	0.039

[a] From Pemco (n.d.).
[b] From Hackler (1980).

magnesium oxide can also be used, but the matting effect is more irregular and uncertain.

An extension of this technology is those glazes that have a surface containing matte islands amid glossy rivers (Hackler, 1980). These glazes are often called tear glazes because the glaze looks as if it had been torn. In these glazes the matte islands are usually crystalline zircon. As can be seen in part C of Table XI, the family of tear glazes is broad, but some common points include low SiO_2, high ZrO_2, relatively high concentrations of other glass formers (B_2O_3, P_2O_5), and moderate amounts of modifiers.

III. Porcelain Enamels

Porcelain enamels are ceramic coatings that are applied to metal substrates. Because metals differ greatly from ceramics in bonding, a new factor is introduced when considering these coatings: the interface between the ceramic coating and the metal and the development of bonding between the two materials. That is, the ceramic coating not only must act as a protective and aesthetically pleasing surface, but also must bond effectively to the substrate metal. Therefore, before proceeding to a discussion of porcelain enamel coatings, it will be well to consider the phenomenon of adherence between enamels and metal substrates and the requirements imposed on the coating thereby.

A. Adherence of Porcelain Enamels to Their Substrates

Several research studies completed since the mid-1930s have converged to give a reasonably coherent picture of the adherence process between ceramic coatings and metals (Kautz, 1936; King et al., 1959; Pask, 1979). The most significant requirements for good adherence seem to be (King et al., 1959) the following: the enamel at the interface must be saturated with an oxide of the base metal and this oxide must be one that, when in solution in the glass, will not be reduced by the metal. In the case of iron, this oxide is FeO. The components in the interfacial region must be arranged in such a way that a continuous electronic structure or chemical bond exists in the interfacial zone (Pask, 1979). On the other hand, surface roughness has been found not to be necessary for excellent adherence and of little value when the chemical bond is weak, although it generally improves adherence.

To create a continuity between the atomic and electronic structures of the metal and the ceramic, a transition zone is needed that is compatible and in equilibrium with both the metal and the glass coating at the interface. Therefore, this zone must include at least a monomolecular layer of the oxide of the metal and is retained only so long as both the metal and the glass coating at the interface are saturated with this metal oxide. Since most commercial enameling requires the heating of the substrate in air, a scale is formed on the substrate during the heat-up period before the coating sinters to a continuous mass. As soon as the enamel fuses, however, it attacks the scale and dissolves it in whole or in part. Although it is theoretically possible that a suitable structure could result with a macroscale oxide coating in the interfacial region, practically, this is not possible. It has been found (Kautz, 1936; King and Stull, 1949). that such oxide layers lack mechanical strength, resulting in flaking off of the coating. Therefore, practically speaking, a monomolecular layer is required (Pask, 1971).

When the glass dissolves the oxide and causes contact with the surface of the metal, a driving force must result, leading to a chemical reaction whose net effect is to oxidize the substrate until both the substrate and the ceramic coating are saturated with the oxide of the substrate metal. This latter effect is sometimes described by saying that the coating must wet the metal.

For many years it has been known that additions to the ceramic coating of certain easily reducible ions, such as cobalt oxide, nickel oxide, and copper oxide, will result in improved adherence between the ceramic coating and the metal. It has been found that these oxides, and in particular cobalt oxide, contribute substantially to the attainment and stabilization of the saturation of both the substrate and the coating with the oxide of the substrate. In the first place, once the initial scale is removed, the presence of an easily reduced oxide such as cobalt oxide in the ceramic coating facilitates the dissolution of the substrate in the coating until saturation is achieved (King *et al.,* 1959). After saturation is initially achieved, further reaction can result in the deposition of cobalt metal dendrites at the surface of the metal extending into the coating (Borum and Pask, 1966). Thus the adherence oxides play a critical role in ground-coat-type enamels by creating and maintaining the saturation at the interface of both the substrate and the coating with the oxide of the substrate. Another technique that is often used in enameling is the application of a nickel flash to the substrate prior to enameling. It has been shown (Cook, 1974) that this nickel flash acts in a similar manner to increase the reaction rate between the substrate and the ceramic coating and to create an interfacial region when saturation in the oxide of the substrate occurs.

In recent years it has become possible to enamel "direct-on." This refers to the application of a titania-opacified cover coat enamel directly onto a metallic substrate without the use of an intervening ground coat. This technique involves the use of a specially prepared extra-low-carbon steel and the use of a nickel flash. The nickel flash (Nedeljkovic, 1974) serves to create an iron–nickel alloy at the interface between the substrate and the metal in such a way as to facilitate the development of saturation of the coating in the oxide of the substrate. In the case of a titania-opacified enamel on a steel substrate, this is evidenced by the precipitation at the interface of iron metatitanate.

B. Ground-Coat Porcelain Enamels

Porcelain enamel ground coats for application to sheet steel and similar metals are basically alkali borosilicate formulations containing small amounts of adherence-promoting oxides, such as cobalt, nickel, and copper (Pemco, 1970) oxides. Some typical examples of these materials are found in Table XII.

These coatings serve three important purposes. First, the ground coat provides satisfactory bond or adherence to the base metal, by complex reactions such as those discussed in the last section. Second, a protective layer of coating is provided, which minimizes surface defects that may be caused either by the substrate itself or by its preparation method. Third, a resistant coating is provided, which, in a few applications, may also be decorative.

Adherence is the role most emphasized in a ground coat. Although satisfactory adherence can in some cases be obtained without a ground coat, the use of special steels and metal surface treatment is required in order to eliminate defects that are normally suppressed by the ground coat.

The other functions of the ground coat are also important. For example, in applications such as home laundries, water tanks, and ranges, thermal-resistant and corrosion-resistant properties are essential.

Ground-coat formulations such as those given in Table XII are normally composed of two, three, or four frit members, with each member contributing some special property to the utility of a ground-coat combination. For example, a hard member frit may provide high resistance to sag or hairline defects and superior corrosion resistance. A medium firing member may assist in preventing the preceding defects and also add resistance to burn-off or copperheading. The soft member frit fuses early in the firing process, seals off the metal, promotes bond, and improves edge coverage. Through the use of frit combinations rather than a single frit, a wider firing range is obtained and/or greater coating smoothness as a result of a given fire. Examples A72 and A73 are typical of general-purpose ground coat enamels. Coating A72 fires for 4 min at 805°C, whereas coating A73 fires for $2\frac{1}{2}$ min at 780°C.

Occasionally a fourth member frit may be used to produce a typical appliance gray-speckled effect. This end member is a titania-opacified,

TABLE XII
Ground-Coat Enamels

No.	Ref.	Li$_2$O	Na$_2$O	K$_2$O	CaO	BaO	MgO	ZnO	CoO	NiO	CuO	MnO$_2$	Cr$_2$O$_3$	Al$_2$O$_3$	B$_2$O$_3$	SiO$_2$	TiO$_2$	ZrO$_2$	Sb$_2$O$_3$	P$_2$O$_5$	F
								A. General-Purpose Ground-Coat Enamels													
A72	a	0.041	0.512	0.039	0.271	0.070	0.005	—	0.017	0.038	0.007	0.005	—	0.144	0.490	1.881	—	—	0.001	0.006	0.245
A73	a	0.065	0.470	0.054	0.244	0.105	—	—	0.014	0.038	0.008	0.005	—	0.138	0.489	1.623	—	0.030	—	0.011	0.316
A74	a	0.036	0.518	0.064	0.268	0.062	0.005	—	0.011	0.032	0.005	0.011	—	0.151	0.507	2.088	0.076	—	0.001	0.011	0.256
								B. Alkali-Resistant Ground-Coat Enamels													
B75	b	0.085	0.638	0.052	0.157	0.015	0.014	0.010	0.015	0.013	—	0.024	—	0.354	0.721	2.171	0.100	0.162	—	0.010	0.381
								C. Hot Water Tank Enamels													
C76	b	0.188	0.549	—	0.112	0.014	—	0.068	0.019	0.006	—	0.017	—	0.067	0.318	2.845	0.087	0.275	—	—	0.331
C77	b	0.135	0.679	—	0.110	0.011	—	0.047	0.019	—	—	0.063	—	0.060	0.330	2.820	—	0.286	—	—	0.349
								D. Continuous Clean Oven Coating													
D78		0.051	0.347	0.046	0.034	—	—	—	0.001	0.001	0.518	0.001	0.024	1.195	0.050	1.186	0.001	0.173	0.003	—	0.112

[a] Pemco (1970).
[b] Eppler et al. (1977).

off-white frit that is compatible with the other member frits. Coating A74 is an example of a coating made with a four-frit combination one of which is a titania-opacified off-white frit. This coating fires for 4 min at 780°C.

Ground-coat frits can also be designed to meet a variety of end use purposes. For example, coating B75 has been formulated for alkali resistance through the addition of substantial quantities of zirconium oxide. It is typical of coatings used in the home-laundry industry. Coatings C76 and C77 are examples of the formulations used when the outstanding thermal and corrosion resistance required of hot water tank systems are needed. The higher concentration of silica and lower concentration of boron oxide in these coatings reflects the substantially higher firing temperature of 7 min at 860°C.

It is instructive to compare these formulations with formulations B6 through B8 in Table II. It can be seen that the two important differences are the addition of adherence oxides to facilitate bonding to a metal substrate and the use of greatly increased concentrations of boric oxide and reduced concentrations of alumina and silica to permit the very much lower firing temperature.

A new and substantially different type of ground coat formulation is given in D78 on Table XII. This coating has been developed in order to provide a means of oxidizing and hence removing food soils from the surfaces of ovens at normal operating temperatures (Monteith *et al.*, 1970). In these materials various active ingredients chosen from a wide variety of metal oxides such as copper, vanadium, niobium, bismuth, chromium, molybdenum, tungsten, manganese, rhenium, iron, cobalt, nickel, cerium, rhodium, palladium, and platinum are incorporated into the formulation (Wilson and Lefort, 1973). In addition, in order that the coating be only partially vitreous and substantially porous, approximately $\frac{1}{2}$ of the silica is replaced with alumina.

C. COVER-COAT ENAMELS

Porcelain enamel cover coats are designed to provide specific color and appearance characteristics combined with resistance to atmospheric and liquid corrosion, surface hardness, abrasion resistance, and resistance to heat and thermal shock, as required (Pemco, 1970). Cover-coat formulations are available to provide a wide variety of appearance properties. They range from opaque whites through pastels and medium-strength colors to strong, dark colors. A wide selection of gloss is also available, ranging from the high-gloss sanitary ware finishes to the full matte architectural enamels.

Porcelain enamel cover coats are classified as opaque, semiopaque, and clear. Opaque enamels are used for white and pastel cover coats, semiopaque enamels are used for most of the medium-strength colors, and clear enamels are necessary to produce bright, strong colors.

Although zirconia and antimony oxides were used for opacification in the past, current opaque enamels are opacified with titanium dioxide. In most cases, all the titanium dioxide is smelted into a clear frit that crystallizes to anatase and rutile during the firing process to provide the required opacification (Eppler, 1971; Eppler and McLeran, 1967; Patrick, 1951; Shannon and Friedberg, 1960). Therefore, the properties of the product are dependent on the concentration of crystals present in the vitreous matrix, which in turn responds to the firing parameters of time and temperature.

Some fully opaque titania opacified porcelain enamels are given in section A of Table XIII. These materials have been formulated to show excellent acid resistance and in most cases fairly good alkaline resistance. Titania-opacified enamels are produced with reflectances ranging from 78%, as with enamels A79 or A80, up to 88%, as with enamel A83. The firing temperatures range from 770°C to 830°C. The five examples shown here are all suitable for application directly to decarburized steel. However, not all titania-opacified porcelain enamels are suitable for application direct-on, because they do not produce a satisfactory surface.

As with the other porcelain enamels, these systems are basically alkali borosilicates. To the basic formulations there have been added large concentrations of titanium dioxide, in the range where it will be soluble at the melting temperature of 1400°C but only partially soluble at the firing temperature of 800°C. In addition, large concentrations of fluoride are added to reduce the stability of the vitreous matrix. Finally, small additions are often made of materials that affect the color of the enamel by altering the crystallization properties. These materials include phosphorus pentoxide (Eppler and Spencer-Strong, 1969), niobium pentoxide, and tungsten oxide (Eppler, 1973).

Two examples of semiopaque cover coat enamels are also given in Table XIII. These materials do not differ in essential constituents from the fully opaque enamels. Rather, the concentration of titanium dioxide is reduced to where the system is compatible with the use of pigments for the production of colors of medium strength.

Clear cover-coat porcelain enamels are used in conjunction with appropriate pigments for the production of strong and medium-strength colors. Some typical examples are given in Table XIV. As can be seen, there is a rather wide variation in formulation of these products, compared with the opaque and semiopaque enamels. Both leadless and leadbearing formulations are given. Although some titanium dioxide is present in order to improve acid resistance, the concentrations are low enough that substan-

TABLE XIII
Opacified Cover-Coat Enamels

No.	Ref.	Li$_2$O	Na$_2$O	K$_2$O	MgO	ZnO	Al$_2$O$_3$	B$_2$O$_3$	SiO$_2$	TiO$_2$	ZrO$_2$	P$_2$O$_5$	Nb$_2$O$_5$	WO$_3$	As$_2$O$_3$	F	Firing
A. Opaque Cover-Coat Enamels																	
A79	a	0.058	0.554	0.342	—	0.047	0.078	0.809	2.606	0.898	0.099	0.036	—	—	0.003	1.172	780°C-3'
A80	a	0.111	0.573	0.303	0.013	—	0.079	0.834	2.702	0.976	0.069	0.038	0.001	—	0.001	1.220	770°C-3'
A81	b	0.127	0.646	0.227	—	—	0.054	0.986	2.902	1.117	—	0.039	0.001	0.001	—	0.710	780°C-3'
A82	a	0.037	0.451	0.450	—	0.062	0.066	0.962	3.097	1.011	0.054	0.038	0.001	—	—	0.617	830°C-3'
A83	a	0.135	0.570	0.277	—	0.018	0.067	0.935	2.772	0.915	0.027	0.038	0.001	0.001	0.003	0.988	770°C-3'
B. Semiopaque Cover-Coat Enamels																	
B84	a	0.129	0.486	0.341	—	0.045	0.046	0.833	2.730	0.582	—	—	—	—	—	0.724	770°C-3'
B85	a	—	0.468	0.304	—	0.227	0.054	0.913	3.405	0.387	0.120	—	—	—	—	0.473	815°C-3'

[a] Pemco (1970).
[b] Eppler (1969).

TABLE XIV
CLEAR COVER-COAT ENAMELS

No.	Ref.	Li$_2$O	Na$_2$O	K$_2$O	BaO	ZnO	CdO	PbO	Al$_2$O$_3$	B$_2$O$_3$	SiO$_2$	TiO$_2$	ZrO$_2$	MoO$_3$	F	Firing
86	a	0.103	0.692	—	0.013	—	—	0.192	0.058	0.418	3.044	0.299	0.055	—	0.469	795°C-3'
87	a	0.182	0.734	0.014	—	0.032	0.014	0.024	0.044	0.438	2.244	0.375	—	—	0.519	770°C-3'
88	a	0.198	0.665	0.137	—	—	—	—	0.090	0.344	3.313	0.151	0.215	0.011	0.417	815°C-3'

[a] Pemco (1970).

TABLE XV
MATTE COVER-COAT ENAMELS

No.	Ref.	Na$_2$O	K$_2$O	CaO	ZnO	Al$_2$O$_3$	B$_2$O$_3$	SiO$_2$	TiO$_2$	ZrO$_2$	Sb$_2$O$_3$	P$_2$O$_5$	F	Firing
89	a	0.220	0.608	0.007	0.165	0.044	0.402	2.026	0.807	0.070	0.008	0.011	0.082	805°C-3'
90	a	0.255	0.561	0.007	0.177	0.039	0.372	2.025	0.664	0.075	0.008	0.014	0.052	805°C-3'

[a] Pemco (1970).

tial crystallization does not occur. Therefore, inclusion of pigment in the mill formulation will permit the development of strong colors.

Particularly in the architectural industry there is the need for cover-coat enamels of lower gloss, but without reduction of the weatherability and durability of the enamel. Traditionally, this need was met through the additions of 5–25% of silica or 2–10% titania to a high-gloss enamel. There are several deficiencies in this technique, however; and recently coatings have been developed specifically for matte finishes. In Table XV will be found examples of these coatings. Example 89 illustrates a fully opaque matte coating, whereas example 90 is typical of a semiopaque coating suitable for use with darker colors.

IV. Summary

In this chapter the various materials that are used to form vitreous ceramic coatings have been described. To a very great extent, they are seen to be variations on a theme in which the basic formulations are varied to meet the requirement of suitable firing temperature, adherence, and coefficient of thermal expansion. At very high temperatures and low expansions, simple variations on silica are suitable. As the temperature is reduced, ingredients such as alkalies, alkaline earths, lead oxide, and boric oxide are added. The particular combination of ingredients is chosen in such a way as to provide the necessary firing temperature, durability, and coefficient of thermal expansion. Finally, when dissimilar materials are to be joined, special ingredients, normally called adherence oxides, may also be required.

References

ASTM (1980). Standard Method for Lead and Cadmium Extracted from Glazed Ceramic Surfaces, ASTM C-738-80. American Society for Testing and Materials, Philadelphia, Pennsylvania.
Booth, F. T., and Peel, G. N. (1959). *Trans. Br. Ceram. Soc.* **58**(9), 532–564.
Borum, M. P., and Pask, J. A. (1966). *J. Am. Ceram. Soc.* **49**(1), 1–6.
Cole, J. S. (1971). *Am. Ceram. Soc. Bull.* **50**(11), 917–919.
Cook, R. S. (1974). *Proc. PEI Tech. Forum* **36**, 1–14.
Eppler, R. A. (1969). *J. Am. Ceram. Soc.* **52**(2), 94–99.
Eppler, R. A. (1971). *In* "Advances in Nucleation and Crystallization in Glassess" (L. L. Hench and S. W. Freiman, eds.). American Ceramic Society, Columbus, Ohio.
Eppler, R. A. (1971a). U.S. Patent 3,561,984.
Eppler, R. A. (1971b). U.S. Patent 3,565,644.
Eppler, R. A. (1972a). U.S. Patent 3,676,204.
Eppler, R. A. (1972b). U.S. Patent 3,679,464.
Eppler, R. A. (1973). *Am. Ceram. Soc. Bull.* **52**(12), 879–881.

Eppler, R. A. (1974a). U.S. Patent 3,840,394.
Eppler, R. A. (1974b). *Proc. Int. Conf. Ceram. Foodware Safety.* Lead Industries Assoc., New York.
Eppler, R. A. (1975). *Am. Ceram. Soc. Bull.* **54**(5), 496–499.
Eppler, R. A. (1977). *Am. Ceram. Soc. Bull.* **56**(2), 213–215, 218, 224.
Eppler, R. A., and McLeran, W. A., Jr. (1967). *J. Am. Ceram. Soc.* **50**(3), 152–156.
Eppler, R. A., and O'Conor, E. F. (1973). *Am. Ceram. Soc. Bull.* **52**(2), 180–184.
Eppler, R. A., and O'Conor, E. F. (1974). U.S. Patent 3,804,666.
Eppler, R. A., and O'Conor, E. F. (1975). U.S. Patent 3,871,890.
Eppler, R. A., and Spencer-Strong, G. H. (1969). *J. Am. Ceram. Soc.* **52**(5), 263–266.
Eppler, R. A., Hyde, R. L. and Smalley, H. F. (1977). *Am. Ceram. Soc. Bull.* **56**(12), 1064–1067.
Granger, A. (1908). "Industrial Ceramics." Julius Springer, Berlin.
Gray, T. J. (1979). *Am. Ceram. Soc. Bull.* **58**(8), 768–770.
Hackler, C. L. (1980). *Am. Ceram. Soc. Bull.* **59**(6), 647–648.
Hinton, J. W. (1978). U.S. Patent 4,084,976.
Hinton, J. W., and Williams, C. L., Jr. (1976). *Am. Ceram. Soc. Bull.* **55**, (11), 986–988.
ILZRO (1970). Lead Glazes for Dinnerware. ILZRO Manual Ceramics No. 1, International Lead-Zinc Research Organization, New York.
Jacobs, C. W. F. (1954). *J. Am. Ceram. Soc.* **37**(5), 216–220.
Kautz, K. (1936). *J. Am. Ceram. Soc.* **19**(4), 93–108.
Kerl, B. (1907). "Handbuch der gesammten Thonwaarenindustrie." p. 1149. Vieweg and Son, Braunschweig.
King, B. W., and Stull, C. W. (1949). *J. Am. Ceram. Soc.* **32**(1), 34–40.
King, B. W., Tripp, H. W., and Duckworth, W. H. (1959). *J. Am. Ceram. Soc.* **42**(11), 504–525.
Kitamura, Y. (1921). *J. Ind. Chem. Tokyo* **2**, 89.
Knapp, R. O. (1978). U.S. Patent 4,120,733.
Koenig, J. H. (1937). Lead Frits and Fritted Glazes. Ohio State Univ. Eng. Experiment Station Bull. No. 95.
Laurs, A. N. (1974). *Proc. Int. Conf. Ceram. Foodware Safety.* Lead Industries Assoc., New York.
Lehman, R. L., Yoon, S. C., McLaren, M. G., and Smyth, H. T. (1978). *Am. Ceram. Soc. Bull* **57**(9), 802–805.
Luttrell, C. B. (1949). *J. Am. Ceram. Soc.* **32**(10), 327–332.
Marquis, J. E. (1952). *Am. Ceram. Soc. Bull.* **31**(5), 161–164.
Marquis, J. E. (1971). *Am. Ceram. Soc. Bull.* **50**(11), 921–923.
Marquis, J. E., and Eppler, R. A. (1974). *Am. Ceram. Soc. Bull.* **53**(5), 443–445, 449.
McCutchen, E. S. (1944). *J. Am. Ceram. Soc.* **27**(8), 233–238.
Mellor, J. W. (1934). *Trans. Br. Ceram. Soc.* **34**, 113–190.
Merwin, B. W. (1971). *Am. Ceram. Soc. Bull.* **50**(11), 915–917.
Merwin, B. W. (1974). *Proc. Int. Conf. Ceram. Foodware Safety.* Lead Industries Assoc., New York.
Monteith, P. G., Linhart, O. C., and Slaga, J. S. (1970). *Proc. PEI Tech. Forum* **32**, 73–79.
Nedeljkovic, A. L. (1974). *Proc. PEI Tech. Forum* **36**, 15–21.
Nordyke, J. S. (1971). *Am. Ceram. Soc. Bull.* **50**(11), 913–915.
Orlowski, H. J., and Marquis, J. E. (1945). *J. Am. Ceram. Soc.* **28**, 343–357.
Orlowski, H. J., and Marquis, J. E. (1946). Lead Replacements in Dinnerware Glazes. Ohio State Univ. Eng. Exp. Sta. Bull. No. 125.
Orth, W. H. (1967). *Am. Ceram. Soc. Bull.* **46**(9), 841–844.

Parmalee, C. W., and Harman, C. G. (1973). "Ceramic Glazes," 3rd ed. Cahners Publ., Boston, Massachusetts.
Parmalee, C. W., and Lyon, K. C. (1934). *J. Am. Ceram. Soc.* **17**(30), 60–66.
Pask, J. A. (1971). *Proc. PEI Tech. Forum* **33**, 1–16.
Patrick, R. F. (1951). *J. Am. Ceram. Soc.* **34**(3), 96–102.
Pemco (n.d.). Pemco Technical Notebook on Ceramic Glazes and Stains. Pemco Ceramics Group, Mobay Chemical Corp., Baltimore, Maryland.
Pemco (1970). Pemco Porcelain Enamel Technical Manual. Pemco Ceramics Group, Mobay Chemical Corp., Baltimore, Maryland.
Pemco (1975). Frit, the Adaptable Glass. Pemco Ceramics Group, Mobay Chemical Corp., Baltimore, Maryland.
Pukall, W. (1910). *Sprechsall* **48**, 1–3, 18–19, 33–35, 47–49.
Rado, P. (1969). "An Introduction to the Technology of Pottery." Pergamon, Oxford.
Rhodes, D. (1973). "Clay and Glazes for the Potter," Rev. ed. Chilton Publ., Radnor, Pennsylvania.
Seger, H. (1902). "Collected Writings," Chemical Publ. Co., Easton, Pennsylvania.
Shannon, R. D., and Friedberg, A. L. (1960). Titania-Opacified Porcelain Enamels. Univ. Illinois Eng. Exp. Sta. Bull. No. 456.
Shaw, K. (1971). "Ceramic Glazes." Elsevier, Amsterdam.
Shteinberg, Y. G. (1971). *Glass Ceram.* **28**(5), 316.
Shteinberg, Y. G. (1974). "Strontium Glazes" (T. J. Gray, trans. ed.). Kaiser Strontium, Div. of Kaiser Aluminum and Chemical of Canada, Halifax, Nova Scotia.
Singer, F., and German, W. L. (1964). "Ceramic Glazes." Borax Consolidated, London.
Singer, F., and Singer, S. S. (1963). "Industrial Ceramics." Chapman and Hall, London.
Steele, E. A. (1970). *Proc. Fall Meeting Mat. Equipment Whitewares Div.* American Ceramic Society, Columbus, Ohio.
Steger, W. (1927). *Ber. Deut. Keram. Ges.* **8**(1), 24–43.
Thorpe, T. E. (1899). The Use of Lead in the Manufacture of Pottery. British Government Paper 8383-150093/1901-wt-32982-Da S-4.
Thorpe, T. E. (1901). Report on the Work of the Government Laboratory on the Question of the Employment of Lead Compounds in Pottery. British Government Paper 9264-1500-61901 wt 6417 Da S-4.
van Gordon, D. V. (1951). *J. Am. Ceram. Soc.* **34**(2), 33–38.
Watts, A. S. (1916). *Trans. Am. Ceram. Soc.* **18**, 631–641.
WHO (1976). Ceramic Foodware Safety, Sampling, Analysis, and Limits for Lead and Cadmium Release. World Health Organization, Geneva, Switzerland.
Wilson, H. H. and Lefort, H. G. (1973). *Am. Ceram. Soc. Bull.* **52**(8), 610–611, 620.
Wood, S., and Blachere, J. R. (1978a). *J. Am. Ceram. Soc.* **61**(7-8), 287–292.
Wood, S., and Blachere, J. R. (1978b). *J. Am. Ceram. Soc.* **61**(7-8), 292–294.
Zachariasen, W. H. (1932). *J. Am. Chem. Soc.* **54**, 3841.

CHAPTER 5

Organic Glasses (Molecular Glasses)

A. Bondi†

SHELL DEVELOPMENT COMPANY
HOUSTON, TEXAS

I.	Introduction	339
II.	Effects of Cooling Rate	342
III.	The Glass Transition Temperature (T_g)	342
IV.	The Glass Transition Temperature of Mixtures	348
V.	Heterogeneous Mixtures	350
VI.	Detailed Effects of Molecular Structure	351
VII.	Equilibrium Properties	356
VIII.	Effect of Admixtures	358
IX.	Transport Phenomena	359
	References	362

I. Introduction

The differentiation of organic as compared to inorganic glasses is wholly artificial. A more appropriate differentiation is molecular as compared to ionically or covalently (including metallically) bonded glasses. This differentiation is reflected in all properties that depend on the strength and directedness of the bonds responsible for the cohesion of a glass. The molecules in molecular glasses are composed of atoms held together—in at least one dimension—by any of the three types of chemical bonds or their hybrids. But the intermolecular forces, responsible for the fact that the particular assembly of molecules is in the liquid or solid state at the temperature of observation, are commonly much weaker, often by an order of magnitude, than the chemical bonding forces.

Assuming, for simplicity, that the so-called glass transition temperature T_g is a corresponding state, we would expect that $T_g \sim E/R$, if E is some suitable measure of the cohesive energy of the system. Derivation of T_g from a free-volume model leads to the same criterion. For instance, Litt's

† Deceased.

free-volume model (Litt, 1976) yields a characteristic temperature for glass transition of $4\pi b^{*2}\epsilon k$, where $4\pi b^{*2}$ is the surface area of a hole the size of a molecular volume at 0 K and ϵ is the corresponding free surface energy. It is easy to show that $\epsilon \sim E/A_\mathrm{w}$, where A_w is the surface area per mole of molecules (Bondi, 1968). Thus $4\pi b^{*2}\epsilon/k \sim E/R$ (q.e.d.).

Hence we would expect that on an absolute temperature scale the T_g of simple molecular glasses is far lower than that of the ionic or covalent glasses. This expectation is confirmed by the data of Table I, where "simple" molecular glasses are glasses composed of small molecules. The economically more important polymeric members of the class of molecular glasses differ from the simple ones by the long extension of their

TABLE I

GENERALIZED GLASS TRANSITION TEMPERATURES OF SIMPLE SUBSTANCES

Substance	$T_\mathrm{g}{}^a$	$T_\mathrm{g}/T_\mathrm{m}{}^a$	$T_\mathrm{g}/T_\mathrm{b}{}^b$	$\dfrac{10RT_\mathrm{g}{}^c}{\Delta H_\mathrm{rb}}$	$\dfrac{mRT_\mathrm{g}{}^d}{H_\mathrm{s}}$	T_g^*
	Molecular Glasses					
Rigid molecules						
S_8	245	0.63	(0.34)		0.123	
Se_6	304	0.63	(0.29)		0.109	
As_4O_6	433	0.74	(0.59)		0.199	
P_4O_{10}	537	0.63	(0.54)		0.185	
Methylbenzene	113	0.64	0.29	0.28	0.135	0.246
Methylcyclohexane	85	0.58	0.23	0.23	0.111	0.194
O-Terphenyl	245	0.71				
H_2O	135	0.50	(0.36)	0.139	0.133	
Methanol	102	0.58	(0.31)		0.114	0.296
t-Butanol	180	0.60	(0.51)		0.157	0.40
Flexible molecules						
3-Methyl heptane	99	0.65	0.25	0.245	0.094	0.232
Di-(2-ethyl hexyl) sebacate	173	0.77				
	Covalent Glasses					
GeO_2	800	0.57				
SiO_2	1463	0.73				
	Ionic Glasses					
$NaPO_3$	563	0.63				
$Na_2Si_2O_5$	732	0.66				

[a] All the T_g, $T_\mathrm{g}/T_\mathrm{m}$ data on inorganic substances are from Sekka and MacKenzie (1971); the data on organic substances are a small sampling from the very extensive tables of Bondi and Tobolsky (1975).

[b] The numbers in parentheses are quite meaningless because, for the substances involved, the species in the vapor phase of $T = T_\mathrm{b}$ differ from those in the liquid phase.

[c] ΔH_rb = heat of vaporization at $T = T_\mathrm{b}$ (atmospheric boiling point).

[d] m = number of external degrees of freedom per molecule.

molecules in one dimension and by the internal mobility of these long threads. As a consequence, the cohesive energy, say, the energy of vaporization per mole, must now be compared with a kinetic energy $3cRT_g$, where $3c$ is the number of external degrees of freedom per molecule, including those due to internal rotation; then $T_g \sim E/3cR$. Since $3c \sim n$, the number of chain links per molecule, T_g of polymers is not very much higher than that of glasses composed of simple molecules. Moreover, since $3c$ is smaller the more rigid the molecule, one would expect that T_g is higher the more rigid the polymeric molecule, and vice versa, as is illustrated by the data in Table II. The balance of this chapter just enlarges on these simple relations in terms of specific structural effects.

Here we shall deal with the dependence of vitrification and of glass properties on cooling rate of the melt through T_g (and therefore on sample geometry). The thermodynamic and the mechanical properties of the glasses can be scaled from the observations on ionic and covalent glasses elsewhere in this text by the cohesive energy and internal rotation effects, if at all. Given the complexity of the glassy state, such scaling can hardly be expected to give better than first-order approximations when we try to understand relations between the properties of a glass and molecular structure.

Cross-linking of molecules to form a single, continuous, three-dimensional network changes the properties of the molecular glass toward those of a covalently bonded glass with increasing cross-link density. When that cross-link density is of the order of one cross-link per thousand chain atoms, the properties of the glass, including T_g, are virtually identical with those of the un-cross-linked material. But at the other limit of one cross-link per five chain atoms, we see the typical covalently bonded glass, with T_g generally above the decomposition temperature of the material because the thermal disruption of covalent bonds is tantamount to pyrolytic conversion of the network into volatile constituents and infusible char.

In the previous examples, each cross-link could be represented as trifunctional. Anionic polymerization techniques permit the preparation of systems with a constant number of chain atoms between cross-links, but with cross-link functionalities ranging up to 12. As one would expect, T_g rises with functionality more steeply the shorter the chains between cross-links (Rietsch et al., 1976).

Organic polymers with saltlike cross-links, such as the multivalent metal ion salts of poly(acrylic acid), behave like covalently bonded glasses because of the highly covalent character of these salts. Their most highly cross-linked (= polyvalent neutralized) derivatives also decompose before reaching a glass transition temperature. The other physical properties of the most highly cross-linked polymers, such as the elastic moduli, approach those of the inorganic covalently bonded glasses (Holliday, 1975).

II. Effects of Cooling Rate

Simple substances that tend to crystallize on cooling must be subcooled rapidly in order to enter the viscous glassy region where crystal nucleation rate becomes negligibly small. Here cooling rate is not a freely disposable independent variable (see Chapter 1 of this volume). Moreover, this "critical" cooling rate is commonly the fastest that can be achieved in simple equipment. Hence the effect of cooling rate on the physical properties of glasses has been studied primarily on noncrystallizing polymers. Here high cooling rates "freeze" the liquid into a glassy state at lower density (higher specific volume) than would be experienced at lower cooling rates. In line with theory (Haward, 1973; Roberts and White, 1973) this excess volume is reflected in lower elastic modulus, lower tensile strengths, etc., as shown by the data of Fig. 1. The low packing density of the polymer glasses is responsible for their comparatively low elastic moduli and strength and equilibrium properties.

This quenched-in low density is not stable. On prolonged aging of glasses, even at quite low temperatures relative to T_g, the density of polymer glasses increases steadily (Kovacs, 1964). The parallel decrease in the internal mobility of glasses is noted as decrease in the creep rate in creep compliance experiments extending over long time intervals (Struik, 1976). Because of their low thermal conductivities, the properties of rapidly quenched molecular glasses vary from the surface to the interior, especially for an article, the interior of which cools slowly, regardless of how rapidly the outside skin has been quenched.

Thus, rapidly quenched molecular glasses have skins that are softer than their interiors. Moreover, because of the proximity of the operating temperature to T_g of most molecular glasses, their creep rate may often suffice to relax the resulting quench stress more rapidly than one observes with ionic and covalent or metallic glasses. For a summary of the effects of rapid cooling on various types of glasses, see Vol. 2 of this treatise.

III. The Glass Transition Temperature (T_g)

The data quoted in the previous section indicated the cooling rate dependence of T_g, at least qualitatively. Numerous authors (Moynihan *et al.*, 1974; Wilkes *et al.*, 1975; Roberts and White, 1973) have presented gener-

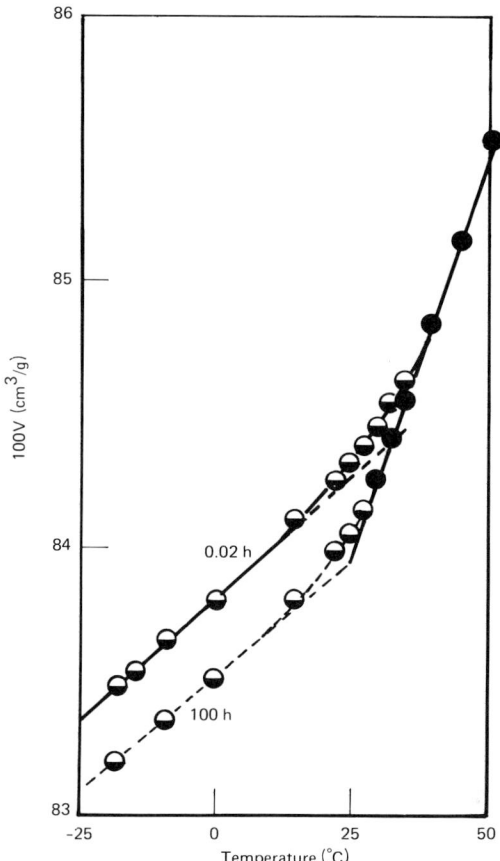

FIG. 1a. Effect of cooling rate on isobaric volume–temperature relationships for poly(vinyl acetate). The two lower temperature lines represent measurements made after 0.02 and 100 h. (From Roberts and White, 1973.)

alized correlations between these two variables so that one can extrapolate the $T_g(t)$ observed at finite cooling rate $\dot{q}$ to the "quasi-equilibrium" $T_g(\infty)$ for $t = \infty$. Such general correlations assume implicitly that the viscosity and the viscosity–temperature function near T_g are identical for all substances, or at least for those for which the general correlation is supposed to hold. The natural experimental validity of this definition of T_g as an isoviscous state is discussed elsewhere in this volume, where it is compared with 15 other theoretical characterizations, such as isentropic, iso(free volume), or iso(packing densities), etc.

Careful experimentation suggests that all such generalizations may have been premature. Indications are that $T_g^{-1} \sim -\ln \dot{q}$, with activation

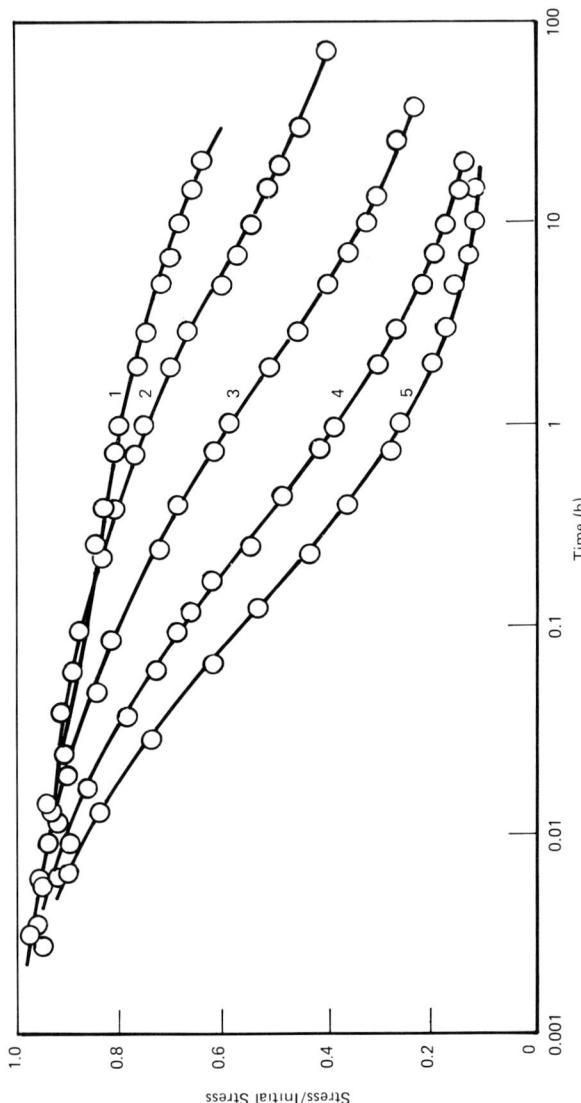

FIG. 1b. Effect of cooling rate on stress relaxation of poly(methyl methacrylate) at 80°C. Sample cooled from 130°C as follows: 1, 5°C/h; 2, 31°C/h; 3, convection in 25°C oil; 4, plunged in 25°C oil; 5, plunged in dry-ice–naphtha bath. (From Bondi, 1968.)

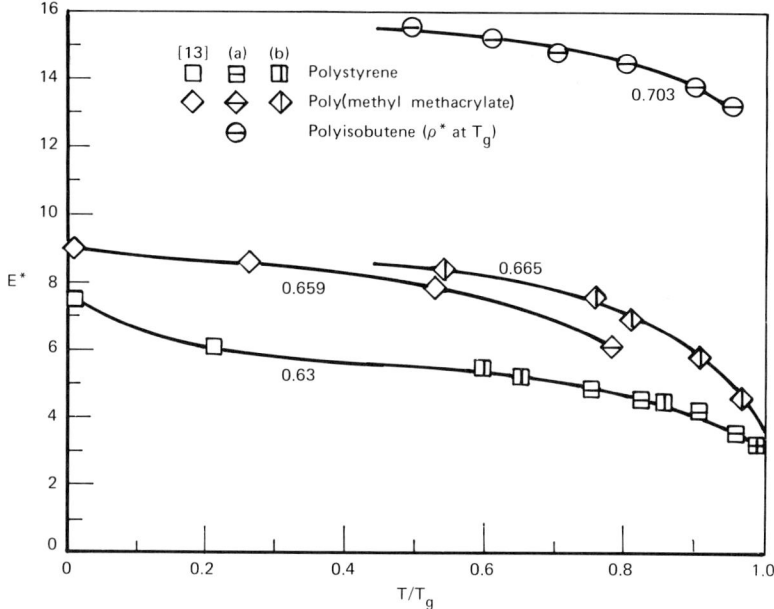

FIG. 1c. Generalized Young's modulus (at 10 kHz) of several polymeric glasses as function of T/T_g over packing density ρ^* (ordering parameter). Note the strong effect of packing density on the modulus curves of poly(methyl methacrylate). (From Bondi, 1968.)

energies $E_a = -R(d(\ln \dot{q})/dT_g)$ of different glassy polymers differing by large factors (Peysen and Bascom, 1977). Most striking is the fact that fillers may raise or lower E_a substantially. Many contradictory observations in the literature on the effect of fillers on T_g may thus turn out to be due to inadequate, or nonexistent, consideration of the cooling rate as an additional independent variable.

Here we shall only be concerned with the effects of molecular structure on T_g. However, to separate the specific effects of a given molecular structure from those simply associated with intermolecular force and molecular flexibility, the absolute magnitude, $T_g(\infty)$, as well as the dimensionless glass transition temperature, $RT_g(\infty)/E$, will be discussed side by side.

The literature data have been obtained by many different techniques, each with a different characteristic time scale. Only in a few cases have enough different data been given to extrapolate securely to $t = \infty$. Hence not all data are exactly comparable, and the method for T_g determination must be specified.

Some authors (Beaman, 1952; Boyer, 1963; Kauzman, 1948; Sakka and MacKenzie, 1971; Van Krevelen, 1976) have claimed a special significance for the ratio T_g/T_m (where T_m is the melting point); this has also been entered on the tables for simple substances. One notes there that this "homologous" glass transition temperature is approximately constant for only very narrowly defined groups of compounds. This result should have been expected from the fact that T_m is generally not a corresponding state and, similarly, that η and $\eta = f(T)$ are far from identical at T_g or at any distance below T_m. The concept of T/T_m as homologous temperature was developed by metallurgists, for whose monatomic systems such a generalization is barely applicable; namely, for solids with equal fractional oscillation amplitude a/r_0 at T_m (Cartz, 1955). The predictive utility of T_g/T_m should be minimal when dealing with the complex polyatomic molecular glasses of concern here. Yet an inspection of the accumulated data shows that the dispersion of T_g/T_m is only slightly wider than that of the theoretically more acceptable dimensionless glass transition temperatures shown in Table I. For solids with first-order solid–solid phase transitions, T_m must be replaced by the more appropriate lower phase-transition temperature T_i.

The vitrification tendency of elemental solids appears to parallel the extent of polymorphism, i.e., the number of solid–solid transitions between T_m and 0 K (Wang and Merz, 1976). No such correlation could be detected among molecular solids. On the contrary, vitrification of such solids appears to be achieved more easily the lower the T_m, i.e., the higher the viscosity of the liquid near that temperature, thus reducing the nucleation rate and crystal growth rate. Among molecular solids one generally finds that polymorphism raises T_m above the temperature where it would be otherwise.

The lack of constancy of the dimensionless glass transition temperature by any of the methods used for defining a corresponding state suggests that T_g cannot be uniquely defined by any current picture of the liquid (or solid) state, i.e., as an equilibrium state. Tammann's old notion that T_g corresponds to a fixed absolute level of the macroscopic viscosity has also not stood the test of time. Perhaps we have to learn to define separate classes of liquids, the vitrification of each of which is dominated by some different molecular mechanism.

Polymeric molecular glasses form that class of compounds for which the absolute value of T_g is of greatest commercial importance because it determines their practical utility. For glassy polymers that are used as solids, the highest T_g consistent with processability is desired, whereas for elastomers a low T_g (far below room temperature) is desired, so that, for instance, automobile tires do not become glassy solids on cold winter nights. Even mild vitrification causes undesirable flat spots, noticeable as bumpy wheels after starting the ride.

TABLE II

Limiting Values for T_g^* at $M \to \infty$, Compared with the
Energy of Rotational Isomerization $(\Delta E_{\text{iso}})^a$

Polymer	T_g^*	ΔE_{iso} (kcal/mole)
Polyisobutene	0.216	0
cis-1,4-Polyisoprene	0.27	
Polyethylene	0.27	0.5–0.6
Poly(dimethylsiloxane)	0.30	0.80
Poly(methyl methacrylate)	0.33	0.8
Poly(n-butyl methacrylate)	0.33	
Polypropylene (isotactic)	0.35	0.6–1.5
Poly(monochlorotrifluoroethylene)	0.36	
Poly(vinyl acetate)	0.39	~1.0
Poly(vinyl chloride)	0.42	1.9
Polystyrene	0.435	1.0–1.5

[a] From Bondi (1968).

As expected from simple substances, we find the generalized glass transition temperature T_g^* of polymers to be dominated by the cohesive energy E^0 and the chain flexibility measure $3c$. But a systematic trend of T_g^* is also noted in Table II, which parallels the commonly accepted measure of chain expansion, the energy of rotational isomerization ΔE_{iso}. Unfortunately, ΔE_{iso} has been measured so far only for few types of backbone chain, and its absolute magnitude is also difficult to determine. Hence, this correlation is only of limited practical utility, even if the basic understanding gained by it is of some value.

As mentioned earlier, rigidity can also be imposed by high degrees of chemical cross-linking. The cross-link between polymeric molecules can take the form of the sulfide (or polysulfide) bridges of classical vulcanization, of either oxygen or C–C bonds, or of the partially ionic bonds of metal ion salts of polymeric acids, best known as ionomers (Jenkins and Duck, 1975; Longworth, 1975).

If one could be sure that the lengths of chains between cross-links is uniform, theoretical relations between T_g and cross-link density (n_c) could even be verified empirically. A model system for such tests of theory would be composed of long-chain dioic acids, cross-linked by trivalent cations. But in more practical systems cross-links are very unevenly distributed, as has been shown many times by degradation experiments. Relations between T_g and cross-link density, as shown in Fig. 2, should

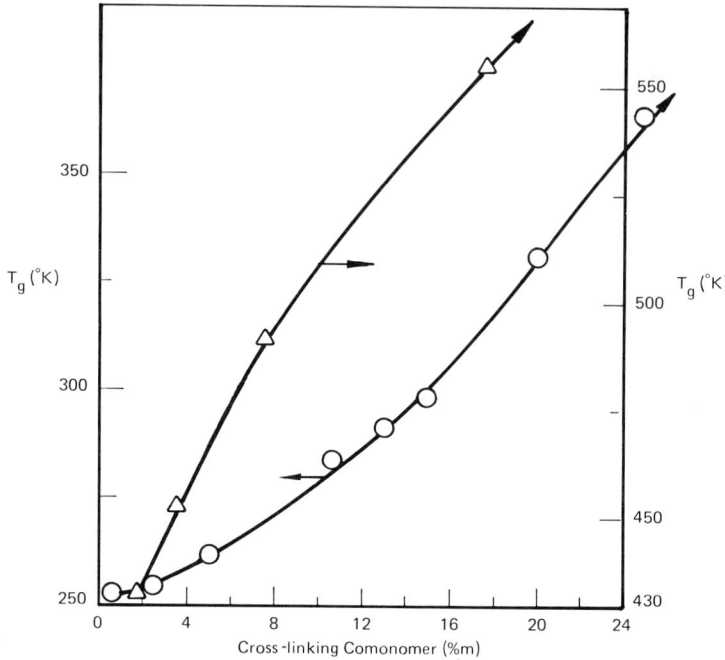

FIG. 2. Effect of cross-link density on T_g. (△) Ethylene (50% naturally neutralized) methacylic acid (Longworth, 1975). (○) Ethylacrylate–ethylene glycol dimethacrylate (Tobolsky and Mark, 1971).

therefore be taken as qualitative guides. In specific cases such relations will be very much affected by the success of the experiments in achieving uniform rather than "lumpy" cross-link density distributions. Obviously, the gradient dT_g/dn_c will be steeper the more uniform the cross-link density distribution over all volume elements of the glass. At present there are no good measures of the uniformity of cross-link density distributions. They can only be inferred from the mechanical behavior of the system at $T > T_g$, especially from stress–strain curves and from mechanical stability data (Tobolsky and Mark, 1971).

IV. The Glass Transition Temperature of Mixtures

We are concerned with several different kinds of mixtures: (a) mutual solutions of substances composed of simple molecules, (b) mixtures of polymers with low-molecular-weight liquids, (c) mixtures of different, but miscible, polymers, (d) random copolymers, and (e) block copolymers, with incompatible blocks.

Mixture rules of varying degrees of sophistication have been developed primarily to predict the T_g of the commercially important cases (b) and (d), namely, plasticized thermoplastics and random copolymers, respectively. With but little adaptation these rules can be used for all microscopically homogeneous systems. Because of the indefiniteness of T_g as a thermodynamic equilibrium property, no mixture rule can be expected to yield precise predictions. Hence the primary purpose of high degrees of approximation is the qualitative understanding of the role of molecular variables, and particularly of the effects of the size and magnitude of the excess free energy of mixing on the T_g of mixtures.

The simplest rule for copolymers, the Fox relation

$$\frac{1}{T_{g(12)}} = \frac{\phi_1}{T_{g(1)}} + \frac{\phi_2}{T_{g(2)}} + f(\phi_1\phi_2 k) \qquad (1)$$

can be derived as a special case from most theoretical treatments. Here the ϕ_i are volume fractions of the components and k is an interaction parameter, which (according to Kovacs and Braun, 1965) is of the order

$$-k = \frac{v^E/v_{12}}{\phi_1\phi_2}, \qquad (2)$$

where v^E is the excess specific volume of mixing and v_{12} is the specific volume of the mixture. The relation (1) should also hold for polymer–plasticizer or polymer–solvent systems. But for mixtures of low-molecular-weight liquids with each other the volume fraction should probably be replaced by the mole fraction x_i. Too few data exist to test that notion.

Some authors treat the first two terms of the Fox relation as an ideal mixing rule and then estimate the correction term from a molecular theory. A typical example is the estimation of a correction term for copolymers by estimating their conformation entropy ΔS_c. Using the assumption that $T_{g(12)}^{-1} \sim \Delta S_{c(12)}$, the difference ΔS_c between ideal and "true" conformational entropy was estimated by rather complex estimation procedures (Tonelli, 1975). A qualitative resemblance between ΔS_c and ΔT_g (relative to "ideal" Fox behavior) is indeed observed.

This appears to be a very inappropriate time to recommend any existing mixture rule. First, none has proved to be widely applicable; second, the more detailed rules contain thermodynamic interaction parameters between polymer and solvent or plasticizer, or between different monomers in a copolymer, including that between center groups and end groups, to account for the molecular weight dependence of T_g; third, these thermodynamic interaction parameters have in only a few cases been measured experimentally in the concentration and temperature ranges of practical interest (Su *et al.*, 1976; Bonner, 1975) and turn out to be far more concentration dependent than had been supposed before. Hence, until these interaction effects are better understood, the more elaborate mixture rules for T_g must be considered as speculative, however appealing their underlying mechanistic hypotheses.

A good mixture rule for T_g must not only permit estimation of the absolute level of $T_{g(12)}$, but also permit the estimation of the vitrification concentration at constant temperature. The steepness of that concentration effect can have very large effects on many technical properties, as will be discussed later.

The general shape of the Fox relation shows that the homologous glass transition temperature T_g/T_m of mixtures cannot be constant. For mixtures that form eutectics T_g/T_m must, and does, go through a fairly steep maximum at the eutectic point.

V. Heterogeneous Mixtures

A class of block copolymers has been developed during the past two decades, the mechanical properties of which are dominated by the ability of the glass-forming blocks to coalesce into small insoluble droplets that act as physical cross-links for the entire system. Thus the elastic modulus and the strength of a bulk sample of such a block copolymer is dominated by the properties of the glassy component. Hence it is important to know whether the physical properties of the glassy droplets are identical with those of their macroscopic 100% bulk equivalent. Obviously, this will depend on the compatibility of the glassy with the rubbery component and on the size. In the limit of full compatibility, the glassy droplets will not segregate. In the limit of total incompatibility, the T_g of the glassy phase will depend on the molecular weight of the glass-forming segment in the block copolymers and to a second approximation on the size of the droplet (Krause *et al.*, 1967). Actually, the compatibility between the phases increases as the block length decreases, so that T_g should decrease more sharply with decreasing block length than predicted from bulk compound data.

There is some evidence that the T_g of very small droplets of the glassy phase is lower than that of the corresponding bulk phase (Bares and Pegoraro, 1971), presumably because of the stress exerted by the rubbery phase blocks that grow heterogeneously out of the glassy droplet. Far more systematic work is required to separate these various effects clearly.

The persistence of the glass transition temperature of the components in a heterogeneous mixture extends also to that of the more mobile component. In the limit of total incompatibility its T_g is that of the pure bulk component. Slight compatibility will raise its T_g slightly.

VI. Detailed Effects of Molecular Structure

The gross effects of molecular structure of polymers (and of monomers) on their glass transition temperature have been outlined in the previous sections as functions of intermolecular forces and of intramolecular flexibility. The deviations of the generalized glass transition temperature from a universal pure number strongly suggest, however, that a more accurate correlation of T_g with molecular structure must be based either on a more complex theory or on straight empiricism. In the absence of such a theory, we turn to the empirical correlations developed by van Krevelen.

For the simple case of polymers without side chains van Krevelen proposes

$$T_g = (S_i Y_{gi})/Z,$$

where the structure increment Y_{gi} is tabulated for different molecular structures (Table III), and Z equals the number of atoms per repeating unit along the backbone chain. If an aromatic ring is part of the backbone, Z is counted differently: phenylene-1,2: $Z = 2$; phenylene-1,3: $Z = 3$; phenylene-1,4: $Z = 4$.

For polymers with side chains,

$$T_g = (Y_{g0} + h(N))/(Z_0 + N),$$

where $Y_{g0} = T_{g0} Z_0$, $h(N) = T_g (Z_0 + N) - T_{g0} Z_0$, N is the number of CH_2 groups (per structural unit) in side chains, and Z_0 and Y_{g0} relate to the structure of the basic polymer (a) and of the second polymer (b).

Typical numerical values for Z_0 and Y_{g0} have been assembled in Table V. The predictive ability of these correlations is quite adequate, the standard deviation being generally of the order of ± 10K. The applicability of this scheme to a number of extreme T_g cases collected from the "Polymer Handbook" (Brandrup and Immengut, 1975) could not be tested because the appropriate group increments had not yet been estimated. A version of Table IV has to be developed for side chains not ending in a methyl group in order to capture some of the most striking molecular structure effects on T_g.

TABLE III
Group Contributions to T_g (kg/mole)

Group	Y_{gi}	Group	Y_{gi}	Group	Y_{gl}	$Y_g(I_x)$
—CH_2—	2,700			—O—	4,000	—
—CH(CH_3)—	8,000	(p-methylphenyl)	32,000	—C(=O)—	27,000	—
—CH(C_2H_5)—	10,500	(chloro-dimethylphenyl)	51,000	—O—C(=O)—O—	8,000	12,000I
CH(C_3H_7)—	13,100	(dimethylphenyl)	35,000	—O—C(=O)—C(=O)—O—	16,000	10,000I
—CH(C_6H_5)—	35,000	(trimethylphenyl)	(55,000)	—S—	(20,000)	?
—CH($C_6H_4CH_3$)—	42,000			—S(=O)(=O)—	7,500	—
—CH(OCH_3)—	11,900				(58,000)	?

Group	Value 1	Structure	Value 2	Value 3	Value 4
—CH(COOCH$_3$)—	21,300	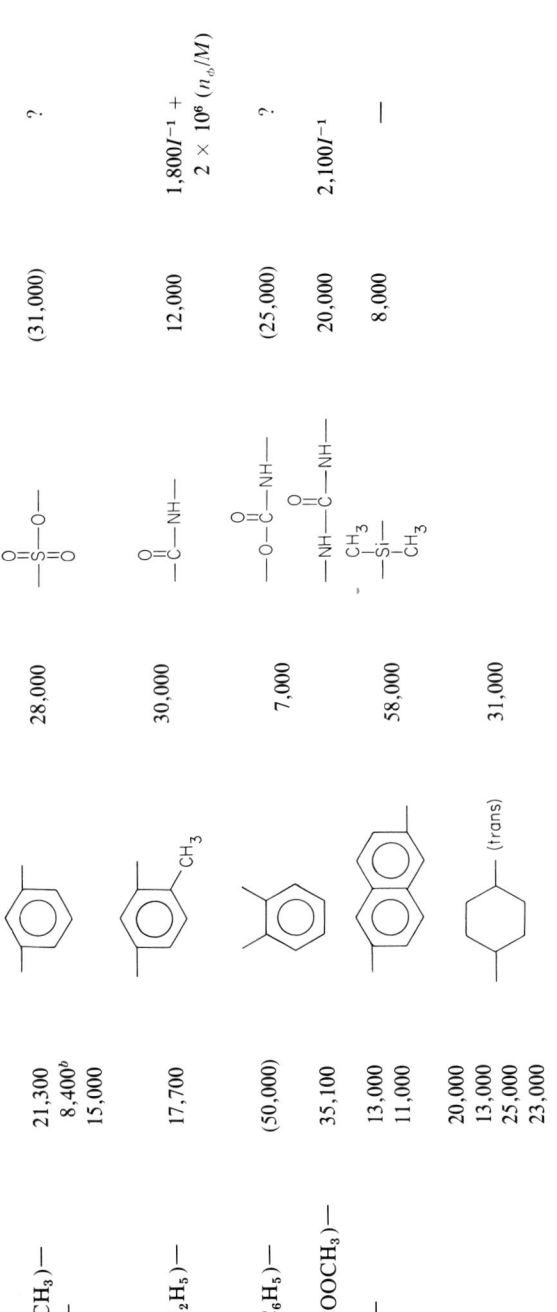	28,000	(31,000)	?
—C(CH$_3$)$_2$—	8,400[b]				
	15,000				
—C(CH$_3$)(C$_2$H$_5$)—	17,700		30,000	12,000	$1,800J^{-1} +$ $2 \times 10^6 (n_{\phi}/M)$
—C(CH$_3$)(C$_6$H$_5$)—	(50,000)		7,000	(25,000)	?
—C(CH$_3$)(COOCH$_3$)—	35,100		58,000	20,000	$2,100J^{-1}$
—CH(OH)—	13,000		31,000	8,000	—
—CHF—	11,000	(trans)			
—CHCl—	20,000				
—CF$_2$—	13,000				
—CCl$_2$—	25,000				
—CFCl—	23,000				

[a] From van Krevelen (1972).
[b] In polyisobutylene only!

TABLE IV Structural Corrections in More Complicated Aromatic Polymers[a]

Case	Group	Correction term	Integral contribution of composed groups
I. Flexible chains containing —CH$_2$— group segments	—C$_6$H$_4$—C$_6$H$_4$—	13,000	77,000
	fused bicyclic (naphthalene)	47,000	61,000
	—C$_6$H$_4$—O—C$_6$H$_4$—	−5,000	63,000
	—C$_6$H$_4$—(CH$_2$)$_1$— Symmetrical	0	34,700
	Asymmetrical	−10,000	25,000
II. Rigid polyester and polyamide chains containing *no* —CH$_2$— group segments	—C$_6$H$_4$— ⎫ In polyamides only	10,000	42,000
	meta-C$_6$H$_4$— ⎭	10,000	38,000
	—C$_6$H$_4$—C$_6$H$_4$—	10,000	87,000
	fused bicyclic (naphthalene)	10,000	71,000
	—C$_6$H$_4$—CH(C$_6$H$_5$)—C$_6$H$_4$—	10,000	109,000
	—C$_6$H$_4$—C(CH$_3$)$_2$—C$_6$H$_4$—	20,000	99,000
	—C$_6$H$_4$—O—C$_6$H$_4$—	20,000	83,000

[a] From van Krevelen (1972).

TABLE V
Basic Data for Vinyl Polymers with Longer Side Chains[a]

Series	Basic polymer	Trivalent group Y	Y_{g0}	Y_{g9}
Polyolefins	Polypropylene	$-\overset{H}{\underset{\vert}{C}}-$	10,700	33,600
Polyalkylstyrenes	Poly(p-methylstyrene)	$-\overset{H}{\underset{\vert}{C}}-$ with phenyl	44,700	48,800
Polyvinyl ethers	Poly(vinyl methyl ether)	$-\overset{H}{\underset{\underset{\vert}{O}}{\underset{\vert}{C}}}-$	14,600	36,800
Polyvinyl esters	Poly(vinyl acetate)	$-\overset{H}{\underset{\underset{\underset{\vert}{C=O}}{\underset{\vert}{O}}}{\underset{\vert}{C}}}-$	26,000	42,400
Polyacrylates	Poly(methyl acrylate)	$-\overset{H}{\underset{\underset{\underset{\vert}{O}}{\underset{\vert}{C=O}}}{\underset{\vert}{C}}}-$	24,000	42,400
Polymethacrylates	Poly(methyl methacrylate)	$-\overset{CH_3}{\underset{\underset{\underset{\vert}{O}}{\underset{\vert}{C=O}}}{\underset{\vert}{C}}}-$	37,800	45,200

[a] For the ester groups, the interaction factor is $I = 2/(2 + 2) = 0.5$. So the group contributions to Y_g are

	Y_{gi}
2 $-CH_2-$	5,400
$-\bigcirc-$	32,000
2 $-COO-2(8{,}000 + 0.5 \times 12{,}000) =$	28,000
	$Y_g = 65{,}400$

VII. Equilibrium Properties

The equilibrium properties of density, thermal expansion, elastic moduli, and heat capacity of molecular glasses share with those of all glasses the indeterminacy of the precise value (for a given composition) because of their dependence on the cooling rate during vitrification. In absolute terms the numerical values of these properties, of course, reflect their low cohesion energy in the same manner as observed earlier for T_g. A typical example for the absolute and the generalized equilibrium properties of molecular glasses side by side with those of covalently and ionicly bound glasses has been assembled in Table VI.

Typical examples for the long relaxation times of volume-related equilibrium properties after heating or compression are shown in Figs. 3 and 4. Such creep phenomena are better known for tensile and torsional deformations, where they are also of far greater technical importance. Since the mechanical properties of glasses are treated in detail in Vol. 5 of this treatise, only a single example is shown, in Fig. 5, demonstrating the extreme dependence of the observed elastic modulus on the rate of deformation that prevailed during its determination. This deformation rate dependence of the modulus and of Poisson's ratio be-

TABLE VI

CUBIC THERMAL EXPANSION COEFFICIENT OF VARIOUS GLASSES

	$10^4 \alpha_g$ (K^{-1})	T_g (K)
Molecular glasses		
1,3,5-Trinaphthylbenzene	1.20	348
Polystyrene	2.3	375
Polyethyleneterephthalate	1.9	353
Glucose	0.9	290
Polyacrylic acid	1.6	378
		400
Covalent glasses		
Calcium polyacrylate	0.7	
Zinc polyacrylate	0.4	473
SiO_2	0.015	1463
Ionic glasses		
$NaPO_3$	1.0	563
$Ca(PO_3)_2$	0.3	

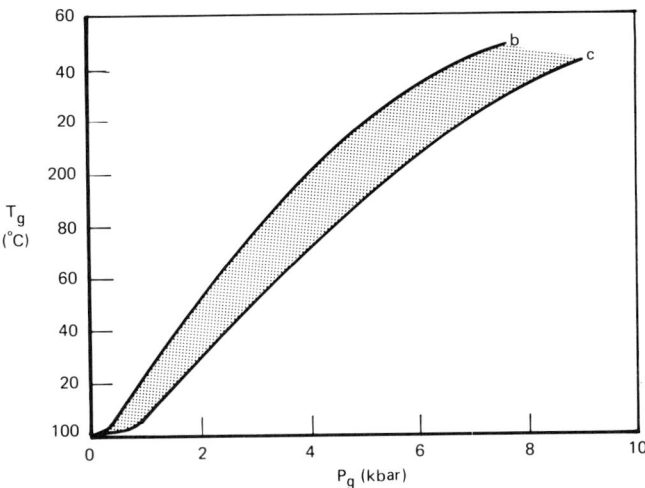

FIG. 3. Vitrification "phase" diagram of polystyrene. Note that vitrification at constant temperature takes place over a pressure range, b = beginning, c = completion of vitrification, or over a temperature range at constant pressure. (From Rehage and Bochard, 1973.)

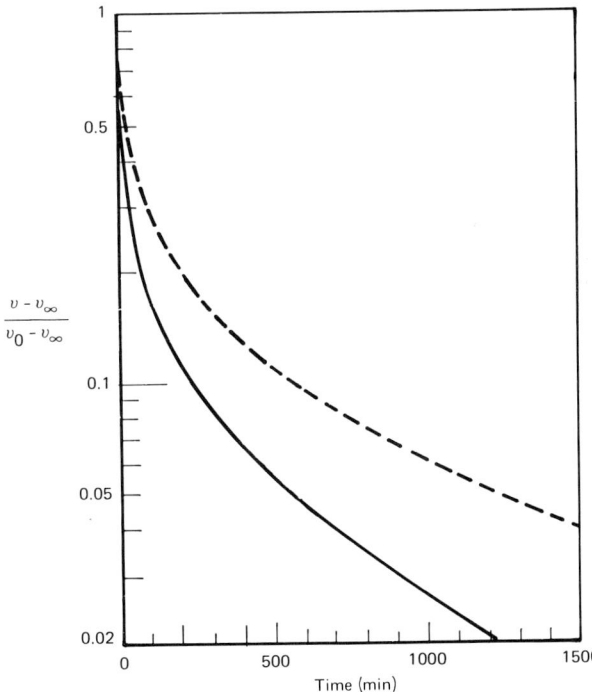

FIG. 4. Function $(v - v_\infty)/(v_0 - v_\infty)$ in polystyrene glass on a logarithmic scale versus time. (– –) Pressure jumps; (———) temperature jumps. (From Rehage and Borchard, 1973.)

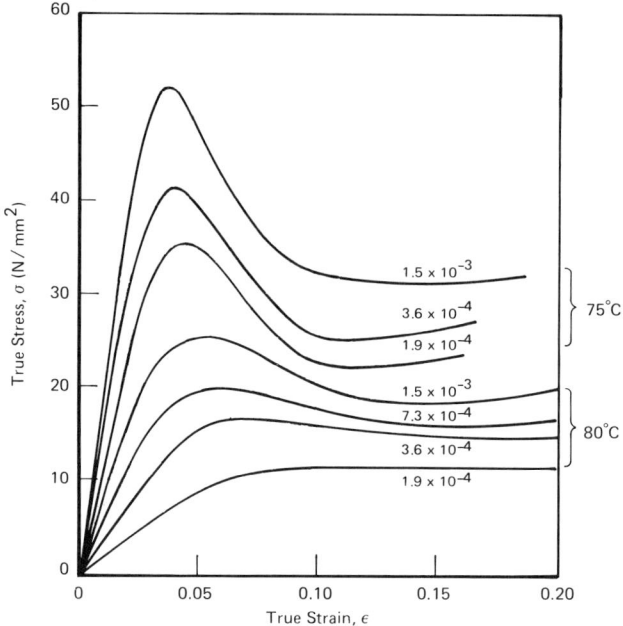

FIG. 5. Stress–strain curves for polystyrene determined in plane strain compression. Strain rate (per second) given for each curve. (From Bowden, 1973.)

comes less and less important the larger is $T_g - T_{exp}$. Hence the room temperature elastic modulus of glasses with very high T_g exhibit insignificant creep effects (see Vol. 2 of this treatise), and creep becomes a "trademark" of the molecular glasses. Their creep is still very much smaller than that of molecular crystals and crystalline polymers at the same $T_m - T_{exp}$ and at comparable load levels.

VIII. Effect of Admixtures

The elastic moduli of polymeric glasses are reduced by the admixture of compatible low-molecular-weight materials, such as plasticizers, just as one would have expected from the corresponding reduction in T_g. Generally, one finds that a plot of the reduced moduli G_χ/G_0 versus $T_{g(\chi)}/T_{g(0)}$ puts all experimental points, if not on a single curve, at least within a narrow band. Here the subscript χ denotes the concentration of a given plasticizer and 0 denotes the unplasticized polymer. However, there are noteworthy exceptions, especially when a very compatible plasticizer at low concentration actually increases the modulus while reducing T_g in the expected manner. This phenomenon is known as antiplasticization, the explanations for which are still controversial and should be examined in the original literature (Bondi, 1968; Petrie et al., 1972).

The effect of heterogeneous admixtures on the elastic moduli of polymeric glasses is of great practical importance in two extreme cases. One is the admixture of small to moderate concentrations of elastomers with low elastic modulus in order to increase the impact resistance of these glasses. The other is the admixture of very high concentrations of generally fibrous materials such as glass fibers or carbon fibers with very high elastic modulus. These can raise the elastic modulus of the resulting composite very appreciably. But the main purpose is to raise the ultimate strength of these materials. The estimation of the physical properties of such heterogeneous "composites" has become a separate field of specialty, with dedicated journals, etc. The extremely high strength–weight ratios that can be achieved by such combinations had their first spectacular demonstration in the rocket casing of the Polaris submarine missile. Commercial applications of composites constitute the fastest growing branch of the polymer industry.

IX. Transport Phenomena

Rotational and mass diffusion phenomena and their congeners are among the most interesting and technically most important properties of molecular, especially polymeric, glasses. Their thermal conductivity, by contrast, is just as one would expect, and thus quite uninteresting. Detailed discussion of these phenomena, from different perspectives, is presented in the books by Bueche (1965) and deGennes (1979).

Rotational diffusion phenomena are observed during elastic and dielectric property measurements as sources of selective energy absorption at frequencies resonating with corresponding molecular rotation frequencies. The relaxation times of these motions have such a wide distribution that the energy absorption versus frequency curves show rather broad peaks (Kolarik and Iavsky, 1970; McCrum et al., 1967; Ferry, 1980). Much sharper spectra are obtained by exploitation of the time–temperature superposition principle, through a plot of energy absorption versus temperature at constant frequency.

The interpretation of these so-called relaxation spectra has proceeded in a largely empirical fashion. Chemists prepared polymers of many different known molecular structures and then observed the changes in the elastic and dielectric relaxation spectra. These showed that energy absorption at the lowest temperatures (as low as 20–50 K) appears to be caused by the rotation of very small molecule fragments, such as the methyl group. The larger the size of the rotating molecule fragments, the higher the temperature of the absorption maximum at a given frequency. Moreover, since dielectric energy absorption can take place only if there are permanent electric dipoles in the system that can oscillate in resonance with the alternating electric field, a comparison between mechanical and dielectric relaxation spectra for polymer molecules with polar and nonpolar groups reveals which group is moving in which part of the spectrum (McCrum et al., 1967; Ferry, 1980).

A plot of the relaxation peaks taken at several frequencies with the coordinates peak frequency versus T^{-1} yields the so-called relaxation map of a system (McCall, 1971). The slope of each curve is the activation energy, a measure of the potential energy barrier hindering the molecular motion associated with that particular relaxation peak. Again the correlation of this information and the absolute magnitude of the limiting frequency involved within the framework of massive amounts of chemical structure information has provided rather good clues as to what components of polymer molecules are moving in which temperature and frequency domain.

Among the practical applications of this relaxation information one finds predictions of the suitability of a given composition as an electrical insulator and, more significantly, the incidence of ductility among glassy polymers. This may sound like a contradiction in terms. It is not in the case of polymeric glasses. Here it happens that the incidence of rigid backbone structure components between highly mobile hinges creates the opportunity for substantial energy absorption during deformation (Argon and Bessonov, 1976). That would explain resistance to shattering at high deformation rates. The same structure elements also permit local realignment during slow deformation. A microscopic theory of this glassy ductility has been developed (Argon, 1973, 1980).

The mass diffusion of gases and vapors through polymer films is of considerable practical importance because of their widespread use as gas and vapor barriers. An increasing number of films are used at $T < T_g$; hence the peculiar differences between the familiar liquidlike diffusion through polymers in the rubbery state and that in the (still somewhat unfamiliar) glassy state should be noted.

That peculiarity can be divided into two parts: the permeant solubility (χ_{ij}) and the permanent diffusion (D_{ij}). The solubility must be considered because the technically significant permeability of gas j through film i is given by the product $P_{ij} = \chi_{ij} D_{ij}$, in the rubbery state, as in the liquid. Here $\chi_{ij} \sim p_i$, where p_i is the partial pressure of the gas in the gas space. However, in the glassy state one portion of the gas is dissolved in the glass and another is adsorbed on the walls of the many microscopic voids that form during vitrification. The resulting combination of Henry's law

5. ORGANIC GLASSES (MOLECULAR GLASSES)

and of a Langmuir adsorption isotherm successfully represents the gas solubility data in the glassy state (Hopfenberg and Stannett, 1973; Koros et al., 1976). The *a priori* estimation of the adsorption coefficient may never be possible, however, because—as expected—they depend somewhat on the thermal and mechanical history of the sample.

The temperature dependency of the diffusion constant D_{ij} also passes through a discontinuity in slope at T_g. In fact, it would not be surprising to find additional slope changes in the temperature ranges in which various polymer molecular motions freeze out as the temperature is reduced further. The detailed interaction between permeant molecules and the motions of polymer molecule segments required for mass diffusion is beginning to be understood, at least qualitatively, as reliable experimental data become available for more and more systems.

There is a fundamental difference between the permeation of slightly soluble gases through polymer films and that of vapors that are quite soluble in the polymer film. The latter group plasticizes the film and can change its properties drastically. Consequently, the permeation rate becomes time and location dependent and, naturally, increases rapidly with increasing permeant concentration. This situation is so complicated that its detailed description is beyond the scope of the present review and experiments should be planned only after careful study of specialized reviews (Hopfenberg and Stannett, 1973).

Symbols

b	Characteristic molecular length dimension
c	One-third of the number of external degrees of freedom per molecule, including those due to internal rotation
E	Cohesion energy
E_a	Activation energy
ΔE_{iso}	Energy of rotational isomerization
$E°$	A standard energy of vaporization†
E^*	$Y \cdot V_w/H_s$, generalized Young's molecules†
H_s	Heat of sublimation
k	Empirical constant
k_B	Boltzmann constant
M	Molal weight
n_c	Number of chain links between cross-links
p	Partial pressure of vapor
P_g	Vitrification pressure
P_{ij}	Permeability of gas i through film j
q	Cooling rate
R	Gas constant

† For detailed discussion and numerical values, see Bondi (1968).

ΔS_c	Conformation entropy
t	(Cooling) time (seconds)
T	Absolute temperature (K)
T_g	Glass transition temperature (K)
T_1	Solid–solid phase transition temperature (K)
T_m	Melting point (K)
T_g^*	$5cRT_g/\Delta E°$, generalized glass transition temperature, dimensionless
v	Specific volume
V_w	Van der Waal's volume†
Y	Young's modulus
ϵ	Molecular free surface energy
η	Viscosity
ρ	Density
ρ^*	$\rho \cdot V_w/M$, generalized packing density†

† For detailed discussion and numerical values, see Bondi (1968).

References

Argon, A. S. (1973). *Phil. Mag.* **28,** 839.
Argon, A. S. (1980). *In* "Glass: Science and Technology," Vol. 5, p. 79. Academic Press, New York.
Argon, A. S., and Bessonov, M. I. (1976). M.I.T. Report 6-57-76.
Bares, J., and Pegoraro, M. (1971). *J. Polym. Sci. Part A-2* **9,** 1287.
Beaman, R. G. (1952). *J. Polym. Sci.* **9,** 470.
Bondi, A. (1968). "Physical Properties of Molecular Crystals, Liquids and Glasses." Wiley, New York.
Bondi, A., and Tobolsky, A. V. (1975). *In* "Polymer Science and Materials" (H. F. Mark and A. V. Tobolsky, eds.), Chapter 6. Wiley, New York.
Bonner, D. C. (1975). *J. Macromol. Sci.-Rev. Macromol. Chem.* **C13,** 263.
Bowden, P. (1973). *In* "The Physics of Glassy Polymers" (R. N. Haward, ed.). Wiley (Halsted), New York.
Boyer, R. F. (1963). *Rubber Chem. Technol.* **36,** 1303.
Brandrup, J., and Immergut, E. H. (1975). "Polymer Handbook." Wiley, New York.
Bueche, F. (1965). "Physical Properties of Polymers." Wiley, New York.
Cartz, L. (1955). *Proc. Phys. Soc. London* **B68,** 163.
deGennes, P.-G. (1979). "Scaling Concepts in Polymer Physics." Cornell University Press, Ithaca, New York.
Ferry, J. D. (1980). "Viscoelastic Properties of Polymers." Wiley, New York.
Haward, R. N. (1973). "The Physics of Glassy Polymers." Wiley (Halsted), New York.
Holliday, L. (1975). "Ionic Polymers." Applied Science Publ., London.
Hopfenberg, H. B., and Stannett, V. (1973). *In* "The Physics of Glassy Polymers" (R. N. Haward, ed.), p. 504. Wiley (Halsted), New York
Jenkins, D. N., and Duck, E. W. (1975). *In* "Ionic Polymers" (L. Holliday, ed.), p. 173. Wiley (Halsted), New York.
Kauzman, W. (1948). *Chem. Rev.* **43,** 219.
Kolarik, J., and Iavsky, M. (1970). *Coll. Czech. Chem. Commun.* **35,** (8), 2286.
Kovacs, A. J. (1964). *Adv. High Polym.* **3,** 394.
Kovacs, A. J., and Braun, G. (1965). "Physics of Non-Crystalline Solids," p. 303. North Holland Publ., Amsterdam.

5. ORGANIC GLASSES (MOLECULAR GLASSES)

Koros, W. J., Paul, D. R., and Rocha, A. A. (1976). *J. Polym. Sci.* **14,** 687.
Kraus, G., Childers, C. W., and Gruner, J. T. (1967). *J. Appl. Polym. Sci.* **11,** 1581.
Litt, M. (1976). *Trans. Soc. Rheol.* **20,** 47.
Longworth, R. (1975). *In* "Ionic Polymers" (L. Holliday, ed.), p. 69. Wiley (Halsted), New York.
McCall, D. W. (1971). *Am. Chem. Soc. Polym. Prepr.* **12** (1), 73.
McCrum, N. J., Williams, G., and Read, B. E. (1967). "Mechanical and Dielectric Relaxation in Polymeric Solids." Wiley, New York.
Moynihan, C. T. *et al.* (1974). *J. Phys. Chem.* **78,** 2673.
Petrie, S. E. B., Moore, R. S., and Flick, J. R. (1972). *J. Appl. Phys.* **43,** 4318.
Peysen, P., and Bascom, W. D. (1977). *J. Macromol. Sci.-Phys.* **B13** (4), 597.
Rehage, G., and Borchard, W. (1973). *In* "The Physics of Glassy Polymers" (R. N. Haward, ed.), p. 54. Wiley (Halsted), New York.
Rietsch, F., Daveloose, D., and Froelich, D. (1976). *Polymer* **17,** 899.
Roberts, G. E., and White, E. F. T. (1973). *In* "The Physics of Glassy Polymers" (R. N. Haward, ed.), p. 153. Wiley (Halsted), New York.
Sakka, S., and MacKenzie, J. D. (1971). *J. Non-Cryst. Solids* **6,** 145.
Struik, L. C. E. (1976). *Ann. N.Y. Acad. Sci.* **279,** 78.
Su, C. S., Patterson, D., and Schreiber, H. P. (1976). *J. Appl. Polym. Sci.* **20,** 1025.
Tobolsky, A. V., and Mark, H. F. (1971). "Polymer Science and Materials." Wiley (Interscience), New York.
Tonelli, A. E. (1975). *Am. Chem. Soc. Polym. Prepr.* **16,** 228.
Van Krevelen, D. W. (1972). "Properties of Polymers." Elsevier, Amsterdam, Holland.
Wang, R., and Merz, M. D. (1976). *Nature (London)* **260,** 35.
Wilkes, I. G. L. *et al.* (1975). *A.C.S. Polym. Prepr.* **16,** 585, 600.

CHAPTER 6

Metallic Glasses

J. B. Vander Sande

DEPARTMENT OF MATERIALS SCIENCE AND ENGINEERING
MASSACHUSETTS INSTITUTE OF TECHNOLOGY
CAMBRIDGE, MASSACHUSETTS

R. L. Freed

ENGINEERING TECHNOLOGY LABORATORY
E. I. DU PONT DE NEMOURS AND COMPANY, INC.
WILMINGTON, DELAWARE

I. Introduction	365
II. Metallic Glass Alloy Types	366
III. Structure of Metallic Glasses	367
IV. Theories for the Formation of Metallic Glasses	369
V. Investigations of Properties and Behavior of Noncrystalline Phases	373
A. Magnetic and Electrical Properties	373
B. Chemical Stability	373
C. Mechanical Properties of Metallic Glasses	376
VI. Theoretical Mechanisms of Deformation and Fracture	381
VII. The Effect of Crystallization on Mechanical Properties	385
VIII. Metallic Glasses at High Temperature	393
IX. Summary and Conclusions	398
References	399

I. Introduction

The discovery of metallic glasses or noncrystalline metallic solids is a direct result of the extension of a standard processing technique in general use by metallurgists to obtain nonequilibrium structures. This technique is the rapid removal of heat from the usually solid metallic alloy, i.e., quenching. This type of processing has long been used to obtain crystalline phases that are thermodynamically stable only at elevated temperatures or to produce entirely metastable crystalline phases that are unattainable through processing that requires thermodynamic equilibrium. The "new" perturbation on this scheme, of interest in this chapter, is the extension of solid-state quenching to liquid-state quenching, allowing the production of metallic solids that exist in a noncrystalline state rather than the normally obtained crystalline state.

Various techniques have been used to obtain glassy metallic alloys. These include vapor deposition, sputtering, electrodeposition, irradiation, laser glazing, and splat cooling. By far the most widely used technique has been splat cooling. With its reintroduction by Duwez *et al.* (1960a,b) and its many subsequent modifications (Predecki *et al.*, 1965; Pietrokowsky, 1963; Moss *et al.*, 1964; Roberge and Herman, 1968; Willens and Buehler, 1966; Pond and Maddin, 1969; Hinesley and Morris, 1970; Jones and Burden, 1971; Thursfield and Jones, 1971) several hundred investigations have been conducted on numerous alloy systems. Basically, the splat-cooling technique involves the very rapid quenching of a small molten droplet or stream of liquid by conduction of heat through a massive substrate. The cooling rates attainable by splat quenching vary with the equipment utilized but generally fall within the range of 10^5-10^8 °C/sec. The gun technique, originally used by Duwez to produce a noncrystalline Au–Si alloy, has been improved repeatedly to create devices that are capable of producing larger, more uniform specimens. Improved techniques are the piston and anvil (Pietrokowsky, 1963; Harbur *et al.*, 1969) melt spinning, roller quench (Chen and Miller, 1970), melt extraction (Maringer *et al.*, 1976), and laser glazing (Breman *et al.*, 1976) methods. The experimental variables that can cause differences in the properties of materials processed by rapid quenching from the melt have been discussed by Jones and Burden (1971). Kavesh (1976) has reviewed the principles of fabrication of metallic glasses from the thermal, hydrodynamic, and flow stability viewpoints of processing.

II. Metallic Glass Alloy Types

The types of metallic alloys that have been found to form metallic glasses as the result of splat quenching generally fall into two categories. The first category consists of metallic alloys that are composed of metal and metalloid constituents. These alloys contain on the average 15–30 at.% metalloid constituents, with the remainder being the metal. Examples from this class are Ni–P (Cargill, 1970; Bagley and Turnbull, 1970; Dixmier, 1965), Au–Si (Klement *et al.*, 1960; Dixmier and Guinier, 1967), Pd–Si (Chen and Turnbull, 1960; Masumoto and Maddin, 1971; Duwez *et al.*, 1965), Fe–P–C (Lin and Duwez, 1969; Rastogi and Duwez, 1970; Rastagi, 1973), Fe–Pd–P (Maitrepierre, 1969), Ni–Pd–P (Maitrepierre, 1970; Sinha, 1970), Ni–Pt–P (Chen and Turnbull, 1968), Au–Si–Ge (Luo and Duwez, 1963), Ge–Te (Luo and Duwez, 1963), Fe–B (Davis *et al.*,

1976), Ni–Fe–P–B (Chen and Polk, 1974) and Ni–Cr–Mo–W–C (Inoue *et al.*, 1979).

The other category of metallic alloys that can form metallic glasses are those composed only of metal constituents. This type of metallic glass-forming alloy is not as restricted to specific composition ranges as the metal–metalloid glasses are. They usually can be formed over a range of composition, often as large as 40 at.%, centered on 50–50 at.% for a binary glass. Typically, at least one of the constituents in these alloys is a transition metal, such as in the glass-forming systems Au–Co (Mader *et al.*, 1967), Au–Cu (Mader *et al.*, 1967), Ni–Nb (Ruhl *et al.*, 1967), Ni–Zr (Ray *et al.*, 1968), Ti–Zr–Be (Tanner and Ray, 1977), and U–Fe (Elliott *et al.*, 1980). Only a few metallic glasses are known to form from metal constituents not within a transition series. Examples of this alloy type are Mg–Zn (Calka *et al.*, 1977) and Ca–Al (Wagner *et al.*, 1969).

III. Structure of Metallic Glasses

Detailed x-ray studies and radial distribution function analyses have been used to describe the noncrystalline structure of metallic glasses. The models that have been proposed include

(1) microcrystalline or distorted crystalline structures for noncrystalline Nb–Ni (Ruhl *et al.*, 1967), Pd–Si, Ag–Cu, Fe–P–C, Ni–P (Wagner, 1969) Ge (Rudee and Howie, 1972), Ni–Pd–P (Maitrepierre, 1969), Fe–P–C (Lin and Duwez, 1969), and Cu–Zr (Waseda and Masumoto, 1975),

(2) randomly stacked close-packed layers for Ni–P alloys (Dixmier *et al.*, 1965),

(3) Bernal dense random packing (DRP) model based on positioning of idealized holes, and

(4) random network model.

Figure 1 shows schematic representations of atomic arrangements in the glass based on some of these models. Examination of this figure illustrates the large conceptual differences between them. Cargill (1970) has critically compared the experimental x-ray diffraction results in several alloy systems with the calculated patterns of various models. He notes many similarities among the experimentally determined pair distribution functions ($W(r)$) but finds poor agreement with calculated microcrystalline or layered structure interference functions. Polk (1970) proposed a Bernal DRP structure (1964) for Ni–P noncrystalline alloys. Cargill (1970) and others (Sinha and Duwez, 1972) have shown that this model agrees well with the $W(r)$ found in many different alloy systems. An example of this

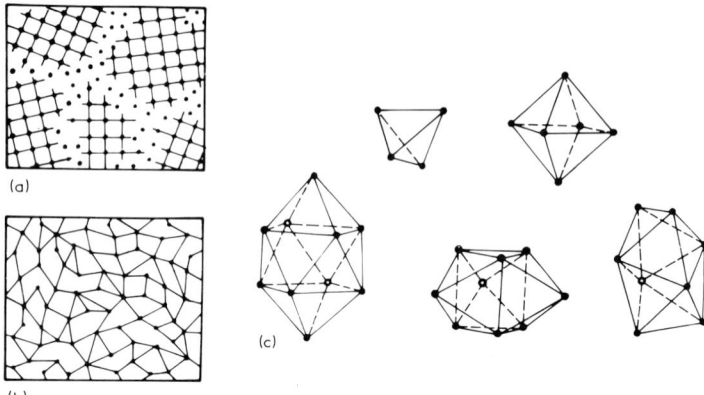

FIG. 1. Schematic representations of proposed models describing noncrystalline or glassy materials: (a) microcrystalline disorder model, (b) random network model, (c) Bernal DPP model based on these idealized hole structures. (Reprinted with permission of Chapman & Hall, Ltd., from Waseda et al. (1977).)

agreement is shown in Fig. 2, where the DRP model is compared to Ni–P glass. For covalently bonded noncrystalline solids, such as silicon and germanium, both random network models (Moss and Graczyk, 1969) and microcrystalline models (Rudee, 1971) appear to agree with experimental x-ray results.

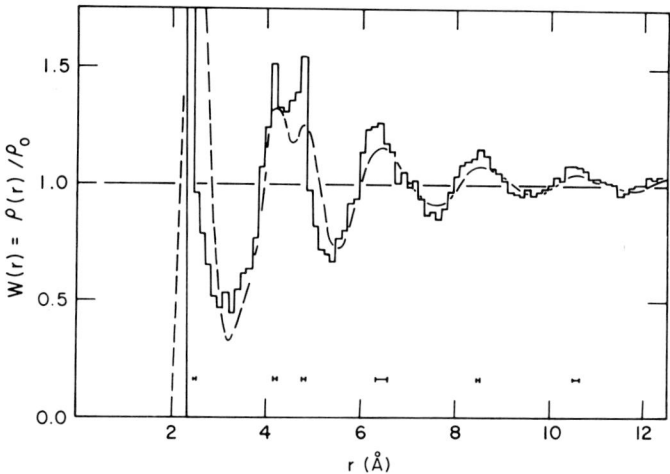

FIG. 2. Comparison of the pair distribution function ($W(r)$) calculated for the DRP structure with sphere of diameter $2\sigma = 2.42$ Å (histogram) with that measured in Ni–P, 76 at. % Ni (--). (Reprinted from Cargill (1970).)

IV. Theories for the Formation of Metallic Glasses

Although numerous approaches have been taken to explain glass-forming tendency in metallic systems, none can be used very successfully to predict, *a priori,* alloy compositions and processing conditions that will result in producing a metallic glass. Theories of metallic glass formation are grouped according to theoretical approach.

Geometric stabilization of the noncrystalline structure is an approach taken by numerous investigators to rationalize glass formation tendency. Polk (1970, 1972) has proposed an explanation for the increased tendency toward noncrystalline phase formation in metal–metalloid systems. His argument is based on the premise that the noncrystalline phases are based on a Bernal DRP structure (Bernal, 1964). Polk notes that a skeleton DRP structure of large metallic atoms would be stabilized, and densified, by the addition of smaller metalloid atoms in the large holes of the structure. These large holes occur frequently enough to allow for an optimum 20–30% metalloid composition that agrees well with experimental observations. Bennett *et al.* (1971) have noted that the free volume model predicts that monatomic glass-forming systems will have large repulsive pair potentials, a characteristic of noble and transition metals. Considering an alloyed amorphous system, Bennett *et al.* suggest that elements with smaller repulsive terms would stabilize the system, in a manner similar to Polk's model of hole filling. Sadoc *et al.* (1973) have calculated radial distribution function (RDF) patterns for two-sphere random dense-packed solids and have concluded that the addition of a smaller minor constituent, as in the case of Ni–P alloys, does not have merely a hole-filling role as Polk suggests. In addition, they provide a special local arrangement or environment for the larger constituent that further stabilizes the structure.

Nowick and Mader (1965) have shown, using a two-dimensional ball model, that two sizes of balls allow for a more dense and more stable amorphous assembly. This is in agreement with their experimental results on vapor-deposited mixtures of metals (Mader *et al.*, 1967; Mader, 1965). They have proposed that a minimum-sized difference of ~30% is necessary for amorphous phase formation. Figure 3 shows an example of pseudo-phase diagrams determined with the model that predicts the range of glass formation. In addition, Mader (1966) notes that noncrystalline vapor-deposited alloys are characterized by close-packed components and insolubility regions in the phase diagrams. However, it has also been pointed out (Turnbull, 1969) that many metal–metalloid systems have Goldschmidt radii (atomic radii of elements in 12-fold coordinates) that are nearly equal. Therefore, the criterion that the atoms in a metallic glass

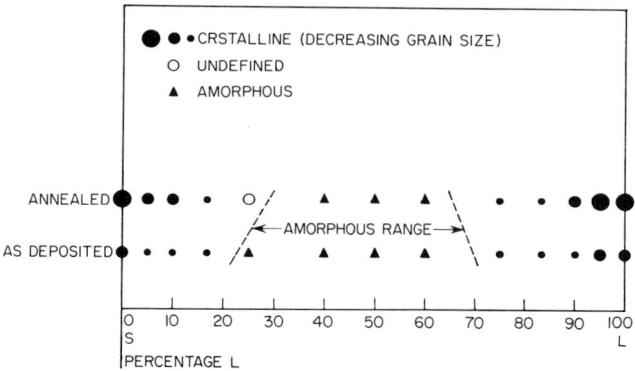

FIG. 3. Pseudobinary phase diagram determined for a hypothetical S–L alloy system. The range of composition where amorphous phases are obtained as vapor deposited and after annealing is shown. (Reprinted with permission of International Business Machines Corporation; copyright 1965 (Nowick and Mader, 1965).)

have a size difference of ~30% might not be a general rule for all metallic glasses. Perhaps it is applicable only to fully metallic systems, i.e., those with no metalloids. Geometric stabilization using nonuniform spheres has also been observed by Spaepen (1972). Chemical bonding (Chen and Park, 1973) and alloy composition (Bennett *et al.*, 1971) have also been presented as factors that have a substantial effect on glass formation.

A different explanation for glass-forming tendencies in metal–metal glass has been proposed by Johnson (1981). Johnson has used self-consistent-field, X-alpha, scattered-wave (SCF-Xα-SW) molecular-orbital calculations for clusters representing local molecular configurations in amorphous Cu–Zr alloys in the composition range $Cu_{60}Zr_{40}$–$Cu_{40}Zr_{60}$. The resulting electronic structures indicate the formation of a network of Zr–Zr bonds through the overlap of Zr d orbitals near the Fermi energy. This is promoted by the interaction with nearest-neighbor Cu d orbitals below the Fermi energy. Zr–Zr bonds bridged by neighboring Cu atoms are evident in the molecular-orbital contour map for a cluster representing $Cu_{60}Zr_{40}$ shown in Fig. 4. This figure illustrates the bonding configuration that may be responsible for the glass-forming abilities of copper–zirconium and similar metal–metal glass alloys. These calculations are in good quantitative agreement with recently measured photoelectron spectra for amorphous $Cu_{40}Zr_{60}$ and $Cu_{60}Zr_{40}$ alloys (Johnson, 1981), which show a predominantly Cu-like d band below the Fermi energy. This approach, similar to the Polk model, also leads to a skeleton formation, explaining glass-forming tendencies.

Calculations based on nucleation and growth kinetics to describe the

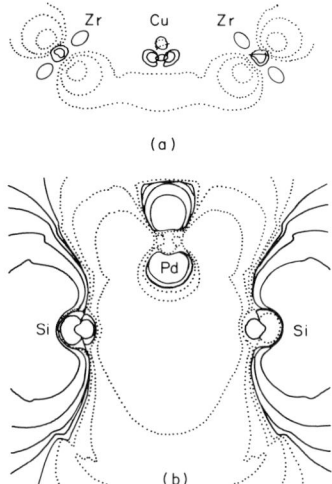

FIG. 4. Bonding configuration, proposed by Johnson, responsible for glass-forming tendency in Cu-Zr and similar metal-metal glass systems. Note the Zr-Zr bonds caused by d-orbital overlap. (Figure from Johnson, 1981.)

conditions of glass formation have been used with some success. For instance, Cohen and Turnbull (1961; Turnbull and Cohen, 1958, 1960) have developed kinetic criteria for amorphous phase formation. They selected minimum nucleation and growth rates for crystallization that must not be exceeded if an amorphous phase is to be retained. These criteria are that I, the nucleation rate, must be below 1/cm³sec; if I reaches this value the corresponding growth rate G must be less than 10^{-5} atom spacings/sec. This criterion requires that the activation energy for either nucleation or growth must be greater than $30RT_m$ (T_m = melting temperature, R = gas constant) in order to avoid crystallization. Cooling rates are not included in these calculations. Qualitatively, the 10^{-5} atom spacings/sec growth rate condition can be relaxed as cooling rate increases. This in turn reduces the $30RT_m$ to possibly $10RT_m$ in cases of extremely high cooling rates. This kinetic treatment is reviewed in more detail in (Rawson, 1967). By relating the activation energies G' and G'' to the bonding energies, and in turn to the heat of vaporization h_v, Cohen and Turnbull (1961) predict increased amorphous phase-forming tendencies with decreasing reduced melting temperature, τ_m (= kT_m/h_v, k = Boltzmann's constant). Thus, in agreement with experimental results, the theory of Cohen and Turnbull predicts increased amorphous phase-forming tendencies near eutectic compositions. Turnbull (1969) reconsidered the problem of noncrystalline formation and determined the variables that would suppress crystallization until the glass transition temperature T_g is reached. Turnbull's analysis is semiquantitative and is summarized as follows: The crystallization rate (either through the nucleation or growth rate) is suppressed by increased cooling rate, decreased liquid volume, and decreased heterogeneous nucleation sites—all laboratory-controlled variables. Crystallization rate is increased by crystal–liquid interfacial energy, increased reduced glass temperature ($= T_g/T_M$), and decreased fraction of atomic sites on the crystal surface where molecules can be attached—all material constants.

Another kinetic approach developed is that which determines the critical cooling rate necessary to prevent the formation of a detectable fraction of crystals from a melt. This approach, developed by Uhlmann (1972, 1977), is used to develop isothermal time–temperature transformation (TTT) and continuous cooling transformation (CCT) curves that define the processing parameters necessary for glass formation. Davies (1976) and co-workers have used this approach to calculate the critical cooling rates to form glass for the Pd–Si and Pd–Cu–Si systems. Tanner and Ray (1979) used the Uhlmann approach to calculate critical cooling rates for the Zr–Be and Ti–Be systems. An example of calculated TTT and CTT diagrams from work of Tanner and Ray is shown in Fig. 5. This is the only model that has been successfully applied to metal–metalloid and metal–

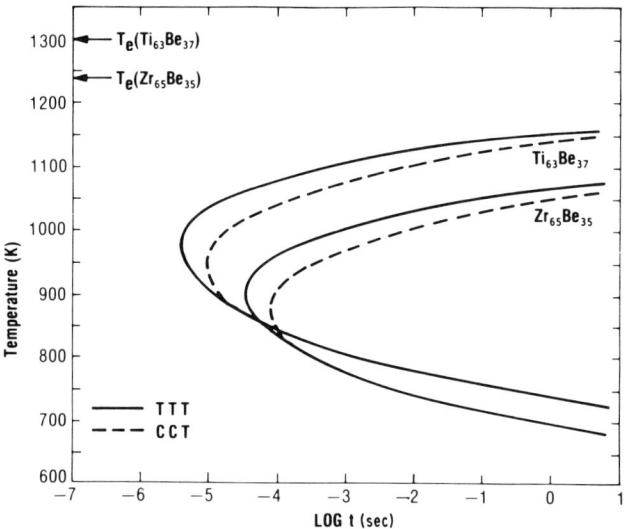

FIG. 5. Computed liquid-to-solid transformation diagrams for eutectic compositions of Zr–Be and Ti–Be with the critical free energy to crystal nucleation (ΔG^*) assumed to be independent of temperature. At $\Delta T_r = 0.2$, $\Delta G^* = 40kT$ and $55kT$ for $Ti_{63}Be_{37}$ and $Zr_{65}Be_{35}$, respectively. (Reprinted with permission from *Acta Metall.* **27**, L. E. Tanner and R. Ray. Copyright 1979, Pergamon Press, Ltd.)

metal glass-forming systems to date. This fact strongly suggests that glass formation tendency is a strong function of the system's nucleation and growth kinetics and is less dependent on geometric stabilization.

Other attempts have been made to predict amorphous phase formation, but these have concentrated on glassy oxide systems. Sarjeant and Roy (1968) developed a qualitative approach incorporating cooling rates. Rawson (1967) and Cahn (1969) have reviewed several different attempts at predicting amorphous phase formation. Many theories used for oxide glasses are not applicable to metallic systems because they rely on ionic and covalent bonding in their arguments. All of these approaches are rather qualitative, pointing out tendencies and noting variables that affect noncrystalline formation. Their ability to predict noncrystalline formation is limited.

V. Investigations of Properties and Behavior of Noncrystalline Phases

Studies of noncrystalline alloys have dealt primarily with (1) the properties (electrical, magnetic, calorimetric, and mechanical) and (2) the transformation to the crystalline state (devitrification). The following sections provide general descriptions of some of these properties of metallic glasses.

A. Magnetic and Electrical Properties

Much information has been collected on the magnetic and electrical properties of metallic glasses. Ferromagnetism has been observed in a variety of noncrystalline alloys (Duwez and Lin, 1967; Hasegawa, 1970). Figure 6 shows that the same ferromagnetic properties—in particular, core loss, impedance, permeability, and excitation power—of certain Fe-base glasses are superior to the conventionally used magnetic alloys. The Kondo effect, that is, a reversal in the temperature coefficient of electrical resistivity below room temperature, has been observed (Hasegawa and Tsuei, 1971; Lin, 1969) in many systems. It has also been reported that, because of the strong effect of structural disorder, the temperature coefficient of electrical resistivity can be minimized and reduced to zero at room temperature by carefully controlling alloy composition in some alloy systems. A review of magnetic and electrical properties of metallic glasses (Giessen and Wagner, 1972) gives additional information.

B. Chemical Stability

Another important property to consider is the chemical stability of metallic glasses in corrosive environments. It might be expected that, because these materials are free of gross heterogeneities (e.g., second

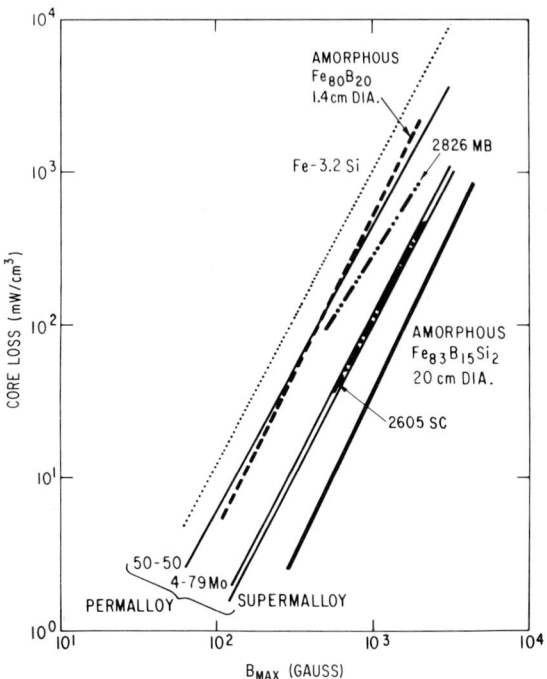

FIG. 6a. Comparison of some ferromagnetic properties of metallic glass alloys with conventional magnetic alloys. Core loss versus induction measured at 50 kHz for 50-μm-thick Permalloys and 30-μm-thick $Fe_{83}B_{15}Si_2$ glass measured as a 20-cm-diameter toroid $Fe_{80}B_{20}$, Metglas 2605 SC, and Metglas 2826 MB. (Reprinted with permission of the General Electric Company, from Luborsky *et al.* (1979).)

phases, grain boundaries, and dislocations), they will be more corrosion resistant than their crystalline counterparts. An example of the better corrosion resistance in 1 M NaCl of Fe–Ni–Cr–P–B glass compared to conventional corrosion-resistant alloys is shown in Fig. 7. A similar result has been shown in Ni–P (Turnbull, 1977), Fe–Cr–P (Hashimoto and Masumoto, 1976), Fe–Cr–Ni–P (Naka *et al.*, 1976; Diegle and Slater, 1976), Fe–Ni–P (Diegle and Slater, 1976), and Co-based systems (Naka *et al.*, 1978). Naka and co-workers (1976, 1978) have compared the general corrosion resistance of Fe–Cr and Fe–Cr–Ni metallic glasses (Cr ≥ 8%) containing various metalloids (P, C, B, and Si) to conventional Fe–Cr and Fe–Cr–Ni stainless steels (Types 304 and 316L). Metallic glasses were found to be superior in NaCl, HCl, and $FeCl_3$ environments. Pitting corrosion resistance was also observed to be superior to the conventional Fe–Cr–Ni alloys in NaCl and H_2SO_4 potentiostatic anodic polarization tests. Naka *et al.* (1978) concluded that the improved resistance was due to the very rapid formation of a highly protective, uniform passive film com-

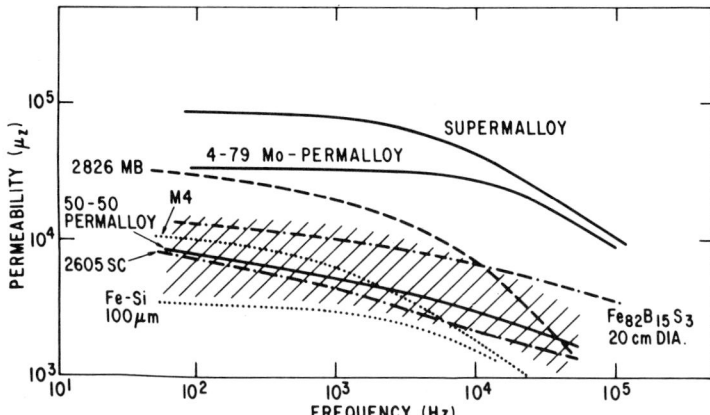

FIG. 6b. Comparison of some ferromagnetic properties of metallic glass alloys with conventional magnetic alloys. Impedance permeability measured at B = 100G versus frequency for various Permalloys, Fe—3¼% Si, Metglas® 2605 SC, Metglas® 2826 MB, and $Fe_{82}B_{15}Si_3$ glass 20-cm-diameter toroid. (Reprinted with permission of the General Electric Company, from Luborsky et al. (1979).)

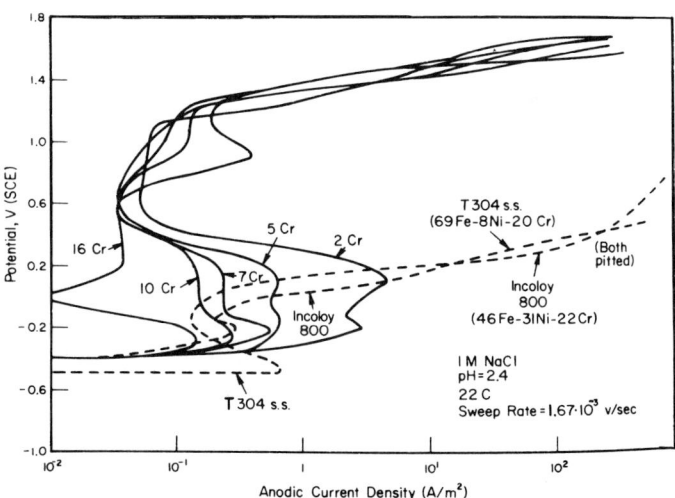

FIG. 7. Potentiodynamic polarization curves for Fe–Ni–Cr–P–B metallic glasses with the indicated Cr contents and for crystalline T304 stainless steel and Incoloy 800 alloys. Expansion of the transpassive region to larger potential demonstrates the better corrosion resistance of these metallic glasses. 1 M NaCl, pH = 2.4, 22°C, sweep rate = 1.67 × 10^{-3} V/sec. (Reprinted with permission of the National Association of Corrosion Engineers, from Diegle (1979).)

posed primarily of chromium oxyhydroxide. More recently, Turn (1979), in a study on the influence of structure on the corrosion behavior of glassy Cu–Zr alloys, showed that structure has a much weaker effect on corrosion resistance than alloy chemistry. He suggests that this could be true for the more complicated alloy glasses studied by others. Therefore, analogous to crystalline materials, both microstructure and alloy chemistry affect the corrosion resistance of metallic glasses but to different degrees in different alloy systems.

C. MECHANICAL PROPERTIES OF METALLIC GLASSES

The study of the mechanical properties of metallic glasses trailed considerably the discovery of noncrystalline metallic solids. This delay is directly related to the development of new splat-quenching techniques, which could produce metallic glasses in a geometry more suitable to mechanical testing. Chen and Turnbull (1968) were the first to report on the mechanics of deformation of metallic glasses by studying an Au–Ge–Si alloy at temperatures close to T_g. Following this initial investigation, the mechanical properties of metallic glasses, often Pd–Si alloys, were explored at temperatures much lower than T_g, generally 20°C (Masumoto and Maddin, 1971; Chen and Wang, 1970; Leamy et al., 1972). The mechanical properties observed for metallic glasses primarily composed of Fe and Ni were first examined by Chen and Polk (1974), with additional studies on similar alloys continuing to the present (Davis et al., 1976; Polk and Pampillo, 1973; Pampillo and Polk, 1974; Davis et al., 1976; Chou et al., 1977).

In general, the features of tensile deformation and fracture of metallic glasses at temperatures close to 20°C have been consistent for the different alloy systems. Comprehensive reviews of this area have been made (Davis, 1976; Pampillo, 1975; Gilman, 1975); however, a brief summary of the details will be presented.

Contrary to the behavior of oxide glasses, metallic glasses do not exhibit extreme brittleness or size effects when tested in uniaxial tension. Compared to crystalline metals, they exhibit remarkable strength when tested in tension, often approaching one-half their theoretical strength. Crystalline metals generally reach strength levels of only one-thousandth to one ten-thousandth of their theoretical strength. Metallic glasses are able to withstand large amounts of elastic deformation. Elastic strain of 1 is seen in bending prior to plastic deformation. This large value of tolerable elastic strain contrasts sharply with that attainable in oxide glasses, ~0.01. The stress–strain curve obtained during a uniaxial tensile test is approximately linear at room temperature. Deviation from linearity is observed only at high stress levels as shown in Fig. 8 (Masumoto and Maddin, 1971). Tensile strain to failure for metallic glasses is ~2%; how-

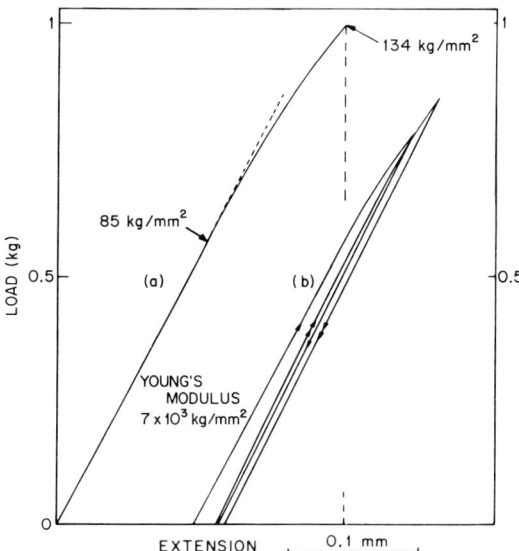

FIG. 8. Typical stress–strain curves obtained for $Pd_{80}Si_{20}$ glass. (Reprinted with permission from *Acta Metall.* **19**, T. Masumoto and R. Maddin. Copyright 1971, Pergamon Press, Ltd.)

ever, careful measurement of actual plastic strain during tensile deformation as shown in Fig. 9 could not be observed within the accuracy of the experimental technique. Plastic strain during tensile deformation was measured to be <0.015% (Megusar *et al.*, 1976). Therefore, it can be

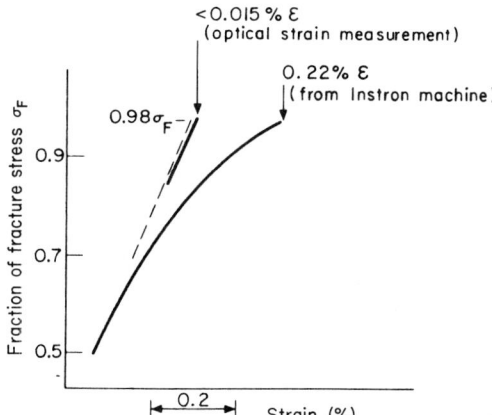

FIG. 9. Portion of stress–strain curve for $Cu_{60}Zr_{40}$ close to the fracture stress showing that the actual strain at failure is extremely small. Very sensitive techniques must be used to measure it. (Reprinted with the permission of MIT Press, from Megusar *et al.* (1976).)

concluded that, in comparison to most crystalline metallic alloys, metallic glasses fail in tension in a macroscopically brittle fashion.

Plastic deformation at temperatures far below the glass transition is completely inhomogeneous. Shear or deformation bands are always observed on the surface of the glass. An example of this is shown in Fig. 10. This large degree of inhomogeneity suggests that metallic glasses do not work-harden. Deformation is confined to bands that are oriented at specific angles to the loading axis, depending on the geometry of the specimen (Davis, 1976; Megusar, 1977). A number of interesting observations have been made regarding the nature of the shear bands. The localized nature of the shear bands is indicative of an absence of strain

FIG. 10. Shear or deformation bands formed on the tension side of a bent metallic glass strip. Lance A. Davis, "Strength, Ductility and Toughness," *Metallic Glasses,* American Society for Metals, 1976, p. 192.

hardening. Argon (1975) has suggested that strain softening is necessary for strain localization. The absence of work hardening was further confirmed in a test by Pampillo and Chen (1974) in which a sample of $Pd_{77.5}Cu_6Si_{16.5}$ glass was compressed to initiate deformation and then repolished. On reloading after polishing, as shown in Fig. 11, shear was observed on the same shear bands. In another test it was observed that shear along bands produced by bending was reversed by bending in the opposite direction (Pampillo, 1975). Such observations are all consistent with an absence of strain hardening.

Polk and Turnbull (1972) have suggested that destruction of short-range ordering is responsible for the strain concentration. It has been shown in a number of glasses (Pampillo and Chen, 1974; Pampillo, 1972; Chen et al., 1973) that shear bands are preferentially etched. This indicates that there is some difference in the chemical nature of material within the shear bands. Evidence of this is shown in Fig. 12. Further, on annealing at temperatures near but below T_g, the etching sensitivity disappears, and on reloading, shear is observed to occur on new bands even though there is

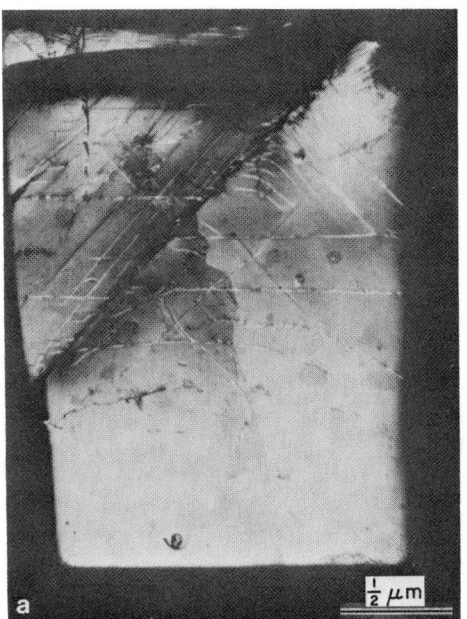
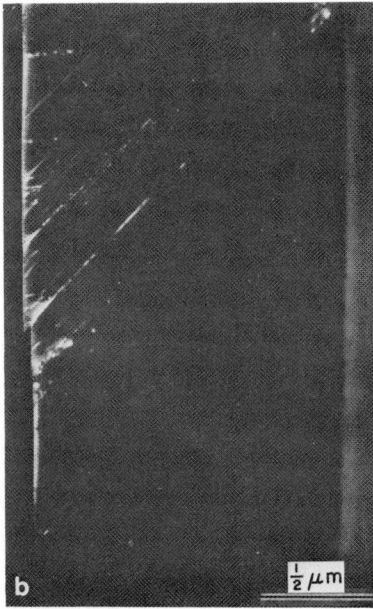

FIG. 11. Demonstration of the lack of work hardening in $Pd_{77.5}Cu_6Si_{16.5}$ glass when deformed in compression. (a) Shear bands produced after 7% compressive deformation. (b) Same area as in (a) after polishing and straining an extra 5%. (Reprinted with permission of Elsevier Sequoia, Ltd., from Pampillo and Chen (1974).)

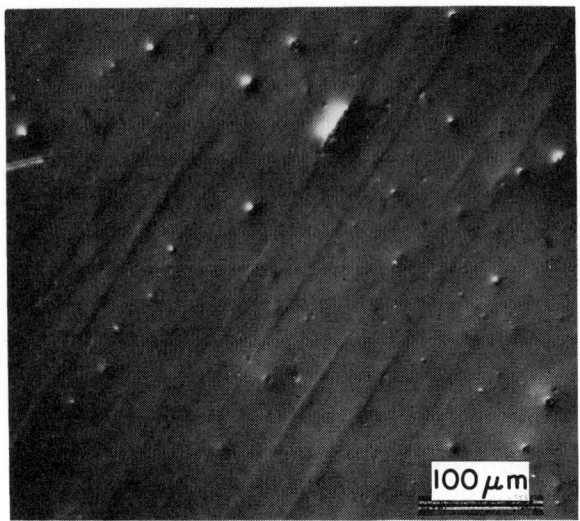

FIG. 12. Selective etching of shear bands produced by compressive deformation of $Pd_{77.5}Cu_6Si_{16.5}$ glass. (Reprinted with permission of Elsevier Sequoia, Ltd., from Pampillo and Chen (1974).)

no apparent change in the undeformed bands. As annealing will tend to increase order, it appears that deformation serves to produce a more disordered structure that is weaker in nature, and this provides an easy path for subsequent deformation.

The final fracture occurs on the shear band planes and the fracture surface is characterized by a featureless zone and a veined zone. An example of a typical fracture surface produced in uniaxial tension is shown in Fig. 13. It would appear that cracks propagate from voids or hard inclusions and expand in the plane of the shear until the sample is held together only by a network of ligaments or veins. The veinlike pattern has been interpreted as indicating that the fracture mechanism is a variant of the Taylor instability observed in viscous materials (Argon and Salama, 1976). On a microscopic scale fracture should be considered ductile because it consists of a final shear rupture of the ligaments in the deformation bands.

Davis and Kavesh (1975) studied the fracture of $Pd_{77.5}Cu_6Si_{16.5}$ under hydrostatic pressure to provide additional details for understanding deformation of metallic glasses. Figure 14 shows their observation of a large increase in the size of the featureless zone. They noted that cracking was initiated at the surface and concluded that internal nucleation of cracks at voids and inclusions involves a dilatational component that is suppressed by hydrostatic pressure.

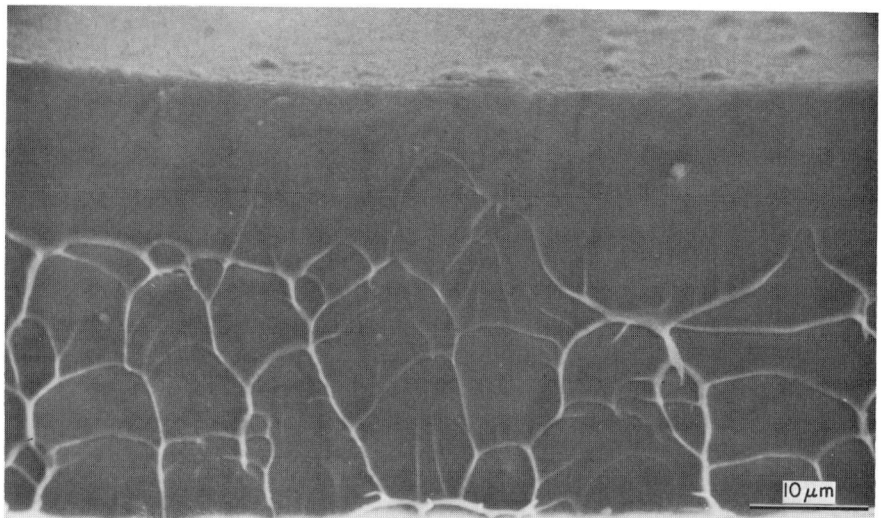

FIG. 13. Tensile fracture surface of as-splat $Cu_{60}Zr_{40}$ glass (Reprinted from Freed and Vander Sande (1979).) Testing was done at room temperature.

VI. Theoretical Mechanisms of Deformation and Fracture

The theoretical mechanisms describing deformation and fracture in metallic glasses fall into a few general categories. There is considerable disagreement among researchers concerning not only the fine points of deformation but also the basic principles of the deformation mechanism. Unfortunately, with current experimental results, it is difficult to prove absolutely the credibility of one model over the others to describe deformation and fracture in metallic glasses accurately.

One view of deformation in metallic glasses is that it is the result of a dislocation mechanism. The basis for this viewpoint is that the slip or deformation bands are the elastic discontinuity existing between sheared and unsheared material, hence, a Volterra dislocation (Davis, 1976). Li (1976) has modeled an amorphous solid as a crystal that has a lattice of closely spaced dislocations within it, depicted schematically in Fig. 15. The distortions in crystal symmetry resulting from the presence of these dislocations cause the solid to appear to have no three-dimensional symmetry. The inhomogeneous shear, which is observed in metallic glasses as the slip bands, is caused by the movement of dislocations within the "lattice" at a critical shear stress. Gilman (1973, 1975) has also proposed a dislocation model for the deformation of glasses that have sufficient viscosity for the viscous relaxation time to be much greater than the time necessary for plastic deformation. The Burgers vector is defined as one

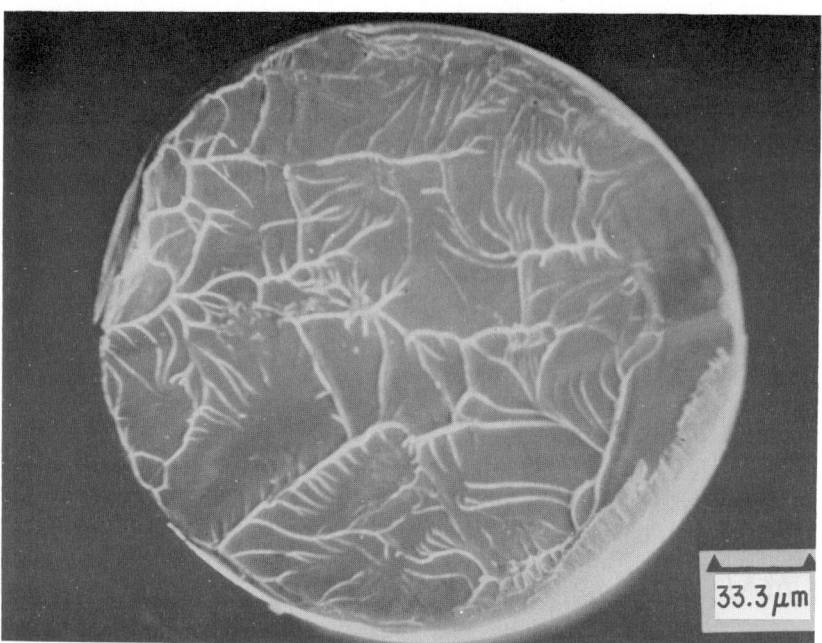

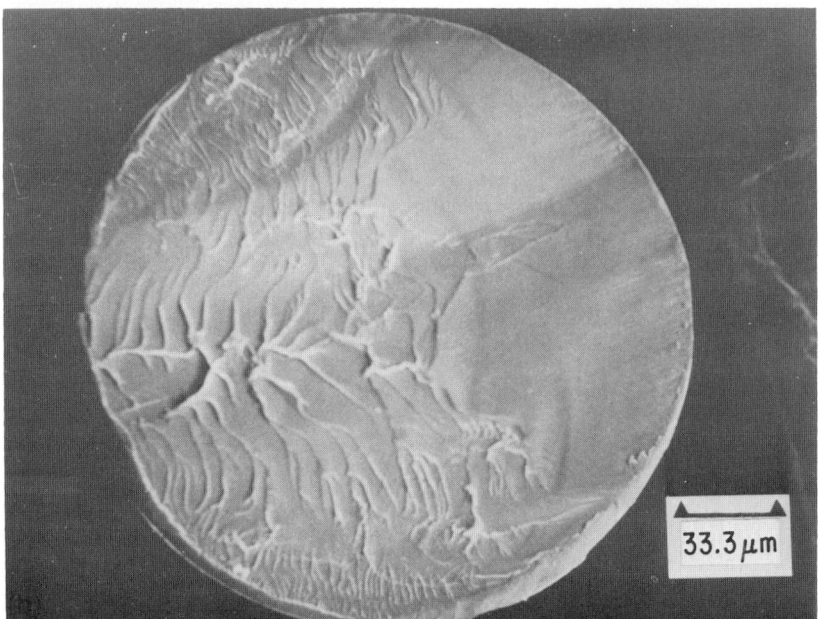

Fig. 14. Tensile fracture surface of $Pd_{77.5}Cu_6Si_{16.5}$ glass produced at (a) 1 atm and (b) 6.9 kbar. In both photos the featureless zone is an arc on the right of the micrograph (the "brush" markings in (a) were caused by handling). The fracture surface slopes in both from left to right at ~45°. (Reprinted with the permission of Chapman & Hall, Ltd., from Davis and Kavesh (1975).)

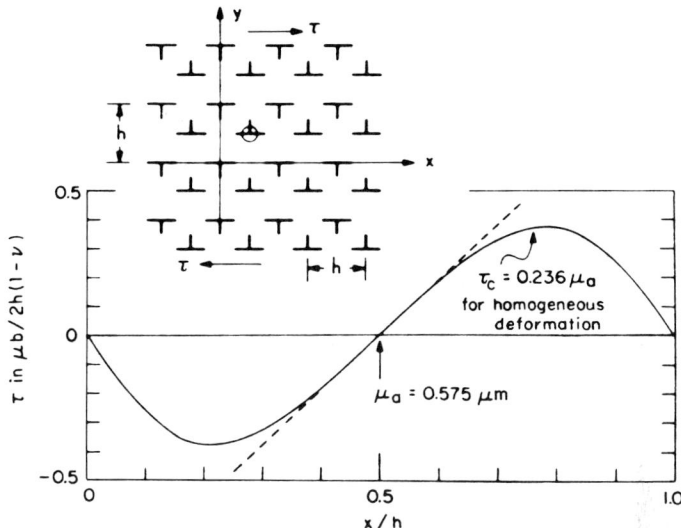

FIG. 15. Dislocation lattice model for a metallic glass. The figure shows the calculated stress–strain curve based on the interaction of the dislocations resulting from an applied shear stress τ. (Reprinted from Li (1976), by courtesy of Marcel Dekker, Inc.)

that fluctuates in direction and magnitude along the dislocation line but has an average value correlated to some structural parameter of the glass, such as the average nearest-neighbor distance. This is shown schematically in Fig. 16. It is postulated that at a critical shear stress these dislocations move through the glass and cause the deformation bands observed.

The dislocation models proposed by Li and Gilman discuss only deformation prior to fracture or crack propagation. One must assume then that fracture initiates when the material reaches a critical cross-sectional area, or stress, at some stress concentration. No explanation is proposed within the framework of these theories for the details of the observed fracture surface morphology.

The remaining proposed deformation mechanisms are more closely related to the actual fracture of the material during tensile deformation. Their strong points are the ability to predict fracture surface morphology. This is a realistic approach, since at temperatures $\ll T_g$ plastic deformation in tension occurs in association with the fracture in the material. For example, no plastic deformations (i.e., slip bands) were found to form prior to stress levels of 97% of tensile fracture stress (σ_F) in $Pd_{80}Si_{20}$ when tested in tension, and then slip bands form and grow only in association with a propagating crack (Megusar et al., 1977). Leamy et al. (1972) have developed a model based on the concept that viscous deformation

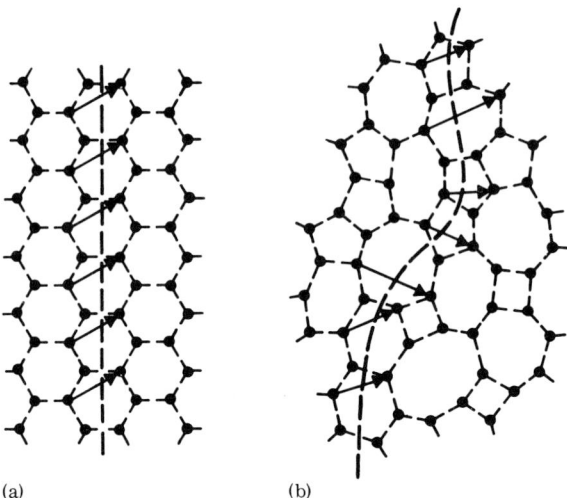

FIG. 16. Schematic comparison of dislocation lines in (a) crystalline and (b) amorphous alloy. Note the change in direction and magnitude of the Burgers vector shown as an arrow. (Reprinted from Gilman (1973).)

proceeds in the slip bands because of the decreased viscosity of the material in these bands. This viscosity decrease is brought about by adiabatic heating, which occurs by the dissipation of plastic work as the glass deforms. After the material reaches a critical stress, the low-viscosity materials fail catastrophically via viscous flow. Pampillo and Reimschessel (1974) use a similar approach in their pseudocleavage mechanism; however, they propose that the material in the slip bands is chemically weakened by some initial slip. Cracks nucleate in this chemically weakened material and final fracture occurs when adiabatically heated material fails by viscous flow.

Spaepen and Turnbull (1974) describe flow and fracture as the result of a viscosity lowering mechanism but not of adiabatic heating. The hydrostatic, stress-assisted dilation, which causes an increase in the free volume of the metallic glass, is experienced in the slip band because of stress considerations. This increase in free volume causes a dramatic decrease in the viscosity of the glass adjacent to such stress concentrators according to free volume models of diffusive transport (Turnbill and Cohen, 1970). This results in essentially a fluidlike layer sandwiched between two rigid, solid layers. On reaching a critical stress level, fracture is initiated at a free surface, and the fluidlike layer fails.

Spaepen (1975) refined the free volume model by adding the formalism of perturbation theory (Mullens and Sekerka, 1964) to model the interface

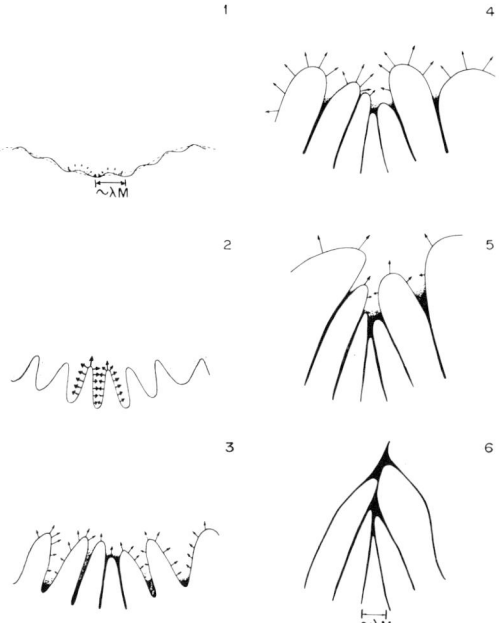

FIG. 17. The development of the "vein" pattern on the tensile fracture of a metallic glass from an initial perturbation λ_M. Shading is qualitatively proportional to density in the glass. (Reprinted with permission from *Acta Metall.*, **23**, F. Spaepen. Copyright 1975, Pergamon Press, Ltd.)

motion in a Laplacian field of the local density gradient at a stress concentration. The final result is a mathematical relationship between fracture stress and fracture-surface morphology. The development of the typical vein fracture pattern as envisioned by Spaepen is schematically shown in Fig. 17. The spacings between veins calculated by Spaepen were found to compare reasonably with experimental measurements.

Argon and Salama (1976) have modeled fracture in pure ductile glassy media as a nonlinear fluid capable of deforming and flowing as a fluid. The rupture of the fluid is discussed in terms of a fluid meniscus instability, which results in a relationship predicting the fracture toughness of the glass and the fracture-surface morphology. Figure 18 shows how Argon and Salama visualize the development of an instability.

VII. The Effect of Crystallization on Mechanical Properties

The effect of crystallization, in particular crystallization brought about by isothermal aging at temperatures below T_g, on the room-temperature mechanical properties of metallic glasses is an area that has received little

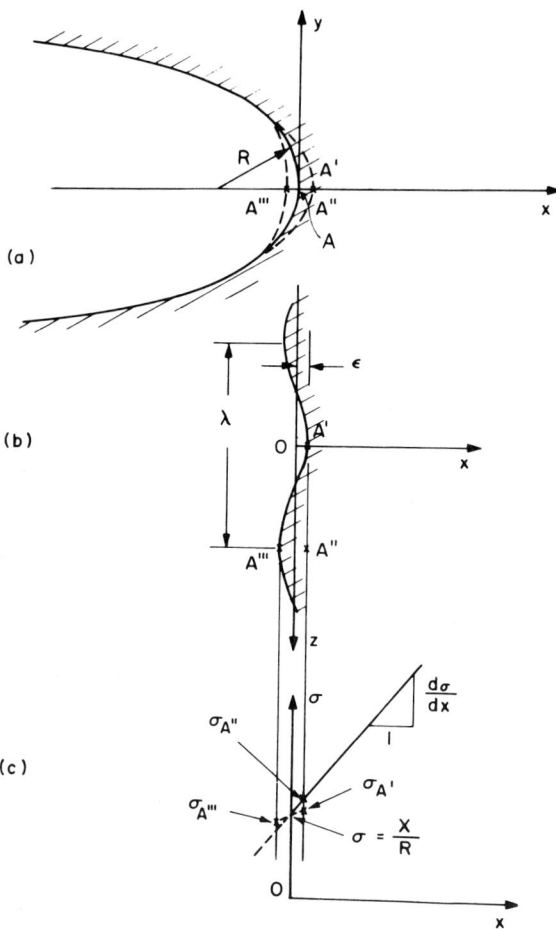

FIG. 18. Development of an instability at a meniscus or blunt crack: (a) Side view of blunted crack at impending instability. (b) Top view showing wavelength λ and strain ε. (c) Alteration of negative pressure distribution in the crack. (Reprinted with the permission of Elsevier Sequoia, Ltd., from Argon and Salama (1976).)

attention. This topic is extremely important if metallic glasses are considered for structural applications where high-temperature excursions are expected. In addition, it is interesting to determine if room-temperature mechanical properties can be improved or enhanced by isothermal aging. Analogous to the effect of very fine precipitates in crystalline solids (Kelly and Nicholson, 1961) or glass ceramics, it is conceivable that metallic glasses could experience similar changes in mechanical properties.

The first experiments concerning the effect of isothermal aging at temperatures below T_g and crystallization on the properties of $Pd_{80}Si_{20}$ metallic glass were conducted by Masumoto and Maddin (1971). As shown in Fig. 19, they observed that σ_F can be increased by isothermal aging in a particular temperature range. Transmission electron microscopy indicated that the increase in σ_F corresponded to the "incipient stages of crystallization." After crystallization has progressed to 1–2%, σ_F was found to decrease dramatically. The conclusion that room-temperature σ_F can be increased as a result of aging is supported by other observations. In high-temperature tensile tests, σ_F increases significantly at temperatures where crystallization just begins after a short tempering period. As the result of higher tempering temperature, the high-temperature σ_F decreases dramatically, presumably after the amount of crystalline material in the metallic glass exceeds 2%. Masumoto and Maddin's observations indicate that isothermal aging and the intervening devitrification noticeably affect the mechanical properties of $Pd_{80}Si_{20}$ metallic glass. The amount of crystalline

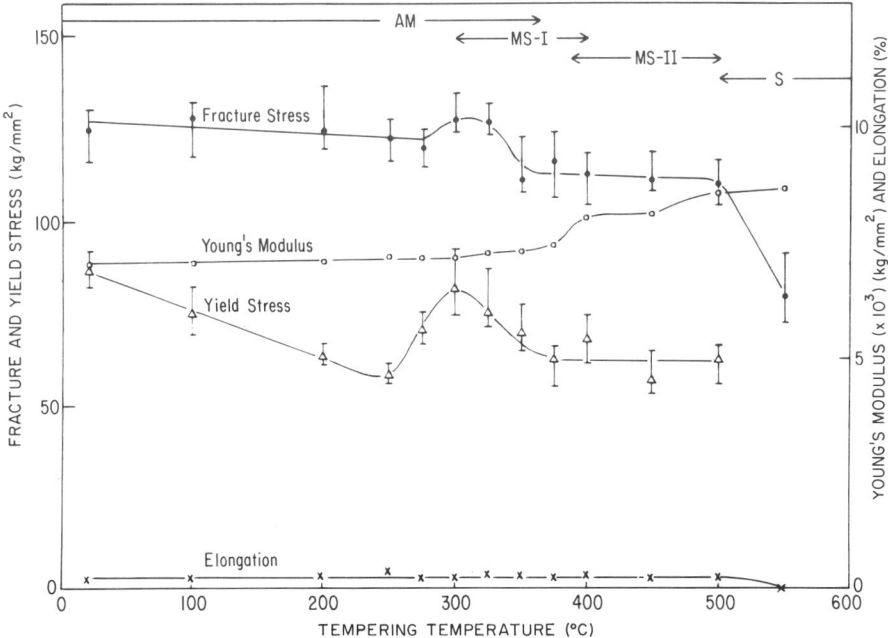

FIG. 19. Mechanical properties of $Pd_{80}Si_{20}$ glass are found to change as a function of isochronal aging at different temperatures. This effect has been related to the volume of crystalline phases produced during devitrification. Increases in tensile yield and fracture stress are maximized at the very early stages of glass devitrification. (Reprinted with permission from *Acta Metall.*, **19**, T. Masumoto and R. Maddin. Copyright 1971, Pergamon Press, Ltd.)

material in the metallic glass determines if properties improve or deteriorate.

Megusar et al. (1976) examined the effects of isothermal aging and the resulting devitrification on the room-temperature σ_F of $Pd_{80}Si_{20}$ and $Cu_{60}Zr_{40}$ metallic glasses. This study also indicated an increase in room-temperature σ_F as a function of aging time; however, no actual determination of the structure of these specimens was provided. Tables I and II summarize the data produced in this study. The fracture-surface morphology was also examined as a function of aging time. Distinct changes were noted in the aged $Cu_{60}Zr_{40}$ fracture-surface morphology relative to that of as-splat $Cu_{60}Zr_{40}$. These changes are similar to those shown in Fig. 20 for aged $Cu_{56}Zr_{44}$ as glass. Interestingly, no changes were observed in the fracture-surface morphology of aged $Pd_{80}Si_{20}$. This study indicates that the mechanical properties of both categories of metallic glass alloys, i.e., either metal–metalloid or metal–metal alloy, are affected by isothermal aging and devitrification. Devitrification seems to be more influential, in fact, in the metal–metal glasses as is indicated by the fracture-surface analysis conducted by Megusar et al. (1976).

Chou and Spaepen (1975) examined the effects of phase separation and crystallization of $Pd_{74}Au_8Si_{18}$ metallic glass on the fracture strength, microhardness, and fracture-surface morphology following aging at 383°C, and also in microhardness after aging at 392°C. Increases in both room-temperature σ_F and microhardness, shown in Figs. 21 and 22, respectively, seem to coincide with early detection of crystallization using differential scanning calorimetry (DSC). The accuracy of determining crys-

TABLE I

TENSILE PROPERTIES OF AMORPHOUS AND ANNEALED $Pd_{80}Si_{20}$ [a]

A.	Annealed at 200°C					
	Annealing time/crystal size (min/Å)		Amorphous	60/20	300/40	6000/80
	Fracture strength	(kg/mm^2)[b]	127	131	135	136
		(ksi)	180	186	191	193
B.	Annealed at 300°C					
	Annealing time/% transferred to MS-1 phase (min/%)		100/25	300/50	600/80	1000/100
	Fracture strength	(kg/mm^2)[b]	138	135	132	125
		(ksi)	196	191	187	177

[a] Reprinted with the permission of MIT Press, from Megusar et al. (1976).

[b] Values are based on a minimum of four tests.

TABLE II

TENSILE PROPERTIES OF AMORPHOUS AND ANNEALED $Cu_{60}Zr_{40}$ [a]

A.	Annealed at 425°C				
	Annealing time/% transferred to equilibrium phase (min/%)		Amorphous	15/30	60/50	150/80
	Fracture strength	(kg/mm^2) [b]	125	140–150	140–150	85–95
		(ksi)	178	200–213	200–213	121–135
	Apparent yield strength as fraction of σ_F		$0.5\sigma_F$	$0.6\sigma_F$	$0.7\sigma_F$	$1.0\sigma_F$
B.	Amorphous $Cu_{60}Zr_{40}$					
	Selected fraction of fracture strength σ_F		0.87	0.91	0.96	0.98
	Strain (%) measured by:					
	Instron machine		0.1	0.12	0.17	0.22
	Optical method		<0.015	<0.015	<0.015	<0.015

[a] Reprinted with the permission of MIT Press, from Megusar et al. (1976).
[b] Values are based on a minimum of four tests.

tallinity using DSC was not verified by other experimental observations, such as transmission electron microscopy. Chou and Spaepen reported a maximum in microhardness at 30% crystallinity; and after 50% crystallinity, both room-temperature microhardness and σ_F decreased rapidly. They determined by fractography that fracture, in partially crystallized material, was initiated in the interior of the specimens, very likely as the result of cracks in the crystallites.

A thorough study of the effect of crystallization on mechanical properties has been completed by Freed and Vander Sande (1979, 1980) in Cu–Zr glasses. The microhardness and fracture stress results for $Cu_{46}Zr_{54}$ are shown in Fig. 23. Glasses of other compositions in this system, $Pd_{80}Si_{20}$ (Masumoto et al., 1976; Chen and Coleman, 1976), $Fe_{80}P_{13}C_7$ (Chen and Coleman, 1976), and $Pd_{74}Au_8Si_{18}$ (Chou and Spaepen, 1975) alloys also show a similar variation of hardness and fracture stress with annealing time.

Although Freed and Vander Sande observed no crystalline phases using transmission electron microscopy of samples aged 240 min just below T_g in $Cu_{46}Zr_{54}$, it is clear from the hardness and fracture stress results that some form of structural relaxation has occurred, producing a small peak in both of these properties at approximately 75 min. X-ray studies indicate (Masumoto and Maddin, 1971; Masumoto et al., 1976) that during this period short-range ordering of atoms resulting in a relaxation of internal

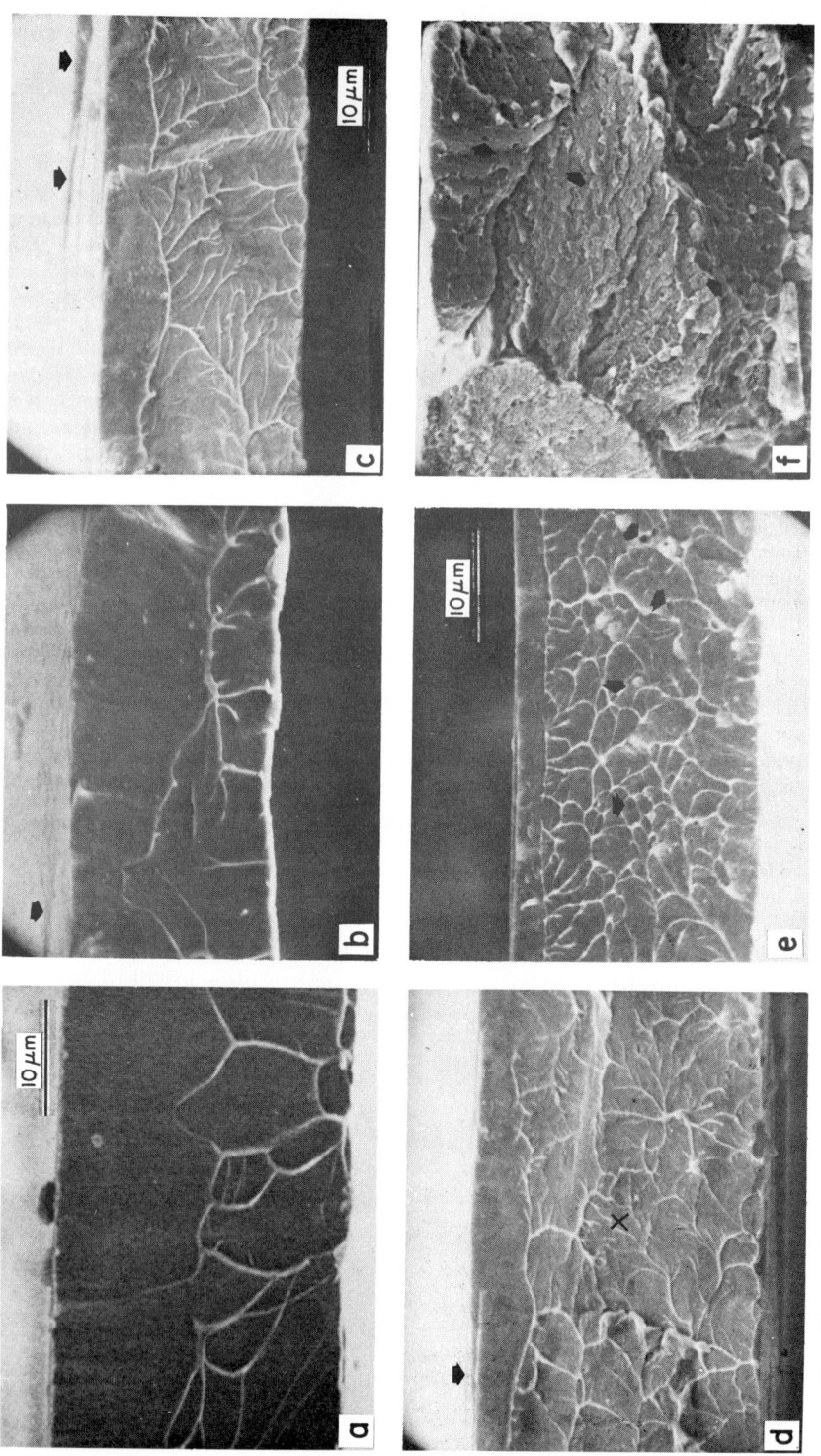

Fig. 20. Tensile fracture surface morphology of $Cu_{56}Zr_{44}$ glass as a function of aging time at 440°C. Note how the extent of the featureless zone (top of each micrograph) and the "cell" size of the vein zone decrease as time at temperature increases. (Reprinted with permission from *Acta Metall.* **28**, R. L. Freed and J. B. Vander Sande. Copyright 1980, Pergamon Press, Ltd.)

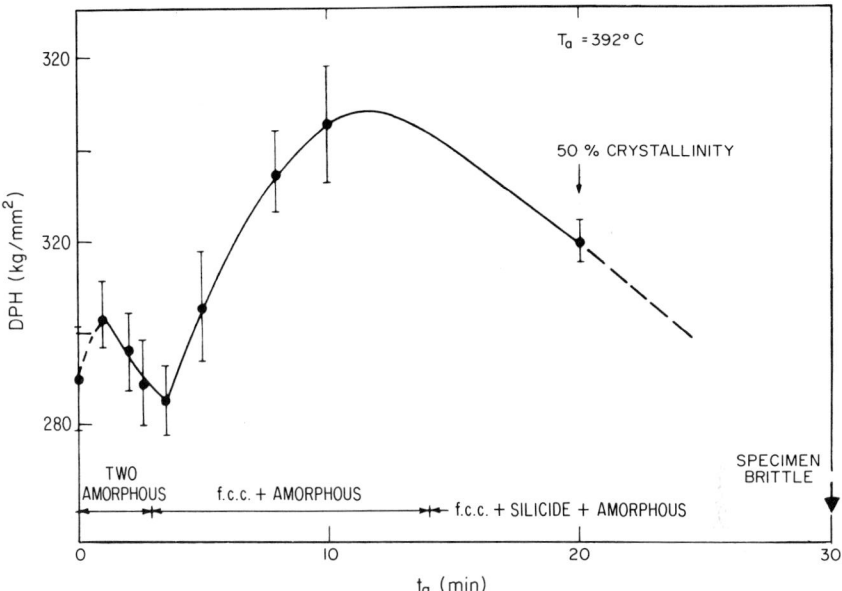

FIG. 21. The effect of aging on $Pd_{74}Au_8Si_{18}$ at 392°C on microhardness. Note that this glass phase separates. The phases present as a function of aging time are shown. (Reprinted with permission *Acta Metall.*, **23**, C. P. Chou and F. Spaepen. Copyright 1975, Pergamon Press, Ltd.)

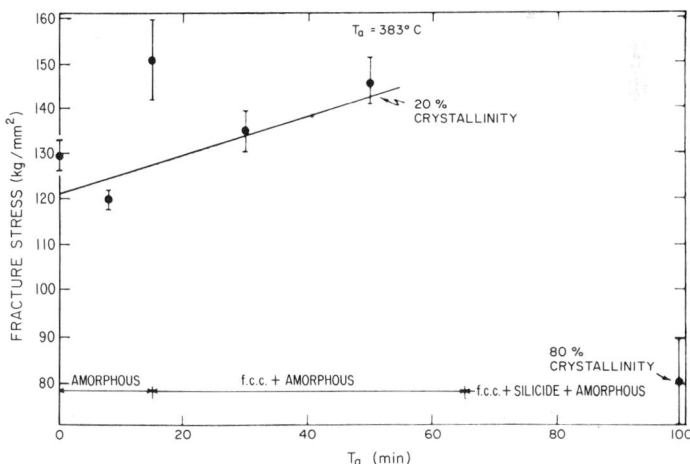

FIG. 22. The effect of aging on $Pd_{74}Au_8Si_{18}$ at 383°C on tensile fracture stress. The phases present as a result of aging are shown. (Reprinted with permission from *Acta Metall.*, **23**, C. P. Chou and F. Spaepen. Copyright 1975, Pergamon Press, Ltd.)

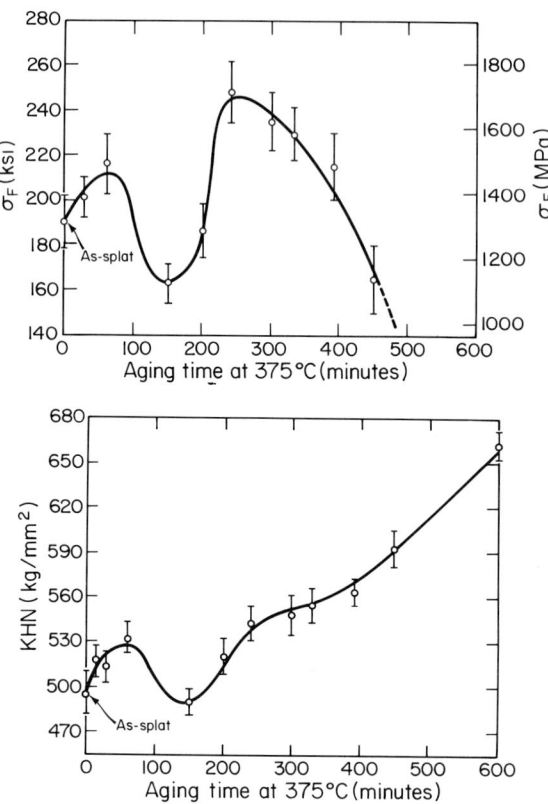

FIG. 23. The effect of aging at 375°C on mechanical properties on $Cu_{46}Zr_{54}$ glass is shown. Crystallization was detected in this glass after an incubation time of ~240 min. Note the correlation of mechanical property changes with the onset of crystallization. (a) Tensile fractures stress (σ_F). (b) Knoop microhardness (KHN) measured with a 50-g load. (Reprinted from Freed and Vander Sande (1979).)

strains occurs. Chen's (Chen and Coleman, 1976; Chen and Lo, 1976) observations of the relaxation spectrum of various glasses indicate that a wide range of different configurational rearrangements occur. He speculates that these may fall into two categories: local atomic groupings and cooperative structural rearrangements. Neither of these has been directly confirmed. Freed and Vander Sande concluded that the initial increase in σ_F and KHN in Cu–Zr glasses is due to such relaxations. The combined effect of relaxation and of nucleation of the crystalline phases produces the decreases at aging times of more than 60 min and less than 150 min near T_g. For times exceeding 150 min, the hardness continually rises because of the increasing presence of the hard crystalline phases. The fracture stress first increases before decreasing again at times in excess of 240 min. The initial strengthening of the material must be due to the presence of a fine dispersion of crystallites in the amorphous phase. The strengthening mechanism may well be similar to that found in a metal containing a fine dispersion of precipitates. The shear resistance is increased by the need of the shear front to circumvent the nondeformable crystallite; thus the observed increase in σ_F. Because they are dispersed and not a continuous phase, as shown in Fig. 24, the crystallites will reduce the load-carrying area and play a role in crack nucleation as well. Based on the fracture surface morphology, Freed and Vander Sande concluded that fracture initiated with void formation due to local stress concentrations at the crystallites. Figure 25 shows evidence for this on the fracture surface. Eventually, as the crystallites increase in size and number, the adverse effects of their presence in the material overwhelm the strengthening mechanism and the material strength deteriorates.

It is clear that the strength properties of metallic glasses may be enhanced through thermal treatments that produce a small degree of crystallization. Overaging, however, will cause a rapid degeneration of strength.

VIII. Metallic Glasses at High Temperature

In the limited work that has been done at temperatures approaching T_g, a transition is observed to a homogeneous mode of deformation as indicated by the results of Masumoto and Maddin (1971). The effect of this change in deformation mode on the stress–strain behavior for $Pd_{80}Si_{20}$ glass is shown in Fig. 26. At increasing temperatures the material becomes highly strain rate dependent. Stress–strain curves show a transition from the near linear behavior observed at low temperatures to curves showing substantial nonlinearity due to the anelasticity of the glass, accompanied by a substantial decrease in fracture stress. At sufficiently low strain rates and high temperatures, extensive homogeneous deformation is observed. A load maximum occurs followed by necking, which results in considerable macroscopic ductility.

Additional information regarding the nature of deformation processes at temperatures immediately below T_g may be obtained from creep data. Chen and Turnbull (1968) first observed Newtonian viscous flow in $Au_{77}Ge_{13.6}Si_{9.4}$ at temperatures within 15°C of T_g. The glass exhibited a viscoelastic behavior, although the viscoelastic strains were small. A long transient region exists between the initial instantaneous deformation and the steady state. They also noted that considerable time elapsed before the steady state was attained after abrupt changes in stress or temperature. The viscosity is highly temperature dependent and, at a given

FIG. 24. Bright-field transmission electron microscope image of $Cu_{46}Zr_{54}$ glass aged 450 min at 375°C. Crystallites are shown that have grown from the glassy matrix. (Reprinted from Freed and Vander Sande (1979).)

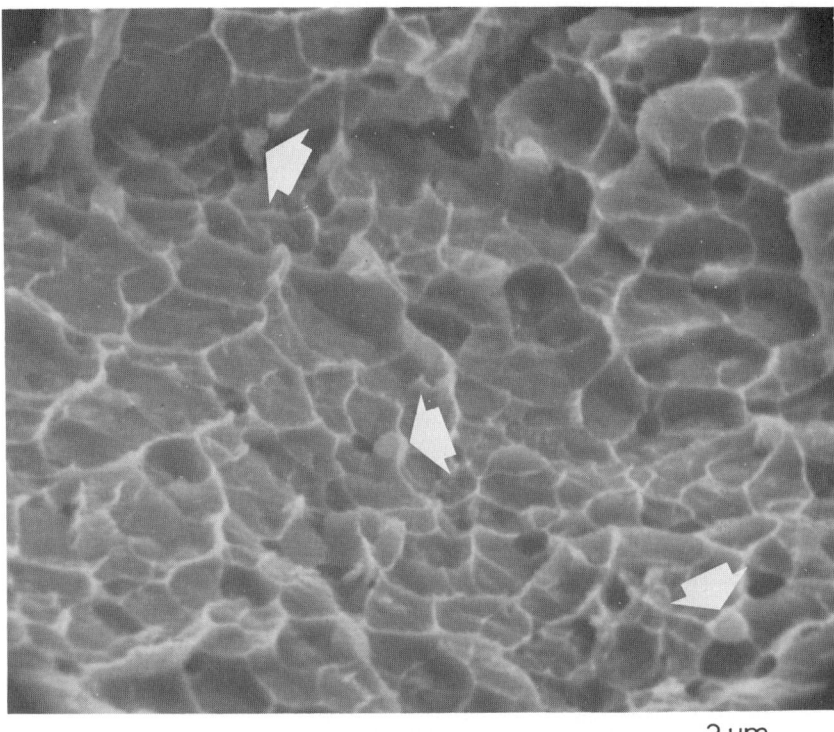

2 μm

FIG. 25. Scanning electron microscope image of the tensile fracture surface of $Cu_{46}Zr_{54}$ aged for 450 min at 375°C. The arrows point out crystallites that have voids situated next to them. As the size and density of the voids increase with applied stress, the ligaments of glass between them rupture, leaving the vein pattern on the surface. (Reprinted from Freed and Vander Sande (1979).)

temperature, independent of stress and the path by which the temperature was attained. This observation rules out the possibility that the viscous flow might be due to Nabarro–Herring creep of a microcrystalline alloy, for which the viscosity would irreversibly increase with temperature. Similar results have been observed in a number of others glasses (Cohen and Turnbull, 1961; Rawson, 1967). The observed temperature dependence of viscosity is in accordance with that observed for several other glass-forming liquids (Polk, 1970; Turnbull, 1969).

Values determined for shear-stress activation volumes are of the order of 50–100 Å^3, increasing with decreasing stresses. These large values indicate that groups of atoms cooperate in atomic rearrangement. At higher stresses (Davis and Kavesh, 1975) or higher temperatures (Logan and Ashby, 1974) non-Newtonian behavior is found. It is supposed that

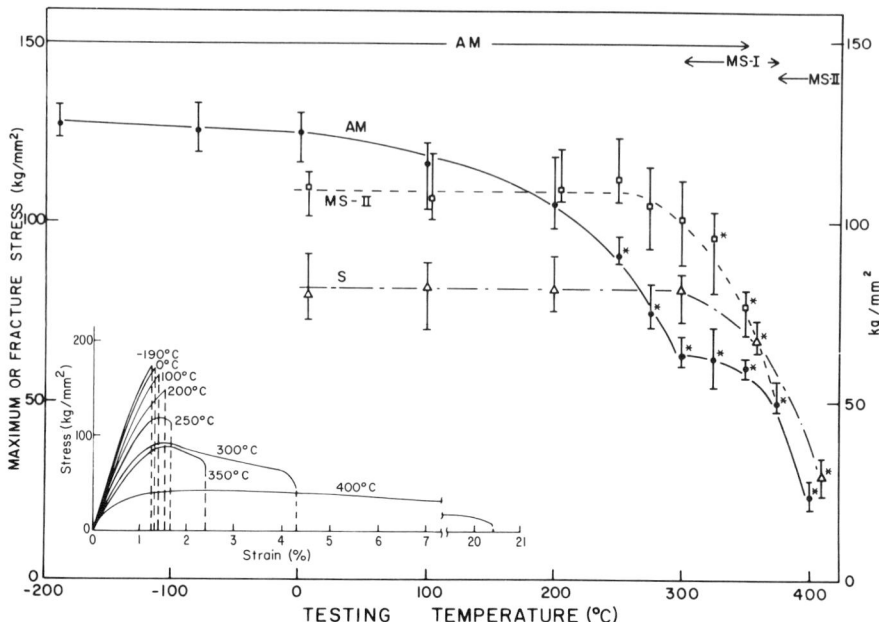

FIG. 26. Temperature dependence of fracture stress of $Pd_{80}Si_{20}$ glass and devitrified crystalline phases. Insert shows the change in the stress–strain with change from inhomogeneous to homogeneous deformation with increasing temperature. Am, amorphous phase; MS-I, metastable phase I; MS-II, metastable phase II; S, stable phase; *, ultimate tensile strength. (Reprinted with permission from *Acta Metall.*, **19**, T. Masumoto and R. Maddin. Copyright 1971, Pergamon Press, Ltd.)

under these conditions the glasses are relaxing to a more stable structure. In their study of $Pd_{80}Si_{20}$, Maddin and Masumoto (1972) determined a strain-rate–stress relation for steady state creep, indicating that it was non-Newtonian. They proposed stress-induced relaxation from the glass to the crystal to account for the observed behavior. Argon (1979) has devised a model based on thermally activated shear transformations around free volume regions, which explains the tendency of the glass to deform homogeneously and inhomogeneously at high temperatures ($0.6T_g < T < T_g$) and low temperatures ($0 < T < 0.6T_g$), respectively. Figure 27 shows the good correlation of the calculated flow stress with that determined experimentally for two glassy alloys.

The increasing ease with which the glass may be deformed and the homogeneous nature of deformation suggest that forming processes or compaction could be performed at temperatures approaching T_g. However, in studying the crystallization of $Pd_{80}Si_{20}$ under a constant applied load, Masumoto and Maddin (1971) observed a substantial decrease in the

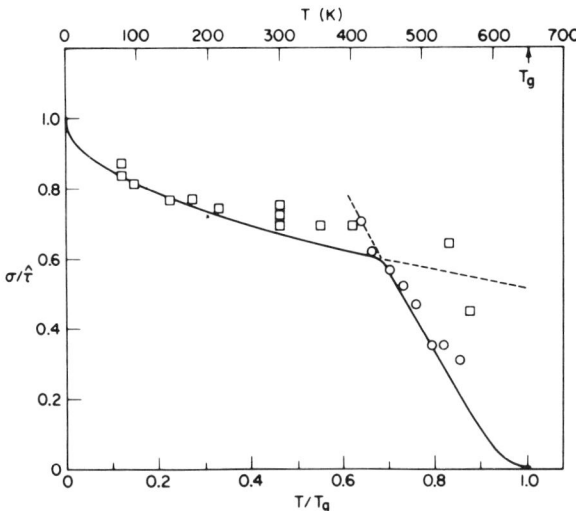

FIG. 27. Computed dependence of the flow stress on temperature compared with low-temperature data for $Pd_{77.5}Cu_6Si_{16.5}$ glass and high-temperature data for $Pd_{80}Si_{20}$ glass. (○) $Pd_{80}Si_{20}$ ($\hat{\tau} = 0.90$ GN/m²), Megusar et al. (1979), (□) $Pd_{77.5}Cu_6Si_{16.5}$ ($\hat{\tau} = 1.03$ GN/m²), Pampillo and Chen (1974). (Reprinted with permission from *Acta Metall.*, **27**, A. Argon. Copyright 1979, Pergamon Press, Ltd.)

time required to initiate crystallization of the low-temperature fcc phase, which apparently forms by continuous growth of preexisting clusters in the as-quenched structure. However, the stress did not greatly affect the beginning of the crystallization of the two high-temperature metastable phases that form by nucleation and growth mechanisms. A similar phenomenon has also been observed in $Fe_{40}Ni_{40}P_{14}B_6$ glass by Patterson and Jones (1975) by crystallization studies and Curie temperature measurements. They determined that crystal growth activation energy during deformation is reduced compared to the unstressed material because of enhanced diffusion under stress. This results in shorter devitrification times, as shown in Fig. 28. Change in crystal structure was not detected.

Exposure to high temperatures may also lead to embrittlement. Catastrophic loss of ductility has been observed on annealing samples at temperatures that are insufficient to cause detectable crystallization (Pampillo and Polk, 1978; Walter et al., 1978). Generally, the embrittlement has been associated with metal–metalloid systems, and Walter et al. (1976) suggested that it is due to the clustering of the metalloid atoms. Chen (1976) has suggested that fine-scale phase separation is responsible; consequently, stability is greatest for those glasses with the fewest constituents and containing elements with electronic configurations closest to the noble elements. Recent work by Walter et al. (1978) provides strong evidence

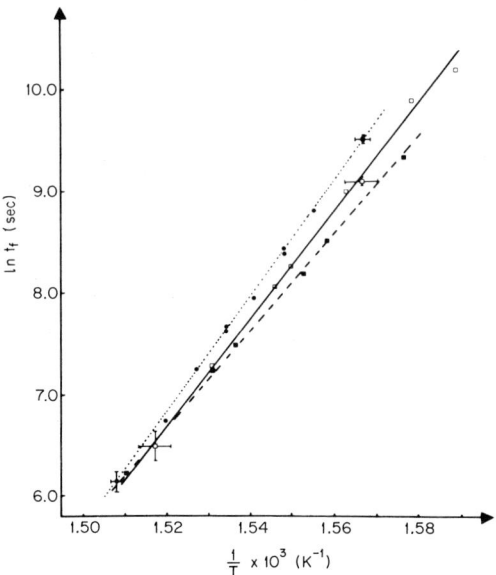

FIG. 28. Plots of *ln* (time for complete crystallization) against reciprocal absolute temperature for stressed and unstressed $Fe_{40}Ni_{40}P_{14}B_6$ glass. A decrease in the time and temperature needed to crystallize this glass was found to correspond to a decrease in crystal growth activation energy. (· · ·) Unstressed (DSC), (———) crystallized under 0.12 GPa, (– – –) crystallized under 0.22 GPa. (Reprinted with permission from *Scripta Metall.*, **13**, J. Patterson and D. R. H. Jones. Copyright 1979, Pergamon Press, Ltd.)

against these arguments. To further complicate the problem, embrittlement has recently been observed in $Ni_{60}Nb_{40}$ glasses (Pratten and Scott, 1978a,b), but not in $Cu_{60}Zr_{40}$. This is somewhat surprising, since copper and nickel and niobium and zirconium are neighbors in the periodic table. In view of this, internal oxidation was suggested as a possible cause of embrittlement in $Ni_{60}Nb_{40}$.

IX. Summary and Conclusions

From a fundamental viewpoint, there are a number of properties of metallic glasses that could be better understood. For example: a generalized theory of glass formation that allows prediction of glass-forming compositions; models that show the subtle differences in glass structure that can be used to understand mechanisms of relaxation, deformation, and crystallization; and, finally, a better appreciation of the effect of structure on magnetic and electrical properties. These would all advance the understanding and technology of metallic glasses significantly. These developments would permit simpler alloy design and exploration of thermomechanical processing, similar to that used routinely for crystalline metallic alloys.

Although the physics and chemistry of metallic glasses may not be totally understood, this has not decreased their potential for commercial application. Some of the most promising applications will use their electrical and magnetic properties. For instance, capitalizing on the "soft" magnetic and mechanical properties of Fe-base alloys, flexible magnetic shielding is produced by weaving strips of melt-spun glass. Also, power transformer cores operating at 50–60 Hz made from metallic glass windings have been made and are in the testing phase of development. It has been demonstrated that power savings of $\sim 30\%$ over conventional crystalline Fe–Si or permalloy alloys will be observed because of their low hysteresis loss. Fabrication of magnetic-bubble memories from metallic glass thin films has been described and is under development. The development of technology to produce these devices could ultimately increase storage density to 1 billion bits of information per square centimeter, or 25 times the density possible in conventional materials. The use of metallic glasses as superconductors is under development. Their very fine scale structure theoretically could improve their performance over crystalline superconducting alloys.

Applications of metallic glasses that use their mechanical attributes have been somewhat slower to occur. Metallic glasses are likely candidates for composite reinforcement applications, such as belts, tires, and high-pressure tubing, because of their high strength and corrosion resistance. Their use in cutlery because of their hardness, strength, and chemical stability is also being considered.

The history of metallic glasses has been a relatively short one compared to their crystalline counterparts. In 10–15 years they have moved from a laboratory curiosity to an article of commerce. The next 10 years will undoubtedly find new applications of these materials and a better understanding of their technology.

References

Argon, A. S. (1975). "Polymer Materials," p. 411. American Society of Metals, Metals Park, Ohio.
Argon, A. (1979). *Acta Metall.* **27,** 47.
Argon, A. S., and Salama, M. M. (1976). *Mat. Sci. Eng.* **23,** 219.
Bagley, B. G., and Turnbull, D. (1970). *Acta Metall.* **18,** 857.
Bennett, C. H., Polk, D. E., and Turnbull, D. (1971). *Acta Metall.* **19,** 1295.
Bernal, J. D. (1964). *Proc. R. Soc. London Ser. A* **280,** 299.
Breman, E. M., Kear, B. H., Banas, C. M., and Greenwald, L. E. (1976). Super-alloys: metallurgy and manufacture, *Int. Symp. 3rd, Seven Springs, Pennsylvania.*

Cahn, J. W. (1969). *J. Am. Ceram. Soc.* **52,** 118.
Calka, A., Madhaua, M., Polk, D. E., Giessen, B. C., Matyja, H., and Vander Sande, J. B. (1977). *Scripta Metall.* **11,** 65.
Cargill, G. S. III (1970). *J. Appl. Phys.* **41,** 12.
Cargill, G. S., III (1970). *J. Appl. Phys.* **41,** 2248.
Chen, H. S. (1976). *Mat. Sci. Eng.* **26,** 79.
Chen, H. S., and Coleman, E. (1976). *Appl. Phys. Lett.* **28,** 245.
Chen, H. S., and Lo, C. C. (1976). "Rapidly Quenched Metals" (N. J. Grant and B. C. Giessen, eds.), p. 369. MIT Press, Cambridge, Massachusetts.
Chen, H. S., and Miller, C. E. (1970). *Rev. Sci. Instrum.* **41,** 1237.
Chen, H. S., and Park, B. K. (1973). *Acta Metall.* **21,** 395.
Chen, H. S., and Polk, D. J. (1974). *J. Non-Cryst. Solids* **15,** 174.
Chou, C.-P., and Spaepen, F. (1975). *Acta Metall.* **23,** 609.
Chen, H. S., and Turnbull, D. (1968). *J. Chem. Phys.* **48,** 2560.
Chen, H. S., and Turnbull, D. (1969). *Acta Metall.* **17,** 1021.
Chen, H. S., and Wang, T. T. (1970). *J. Appl. Phys.* **41,** 5338.
Chen, H. S., Leamy, H. J., and O'Brien, M. J. (1973). *Scripta Metall.* **7,** 415.
Chou, C.-P., Davis, L. A., and Narasinham, M. C. (1977). *Scripta Metall.* **11,** 417.
Cohen, M. H., and Turnbull, D. (1961). *Nature (London)* **189,** 131.
Davies, H. A. (1976). *Phys. Chem. Glasses* **17,** 159.
Davis, L. A. (1976). "Rapidly Quenched Metals" (N. J. Grant and B. C. Giessen, ed.), p. 369. MIT Press, Cambridge, Massachusetts.
Davis, L. A. (1978). "Metallic Glasses," p. 192. American Society of Metals, Metals Park, Ohio.
Davis, L. A., and Kavesh, S. (1975). *J. Mat. Sci.* **10,** 453.
Davis, L. A., Ray, R., Chou, C.-P., and O'Handley, R. C. (1976). *Scripta Metall.* **10,** 541.
Davis, L. A., Chou, C.-P., Tanner, L. E., and Ray, R. (1976). *Scripta Metall.* **10,** 937.
Diegle, R. G. (1979). *Corrosion* **35,** 250.
Diegle, R., and Slater, J. (1976). *Corrosion* **32,** 155.
Dixmier, J., and Guinier, A. (1967). *Rev. Metall.* **64,** 53.
Dixmier, J., Doi, K., and Guinier, A. (1965). "Physics of Non-Crystalline Solids (J. A. Prins, eds.), p. 67. North-Holland Publ., Amsterdam.
Duwez, P., and Lin, S. C. H. (1967). *J. Appl. Phys.* **38,** 4096.
Duwez, P., Willens, R. H., and Klement, W. (1960a). *J. Appl. Phys.* **31,** 1136.
Duwez, P., Willens, R. H., and Klement, W. (1960b). *J. Appl. Phys.* **31,** 1137.
Duwez, P., Willens, R. H., and Crewdson, R. C. (1965). *J. Appl. Phys.* **36,** 2267.
Elliott, R. O., Koss, D. A., and Giessen, B. C. (1980). *Scripta Metall.* **14,** 1061.
Freed, R. L., and Vander Sande, J. B. (1979). *Metall. Trans.* **10A,** 1621.
Freed, R. L., and Vander Sande, J. B. (1980). *Acta Metall.* **28,** 103.
Giessen, B. C., and Wagner, C. N. J. (1972). "Liquid Metals" (S. Bear, ed.), p. 633. Dekker, New York.
Gilman, J. J. (1973). *J. Appl. Phys.* **44,** 675.
Gilman, J. J. (1975). *J. Appl. Phys.* **46,** 1625.
Harbur, D. R., Anderson, J. W., and Maraman, W. J. (1969). *Trans. AIME* **245,** 1055.
Hasegawa, R. (1970). *J. Appl. Phys.* **41,** 4096.
Hashimoto, K., and Masumoto, T. (1976). *Mat. Sci. Eng.* **23,** 285.
Hasegawa, R., and Tsuei, C. C. (1971). *Phys. Rev. B* **3,** 214.
Hinesley, C. P., and Morris, J. G. (1970). *Metall. Trans.* **1,** 1476.
Inoue, A., Naohara, T., Masumoto, T., and Kumada, K. (1979). *Trans. Jpn Inst. Met.* **20,** 577.

Johnson, K. (1981). Personal communication.
Jones, H., and Burden, M. H. (1971). *J. Phys. E Sci. Instrum.* **4,** 671.
Kavesh, S. (1976). *ASM Mat. Sci. Seminar Metall. Glasses, Niagara Falls, New York.*
Kelly, A., and Nicholson, R. B. (1961). *Prog. Mat. Sci.* **10,** 149.
Klement, W., Jr., Willens, R. H., and Duwez, P. (1960). *Nature (London)* **187,** 869.
Leamy, H. J., Chen, H. S., and Wang, T. T. (1972). *Metall. Trans.* **3,** 699.
Li, J. C. M. (1976). "Frontiers in Material Science" (L. E. Murr and C. Stein, eds.), p. 527. Dekker, New York.
Lin, S. C. H. (1969). *J. Appl. Phys.* **40,** 2173.
Lin, S. C. H., and Duwez, P. (1969). *Phys. Status Solidi* **34,** 469.
Logan, J., and Ashby, M. F. (1974). *Acta Metall.* **22,** 1047.
Luborsky, F. E., Frishmann, R. G., Johnson, L. A. (1979). General Electric Company, Rep. No. 79CRD209.
Luo, H. L., and Duwez, P. (1963). *Appl. Phys. Lett.* **2,** 21.
Mader, S. (1965). *J. Vac. Sci. Technol.* **2,** 35.
Mader, S. (1966). "Recrystalization, Grain Growth, and Textures." 1965 Seminar, p. 523. American Society of Metals, Metals Park, Ohio.
Mader, S., Nowick, A. S., and Widmer, H. (1967). *Acta Metall.* **15,** 203.
Maddin, R., and Masumoto, T. (1972). *Mat. Sci. Eng.* **9,** 153.
Maitrepierre, P. L. (1969). *J. Appl. Phys.* **40,** 4826.
Maitrepierre, P. L. (1970). *J. Appl. Phys.* **41,** 498.
Maringer, R. E., Mobley, C. E., and Collings, E. W. (1976). "Rapidly Solidified Metals—Section I" (N. J. Grant and B. C. Giessen, eds.), p. 29. MIT Press, Cambridge, Massachusetts.
Masumoto, T., and Maddin, R. (1971). *Acta Metall.* **19,** 725.
Masumoto, T., Waseda, Y., Kimura, H., and Inoue, A. (1976). *Sci. Rep. Res. Inst. Tohoku Univ. (A)* **26,** 21.
Megusar, J., Vander Sande, J. B., and Grant, N. J. (1976). "Rapidly Solidified Metals—Section I" (N. J. Grant and B. C. Giessen, ed.), p. 401. MIT Press, Cambridge, Massachusetts.
Megusar, J., Clay, C., and Grant, N. J. (1977). Personal communication.
Megusar, J., Argon, A. S., and Grant, N. J., (1979). *Mat. Sci. Eng.* **38,** 63.
Moss, S. C., and Graczyk, J. F. (1969). *Phys. Rev. Lett.* **23,** 581.
Moss, M., Smith, D. L., and Lefever, R. A. (1964). *Appl. Phys. Lett.* **5,** 120.
Mullens, W. W., and Sekerka, R. F. (1964). *J. Appl. Phys.* **35,** 444.
Naka, M., Hashimoto, K., and Masumoto, T. (1976). *Corrosion* **32,** 146.
Naka, M., Hashimoto, K., Asami, K., and Masumoto, T. (1978). "Rapidly Solidified Metals III" (B. Cantor, ed.), Vol. II, p. 449. The Metals Society, London.
Naka, M., Hashimoto, K., Masumoto, T. (1978). *J. Non-Cryst. Solids* **28,** 403.
Nowick, A. S., and Mader, S. (1965). *IBM J. Res. Dev.* **9,** 358.
Pampillo, C. A. (1972). *Scripta Metall.* **6,** 915.
Pampillo, C. A. (1975). *J. Mat. Sci.* **10,** 1194.
Pampillo, C. A., and Chen, H. S. (1974). *Mat. Sci. Eng.* **13,** 181.
Pampillo, C. A., and Polk, D. E. (1974). *Acta Metall.* **22,** 741.
Pampillo, C. A., and Polk, D. E. (1978). *Mat. Sci. Eng.* **33,** 275.
Pampillo, C. A., and Reimschuessel, A. C. (1974). *J. Mat. Sci.* **9,** 718.
Patterson, J., and Jones, D. R. H. (1979). *Scripta Metall.* **13,** 947.
Pietrokowsky, P. (1963). *Rev. Sci. Instrum.* **34,** 445.
Polk, D. E. (1970). *Scripta Metall.* **4,** 473.
Polk, D. E. (1972). *Acta Metall.* **20,** 485.
Polk, D. E., and Pampillo, C. A. (1973). *Scripta Metall.* **7,** 1161.

Polk, D. E., and Turnbull, D. (1972). *Acta Metall.* **20**, 493.
Pond, R., and Maddin, R. (1969). *Trans. Metall. Soc. AIME* **245**, 2475.
Pratten, N. A., and Scott, M. G. (1978). *Scripta Metall.* **12**, 137.
Pratten, N. A., and Scott, M. G. (1978). *In* "Rapidly Solidified Metals III" (B. Cantor, ed.), Vol. I, p. 387. Metals Society, London.
Predecki, P., Mullendore, A. W., and Grant, N. J. (1965). *Trans. Met. Soc. AIME* **233**, 1581.
Rastogi, P. K. (1973). *J. Mat. Sci.* **8**, 140.
Rastogi, P. K., and Duwez, P. (1970). *J. Non-Cryst. Solids* **5**, 1.
Rawson, H. (1967). "Inorganic Glass Forming Systems." Academic Press, New York.
Ray, R., Giessen, B. C., and Grant, N. J. (1968). *Scripta Metall.* **2**, 357.
Roberge, R., and Herman, H. (1968). *Mat. Sci. Eng.* **3**, 62.
Rudee, M. L. (1971). *Phys. Status. Solidi* (6), **46**, K1.
Rudee, M. L., and Howie, A. (1972). *Phil. Mag.* **25**, 1001.
Ruhl, R. C., Giessen, B. C., Cohen, M. C., and Grant, N. J. (1967). *Acta Metall.* **15**, 1693.
Sadoc, J. F., Dixmier, J., and Guinier, A. J. (1973). *J. Non-Cryst. Solids* **12**, 46.
Sarjeant, P. T., and Roy, R. (1968). *Mat. Res. Bull.* **3**, 265.
Sinha, A. K. (1970). *Phys. Rev. B* **1**, 4541.
Sinha, A. K., and Duwez, P. (1972). *J. Appl. Phys.* **43**, 431.
Spaepen, F. (1972). Harvard Univ., private communication.
Spaepen, F. (1975). *Acta Metall.* **23**, 615.
Spaepen, F., and Turnbull, D. (1974). *Scripta Metall.* **8**, 563.
Tanner, L. E., and Ray, R. (1977). *Scripta Metall.* **11**, 783.
Tanner, L. E., and Ray, R. (1979). *Acta Metall.* **27**, 1727.
Thursfield, G., and Jones, J. (1971). *J. Phys. E Sci. Instrum.* **4**, 675.
Turn, J. C., Jr. (1979). Ph.D. Thesis, MIT, Cambridge, Massachusetts.
Turnbull, D. (1969). *Contemp. Phys.* **10**, 473.
Turnbull, D. (1975). *J. Electr. Mat.* **4**, 771.
Turnbull, D., and Cohen, M. H. (1958). *J. Chem. Phys.* **29**, 1049.
Turnbull, D., and Cohen, M. H. (1960). "Modern Aspects of the Vitreous State" (J. C. Mackenzie, ed.), p. 38. Butterworths, London.
Turnbull, D., and Cohen, M. H. (1970). *J. Chem. Phys.* **52**, 3038.
Uhlmann, D. R. (1972). *J. Non-Cryst. Solids* **7**, 33.
Uhlmann, D. R. (1977). *J. Non-Cryst. Solids* **25**, 43.
Wagner, C. N. J. (1969). *J. Sci. Technol.* **6**, 650.
Walter, J. L., Bacon, F., and Luborsky, F. E. (1976). *Mat. Sci. Eng.* **24**, 239.
Walter, J. L., Bacon, F., and Luborsky, F. E. (1978). *Mat. Sci. Eng.* **33**, 91.
Waseda, Y., and Masumoto, T. (1975). *Z. Phys.* **B21**, 235.
Waseda, Y., *et al.* (1977). *J. Mat. Sci.* **12**, 1927.
Willens, R. H., and Buehler, E. (1966). *Trans. Metall. Soc. AIME* **236**, 171.

CHAPTER 7

Glass-Ceramic Technology

George H. Beall

TECHNICAL STAFFS DIVISION
CORNING GLASS WORKS
CORNING, NEW YORK

David A. Duke

TECHNICAL PRODUCTS DIVISION
CORNING GLASS WORKS
CORNING, NEW YORK

I. Introduction	404
II. Nucleation of Crystals in Glass	405
A. The Nature of Glass	405
B. Homogeneous and Heterogeneous Nucleation	406
C. Nucleation through Microimmiscibility	409
III. Crystallization Phenomena in Glass-Ceramics	413
A. Metastable Phases and Transformation	416
B. Crystallization of an Li_2O–Al_2O_3–SiO_2–TiO_2 Glass-Ceramic	418
IV. Properties of Glass-Ceramics	422
A. Ultralow Thermal Expansion	422
B. Transparency	424
C. Mechanical Strength	425
D. Machinability	429
E. Miscellaneous Properties	433
V. Glass-Ceramic Processing	436
A. Raw Materials	436
B. Handling and Mixing	436
C. Glass Melting	437
D. Forming	439
E. Finishing	440
F. Crystallization—Heat Treatment	441
VI. Applications	441
References	444

I. Introduction

Glass-ceramics are crystalline materials formed through the controlled devitrification of glass. Glasses are melted, fabricated to shape, and then converted to a ceramic by a specific heat treatment. Hot-glass-forming techniques such as pressing, blowing, spinning, rolling, and casting are used to rapidly produce a variety of articles. Their shapes are preserved with only minor shrinkage or deformation during the heating cycle that converts glass to glass-ceramic. Glass-ceramic structure is characterized by fine-grained, randomly oriented crystals with some residual glass but no voids, microcracks, or other porosity. As a result of this unique microstructure, properties such as translucency, high strength, and very low and uniform thermal expansion can be routinely produced.

Controlled glass crystallization has a number of key advantages over conventional ceramic processing. These are:

1. Flexibility and ease of forming.
2. The ability to inspect the transparent glass article prior to crystallization to eliminate obvious defects.
3. Uniformity and reproducibility of properties.
4. Ability to produce unique properties inherent in extremely fine-grained crystalline materials; i.e., transparency, homogeneous thermal expansion, etc.
5. Lack of porosity; suitability for subsequent strengthening by prestressing techniques.
6. Process economy in high-volume manufacturing.

Certain restrictions or disadvantages are inherent in glass crystallization as a ceramic-forming technique:

1. Compositions are restricted to those that can form glasses.
2. Melting and glass-forming steps may sometimes be prohibitive in cost, particularly where volume is low. With sufficient volume, however, the capital investment required for a complex glass tank and associated forming equipment will generally be more than balanced by the process economy inherent in a high-speed and continuous operation.

The basis for the controlled crystallization of glass lies in efficient nucleation. Nucleation of crystals from the surface of a glass article or from a small number of sites in the interior results in a coarse-grained, oriented form of crystallization. This structure is generally accompanied by voids, planes of weakness, and gross distortion of the material, and consequently results in low strength. The discovery of the role of nucleating agents in initiating glass crystallization from a multitude of centers was the major factor allowing the introduction of glass-ceramics (Stookey, 1959).

II. Nucleation of Crystals in Glass

A. The Nature of Glass

To understand the process of nucleation in glasses, it is useful to consider what is known of their structure. Silicate glasses have a random structure based on an irregular arrangement of SiO_4 tetrahedra linked through corner sharing into a three-dimensional network (Zachariasen, 1932). The difference in structure between vitreous silica and the crystalline polymorphs of silica lies primarily in long-range order (beyond about 8 Å). Both are composed structurally of the same cation–anion polyhedra with similar linkage, but the crystalline forms show continuous spatial pattern repetition; the vitreous form is random. In Fig. 1a,b this difference is illustrated.

The chemical composition of glasses may vary widely and can include more than 60 of the elements of the periodic table. Certain cations have been classified according to their role in the structure of oxide glasses (Zachariasen, 1932): network formers, such as Si, B, P, Ge, and As, have oxygen coordination numbers of 3 or 4 and tend to produce the basic cross-linked polymeric glass structure; network modifiers, such as Na, K, Ca, and Ba, have coordination numbers of 6 or more and generally tend to

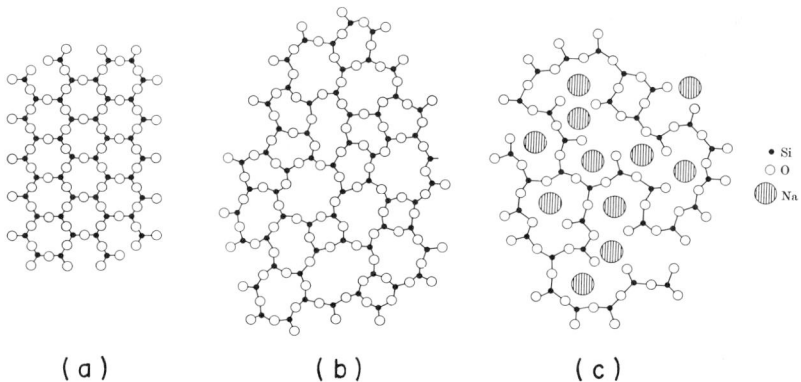

FIG. 1. Schematic two-dimensional representation of the structures of (a) crystalline silica. (b) silica glass, and (c) sodium silicate glass. (Reprinted from Warren (1938) and Zachariasen (1932), by courtesy of the American Chemical Society.)

reduce the polymerization and viscosity of the glass; intermediate oxides with cations such as Al, Zn, Mg, Pb, and Be have intermediate coordination numbers of 4 to 6 and may either act as modifiers or replace network cations, depending on the glass composition. Figure 1c shows the effect of modifying sodium ions on the structure of silicate glass (Warren, 1938).

Although glass has often been referred to as the ultimate solvent, certain oxides have limited solubility and may enhance immiscibility in glass, particularly at lower temperatures. Such oxides are usually based on ions of strong polarizing power. These ions coordinate tetrahedrally with oxygen at high temperatures (near the melting point of the glass) and therefore enter the silicate melt structure, but are larger than silicon and prefer octahedral or cubic coordination at lower temperatures near the annealing point of the glass. They include Ti^{4+}, Zr^{4+}, Hf^{4+}, Nb^{5+}, Ta^{5+}, Mo^{6+}, W^{6+}, and to some extent Cr^{3+}. Such cations can effect immiscibility in a large number of glasses.

Until recently, most commercial glass compositions were specifically designed on the basis of resistance to devitrification so that they could be easily formed into useful shapes in the plastic state. With the advent of glass-ceramics, glasses with intermediate stability became increasingly important, since it was necessary to form these glasses in a viscous state without devitrification, yet cause them to crystallize during subsequent thermal treatment.

B. Homogeneous and Heterogeneous Nucleation

When a homogeneous viscous liquid is cooled below the equilibrium solubility point of its most insoluble species (liquidus temperature), it enters a metastable zone in which nuclei do not form at a detectable rate but where crystals, once nucleated, can easily grow (Tamman, 1925). Below this temperature zone, nuclei may spontaneously and uniformly form, but with further cooling, the liquid reaches a high viscosity that impedes both the formation and growth of nuclei.

Figure 2 shows the observed behavior of both homogeneous nucleation and growth rates as a function of temperature for a typical undercooled liquid of high viscosity (Stookey, 1959). These rates follow relationships of the type (Becker, 1938; Uhlmann, 1972; McMillan, 1979)

$$I = A \exp \frac{-\Delta F^* + Q}{kT},$$

where I is the homogeneous nucleation rate, A is a constant approximately equal to NkT/h (N is Avogadro's number, k is Boltzmann's constant, T is the absolute temperature, and h is Planck's constant), ΔF^* is the maximum free energy change at the critical radius of a spherical nucleus (acti-

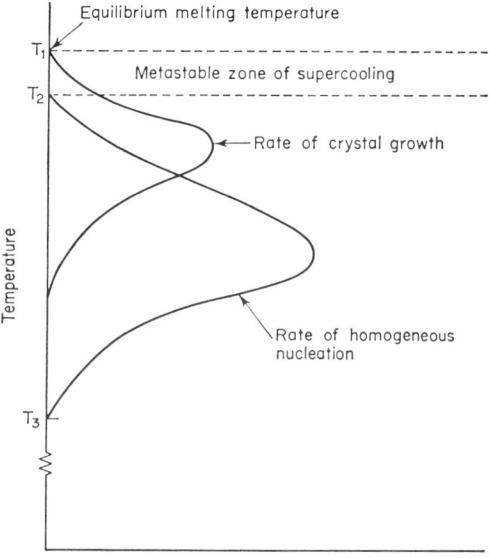

FIG. 2. Effect of temperature on the rates of homogeneous nucleation and crystal growth. (Reprinted from McMillan, 1976.)

vation barrier for nucleus formation), and Q is the activation energy for short-range diffusion of atoms or molecules across the interface;

$$U = fRT(1 - \exp \Delta G/RT)3\pi N a_0^2 \eta,$$

where U is the rate of crystal growth, f is the fraction of the total number of sites available for growth, ΔG is the bulk free energy of crystallization, R is the gas constant, a_0 is the interatomic separation, and η is the viscosity of the liquid.

Most glass-forming liquids, when supercooled, do not crystallize according to the simple laws of homogeneous nucleation. Nucleation usually occurs at the surface of glass in contact with air or other foreign substances where abundant nuclei are already present. When internal crystallization does occur, it almost always results from the nucleation of the major crystalline components upon nuclei of foreign, highly insoluble particles, i.e., metallic particles, halides, sulfides, and certain oxides. These forms of induced nucleation are controlled by irregularities in the glass structure or foreign particles within or on the glass surface and are referred to as heterogeneous nucleation.

Actual examples of homogeneous nucleation of crystals in glass are rare. Binary baria–silica glasses, however, do appear to crystallize in this manner (MacDowell, 1965). Thus, a glass of the barium disilicate composition ($BaSi_2O_5$) will spontaneously and internally crystallize upon heat treatment without the aid of a nucleation catalyst.

Most glasses, if held for a long period at temperatures below their liquiduses, but considerably above their annealing point, will crystallize from the surface by heterogeneous nucleation. At the surface, coordination of certain ions is incomplete and deviations from bulk structure are locally large. This creates a high-energy state where devitrification can readily occur. An abundance of foreign nuclei doubtless enhances this phenomenon. Crystallization in this case proceeds generally toward the interior of the glass in a more or less oriented dendritic pattern. This form of heterogeneous nucleation and devitrification generally results in weak ceramic bodies, with coarse and oriented crystals usually accompanied by pits and voids.

Useful heterogeneous nucleation was first achieved using metallic nuclei precipitated throughout the body of glass (Stookey, 1959). For centuries metals such as gold, copper, and silver have been used to produce colored glass for decorative purposes. Initially dissolved in the melt in the form of ions, they are reduced as the glass cools and finally precipitate as extremely fine colloidal particles. It was found that gold and silver, in particular, could be precipitated photochemically by the action of ultraviolet radiation on a glass containing a readily available source of electrons such as the cerous ion. The photosensitive reaction is

$$Au^{1+}, Ag^{1+} + Ce^{3+} \rightarrow Au^0, Ag^0 + Ce^{4+}.$$

Metallic nuclei formed in this manner were found to act as heterogeneous nuclei for crystallization of lithium metasilicate and in glasses of the basic $SiO_2-Li_2O-Al_2O_3$ system (Stookey and Maurer, 1961). Unfortunately, this phenomenon of heterogeneous nucleation of silicate crystals from metallic particles has not found general applicability in crystallization. This is because a similarity in structure must exist between the nucleating particle and the crystallizing phase for effective nucleation. Experimental evidence indicates that effective nucleation catalysts in supercooled melts generally show a maximum disregistry of about 15% between spacings in some low index planes of the catalyst and the nucleated phase. The disregistry between the 111 planes of gold and silver and the 001 plane of lithium metasilicate is 0.5%, an excellent match, but a rare one between metals and silicates. This may explain the particular effectiveness of these two metals in the nucleation of lithium metasilicate from glass.

The major breakthrough allowing general nucleation and widespread glass-ceramic formation in silicate and aluminosilicate glasses came when a method of internal precipitation of various complex oxide crystals more structurally akin to the silicates was discovered (Stookey, 1959). This involved the role of immiscibility-producing oxides, in particular titania.

C. Nucleation through Microimmiscibility

A large number of glass-forming silicate systems, particularly binary systems, are stable above their liquiduses in the molten state as two immiscible liquids (Grieg, 1927). In recent years, the electron microscope and low x-ray diffraction camera have revealed that many glasses, formerly considered homogeneous, have actually spontaneously separated metastably into two amorphous phases upon cooling. Others, homogeneous on cooling, can be observed to separate on a microscopic scale during reheating near their annealing point (McMillan, 1979).

The binary Al_2O_3–SiO_2 system presents examples of such glasses (MacDowell and Beall, 1969). Liquids over most of the range between the compositions of mullite ($3Al_2O_3 \cdot 2SiO_2$) and silica are homogeneous at temperatures above their liquiduses (~1850°C) but spontaneously separate into two liquids on cooling (Figs. 3 and 4). The alumina-rich glass is far more unstable than the siliceous portion and readily crystallizes on subsequent thermal treatment of the emulsion. The high surface energies associated with droplet interfaces afford abundant nuclei for crystallization of the unstable glass to mullite. A variety of fine-grained crystal-glass structures and textures result, based on the form of the original amorphous separation. If the aluminous phase is minor in volume fraction, forming dispersed spherical particles, the resulting mullite reflects this spherical form (Fig. 4b). If the aluminous phase is similar in volume fraction to the

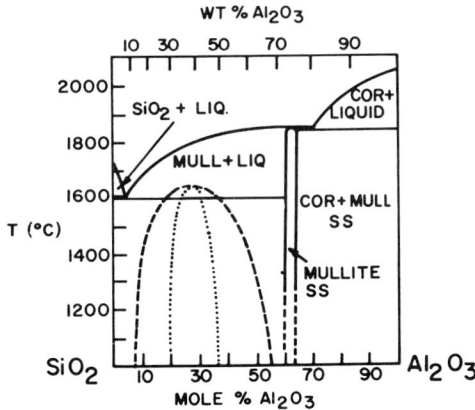

Fig. 3. The Al_2O_3–SiO_2 phase diagram showing schematic region of metastable immiscibility. (– – –) metastable boundary, (· · ·) co-connectivity limit on rapid quenching. (Reprinted from MacDowell and Beall (1969), by courtesy of the American Ceramic Society.)

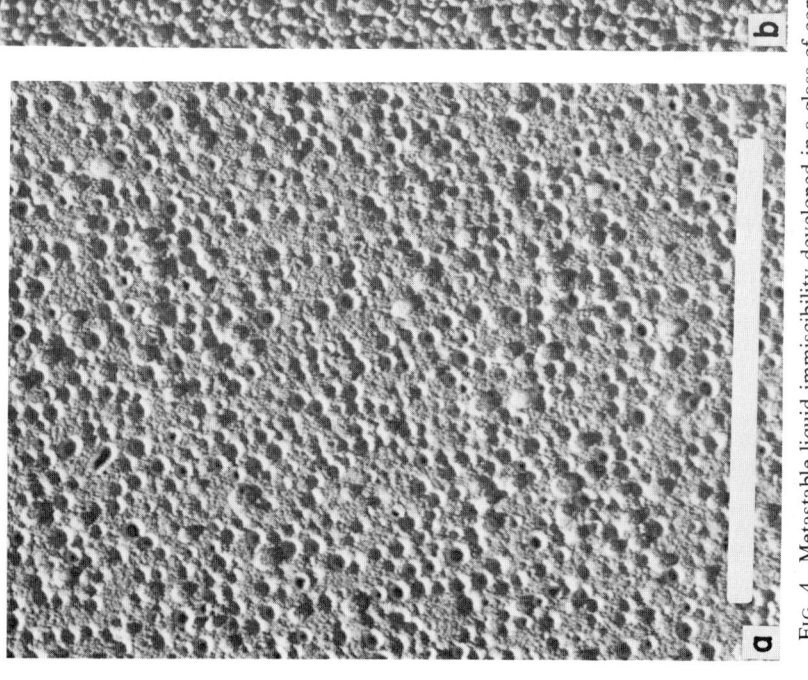

FIG. 4. Metastable liquid immiscibility developed in a glass of composition Al_2O_3, 15 mole %; SiO_2, 85 mole % (replica, bar = 1 μm). (a) Glass after initial cooling from the melt. Droplets represent high-alumina regions (negative relief), some of which have crystallized to mullite (positive relief). (b) Glass after heat treatment (950°C for 10 h). Aluminous droplets have crystallized producing a mullite glass-ceramic. (Reprinted from MacDowell and Beall (1969), by courtesy of the American Ceramic Society.)

siliceous portion, the separation produces wormy co-continuous phases. A major portion of mullite results, reflecting the original interconnected separation (Fig. 5). Further heat treatment causes crystal growth and eventual crystallization of the siliceous phase to cristobalite, but the early stages of crystallization faithfully reflect the geometry of the original amorphous phase separation.

The cause of this metastable phase separation in systems such as Al_2O_3–SiO_2 is not completely understood, but it is believed due to structural incompatibilities occurring during cooling from the melt. At high temperatures, the aluminosilicate liquid is stable as a single phase probably composed structurally of partially linked SiO_4 and AlO_4 tetrahedra. On cooling, the melt polymerizes toward a corner-shared tetrahedral

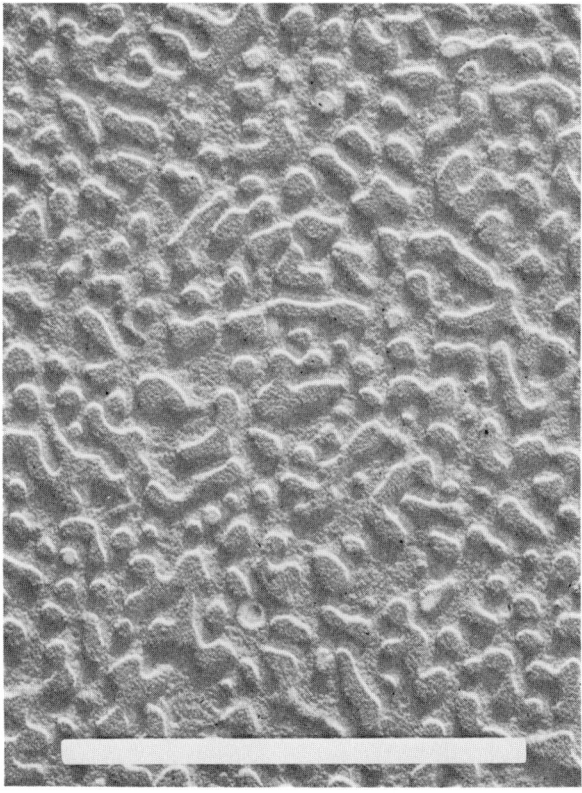

FIG. 5. Mullite glass-ceramic with microstructure reflecting the original co-connective immiscibility. Raised particles are mullite, background is siliceous glass (replica, bar = 1 μm). (Reprinted from MacDowell and Beall (1969), by courtesy of the American Ceramic Society.)

arrangement. For alumina to enter such a structure, an oxygen ion must tribridge three tetrahedra for every Al for Si replacement (MacDowell and Beall, 1969). This triclustering of tetrahedra, illustrated in Fig. 6, produces excessive densification of the network of normal double-bridging oxygen ions. As a result, only limited miscibility is permitted. An alumina-rich amorphous phase close to mullite in composition and presumably reflecting a mixed 4–6 oxygen coordination of aluminum (as in mullite) therefore separates.

More complex aluminosilicates glasses modified by common oxides like those of the alkalies and alkaline earths do not generally show such structural incompatibilities on cooling. This is due to the fact that the larger modifying cations (Na, K, Ca, etc.), which reside in interstitial structural positions, provide charge balance for aluminum-for-silicon network replacement and allow normal cross-linking as in silica glass. Additions of an oxide such as TiO_2, which can neither modify the glass structure nor replace SiO_2 in the network, are therefore necessary to produce microimmiscibility in such glasses. Relatively small amounts (less than 10 mole %) of oxides such as TiO_2 generally induce separation of a secondary amorphous phase on cooling the glass. These oxide species are concentrated in the separating phase and usually promote crystalline nuclei when the glass is reheated. Titanates form from TiO_2-rich amorphous particles in the same manner as mullite forms from the aluminous droplets of Al_2O_3–SiO_2 emulsions. The primary crystalline nuclei, generally metastable oxide phases, presumably reflect the basic structure of the original

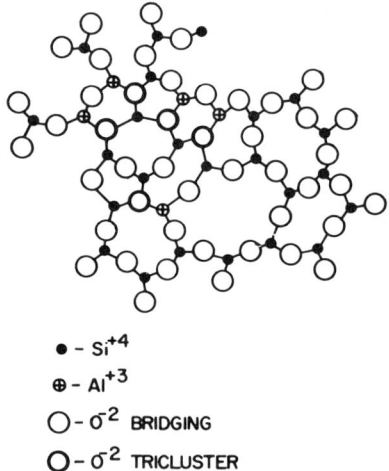

- ● – Si^{+4}
- ⊕ – Al^{+3}
- ○ – O^2 BRIDGING
- ○ – O^2 TRICLUSTER

FIG. 6. Two-dimensional structural model showing the possible effect of additions of alumina to silica glass. (Reprinted from MacDowell and Beall (1969), by courtesy of the American Ceramic Society.)

separating amorphous droplets and are observed to differ in basic structural elements from the major glassy phase. These crystalline nuclei, however, typically form effective heterogeneous nuclei for the subsequent crystallization of the major aluminosilicate components. Table I lists for a series of glass-forming systems the nucleating agents that are added in order to produce effective internal nucleation and glass-ceramic formation. In addition, for each of these systems, the structure and composition of the primary crystalline nuclei have been determined by x-ray diffraction and/or chemical extraction techniques. The structural differences in coordination and linkage between these nuclei and the major glass phase is presented. By assuming that these crystals are similar in coordination and linkage to their parent amorphous phase, the structural differences effecting the original immiscibility are in evidence.

Internal crystallization by the route of microimmiscibility has several benefits in terms of producing a strong ceramic body from a glass article without disruption. Since amorphous phase separation occurs typically on a very fine scale, usually producing particles less than 1000 Å in diameter, there is an enormous surface area available for the primary crystalline nuclei to form. This normally results in very effective nucleation of the major crystals and a fine crystalline microstructure. Stresses due to differences in thermal expansion between phases, anisotropic expansion of crystals, and densification during crystallization are thereby minimized, allowing the development of high body strength. Moreover, the unstable amorphous droplets from which crystallization initiates are generally less stable than the host glass and begin to crystallize at temperatures where the host glass is very viscous. Major crystallization of the host glass on abundant nuclei generally occurs at viscosities of $10^{10}-10^{14}$P, just above the annealing temperature. Distortion of the glass during crystallization is therefore eliminated or minimized. This contrasts with crystallization of glass from surface nuclei, which generally occurs near the softening point of the glass or at least below 10^{10} P viscosity, and results in severe distortion.

In practice, the nucleation of the major crystalline phases on the primary oxide nuclei follows a qualitatively similar function with respect to temperature as does homogenous nucleation (Fig. 2). Therefore, nucleation holds of several hours at an optimized temperature are often employed to assure minimum distortion.

III. Crystallization Phenomena in Glass-Ceramics

Although the phenomena governing the internal crystallization of glass are extremely varied and complex, there is a general pattern or sequence observed during the crystallization cycle. For aluminosilicate glass-

TABLE I
CRYSTALLINE NUCLEI IN MICROIMMISCIBLE GLASS-FORMING SYSTEMS

Glass system	Structural cation coordination	Basic polyhedral linkage	Nucleating agent	Primary crystalline nuclei	Structural cation coordination (nuclei)	Polyhedral linkage (nuclei)
Alkali silicates						
SiO_2–Na_2O	4	Corner-shared tetrahedra	NaF	NaF	6	Face-shared cubes
SiO_2–Li_2O	4	Corner-shared tetrahedra	P_2O_5[a]	?	—	—
Aluminosilicates ($R_2O : Al_2O_3 \gtrsim 1$)						
SiO_2–Al_2O_3	4	Corner-shared tetrahedra	—	$Al_6Si_2O_{13}$[a]	4, 6	Edge-shared octahedra
SiO_2–Al_2O_3–Li_2O	4	Corner-shared tetrahedra	TiO_2	$Al_2Ti_2O_7$[c,d]	6	Edge-shared octahedra
SiO_2–Al_2O_3–Li_2O	4	Corner-shared tetrahedra	ZrO_2	ZrO_2[c]	8	Edge-shared cubes
SiO_2–Al_2O_3–Li_2O	4	Corner-shared tetrahedra	Ta_2O_5	$LiTaO_3$	6	Corner-shared octahedra

System					
SiO_2–Al_2O_3–Na_2O	4	Corner-shared tetrahedra	TiO_2	TiO_2	6	Edge-shared octahedra
SiO_2–Al_2O_3–MgO	4	Corner-shared tetrahedra	TiO_2	$MgTi_2O_5$[c]	6	Edge- and corner-shared octahedra
SiO_2–Al_2O_3–MgO	4	Corner-shared tetrahedra	ZrO_2	ZrO_2[c,e]	8	Edge-shared cubes
SiO_2–Al_2O_3–MgO	4	Corner-shared tetrahedra	WO_3	$MgWO_4$	6	Edge- and corner-shared octahedra
SiO_2–Al_2O_3–CaO	4	Corner-shared tetrahedra	TiO_2	$CaTiO_3$[c]	6	Corner-shared octahedra
SiO_2–Al_2O_3–CaO–MgO	4	Corner-shared tetrahedra	Fe_2O_3	Fe_3O_4	6	Edge-shared octahedra
Aluminates						
Al_2O_3–CaO	6, 4	Edge-shared octahedra	ZrO_2	$ZrO_2 \cdot CaO$	8	Edge-shared cubes

[a] McMillan (1979).
[b] Grieg (1927).
[c] Metastable phase.
[d] Lacy (1963).
[e] Neilson (1970).

ceramics, where nucleation is initiated by amorphous phase separation, this normally includes three stages: (1) precipitation of primary crystalline nuclei, (2) nucleation and growth of metastable crystalline phases, and (3) approach to the stable crystalline assemblage.

A. METASTABLE PHASES AND TRANSFORMATION

Metastable crystalline phases are particularly common as the first major silicates to form on the primary oxide nuclei. These metastable crystals are usually solid-solution phases that can incorporate the major chemical species of the aluminosilicate glass in roughly the same proportions as they are present in the glass. Structurally, these metastable aluminosilicates are usually related to the polymorphs of silica. They are referred to as stuffed derivatives, because they can be considered derived from silica polymorphs through replacement of silicon by aluminum in conjunction with stuffing interstitial vacancies by a modifying cation to maintain charge balance (Buerger, 1954). This is similar to the presumed modifying effect of cations in preserving the cross-linked silica network in aluminosilicate glasses, and the primary crystallization of such glasses to these phases supports the structural similarity.

For example, stuffed derivatives of β quartz, the hexagonal trapezohedral polymorph of silica, first crystallize as metastable forms from a wide variety of Li, Mg, Zn, Mn, and mixed-cation aluminosilicates where the size of the modifying cation is between 0.5 and 0.8 Å ionic radius (Buerger, 1954; Schereyer and Schairer, 1961; Beall et al., 1967). The interstitial vacancies in the quartz structure tolerate ions of this size. Stuffed derivatives of the more open, high-temperature silica polymorph, cristobalite, similarly crystallize metastably from many Na, Ca, K, and mixed-cation aluminosilicates where the radii of the modifying cations are over 0.9 Å (Duke et al., 1967).

These silica derivatives, unlike the pure polymorphs to which they are related, are generally stable toward thermal inversions involving displacive motion of the lattice. For example, the β-quartz form of pure silica is stable only above 573°C, below which it undergoes a displacive or second-order transition to the trigonal polymorph α quartz (Fig. 7). No bonds are broken, but the trigonal form is characterized by high thermal expansion. Moreover, there is a 1% volume decrease accompanying this collapse of the quartz structure. The stuffed derivatives of β-quartz, containing interstitial ions that serve to buttress the tetrahedral framework, cannot invert to the trigonal structure, and hence maintain their low-expansion behavior over the complete cooling range.

With increasing and prolonged thermal treatment, metastable solid-

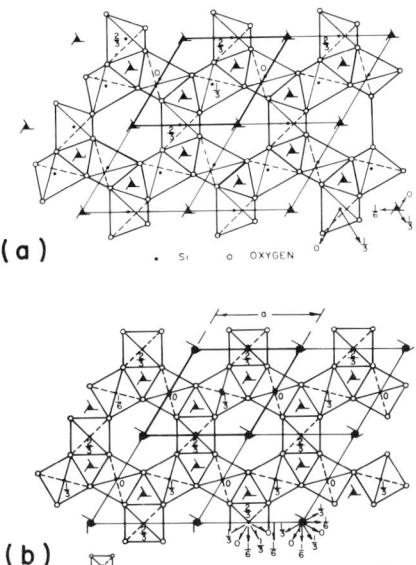

FIG. 7. The structure of (a) α and (b) β quartz. (Reprinted from Beall et al. (1967), by courtesy of the American Ceramic Society.)

solution phases that may persist at temperatures of 1000°C for many hours or even days break down to the stable crystalline assemblage for the particular bulk composition. This irreversible reaction may take the form of an isochemical phase transformation, a reaction between metastable phases, or an exsolution phenomenon. Commonly, some residual glass of a particularly stable or persistent composition remains along crystal boundaries or at grain boundary nodes. This glass may facilitate or impede secondary grain growth of the stable phases, depending on its composition and structure.

In certain cases, the properties of the metastable crystalline phases are unique and desirable in influencing the bulk properties of a glass-ceramic. In this case the heat treatment cycle is designed to operate at an intermediate temperature and a time sufficient to crystallize the glass but insufficient to promote chemical equilibrium. Stable silicate structures, however, because of their superior thermal stability, form the basis of most glass-ceramic materials. These phases are normally three-dimensional-framework silicates and aluminosilicates of various stoichiometries and structural complexities. In some cases, however, they may be sheet silicates, that is, silicate phases based on stacking of discrete two-dimensional layers of anion polyhedra.

B. Crystallization of an $Li_2O-Al_2O_3-SiO_2-TiO_2$ Glass-Ceramic

It is useful to examine the crystallization cycle of a typical glass-ceramic based on a commercially important framework silicate, the stable lithium aluminosilicate, β spodumene. The sequence of events and resulting microstructures are complex but nevertheless characteristic of the crystallization behavior of glass-ceramics.

Beta-spodumene solid solution is a stable low-expansion crystal phase of composition varying from $Li_2O \cdot Al_2O_3 \cdot 4SiO_2$ to $Li_2O \cdot Al_2O_3 \cdot 10SiO_2$ (Ostertag *et al.*, 1968). It is a stuffed derivative of the polymorph of silica keatite. Figures 8a–e show a sequence of micrographs depicting the structure of a $Li_2O-Al_2O_3-SiO_2$ glass approaching the 1:1:7 stoichiometry* (approximately 4 mole % TiO_2 addition as nucleating agent) as a function of heat treatment during the crystallization cycle (Doherty *et al.*, 1967; Chyung, 1969; Beall, 1972).

Figure 8a is a transmission electron micrograph showing the amorphous phase separation that develops on cooling of the glass. No phase separation can be observed in a similar glass lacking titania. Figure 8b shows the early crystalline nuclei that develop near 825°C, some 100°C above the annealing point of the glass. These crystallites have been extracted and identified as aluminum titanate ($Al_2Ti_2O_7$) with the pyrochlore structure (Doherty *et al.*, 1967). Figure 8c shows the results of the crystallization of the major aluminosilicate glassy phase near 900°C to a metastable stuffed β-quartz solid solution. The crystals are roughly 0.1 μm in diameter. The glass-ceramic is highly crystalline and, because of the extremely fine grain size, is transparent at this stage of development. The metastable β-quartz phase breaks down above 950°C to the stable β-spodumene solid-solution crystal. Figure 8d shows the resulting microstructure. Note the increase in grain size of the silicate phase (by a factor of about 7) associated with this almost isochemical phase change. Accompanying the silicate transformation, the stable form of TiO_2, rutile, emerges as a major phase. Rutile was evidently far more soluble in the metastable β-quartz phase than in the β-spodumene structure, and it therefore exsolved during the transformation. A significant result of the appearance of rutile, a dense phase of high refractive index, is the increasing opacity of the glass-ceramic. Figure 8e shows the effect of severe thermal treatment at 1200°C on the microstructure. An increase in both the β-spodumene solid-solution and rutile grain size is apparent.

The growth behavior of the β-spodumene crystals as a function of temperature and time is illustrated in Fig. 9. The secondary grain growth after

---

* Actual composition contains some MgO and ZnO, which replace Li_2O within the solid-solution crystals.

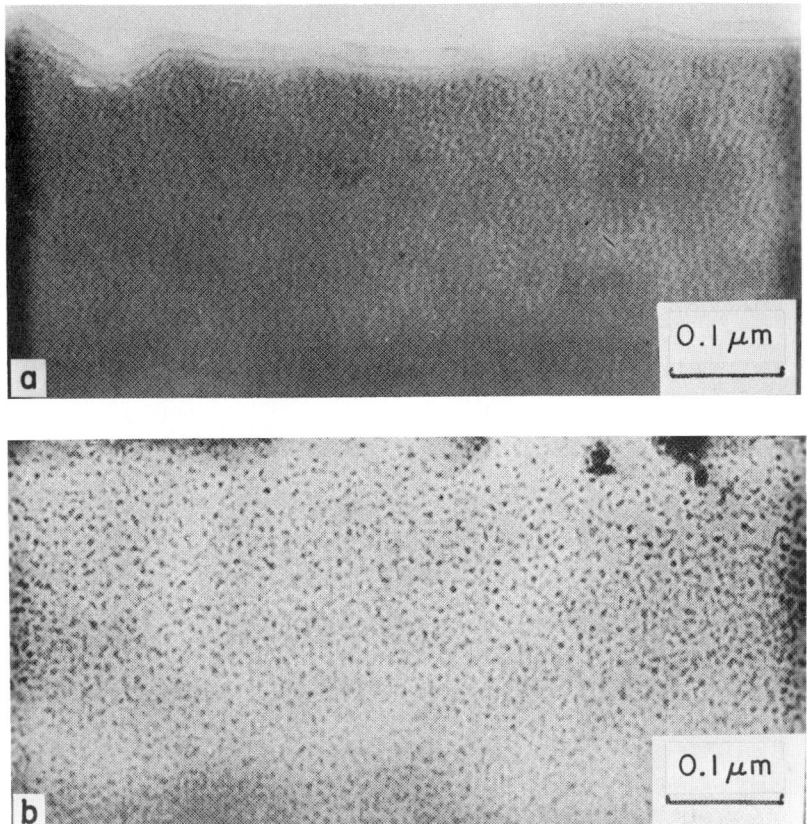

FIG. 8a,b. Microstructure of a $Li_2O-Al_2O_3-SiO_2-TiO_2$ glass during the crystallization sequence. (a) "Green" glass prior to heat treatment showing phase separation (bright field transmission). (b) Glass heated to 825°C showing crystalline nuclei (bright field transmission).

phase transformation and impingement of the grains is linear with the cube-root of time. The activation energy associated with this growth is roughly 55 kcal/mole. This value resembles that for diffusion of trivalent aluminum ions in silica glass. Indeed, the slow diffusion of aluminum through the siliceous residual glass is likely the rate-limiting factor in grain growth in these glass-ceramics (Chyung, 1969).

Grain size stability is one of three factors that provide unique thermal and dimensional stability in β-spodumene glass-ceramics. The others are the ultralow expansion of the solid-solution crystals and the matching low thermal expansion of the siliceous residual glass. The resistance to grain growth is particularly important in view of the marked anisotropy in

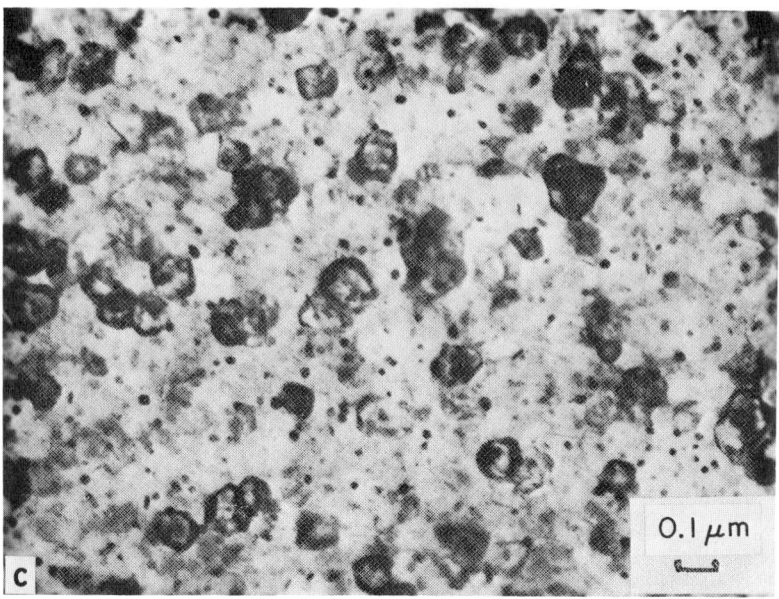

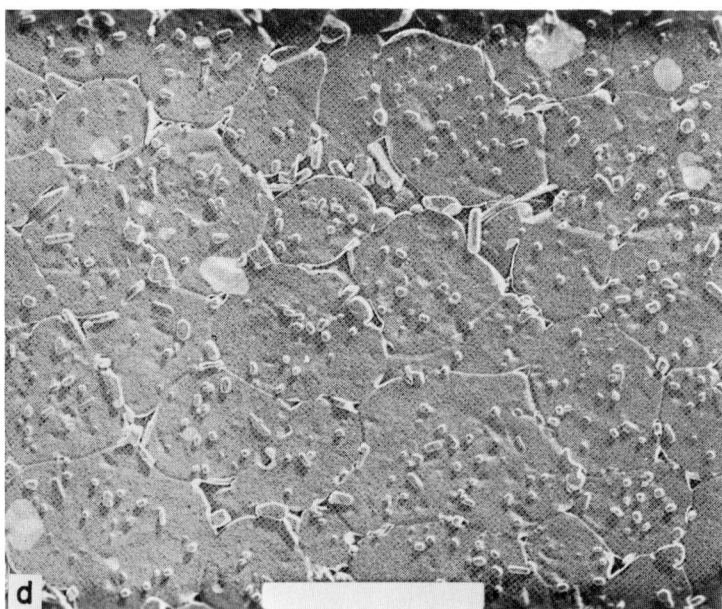

FIG. 8c,d. $Li_2O-Al_2O_3-SiO_2-TiO_2$ glass. (c) Crystallized sample heated to 950°C that contains β-quartz solid solution as the major phase (bright field transmission). (d) Typical microstructure of β-spodumene solid-solution glass-ceramic after 45 min at 1000°C, showing general impingement of grains (replica, bar = 1 μm).

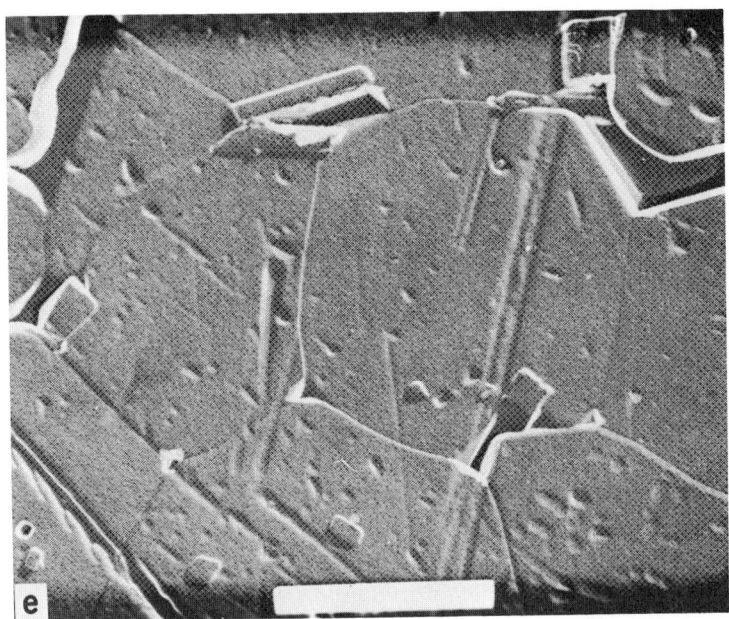

FIG. 8e. Li$_2$O–Al$_2$O$_3$–SiO$_2$–TiO$_2$ glass. Microstructure of β-spondumene solid-solution glass-ceramic after 20 hr at 1200°C showing the results of secondary grain growth (replica, bar = 1 μm). (All reprinted from Beall (1972), by courtesy of the American Ceramic Society.)

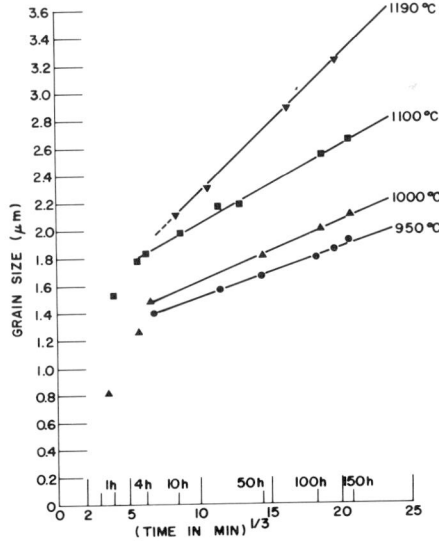

FIG. 9. Grain size as a function of temperature and time for a typical β-spodumene solid-solution glass-ceramic. (Reprinted from Chyung (1969), by courtesy of the American Ceramic Society.)

thermal expansion of β-spodumene crystals (Ostertog *et al.*, 1969) (see also Section IV).

IV. Properties of Glass-Ceramics

The properties of glass-ceramics are limited only by the spectrum of properties of crystals capable of being formed from glass. Some of the most useful characteristics that can be designed into a glass-ceramic include ultralow thermal expansion, optical transparency, high mechanical strength due to surface compression, and physical machinability. The choice of a glass-ceramic system, the development of the desired crystalline assemblage and microstructure, and the various processes required to control these properties will be now considered.

A. Ultralow Thermal Expansion

Of the crystalline polymorphs of silica, β quartz and keatite possess the unique characteristic of ultralow volume thermal expansion. This property is based on the unique helical structure of these phases. The silica tetrahedra are arranged around hexagonal and tetragonal screw axes in β quartz and keatite, respectively, in such a manner that when one crystal direction expands the other contracts. In β quartz, the a axis expands with increasing temperature while the c axis contracts. The opposite is true of keatite. Stuffed derivatives of these two polymorphs can be crystallized from glasses in the lithium aluminosilicate system (see Section III). Beta-spodumene solid solution is a stuffed derivative of keatite, whereas β eucryptite and related solid solutions are stuffed derivatives of β quartz. Figure 10 shows the unique thermal expansion characteristics of the β-spodumene solid-solution phase, which allows the development of glass-ceramics of exceptional thermal stability and shock resistance. The thermal expansion decreases with increasing silica but remains very low across the solid-solution region. The behavior over a wide range of silica levels allows considerable flexibility in the base composition of a β-spodumene glass-ceramic. As a result, such materials can be tailored to meet other property and glass-forming requirements. Figure 11 shows the crystallographic expansion in the two lattice directions of the same solid-solution compositions. The anisotropy is evident. Beta-spodumene actually has a very high thermal expansion along the c axis. It is therefore necessary for a ceramic material based on this phase to have a fine-grained microstructure so that stresses due to this anisotropy can be minimized.

The β-quartz phase represents an even more complex series of solid solutions, combining silica with $AlPO_4$, BPO_4, $Li(AlO_2)$, $Mg(AlO_2)_2$, $Mn(AlO_2)_2$, and $Zn(AlO_2)_2$ (Schereyer and Schairer, 1961; Beall *et al.*,

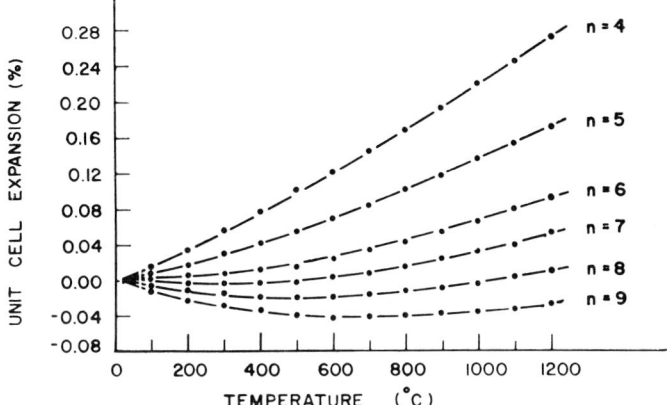

FIG. 10. Volume thermal expansion of β-spodumene solid solutions: $Li_2O-Al_2O_3-n\,SiO_2$. (Reprinted from Ostertag et al. (1968), by courtesy of the American Ceramic Society.)

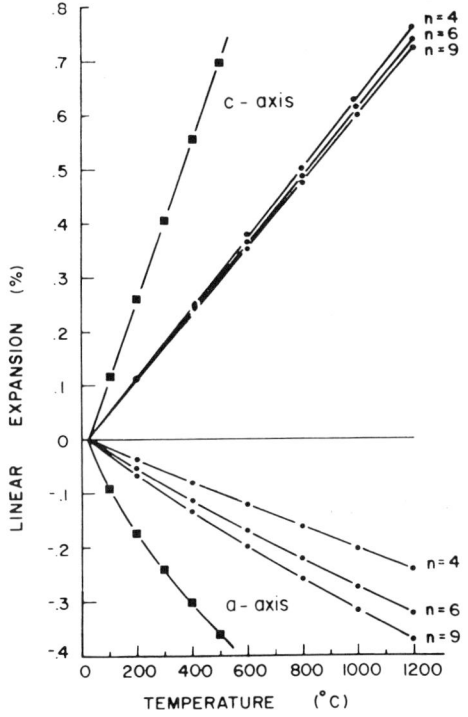

FIG. 11. Linear thermal expansion in direction of a and c axis in β-spodumene solid solution: $Li_2O-Al_2O_3-n\,SiO_2$. (●) β-spodumenes. (■) Keatite. (Reprinted from Ostertag et al. (1968), by courtesy of the American Ceramic Society.)

1967; Petzoldt, 1967). Most of these crystals possess low thermal expansion, but unlike the β-spondumene or keatite series, most are metastable, breaking down to other phases when heated above 1000°C for relatively short periods.

Applications of glass-ceramics exhibiting ultralow thermal expansion include transparent and opaque baking and top-of-stove ware, flat cooking surfaces, telescope mirror blanks, and heat-resistant windows.

B. Transparency

Most ceramic bodies appear white because they are composed of crystalline particles that act as scattering centers for visible light. The intensity of light scattering by homogeneous spherical particles whose diameter is small in comparison to the wavelength of light was derived by Rayleigh (1871, 1899) as

$$I(\theta) = \left(\frac{1 + \cos^2 \theta}{r^2}\right) \frac{8\pi^4}{\lambda^4} \alpha^6 \left|\frac{M^2 - 1}{M^2 + 2}\right|^2 I_0,$$

where I is the specific intensity, θ the scattering angle, r the distance from the scattering center, I_0 the intensity of the incident beam, λ the wavelength of light, α the radius of the particle, and M the ratio of the refractive index of the particle to that of the surrounding medium.

From this equation the criteria for complete transparency in a light-transmitting material are (i) $\alpha \ll \lambda$, where the particles are much smaller than the wavelength of light, or (ii) $M \cong 1$, where the refractive indices of the particles and surrounding medium approach equality. Either condition is sufficient for transparency, providing, of course, that there is insufficient absorption of light to cause serious attenuation. In the more common case of scattering particles where the size of the particle may approach or exceed the wavelength of light, say, where $2\alpha \gtrsim \lambda$, Mie (1908) showed that for transparency the refractive index of the particles must approach that of the surrounding medium.

It is therefore clear that two possibilities exist for producing a transparent glass-ceramic system (Beall and Duke, 1968). In the first case, the dispersed crystallites must be sufficiently small to produce no effective scattering in the visible spectrum. In the second case, where larger crystals are allowed, optical isotropy must be achieved within the glass-ceramic, i.e., the refractive index of the crystals and glass must be virtually equal, and the birefringence within any crystal must be very small.

An example of the first case occurs in the $Al_2O_3-SiO_2$ system, where glass-ceramics containing a fine dispersion of mullite in siliceous glass can be formed (Fig. 4b). The mullite crystallites form with the relict spherical structure of the original amorphous phase separation, less than 1000 Å in diameter. If the mullite crystals are allowed to grow beyond 2000 Å through prolonged thermal exposure above 1150°C, light scattering ensues, because mullite is not optically isotropic, nor does it nearly match the glass in refractive index.

Highly crystalline glass-ceramics composed of low-expansion β-quartz solid solution can be made transparent through special heat treatment of a series of glass compositions in the system $SiO_2-Al_2O_3-Li_2O-MgO$. Zirconia and/or TiO_2 are used for effective nucleation of this metastable phase. In the composition region where the β-quartz solid solution is primarily stuffed with magnesium, the birefringence of the quartz solid solutions can be made very small. With little residual glass and a close index match, excellent transparency can result even with crystals exceeding 1 μm in size. Some of these transparent materials can withstand temperatures over 1100°C for several hours without suffering opalization.

The transparent β-quartz solid-solution glass-ceramics combine low thermal expansion and excellent chemical durability with good forming characteristics in the glassy state. Fine-grained compositions can also take a very fine optical polish. For these reasons they have been applied as materials for transparent cookware, heating coil housing, oven windows, and telescope mirror blanks.

C. MECHANICAL STRENGTH

Mechanical strength superior to glass is one of the most important characteristics of glass-ceramics. The theoretical strengths of homogeneous nonporous brittle silicate materials such as glass has been estimated at about 10^6 psi (Hutchins and Harrington, 1966). The measured strength of fibers of fused silica, protected in vacuum, has reached this value (Ernsberger, 1964). Commercial glass fibers and acid-etched glass rods have shown measured bending strength values of 250,000 psi (Hutchins and Harrington, 1966). Moderate abrasion, however, quickly reduces the practical strength of glass to the typical range of 5,000–10,000 psi. Polycrystalline glass-ceramics provide more resistance than glass to the propagation of fractures. While their inherent strength in a pristine and flawless state may not be as great, their practical or abraded strength is almost always better. Typical values range from 10,000 to 30,000 psi.

Significant progress has been made in recent years developing techniques to increase the practical strength of glass-ceramic articles. The most successful involve inducing a thin layer of compression, generally several mils in thickness, at the surface of the material. Two phenomena that can be used to effect this surface compression are ion exchange and differential surface crystallization.

1. Surface Strengthening from Ion Exchange

Compression can be developed at the surface of glass-ceramic articles through ion exchange between crystals in the glass-ceramic and molten salts. Stress may result directly from crowding of large ions in positions previously held in the crystal by smaller ions or indirectly through differential thermal expansion between the exchanged surface and the body of the glass-ceramic.

The ion-exchange phenomena can be divided into two types: (i) where the surface phases resulting from the exchange are structurally similar to the original and interior phases (i.e., where solid solution is permitted) and (ii) where the surface phases are structurally distinct from the original and interior phases (i.e., where phase transformation has occurred). The first type is typified by $Na^+ \rightleftarrows Li^+$ exchange in β-spodumene solid-solution glass-ceramics (Karstetter and Voss, 1967). Replacement of the small Li^+ ion by the larger Na^+ ion is believed to elastically deform the cation–oxygen bonds in the crystal lattice. The distortion caused by this ionic crowding directly produces a surface compression that typically results in modulus-of-rupture values on abraded samples of around 50,000 psi, an increase of 35,000 psi above the normal body strength. Exchange normally takes place at relatively low temperatures (500–600°C) in a salt bath composed largely of $NaNO_3$. A compression layer typically a few mils in thickness is developed within several hours. The thermal stability of the resulting strength is not great because of the low exchange temperature.

Another example of ion exchange involving solid solution in a glass-ceramic is the $2Li^+ \rightleftarrows Mg^{2+}$ exchange, which can be achieved in β-quartz solid-solution compositions. Here glass-ceramics containing Mg-stuffed derivative of β quartz as the major phase are placed in a high-temperature Li_2SO_4 bath (800–900°C) (Beall et al., 1967). Surface compression in this case results from both ionic crowding and differential thermal expansion. The $2Li^+ \rightleftarrows Mg^{2+}$ exchange results in an increase in the unit cell volume and a decrease in the thermal expansion coefficient of the β-quartz solid-solution crystals. Typical abraded strengths range from 40,000 to 120,000 psi for bodies of thermal expansion coefficient from 15 to 45×10^{-7}/°C. Because this exchange is controlled kinetically by the mobility of the sluggish Mg^{2+} ion, the thermal stability of strengthened articles is considerable.

An example of ion exchange that results in compression due to a different phase forming in contact with the molten salt (i.e., surface phase transformation) occurs when nepheline ($Na_2O \cdot Al_2O_3 \cdot 2SiO_2$) glass-ceramics are ion-exchanged in a K_2SO_4–KCl salt bath at 750°C (Duke et al., 1967). In this case, crowding of the larger K^+ ions in the smaller Na^+

sites in the nepheline structure results in a displacive transformation to kalsilite ($K_2O \cdot Al_2O_3 \cdot 2SiO_2$). This transformation normally involves a significant increase in specific volume (~10%), and it produces great surface compression in the glass-ceramic, despite significant compensating viscous relaxation. Modulus-of-rupture values on rods ion-exchanged in this manner and then abraded may average in excess of 200,000 psi. Figure 12 shows an electron microphotograph of the sharp phase profile due to this ion-exchange-induced transformation. The chemical and stress profiles follow similar step functions in these materials, in contrast to the smooth profiles characteristic of ion exchange involving solid-solution phases, where the standard diffusion laws governing ideal solutions (Fick's laws) prevail.

Applications of glass-ceramic materials strengthened by ion exchange include high-strength cooking utensils, strengthened glass-ceramic panelling, and frangible material devices. The first two applications require a combination of high-strength and low-thermal-expansion behavior. The last application requires an article in a highly stressed condition such that, if the surface is penetrated, it will fail violently, producing minute fragments.

2. Surface Compression from Differential Crystallization

When a glass is converted to a glass-ceramic, there are frequently phenomena occurring at or near the surface that alter the kinetics of nuclea-

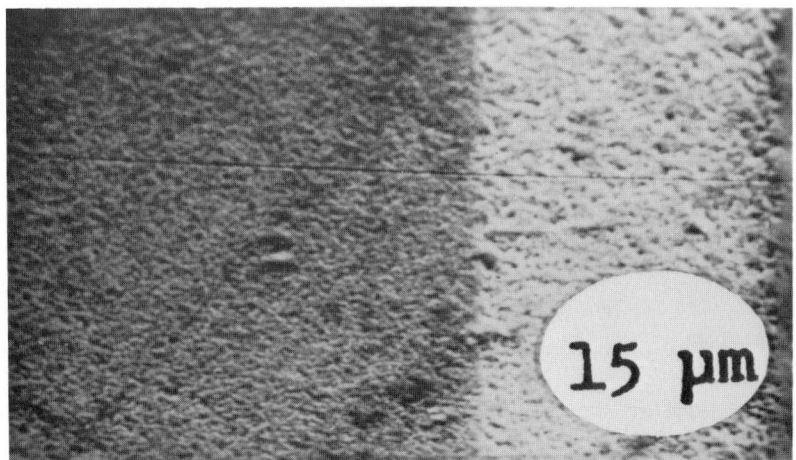

FIG. 12. Secondary electron image of a $K^+ \rightarrow Na^+$-exchanged ($\frac{1}{2}$ h at 720°C) glass-ceramic showing nepheline (dark)–kalsitite (light) interface. The surface of the glass-ceramic is at the extreme right. (Reprinted from Duke *et al.* (1967), by courtesy of the American Ceramic Society.)

tion or crystal growth. These may lead to a different phase assemblage at the surface as compared to the interior and may, on cooling the glass-ceramic, result in the spontaneous development of surface compression. Such glass-ceramics are sometimes referred to as self-strengthening, in the sense that a glass article on transformation to a glass-ceramic is strengthened to a degree far greater than that due to the difference in body strength between the glass and glass-ceramic.

One example of strengthening from differential surface crystallization occurs when a crystal phase of low thermal expansion nucleates selectively at the surface of a glass-ceramic. The formation of cordierite, $2MgO \cdot 2Al_2O_3 \cdot 5SiO_2$, commonly shows such preferential surface development (MacDowell, 1966). High-alumina glasses in the $MgO-Al_2O_3-SiO_2$ system develop fine-grained mullite internally on heat treatment above 900°C. At higher temperatures, near 1100°C, cordierite is nucleated at the surface. A layer several mils in thickness may be developed, which, on cooling, is pulled into compression. Differential thermal expansion between the cordierite skin and the higher-expansion mullite-glass interior causes this compression and consequent strengthening. Modulus-of-rupture values of 50,000 psi can be measured on abraded samples of these materials.

Zirconia-nucleated $MgO-Al_2O_3-SiO_2$ glass-ceramics develop β-quartz solid solution as the initial major phase. This metastable phase breaks down on further treatment through the exsolution of spinel. The spinel, higher in thermal expansion than the β-quartz solid solution, is selectively nucleated in the body of the article. The surface, on which spinel development is retarded, develops compression during cooling of the article (Tucker and Stewart, 1971). Modulus-of-rupture measurements as high as 75,000 psi have been measured on abraded samples of glass-ceramics of this type.

Selective volatilization of halide species during the crystallization of glass-ceramics is another phenomenon that can result in self-strengthening through surface compression. Titania-nucleated $Li_2O-Al_2O_3-SiO_2$ glass-ceramics can be doped with fluorides and heat-treated to develop β-spodumene solid solution as the major phase (Chyung, 1972). The fluoride is incorporated in the β-spodumene structure, presumably replacing oxide ions in the anion framework. On heat treatment above 1000°C, however, fluoride species volatilize from the surface. On further heating of the sample above 1100°C, a dense fluoride phase, in this case topaz ($Al_2SiO_4F_2$), is nucleated in the interior of the body, resulting in shrinkage of the body relative to the fluoride-depleted skin. Surface compression develops accompanying this differential densification of the interior. Modulus-of-rupture measurements as high as 40,000 psi have been measured on β-spodumene glass-ceramics strengthened by this technique. Unlike compression developed through a differential expansion mechanism, the stress due to this volatilization–differential densification phenomenon is retained at high temperatures.

The obvious practical advantage in producing surface compression in glass-ceramics through differential crystallization is that no postcrystallization process, such as in the case of ion exchange, is required. Self-strengthening techniques have found widespread application in developing useful strength in glass-ceramic sheet used in countertop and top-of-stove applications.

D. Machinability

Certain glass-ceramics containing major proportions of sheet silicate minerals of the mica group possess the unusual property of machinability; that is, they can be machined to precise tolerances and surface finish with conventional metal-working tools (Beall, 1972). These materials show less sensitivity to surface damage and greater resistance to brittle fracture than do most other ceramic materials.

Good machinability is developed in glass-ceramics whose microstructure is composed of mica crystals as a continuous phase. If the aspect ratio of these flaky crystals is large, interlocking of crystals usually develops with over 40 vol % mica. Greater crystallinity is required for ease of machining if the aspect ratio is small. Machinability is based on the unusual mechanical characteristics of mica flakes that result from weak bonding between structural layers. Mechanical rotation and/or displacement is easily effected along the cleavage planes, and fracture propagation across these planes is very difficult. This induces fractures to follow cleavage planes or crystal boundaries, causing detachment of individual crystallites or small groups of crystallites and enclosed glass. It is therefore easy to cause local damage but difficult to propagate large cracks. This explains the ease of machining and insensitivity of strength to abrasion that typify mica glass-ceramics.

The atomic structure of mica explains its two-dimensional habit and two-dimensional cleavage behavior. Figure 13 represents the structure of fluorophlogopite, a fluorine mica easily crystallized from glass. The basal section (001 plane) shows the infinite sheet network of corner-shared SiO_4 and AlO_4 tetrahedra. Monovalent fluoride ions, or in the case of most natural micas, hydroxyl ions, are essential to the structure and alternate with oxygen in forming an edge-shared octrahedral sheet of $Mg(O,F)_6$ sandwiched between two tetrahedral layers. The 010 section shows the weakly cross-bonded nature of the crystal. The two tetrahedral sheets are tightly bonded by the octrahedral layer forming a complex plate of

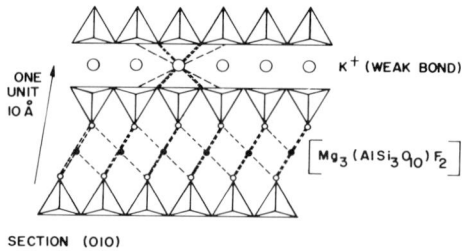

Fig. 13. The structure of fluorophlogopite. (Reprinted from Beall (1972), by courtesy of the American Ceramic Society.)

composition $(Mg_3AlSi_3O_{10}F_2)^-$. These plates are weakly attached by 12-coordinated K^+ ions, giving rise to the basal cleavage phenomenon.

Considerable solid solution has been recognized in the phlogopite phase as it is crystallized from glass. Of particular interest, potassium deficiency is possible as long as the charge balance is maintained by cation substitution, that is, Si for Al in the tetrahedral layer or Al for Mg in the octrahedral layer. Thus it is possible to decrease further the weak cross bonding of mica crystals as precipitated from glass.

Figure 14 shows three different microstructures typical of machinable glass-ceramics. The aspect ratio of the mica crystals is seen to vary from approximately 2 (thick books) to over 10 (thin sheets). The aspect ratio depends on several factors, the most important of which are (1) cross bonding in the mica crystals and (2) viscosity of the glassy medium in which the crystals grow. Weak cross bonding favors a more two-dimensional habit because of surface energy considerations. Similarly, growth in a fluid medium would favor the natural flaky development, because mica-forming species could be transported easily. Conversely, strong cross bonding and a viscous growth medium would tend to hinder two-dimensional growth and aid in the development of thick crystals.

Figures 14a–c illustrate these effects of cross bonding and growth media variation on the microstructure. It should be noted that machinability is increased with increasing aspect ratio and is generally optimized with the development of the "house of cards" structure illustrated in Fig. 14c.

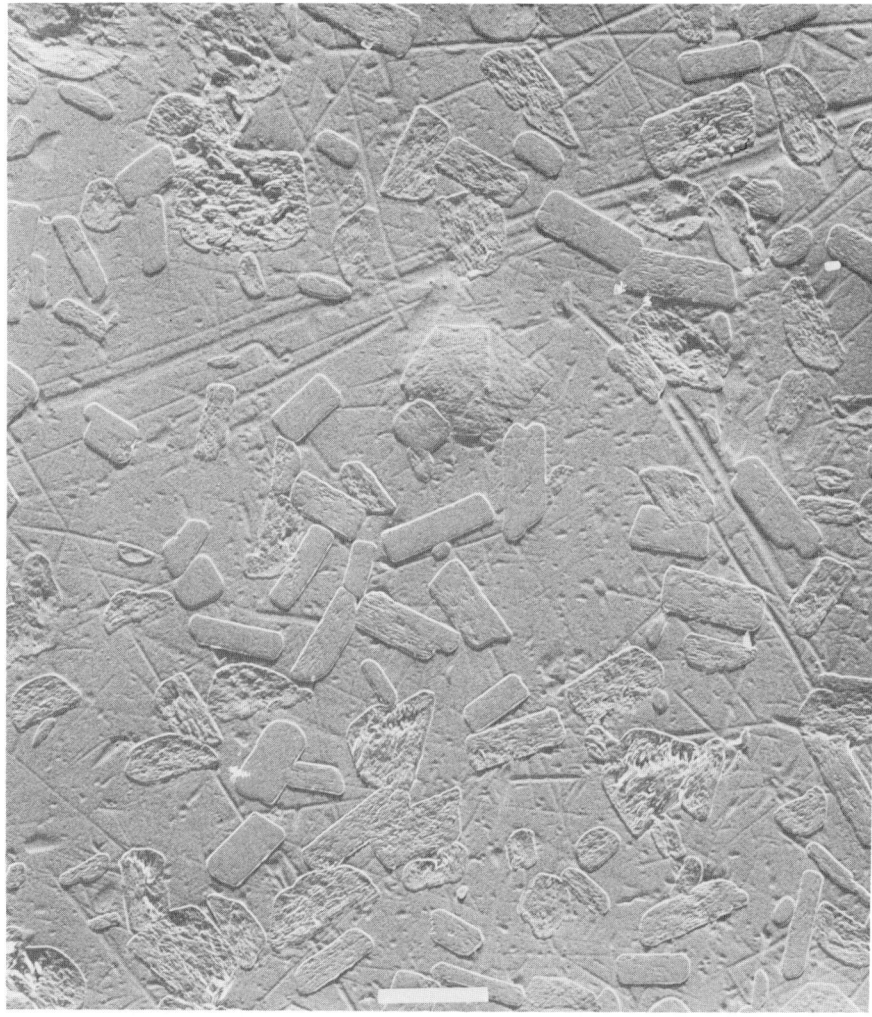

FIG. 14a. Variable microstructure in mica glass-ceramics. Saturated phlogopite solid-solution crystals of low aspect ratio in highly siliceous residual glass (replica, bar = 1 μm). Saturated phlogopite solid-solution crystals of moderate aspect ratio in modified silicate residual glass (replica, bar = 1 μm). Potassia-deficient phlogopite solid-solution crystals of high aspect ratio showing interlocking "house of cards" structure. (a), (b), and (c) reprinted from Beall (1972), by courtesy of the American Ceramic Society.]

Fig. 14b.

Aside from machinability, other properties of interest that can be obtained from mica glass-ceramics are thermal shock resistance and excellent dielectric characteristics. Although thermal expansion coefficient vary with composition and are usually moderate to high—matching a variety of metals—thermal shock resistance is often excellent. This behavior is related to the difficulty of fracture propagation. Dielectric properties are controlled by the inherent electrical characteristics of mica: high

FIG. 14c.

resistivity, high dielectric strength, and low dielectric loss. Actual measurement of volume resistivity and dielectric strength of machinable glass-ceramics have been as high as 10^{12} ohm-cm (500°C) and 3000 V/mil (25°C), respectively.

Possible applications for machinable glass-ceramics include precision dielectric components, high-quality insulators, and containers and conduits for molten metals.

E. MISCELLANEOUS PROPERTIES

Nucleation and internal crystallization of glass can be catalyzed by light or x radiation producing photosensitive glass-ceramics. In the SiO_2–Li_2O system, gold and silver particles act as nuclei produced through ultraviolet irradiation (Stookey and Maurer, 1961). Selected portions of a glass plate are irradiated and on heat treatment converted to a glass-ceramic containing dendritic crystallites of lithium metasilicate (Fig. 15a) (Stookey, 1949). Because this phase is easily attacked by acid, the glass-ceramic portion can be etched away while the more resistant glassy portion remains. The glass can subsequently be flood-exposed and heat treated at high temperature, where it is converted to a highly crystalline

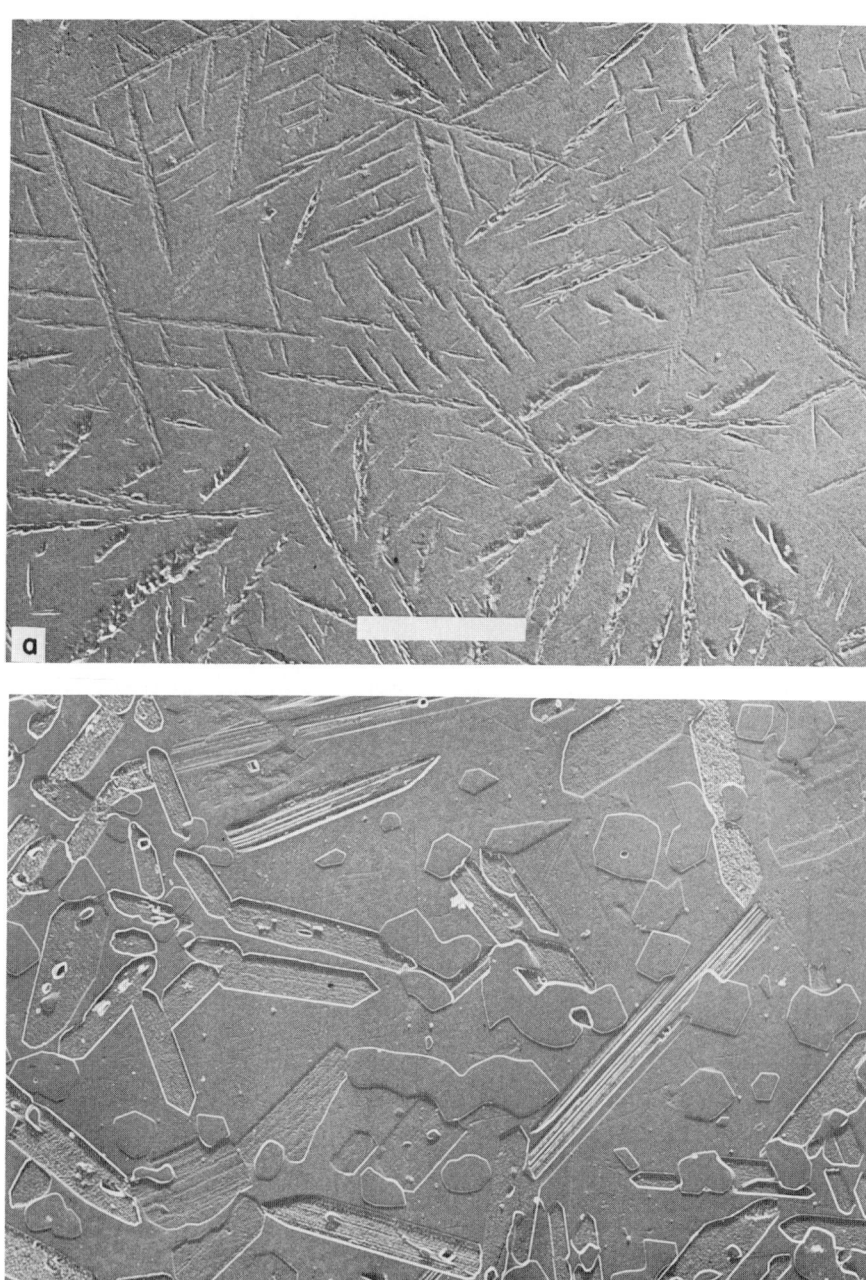

and tough lithium disilicate glass-ceramic (Fig. 15b) (MacDowell, 1966). By stacking patterened and etched plates together and fusing during crystallization, channelled fluid amplifier circuits can be produced.

Chemical durability is an important accessory property of glass-ceramics. Resistance to etching and staining by most corrosive liquids is exceptional in many aluminosilicate glass-ceramics, particularly those containing siliceous β-spodumene solid solution and mullite as major phases. Such glass-ceramics are used in top-of-stove applications, where outstanding acid and alkali resistance is essential. Sometimes, a glassy skin or glaze can be applied to a glass-ceramic of average or poor durability to improve its chemical resistance. If the glaze is fired on at high temperature, and if the thermal expansion coefficient is significantly lower than that of the glass-ceramic, the strength of the resulting article may also be increased. Such is the case with Centura tableware, where high strength and good durability result from the application of such a compressive glaze (Duke et al., 1966).

Hardness and abrasion resistance is another property that is far more variable in glass-ceramics than in their parent glasses. Machinable glass-ceramics containing mica are obviously very soft. Glass-ceramics containing major phases of high abrasion resistance, however, such as spinel ($MgAl_2O_4$), magnetite (Fe_3O_4), and sapphirine ($4MgO \cdot 5Al_2O_3 \cdot SiO_2$) often show Knoop hardness values in excess of 1000. Because of the fine-grained nature of these materials, they can be polished to a high degree. Inexpensive slag and basalt glass-ceramics containing magnetite and related minerals have been widely used in eastern Europe in applications where abrasion resistance is of prime importance (Kopecky and Voldan, 1964; Berzhnoi, 1970; Beall and Rittler, 1976).

Unique electrical properties can be designed into glass-ceramic materials primarily through controlling the composition of the constituent crystalline phases. High electrical resistivity at high temperature ($>10^{10}$ ohm-cm at 500°C) is characteristic of glass-ceramics composed of high-MgO phases such as cordierite and subpotassic fluorophlogopite. Low dielectric loss is also typical of these materials. The transmission of microwave signals through a glass-ceramic with minimum loss is critical for commercial applications. For example, cordierite glass-ceramics,

---

FIG. 15. Replica electron micrographs of Corning Code 8603 photonucleated glass. White line = 1 μm. (a) Lithium metasilicate crystals after exposure to x rays (10,000 roentgens) and heat treatment at 600°C for 1 hr. Crystals have been etched 5 sec in 0.5% HF, showing beginning of chemical machining process. (b) Lithium disilicate crystals after exposure to x rays (160,000 roentgens) and heat treatment of 600°C and 850°C for 1 hr each; polished, etched for 1 min in 0.5% HF. (Reprinted from MacDowell (1966), by courtesy of the American Chemical Society.)

combining this property with thermal-shock resistance and mechanical strength, are useful as missile nose cones.

Ferroelectric glass-ceramics have been developed for application as capacitor dielectrics (McMillan, 1979; Herczog, 1964). Certain niobate glass compositions form glass-ceramics composed of perovskite phases similar in structure to barium titanate. These materials are easily manufactured as glasses into thin sheet form, and on crystallization combine very high dielectric constants with low loss factors. Ferromagnetic crystals such as hexagonal and cubic ferrites have also been crystallized from glasses, but unique properties not attainable from conventional ceramic syntheses have not been observed.

V. Glass-Ceramic Processing

The unique properties of glass-ceramics depend not only on their composition but on the uniform, nonporous microstructure that results from their processing. In this section the process of glass manufacture is briefly described with emphasis on the special care required for those glasses designed for conversion to glass-ceramics.

A. Raw Materials

There are many raw materials used in the manufacture of glass. Most are inexpensive and simple in composition, i.e., sand (SiO_2), soda ash (Na_2CO_3), limestone ($CaCO_3$), clay etc. The ultimate choice is dependent primarily on cost and purity, but other factors, such as handling, mixing, and melting rate, often come into play. Since impurity levels and concentrations can often radically affect the crystallization sequence of glass, extra care in analysis and testing of batch is required.

Conventional glass systems are composed of oxides of the common elements silicon, aluminum, boron, sodium, calcium, etc. Nonoxide species such as fluorides, chlorides, and sulfides may be included. Because of the multicomponent nature of glass, complex natural minerals containing several chemical species common to the particular glass formulation may be used in the batch. In fact, the use of such minerals generally aids in the mixing and melting process. Important examples include feldspar (($Na,K)AlSi_3O_8$), dolomite ($CaMg(CO_3)_2$), petalite ($LiAl(Si_4O_{10})$), spodumene ($LiAlSi_2O_6$), and nepheline (($Na,K)AlSiO_4$).

B. Handling and Mixing

The handling and mixing of raw materials used in glass melting is a very important, but often neglected, part of the manufacturing process. Uniform mixing of the batch materials helps to facilitate melting and insure homogeneity. Care must be exercised to minimize impurities, especially iron and other transition metals, which may be introduced during mixing or handling.

Batch handling may be a small-scale manual operation or may be totally automated. Materials are shipped in packages or in bulk, depending on cost and plant requirements. Major batch ingredients such as sand are commonly stored in large silos, whereas minor chemical components are kept sealed in their original containers.

Mixing systems range from simple hand operations to complex computer-controlled systems. In a typical automated system, a belt conveyor collects the preweighed ingredients from electrically operated feeders and conveys them to a mixer. Power-driven mixers of a revolving drum or a double-cone configuration tumble the batch upon itself. Large blades act as shovels to lift and spread the materials more effectively.

The ingredients of a glass batch often vary in grain size and density, and one requirement of good mixing is to minimize batch segregation, which causes subsequent problems in melting. Once the batch has been thoroughly mixed, it is ready to be delivered to a melting furnace. In some cases, however, additional preparation of the batch is required. This may involve compaction of the batch into briquettes using some suitable bonding agent. This precludes segregation, eliminates batch loss, minimizes the inconvenience and hazards caused by "dusting" of fine materials, and facilitates storage and handling.

C. Glass Melting

Melting is the thermal process by which mixed batch materials are converted into a liquid free of any crystalline matter. Many of the individual batch components have melting points far in excess of any temperature that can be achieved in a glass furnace, but reaction of batch materials and subsequent dissolution allows melting at reasonable temperatures. On initial heating, various components of the batch react to form a liquid. Unreacted oxides then become part of the melt by a process of dissolution, where the reaction rate is controlled by diffusion of the components. Dissolution can be accelerated by increasing the melting temperature or through agitation techniques.

When the batch is thoroughly melted and fairly homogeneous, gaseous inclusions may remain, especially where the melt is viscous. Such inclusions often affect the appearance, light transmission, mechanical strength, and other properties of the final product and must therefore be removed. Fining is the name given to the process of removing gas bubbles from a melt. Sources of these bubbles are (1) gases evolved during melting, (2) air entrapped within the batch, or (3) gases evolved from the porous refractory materials lining the melting chamber.

The most common method of fining glass is to add an extra component to the batch that will accelerate fining by causing the existing bubbles to become larger. Gas evolution is a chemical characteristic common to these fining agents. For example, arsenic and antimony pentoxides release oxygen, and sulfates decompose, releasing SO_2. These species enter the entrapped bubbles and cause them to grow larger. They then rise to the surface and eventually break, leaving a high-quality glass melt. Other techniques used to eliminate bubbles include decreasing the melt viscosity by increasing temperature and the use of external gas pressure or vacuum. The fining of glass is a complex and well-researched phenomenon (Hutchins and Harrington, 1966).

The final stage of the melting process, termed conditioning, allows the glass to achieve uniform physical and chemical properties. This may be accomplished simply by holding the glass at temperature for sufficient time for diffusion to eliminate composition gradients. Agitation techniques include passing large gas bubbles vertically through the melt or mechanically mixing the glass in a stir chamber.

There are numerous types of furnaces used in glass melting. These range from small electrical units that melt experimental batches to large, continuous glass-melting tanks. There are two basic types of furnaces: batch and continuous. A batch furnace or "day tank" is one in which all phases of melting occur in the same enclosure over a period of time. Batch is introduced into a cooled furnace area, which is then heated to melt, fine, and condition the glass. A continuous tank is designed with different sections to accomplish various facets of melting and fining. Batch is introduced into a melting section (back end); the melt flows into a fining section; and finally into a forehearth, where conditioned glass is delivered (front end). All phases of melting occur simultaneously, but at different locations. One of the main advantages of the continuous tank is the reproducibility and homogeneity of the glass, that can be achieved piece after piece, day after day.

Glass tanks are normally lined with refractory oxides that will not react excessively with the melt and that will withstand the high temperatures employed in the glass-melting process. For certain types of special glasses, such as those used in optics, even slight reaction between the melt and the refractory may produce intolerable variations in composition and refractive index of the glass. In such cases the tank is lined with nonreactive platinum-group metals. Most glass tanks utilize natural gas as the prime method of heating. Other sources of heat are the burning of fuel oil and electric heating by means of direct current passed through the glass from a series of electrodes. Because of the natural gas shortage in many areas, fuel oil heating and electric boosting are becoming more common.

D. FORMING

Once a glass is melted and is ready for delivery, there is a wide variety of forming or shaping procedures that can be used. These include spinning, pressing, blowing, rolling, and casting. The choice of a forming method is limited by the liquidus temperature of the glass and the viscosity of the glass at this temperature. If forming is allowed to take place below the liquidus, uncontrolled devitrification can occur.

Figure 16 shows typical viscosity–temperature curves for three representative glass-ceramic compositions. The normal range of viscosities where melting occurs and where some of the different forming methods are used is also shown. The melting process takes place at low viscosities in the range of 70–500 P. For the three glasses shown, this would translate to temperatures ranging from about 1400°C for glass B to 1600°–1700°C for glasses A and C. Glass B is representative of glasses in the $MgO-Al_2O_3-SiO_2$ system designed to crystallize to a major portion of cordierite, whereas glasses A and C are from the β-spodumene solid-solution region of the $Li_2O-Al_2O_3-SiO_2$ system.

Forming operations are normally carried out in the viscosity ranges shown. Glass is delivered from the tank in a fluid condition of about 400–1000 P. Centrifugal casting is possible only with very fluid glasses,

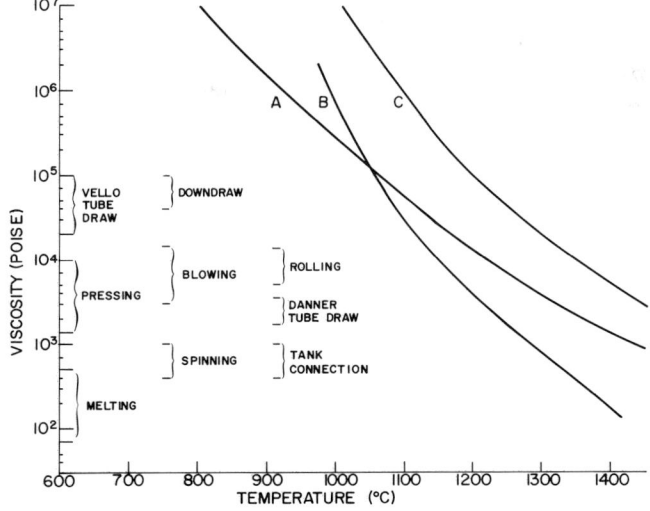

FIG. 16. Viscosity-forming relationships for three representative glass-ceramic compositions.

whereas a Vello tube draw or a downdraw process must be carried out with very stiff glass. The techniques used in forming of glass-ceramic compositions are limited not only by temperature and therefore viscosity, but also by the liquidus temperature of the glass. Since these compositions are specifically designed to crystallize, great care must be exercised during forming to minimize or eliminate uncontrolled crystallization. Forming operations must, therefore, take place at temperatures above the liquidus.

For illustration, let us consider the liquidus temperatures of the three glasses in Fig. 14 as 1220°C for A, 1300°C for B, and 1250°C for C. The viscosity of glass A at its liquidus is 10^4 P. This glass will not crystallize at temperatures above 1200°C, so that pressing, blowing, and certain rolling operations could be performed. Glass C has a viscosity at its liquidus of 5×10^4 P, which would make it suitable for forming by any process except a downdraw, although casting of such a viscous glass would be impractical because of the very high temperatures necessary to reach the required fluidity. Glass B has a viscosity of only 800 P at its liquidus, which would restrict it to being formed by casting. This limitation in forming is often of great commercial significance and is a major parameter to consider when working with glass-ceramics.

For details of the forming processes themselves, the reader is referred to Hutchins and Harrington (1966).

Once glass articles have been formed, they are commonly annealed at lower temperatures to remove residual stresses caused by nonuniform cooling. The viscosity of glass at the annealing point is about 10^{13} P, the viscosity where stress can be removed through viscous flow in practical time periods.

E. Finishing

There are a number of operations subsequent to forming that may be performed on a glass article to ensure the desired shape. Secondary hot-forming operations, such as sagging or redrawing, involve reheating the glass until it softens and then reworking the article. Fire polishing is commonly used to remove rough or sharp edges that result from the forming operations. These rough areas are heated until they soften while the rest of the article is cool enough to retain its shape. Fire polishing is often difficult in the case of glass-ceramics because premature crystallization may develop.

Mechanical grinding of glass articles is another common finishing technique and is necessary when close dimensional tolerances not achievable in hot-forming processes are required.

F. Crystallization—Heat Treatment

Glasses that have been especially formulated for conversion to glass-ceramics require a heat treatment step to effect their crystallization. The time–temperature releationship is often critical. Heating rates and the time held at different temperatures play an important role in the nucleation and development of the desired crystalline phase and resulting microstructure (National Academy of Science, 1968).

In a normal heat treatment, the glass article is heated quite rapidly to a temperature about 50–100°C above its annealing point and then held for a period of time. This hold ensures adequate nucleation. The temperature is then raised more slowly until a top temperature is reached. The bulk of the glass crystallizes and the crystallites grow until the desired microstructure is achieved. Depending on the composition and the crystalline assemblage desired, the total heating cycle can range anywhere from a few hours to several days.

As the glass crystallizes, there is usually an increase in the density of the body and consequent shrinkage of the glass-ceramic. Although the shrinkage is normally small, of the order of 1–5 vol %, special care must be taken during the processing to ensure that it is uniform. Maintaining uniform temperature in the body is therefore essential. This can be difficult, however, in the case of crystallization of large articles. Crystallization is an exothermic reaction in which is generated by the transformation from an amorphous to an ordered structure. In some glasses, the heat of crystallization has been estimated to be as high as 70 cal/g. Assuming a specific heat of 0.3 cal/g °C, temperature increases within the body of up to 200°C can be achieved by instantaneous crystallization. In actual practice, different parts of large glass-ceramic castings have been shown to have temperature differentials up to 100°C as a result of this heat of crystallization. Very long crystallization schedules with slow evolution of heat are therefore required to minimize thermal and mechanical stresses and to prevent failure of large articles.

VI. Applications

Numerous glass-ceramic products have been introduced to the marketplace in the last 15 years. Table II lists some of the most important commercial materials. Major crystalline phases, general properties, and applications are indicated. Where surface strengthening techniques are used to augment the body strength of the glass-ceramic for the specific product application involved, the strengthening process is noted.

TABLE II

SOME COMMERCIAL GLASS-CERAMICS

Commercial designation	Major crystalline phases	Surface strengthening process	Properties	Applications
Corning Glass Works				
8603	$Li_2O \cdot SiO_2$ $Li_2O \cdot 2SiO_2$		Photochemically machinable	Fluidics devices, ink-jet-printing, numeric displays
9606	$2MgO \cdot 2Al_2O_3 \cdot 5SiO_2$ (cordierite) $MgO \cdot SiO_2$ (enstatite) TiO_2 (rutile)		Transparent to microwaves	Radomes
9607	β-Quartz s.s. (vanadium-doped)		Resistant to thermal shock and erosion	Range-top wear
9608	β-Spodumene s.s.		Low expansion, chemically durable, transparent tinted	Cooking utensils
0303	$Na_2O \cdot Al_2O_3 \cdot 2SiO_2$ (nepheline) $BaO \cdot Al_2O_3 \cdot 2SiO_2$ (celsian) TiO_2 (anatase)	Low expansion glaze	Low expansion, chemically durable High strength	Tableware
0330	β-Spodumene s.s. TiO_2 (rutile)	Ion exchange $Na^+ \rightleftarrows Li^+$	Mechanical, chemical, and thermal stability	Exterior, interior cladding, laboratory bench tops
9455	β-Spodumene s.s. $3Al_2O_3 \cdot 2SiO_2$ (mullite)		Low expansion, high thermal and mechanical stability	Heat exchangers
9611	α-Quartz s.s. $MgO \cdot Al_2O_3$ (spinel) $MgO \cdot SiO_2$ (enstatite)	Ion exchange $2Li^+ \rightleftarrows Mg^+$	Very high strength	Classified
9615	β-Spodumene s.s. TiO_2 (anatase)	Differential crystallization	Low expansion, high strength, thermal stability, chemical durability	Electric range tops

Name	Phase	Mechanism	Properties	Applications
9658	Phlogopite s.s.		Machinable, high dielectric strength, thermal stability	Precision dielectric components, vacuum feedthroughs
High K Glass-Ceramic	$(Ba,Sr,Pb)Nb_2O_6$		High dielectric constant	Capacitors
English Electric	$Li_2O \cdot 2SiO_2$		Sealable to metals	Bushings
Fuji HEATRON	β-Quartz s.s.		Low expansion, transparency, thermal stability	Heater tubes
General Electric Re-X	$Li_2O \cdot 2SiO_2$		Sealable to metals	Housings, bushings
Narumi Seito Neoceram 11	β-Spodumene s.s.		Low expansion, chemically durable	Cooking ware
Owens-Illinois Cer-Vit C101	β-Quartz s.s.		Low expansion, polishability	Telescope mirrors
C106	β-Spodumene s.s.		Translucent, low expansion	Covers for heat sources
C126	β-Spodumene s.s.	Differential crystallization	Opaque, higher strength, low expansion	Cooking surfaces pipes, valves
Pfaudler Nucerite	Predominantly alkali silicates		Coating of steel, chemical durability, impact resistance, abrasion resistance	Chemical process equipment, heating devices
PPG Hercuvit 106	β-Spodumene s.s.	Differential crystallization	Low expansion, translucent	Cooking surfaces
101	β-Quartz s.s.		Transparent	High-temperature windows, infrared transparencies
Schott Zerodur	β-Quartz s.s.		Transparent, low expansion, chemically durable	Telescope mirrors, cooking surfaces
Schott Ceran[a]	β-quartz s.s. (cobalt-doped)		Low expansion, dark red-violet	Cooking surfaces
Schott Robax[a]	β-quartz s.s.		Low expansion, transparent	Wood-burning stove and furnace windows

[a] Scheider and Kristen (1980).

The majority of these materials are aluminosilicate glass-ceramics nucleated with titania and/or zirconia. Beta-spodumene and β-quartz solid solutions are the most common aluminosilicate phases. The reason for this is that thermal stability, thermal shock resistance, and chemical durability are the unique properties relied on in the majority of applications.

It is clear from the diversity of unique properties and applications that glass-ceramics have taken their place with glasses and ceramics as a major family of inorganic nonmetallic materials.

References

Beall, G. H. (1972). *In* "Advances in Nucleation and Crystallization in Glasses" (L. L. Hench and S. W. Frieman, Eds.), American Ceramic Society Special Publ. No. 5.
Beall, G. H., and Duke, D. A. (1968). *J. Mat. Sci.* **4**, 340–352.
Beall, G. H., and Rittler, H. L. (1976). *Bull. Am. Ceram. Soc.* **55**(6), 579–582.
Beall, G. H., Karstetter, B. R., and Rittler, H. L. (1967). *J. Am. Ceram. Soc.* **50**(4), 181.
Becker, R. (1938). *Ann. Phys.* **32**, 128–140.
Berzhnoi, A. I. (1970). "Glass-Ceramics and Photo-Sitalls" Plenum Press, New York.
Buerger, M. J. (1954). *Am. Min.* **39**, 600.
Chyung, C. K. (1969). *J. Am. Ceram. Soc.* **52**(5), 342–345.
Chyung, C. K. (1972). *In* "Equilibria and Kinetics in Modern Ceramic Processing," pp. 110–121. U.S.–Japan Seminar Reports on Basic Science of Ceramics.
Doherty, P. E., Lee, D. W., and Davis, R. S. (1967). *J. Am. Ceram. Soc.* **50**(2), 77–81.
Duke, D. A., MacDowell, J. F., and Karstetter, B. R. (1967). *J. Am. Ceram. Soc.* **50**(2), 67–74.
Duke, D. A., Megles, J. E., MacDowell, J. F., and Bopp, H. (1968). *J. Am. Ceram. Soc.* **51**(2), 98–102.
Ernsberger, F. M. (1964). *Glass Ind.* **45**, 349–353, 384–388.
Grieg, J. W. (1927). *Am. J. Sic.* **13**, 1.
Herczog, A. (1964). *J. Am. Ceram. Soc.* **47**, 107–115.
Hutchins, J. R. III, and Harrington, R. V. (1966). *Ency. Chem. Technol.* **10**, 533–604.
Karstetter, B. R., and Voss, R. O. (1967). *J. Am. Ceram. Soc.* **50**(3).
Kopecky, L., and Voldan, J. (1964). *Ann. N.Y. Acad. Sci.* **266**, 1086–1105.
Lacy, E. D. (1963). *Phys. Chem. Glasses* **4**(6), 234–238.
MacDowell, J. F. (1965). *Proc. Br. Ceram. Soc.* **3**, 229–240.
MacDowell, J. F. (1966). U.S. Patent 3,275,494, September 27.
MacDowell, J. F. (1966). *In* "Nucleation Phenomena," pp. 68–75. Symposium sponsored by the Division of Industrial and Engineering Chemistry, American Chemical Society Publ., Washington, D.C.
MacDowell, J. F., and Beall, G. H. (1969). *J. Am. Ceram. Soc.* **52**(1), 17–25.
McMillan, P. W. (1979). "Glass-Ceramics," 2nd ed. Academic Press Inc., New York.
Mie, G. (1908). *Ann. Phys.* **25**, 377.
National Academy of Science (1968). "Glass Crystallization, Ceramic Processing," pp. 81–83. Prepared by the Committee on Ceramic Processing, Materials Advisory Board, Div. of Eng., Nat. Res. Council, Publ. 1576. Washington, D.C.
Neilson, G. F. Jr. (1970). Small Angle X-Ray Scattering Study of Nucleation and Devitrification in a Glass Ceramic Material. AFOSR Contract/Grant Number AF49 (638)-1701.

Ostertag, W., Fischer, G. R., and Williams, J. P. (1968). *J. Am. Ceram. Soc.* **51**(11), 651–654.
Petzoldt, J. (1967). *Glastech. Ber.* **40**(10), 385–396.
Rayleigh, Lord (1871). *Phil. Mag* **41**, 274, 447.
Rayleigh, Lord (1899). *Phil. Mag.* **47**, 375.
Scheider, H., and Kristen, K. (1980). Schott Information, Schott Jena Glaswerk, Mainz, No. 1.
Schereyer, W., and Schairer, J. F. (1961). *Z. Krist.* **116**, 60.
Stookey, S. D. (1949). *Ind. Eng. Chem.*, **41**, 856.
Stookey, S. D. (1959). *Ind. Eng. Chem.* **51**(7), 805–808.
Stookey, S. D. (1959). *Glastech. Ber. Sonderband* **32K**, 1.
Stookey, S. D., and Maurer, R. D. (1961). *In* "Progress in Ceramic Science," Vol. 2. Pergamon, Oxford.
Tamman, G. (1925). "The States of Aggregation." Van Nostrand Reinhold, New York.
Tucker, R. W., and Stewart, D. R. (1971). *In* "Scientific and Technical Communications," **2**, IX Int. Cong. on Glass, pp. 1119–1132.
Uhlmann, D. R. (1972). *In* "Advances in Nucleation and Crystallization of Glasses" (L. L. Hench and S. W. Frieman, Eds.), American Ceramic Society Special Publ. No. 5.
Warren, B. E. (1938). *J. Am. Ceram. Soc.* **21**, 259–265.
Zachariasen, W. H. (1932). *J. Am. Chem. Soc.* **54**, 3841–3851.

Materials Index*

A

A, 65
Ag, 191, 256–258
AgI, 181, 195, 231
$Ag_7I_4AsO_4$ glass, 201–202
Ag_2O, 181, 196
$AgPO_3$, 196
Ag_2SO_4, 195, 231
Al, 3, 64, 116, 133, 191
Albite–anorthite system, crystallization temperatures, 39
$AlCl_3$, 121
AlF_3, 97, 160, 227
Alkali disilicates, crystallization temperatures, 39
Alkali metals, with oxides, 56
Alkaline earths, with oxides, 56
Alkali nitrates, 62
Al_2O_3, 23, 56, 61, 62, 64, 72, 75, 78, 88, 94, 122, 153, 154, 170, 205, 206, 410
  radiation effects on, 90
Al_2O_3–CaO–SiO_2, 58
Al_2O_3–Ga_2O_3, 57
Al_2O_3–Gd_2O_3, 51
Al_2O_3–Nb_2O_5, 57
Al_2O_3–PbO, 56
Al_2O_3–SiO_2, 56, 68, 181–183, 411, 412
  fibers, 89
  phase diagram, 182, 409
Al_2O_3–Ta_2O_5, 57
Al_2O_3–TeO_2, 205, 206
Al_2O_3–TiO_2, 57
Al_2O_3–UO_2, radiation effects on, 90
Al_2O_3–Y_2O_3, 51
$Al_2(OH)_2SiO_4$ (topaz), 89
$Al(PO_3)_3$, 193, 194
$AlPO_4$, 191
$Al_2SiO_4F_2$, 428
Alumina silicates, 73
Aluminum borosilicate, 68, 73

Anorthite, 21, 26, 27, 33–35, 40
Arsenosilicates, 73
As^{5+}, 3, 7
As–S, 62
As–S–Cl, 247–248
As–S–Se, 247
As–Se–Sb, 62
As–Se–Te, 247
As_2O_3, 18–19
AsO_5, 152
As_2S_3, 235, 241, 251
As_2Te_3, 235, 242
As_2Se_3, 233, 234, 241–244, 251
As_2Te_3 compositions, 38
Au, 64
Au–Ge–Si alloy, 31, 376
Au–Si alloy, 53

B

B, 3, 69, 73
Ba, 3, 133, 200
$Ba(Ac)_2$, 213
BaO, 149, 154, 206
BaO–Nb_2O_5, 57
BaO–TiO_2, 57
$BaPO_3F$, 195
Barium silicate, 64
$BaSi_2O_5$, 408
BBr_3, 79
BCl_3, 73, 79
Be, 3, 190, 200
BeF_2, 209, 223, 224, 226–228, 230
$Be(NO_3)_2$, 219
BeO, 153, 154, 206
$3BeO \cdot Al_2O_3 \cdot 6SiO_2$ (beryl), 89
Bi, 65, 169, 208
Bi–Se, 62
$BiCl_3$, 225–226
$Bi_{12}GeO_{20}$, 64
BiO_3, 205

* Formulas given only. For given or trivial name, see Subject Index.

B_2O_3, 4, 19, 50, 62, 68, 73, 75, 78, 97, 121, 149, 161–163, 207, 231
B_2O_3–NaF, 231
B_2O_3–SiO_2, 72, 83, 171–172
BO_4, 164–166

C

C, 64, 65
Ca, 3, 133, 191
$CaAl_2Si_2O_8$ (anorthite), 91, 94, 95
$Ca_3Al_2Si_3O_{12}$ (garnet), 89
$CaCO_3$, 59, 213
Cadmium stannate, 64
CaF_2, 97, 227, 229, 230
Calcium aluminosilicate glass, 64, 65
$Ca(NO_3)_2$, 211, 212
$Ca(NO_3)_2$ + KSCN, 215
$Ca(NO_3)_2 \cdot 4H_2O$, 217, 218, 222
0.4 $Ca(NO_3)_2$–0.6KNO_3[7], 1
CaO, 23, 153, 154, 206
CaO $\cdot$ Al_2O_3 $\cdot$ 2SiO_2, 21
CaO–Fe_2O_3, 57
CaO–Nb_2O_5, 57
CaO–NiO, 57
CaO–ZrO_2, 57
$Ca(PO_3)_2$, 193
$Ca(SCN)_2$ + KSCN, 215
CCl_2F_2, 78
Cd, 3, 59, 64, 133
$Cd(Ac)_2$, 214
$Cd(NO_3)_2 \cdot 4H_2O$, 217
CdO, 199
CdS, 259
Cd–Sb–S–I, 62
CdSe, 259
Ce, 190
CeO_2, 154, 159, 194
CeO_2–Al_2O_3, 57
CeO_2–Ga_2O_3, 57
CeO_2–Nb_2O_5, 57
CeO_2–NiO, 57
CeO_2–Ta_2O_5, 57
CeO_2–TiO_2, 57
CeO_2–ZrO_2, 57
$CH_3CO_2^-$, 210
$(CH_4)_3B$, 73
Cl, 238
Co, 159
Co–Au alloys, 62

$CoCl_2$, 216, 218
CoO, 122, 328
Cristobalite, 15
Cr_2O_3, 121
Cs, 3, 59, 124, 229
Cs_2O, 134, 204
  density, 137
Cu, 159, 180, 256–258
CuO, 121, 181, 194, 205
Cu_2O, 171
Cu–Sb–S–I, 62
$CuSO_4$, 216
Cu–Zr alloys, 54, 370–371, 376
Cu–Zr glasses, 377, 381, 388, 389, 390, 392, 395, 398
Cyclopentane, 43

D

Dy, 64
Dynasil, 77

E

Er, 64
Er_2O_3–Al_2O_3, 57
Er_2O_3–Ga_2O_3, 57
Eu_2O_3, 122

F

F, 70, 180, 187, 188–189
Fe, 64, 159, 185
$Fe_{82}B_{15}Si_3$ glass, 375
$Fe_{83}B_{15}Si_2$ glass, 374
$Fe_{40}Ni_{40}P_{14}B_6$ glass, 1, 397, 398
FeO, 23, 121, 159, 328
Fe_2O_3, 170, 200
Fe_2O_3–BaO, 56, 209
Fe_2O_3–CaO, 56
Fe_2O_3–PbO, 56
Fe–P–C metallic glasses, 389
FeW, 64

G

Ga, 3, 133
GaAs, 70
GaAsP, 70
Ga_2O_3, 57, 122, 194, 200
Ga_2O_3–CaO + SiO_2, 58

MATERIALS INDEX

Ga_2O_3–GeO_2–P_2O_5, 85, 86
Ga_2O_3–Nb_2O_5, 57
Ga_2O_3–PbO, 56
Ga_2O_5–Ta_2O_5, 57
GaP, 70
GaTe, 248
Gd_2O_3, 58, 200
Gd_2O_3–Al_2O_3, 57, 58
Gd_2O_3–Ga_2O_3, 57
Ge, 3, 31
Ge–As–Se system, 250, 252
$GeCl_4$, 73, 79, 88, 121
GeO_2, 11, 12, 17, 19, 42, 68, 73, 78, 111, 121, 177, 181, 196–197, 203, 207, 210, 340
GeO_2–P_2O_5–V_2O_5, 64
GeS_2, 235
Ge–S system, 249–251
Ge–Sb–Se system, 252
Ge–Sb–S–I, 62

H

H, 65
HCO_3^-, 210
He, 65
HfF_4, 227
HfO_2, 207, 208
HfO_2–CaO–SiO_2, 58
HfO_2–La_2O_3, 57
Hg, 3
Ho_2O_3–Al_2O_3, 57

I

In, 3
InAs, 70
In_2O_3–CaO–SiO_2, 58
InSb, 70
Ir, 64

K

K, 3, 59, 176, 229
KAc, 215
KAc–$CaAc_2$ glasses, 213
$KAlSi_3O_8$, 91, 94
K_3AsO_4, 152, 153
K_2BeF_4, 209
K_2CO_3, 59
KF, 227
$K(Mg, Fe^{2+})_3Si_3AlO_{10}(OH)_2$ (biotite), 91, 94
K_2O, 23, 136–142, 206
K_2O–SiO_2, 30, 59
$K_2O \cdot 3SiO_2$, 4
$K_2O \cdot 4SiO_2$, 59
$0.10K_2O$–$0.90SiO_2$, 14
K_2O–SiO_2–TiO_2, 51
$K_2S_2O_3 \cdot 5H_2O$, 218
K_2SO_4, 216
Kr, 65

L

La, 3, 190, 205
LaF_3, 97
La_2O_3, 97, 159, 160, 170–171, 194, 206–208
La_2O_3–Al_2O_3, 57
La_2O_3–CaO–SiO_2, 58
La_2O_3–Fe_2O_3, 57
La_2O_3–Nb_2O_5, 57
La_2O_3–NiO, 57
La_2O_3–TiO_3, 57
Lead borosilicate glass, 64
Lead silicates, 73
Li, 3, 59, 176
LiAc, 6, 214
LiCl, 221–222
$Li_2Cr_2O_7$, 216
LiF, 227
Li_2O, 136–144, 154, 167, 206
Li_2O–Al_2O_3–$nSiO_2$, 422–423
Li_2O–Al_2O_3–SiO_2–TiO_2 glass–ceramic, crystallization, 418–422
Li_2O–Bi_2O_3, 209
Li_2O–GeO_2, 198
Li_2O–SiO_2, 30
Li_2O–$2SiO_2$, 32, 34
Lu_2O_3, 58
Lu_2O_3–Al_2O_3, 57

M

Mg, 3, 59, 133, 168, 185, 191, 200
MgAl spinel, 88
$MgCO_3$, 59
MgF_2, 227
$(Mg, Fe^{2+})_2SiO_4$ (olivine), 91
MgO, 149, 153, 154, 200, 205, 206
   in lunar samples, 23

$Mg(O,F)_6$, 429
$MgO-La_2O_3$, 57
$MgO-SiO_2$, 51, 56
$MgO-Ta_2O_5$, 57
$MgO-Y_2O_3$, 57
$MgO-ZrO_2$, 57
$Mg(PO_3)_2$, 193
$MgSiO_3$ (enstatite), 94, 95
Mg_2SiO_4 (forsterite), 91, 93, 94
Mn, 159, 190
$MnCO_3$, 59
MnO_2, 205
MoCo, 64
MoO_3, 56, 78, 176
MoS_2, 64
$MoSi_2$, 110
Mullite, 68

N

N, 65, 121
Na, 3, 59, 238
NaAc, 214
$NaAl(SiO_3)_2$ (jadeite), 91
$NaAlSi_3O_8$ (albite), 91, 94, 95
Na_2BeF, 227
NaF_2, 230
$NaMgB_3Si_6O_{27}(OH)_4$ (tourmaline), 89
Na_2O, 15, 23, 78, 136–144, 206
$Na_2O-B_2O_3-SiO_2$, 59
$Na_2O-CaO-SiO_2$, 32, 33, 59
$Na_2O-K_2O-ZnO-Al_2O_3-SiO_2$, 147
$Na_2O \cdot 2SiO_2$, 27
Na_2O-SiO_2, critical cooling rates in, 30
$Na_2O-2SiO_2$, crystal nucleation in, 33, 34
$Na_2O \cdot 2SiO_2$, 15
$Na_2O \cdot 2SiO_2$, 11
   crystallization and melting properties of, 12
$NaNO_3$, 152
$NaPO_3$, 97
Na_2SiO_3, 192
Na_4SiO_4, 192, 209
$Na_2Si_2O_5$, 192
$Na_2Si_4O_9$, 192
$([NaSi]_{9-7}[CaAl]_{1-3})AlSi_2O_8$ (oligoclase), 91
Nb, 205

Nb_2O_3, 97, 122, 194
Nb_2O_5, 56, 78, 206, 333
$Nb_2O_5-CaO-SiO_2$, 58
Nd, 159–160, 195, 226, 227, 230
Nd_2O_3, 121
$Nd_2O_3-Al_2O_3$, 57, 200
Ne, 65
$(NH_4)_2SO_4$, 199, 215
Ni, 31, 159, 329
$NiCl_2$, 218
$Ni_{60}Nb_{40}$ glass, 398
NiO, 122
$NiO-BaO$, 57
Nitride layers, 61

O

O_2, 70
OH, 112, 113, 114, 117, 188–189

P

P, 3, 59, 69, 198
Pb, 3, 133, 190
$PbAc_2$, 213
$PbCO_3$, 59
PbF_2, 227
PbO, 43, 155, 198–199, 206
$PbO-Bi_2O_3$, 209
$PbO-B_2O_3-Al_2O_3$, 170
$PbO-SiO_2$, 68
Pb-Sb-S-I, 62
$PbSO_4$, 199
$PbTeO_3$, 68
PCl_3, 121
Pd-Au-Si metallic glasses, 388, 389, 391
0.775Pd-0.06Ca-0.165Si alloy, critical cooling rate, 31
Pd-Cu-Si glasses, 379, 380, 382, 397
Pd-Si alloys, 90, 376
0.82Pd-0.18Si alloy, critical cooling rate, 31
$Pd_{80}Si_{20}$ glass, 377, 383, 387, 388, 389, 396, 397
PH_3, 70
Phosphosilicate glass, 61, 68, 72
P_2O_5, 19, 69, 78, 83, 88, 121, 176, 188, 196, 228–229
$POCl_3$, 69, 70, 73, 79

MATERIALS INDEX

P_2O_5–SiO_2, 71, 72
P_2S_5, 248
Pt, 64

R

Rb, 3, 59, 124, 229
Rb_2O, 134, 136–142, 167, 206

S

S, 231, 236, 246–247, 259
Sb, 3, 7, 198
Sb_2O_3, 152
Sc, 3
Sc_2O_3, 58
Se, 231, 237–240, 259
Si, 3, 110
Si–As–Te, 62, 252
$SiCl_4$, 70, 73, 74, 78, 79, 119, 121, 204
  glasses from, 120
SiF_4, 122
SiH_4, 70, 73
Si_3N_4, 61, 64, 68, 72, 160
$Si_xN_yH_z$, 72
SiO, 61, 62, 110
SiO_2, 1, 6, 17, 19, 23, 41, 50, 61, 62, 64, 65, 68–70, 72, 73, 83, 84, 94, 95, 312, 313, 340
  energy-level models, 111
  in glass-forming systems, 107–147
SiO_4, 405
SiO_6, 108
$86.85SiO_2$–$5.91B_2O_3$–$2.62Al_2O_3$–$3.91Na_2O$–$0.66K_2O$, 88
$Si(OCH_3)_4$, 88
$Si(OC_2H_5)_4$, 88, 119
$Si(OEt)_4$, 88
SiO_2–GeO_2, 88
SiO_2–Li_2O–Al_2O_3, 408
$Si(OMe)_4$, 119
SiO_2–TiO_2, 88
$SiSiO_4$, 191, 194
Sm_2O_3, 58
Sn, 3, 64
$SnCl_4$, 121
SnO_2, 62, 230
Sn–Sb–S–I, 62
SnTe, 248

SO_4, 149
Sr, 3, 200, 315
SrF_2, 97, 227, 230
SrO, 154, 206
$Sr(PO_3)_2$, 193

T

$TaCl_5$, 121
TaN, 61
Ta_2O_5, 56, 61, 62, 70, 78, 97, 180, 206
Tantalum aluminates, 73
Ta_2O_5–CaO–SiO_2, 58
Ta_2O_5–Nb_2O_5, 57
Te, 31, 231, 239, 240–241, 254–255
$Te_{48}As_{30}Ge_{10}Si_{12}$, 73
TeO_2, 56, 194, 204–207, 231
o-Terphenyl, 35–36, 39
Th, 3, 205
$Th(NO_3)_4$, 211
ThO_2, 97, 206, 208
Ti, 3, 185, 198, 205
0.63Ti–0.37 Br alloy, critical cooling rate, 31–32
$TiCl_4$, 73, 88, 121
Ti_xO_y, 73
TiO_2, 23, 56, 75, 76, 78, 94, 149, 154, 158, 160, 187, 188, 194, 203–204, 206, 230, 333, 412
TiO_2–BaO, 51
TiO_2–CaO, 51
TiO_2–Nb_2O_5, 57
TiO_2–SiO_2, 76
  fibers, 89
TiO_2–SrO, 51
TiO_2–Ta_2O_5, 57
Tl, 59, 238, 256–258
Tl_2O, 171, 199, 206
TlSe, 238, 258
Tl_2SO_4, 216

U

UO_2, 89–90

V

V, 3
V_2O_5, 56, 122, 176, 181, 199, 202–203

W

WO$_3$, 56, 122, 176, 205, 206, 231, 333

X

Xe, 65

Y

Y, 3
U–Al–Si-O-N glasses, 160
Yb$_2$O$_3$, 58
Yb$_2$O$_3$–Al$_2$O$_3$, 57
Y$_2$O$_3$, 58, 153, 159–160
Y$_2$O$_3$–Al$_2$O$_3$, 57
Y$_2$O$_3$–Fe$_2$O$_3$, 57
Y$_2$O$_3$–Ga$_2$O$_3$, 57
Y$_2$O$_3$–Nb$_2$O$_5$, 57
Y$_2$O$_3$–NiO, 57
Y$_2$O$_3$–Ta$_2$O$_5$, 57
Y$_2$O$_3$–TiO$_2$, 57

Z

Zinc silicates, 73
Zn, 3, 59, 133, 190, 259
Zn(Ac)$_2$, 214
ZnCl$_2$, 210, 218
  glass formation, 43, 223–225
ZnO, 154–155, 168, 200, 206
Zn(PO$_3$)$_2$, 193
ZnS, 62
ZnSO$_4$, 209, 215, 216
Zr, 3
0.65Zr–0.35Be alloy, critical cooling rate, 31
ZrCl$_4$, 121
ZrF$_4$, 227
ZrO$_2$, 78, 95, 154, 158, 188, 207, 208, 230, 328
ZrO$_2$–Al$_2$O$_3$, 56, 57
ZrO$_2$–BaO, 57
ZrO$_2$–CaO–SiO$_2$, 58
ZrO$_2$–Fe$_2$O$_3$, 57
ZrO$_2$–Ga$_2$O$_3$, 57
ZrO$_2$–La$_2$O$_3$, 57
ZrO$_2$–Nb$_2$O$_5$, 57
ZrO$_2$–SiO$_2$, 56
ZrO$_2$–Ta$_2$O$_5$, 57
ZrO$_2$–TiO$_2$, 57
ZrO$_2$–ZrN, 56
ZrSiO$_4$ (zircon), 89, 91, 93–95

Subject Index

A

Acetate glasses, formation and properties, 213–214
Alamosite structure, 156
Albite, 186
  composition, 91
  melting point, 185
  shock effects on, 94, 95
Alkali acetate mixtures, glass formation in, 214
Alkali aluminosilicate glasses, 183–187
Alkali borate glasses
  selenium in, 259
  tellurium in, 259
Alkali feldspar
  composition, 91
  shock effects on, 93
Alkali indium germanate glasses, 199
Alkaline earths, in borate glasses, 168
Alkaline earth tungstate glasses, 208
Alkali silicate glasses, 122–147
  alkali distribution in, 122–131
  alkali species role, 134–135
  atomistic models, 135
  chemical resistance, 140–141
  coefficient of expansion, 135–136
  dielectric constant, 139, 141–142
  diffusion in, 144–145
  elastic properties, 136–137
  electrical conductivity, 137–138
  $H_2O-R_2O-SiO_2$ systems, 145–147
  mixed alkali type, properties, 142–144
  refractive index, 136
  soluble type, 145–146
  structure and properties, 142
  viscosity, 137
  water in, 146–147
Alkali silicates
  crystalline nuclei in, 414
  glass formation in, 30

Alkali titanate glasses, 203, 204
Alkoxides
  glass films, fibers, and monoliths from, 89
  multicomponent glasses from, 87
Alumina
  as fluxing agent for glazes, 306, 309
  raw materials for, 310
Alumina hydrate, as glaze raw material, 310
Aluminate (gallate and beryllate) glasses, properties and composition, 200–201
Aluminates, crystalline nuclei in, 415
Aluminogermanate glasses, 199
Aluminosilicate glasses, 159, 181–191
  alkali-free, 189–190
  alkali role in, 185–187
  aluminum role in, 181
  complex type, 185
  F and OH in, 188–189
  Lacy's model for, 187–188
  structure, 184
  tricluster model, 182
Aluminosilicates, crystalline nuclei in, 414–415
  melting points compared to borosilicates, 190
Aluminum, role in silicates, 181
Aluminum alloys, in phosphate glass, 196
Amorphous metal films formation, 62
Aluminum metaphosphate, in phosphate glasses, 196
Anatase, 186
Anodic oxidation
  film formation by, 69–70
  apparatus, 70
Anodization glass formation by, 1
Anorthite
  composition, 91
  shock effects on, 94, 95

SUBJECT INDEX

Antimonate glasses, properties and compositions, 201–202
Antimony, in chalcogenide glasses, 241–248
Antimony germanate glasses, 199
Arsenate glasses, properties and composition, 201–202
Arsenic, in chalcogenide glasses, 241–248
Arsenic germanate glasses, 199
Arsenolite, 201
Aventurine glasses, 159

B

Baria (barium oxide), as glaze raw material, 309, 310
Barium, in silica glasses, 154
Barium carbonate, as glaze raw material, 310
Barium crowns, composition, 180
Batteries, fluoroborate glasses in, 231
Bentonite, borosilicate glass from, 180
Berlinite, 194
Bernal dense random packing (DRP) model for metallic glasses, 367, 369
Beryl, radiation effects on, 89
Beryllate aluminate glasses, properties, 200–201
Beryllium in silicate glasses, 153
Binary oxides, quenches, 57
"Bioglass," description, 178
Biotite
    composition, 91
    shock effects on, 91
Bismuth
    in borate glasses, 169
    in silicate glasses, 155–158
Bismuth chloride, glasses based on, 225–226
Bismuth germanate glasses, 199
Bisulfate glass, 215–216
Bloch theory, 232
Blood analyzers, chalcogenide glass use in, 236
Blue glasses, titanate-type, 204
Bonding in glasses, 1
Borate glasses, 160–171
    acoustic losses in, 168
    alkali addition to, 163–168
    alkaline earths in, 168

electronic structure of, 167
structure of, 160–163
Boric oxide, as fluxing agent for glazes, 306, 308
Boroaluminosilicate glasses, properties and compositions, 190–191
Borogermanate glasses, 199
Borosilicate crowns, composition, 180
Borosilicate glasses, 171–181
    alkali-free, 180–181
    chlorine and bromine in, 230
    leaching and sintering processes, 177
    melting points, 190
    phase separation, 174–175
    Raman spectra, 173–174
    reconstructed, 175–178
    systems and structures, 171–172
    viscosity and density, 173
Borovanadate glasses, 202
Boroxol ring in borate glasses, 162
Boule process for fused silicas, 76
Bristol glazes
    compositions of, 314
    description of, 313
Bromine in borosilicate glasses, 230
Bulk glass
    processes for, 73–86
    from silica gel, 87
Burgers vector, 384
    definition, 381–382

C

Cabal glasses, properties, 170
Cadmium borate glasses, 168
Cadmium selenide, in oxide glasses, 259
Cadmium stannate, conducting glass films of, 64
Cadmium sulfide, in oxide glasses, 259
Calcium, in silicate glasses, 153
Calcium aluminate glasses, properties, 200
Calcium barium aluminophosphate glasses, properties, 195
Calcium carbonate, as glaze raw material, 310
Calcium oxide (lime)
    as fluxing agent for glazes, 307
    raw materials for, 310
    in silicate glasses, 147–153

SUBJECT INDEX

Calcium phosphate glass, thermal expansion coefficient, 356
Calcium polyacrylate glass, thermal expansion coefficient, 356
Carbide glasses, production, 64
Carbonate glasses, structure and composition, 212–213
Carnegieite, 184
  melting point, 186
Cations, classification of, in glass formation, 3, 7
Celsian, 326
Ceramic glazes, 303–328
  general description, 303–304
  lead-containing, 317–322
    formulas, 319
  lead release from, 320–322
  leadless gloss type, 311–317
  low-expansion type, formulas, 318
  minerals in, 309
  oxide role in, 305–309
  raw materials and frits for, 309–311
  satin and matte type, 326–328
Ceramics
  multicomponent glass coating of, 59
  solution methods for, 88
Cervit, as lithium aluminosilicate glass, 187
Chalcogenide glasses, 112, 231–259
  applications, 235–236
  arsenic and antimony in, 241–248
    formation and structure, 241–245
    properties, 245–246
    ternary systems, 246–248
  devitrification, 38
  electrical properties, 239–240
  electronic states in, 232–235
  germanium and silicon in, 249–256
    formation, 249
    germanium–selenium glasses, 251–254
    germanium–sulfur system, 249–251
    germanium–tellurium glasses, 254–255
    silicon–tellurium glasses, 255–256
  optical properties, 240
  silicon in, 255–256
  silver, copper, and thallium in, 256–258
  sulfur in, 236
  switching in, 235–236
  tellurium in, 240–241

Chalcogenides, structures, 4
Chalcogens in oxide glasses, 258
Chemical resistance of phosphate glasses, 195
Chemical vapor deposition
  glass film formation by, 70–73
  method comparison, 72
  reactors, 71
Chloride systems, zinc chloride–based, glass formation, 223–225
Chlorine in borosilicate glasses, 230
Claudedite, 201
Clay, as glaze raw material, 310
Coating techniques, comparison, 66
Coesite temperature range, 109
Computer, use in crystal growth studies, 11
Continuous cooling (CT) curves
  for glasses, 22
  uses, 25, 26
Copper in chalcogenide glasses, 256–258
Copper aluminosilicate glasses, compositions and properties, 190–191
Cordierite, 189
  formation, 428
Cordierite bodies, glazes for, 317
Cordierite magnesium aluminosilicate glasses, properties, 189
Corning Code 7940, 77, 118
Corning Code 7943, 77, 118
Corning Code 7971, 77
Corning Code 8603 photonucleated glass, 435
Corundum, as glaze raw material, 310
Covalent bonding in glasses, 1
Cover-coat enamels, description and formulas, 332–336
Cristobalite, 114, 184, 194
  formation, 110
  phases, 109, 110
  structure, 107
  radiation effects on, 90
Critical cooling rate
  for glass formation, 26, 28
    dependence on material parameters, 26–28
  in metallic systems, 53
  theory compared to experimental data, 28–32

Crystal(s)
  growth, in glass formation, 10–12
    kinetics, 14–20
  nucleation, in glass formation, 8–10
    kinetics, 14–20
Crystal bombardment, glass formation by, 1–2
Crystallization of reheated glass, 37–41

D

Dangling bonds, 234
Deposition methods
  for glass formation, 60–86
  comparison, 61, 66
Devitrification, differential thermal analysis, 38–39
Devitrite, soda–lime glasses compared to, 147–148
Diaplectic glasses, 91
Dichromate glasses, formation, 216
Differential thermal analysis (DTA), in devitrification studies, 38–39
Dinnerware glazes, 313–317
  formulas, 316, 319
  with lead, 308
Dolomite
  as glass-ceramic raw material, 436
  as glaze raw material, 310
Dynasil, 118, 120

E

ECD, as switching glass, 235
E glass
  composition, 190
  production, 191
Electrical properties of phosphate glasses, 196
Enamels, 301–338
  definition and description, 301–302
Enstatite shock effects on, 94, 95
Entropy of fusion in crystal growth, 11
Equilibrium properties of organic glasses, 356–358
Ethylammonium ion, in glasses, 210
Ethylene, glass-transition temperature, 348
Eucryptite, melting point, 186
Evaporation, glass formation by, 60–62
Explosive compaction of glassy materials, 97

F

Faraday rotators, fluorophosphate glasses for, 229
Feldspars, 312
  composition, 91
  as glass-ceramic raw material, 436
  as quartz raw material, 310
  shock effects on, 93
Ferrite glasses, precursors, 186
Fiber glasses, E glass as, 190–191
Films, deposited, structure of, 65
Flint glasses, composition, 157
Fluor crowns, composition, 189
Fluoride glasses
  formation and properties, 226–228
  shock-produced, 97
Fluoride-oxide glasses, 228
Fluorides, in glazes, 305, 309
Fluorine, in fused silica, 78
Fluoroaluminosilicates in glasses, 230
Fluoroborate glasses, 170
  formation and properties, 231
Fluoroborosilicate glasses, formation, 230
Fluorophlagopite, 189, 430
Fluorophosphate glasses for lasers, 229
Food-contact surface, glazes for, 315–316
Formate glasses, formation and properties, 213–214
Forsterite
  composition, 91
  shock effects on, 93
Free volume model of liquids, 6–7
Frenkel excitons, 240
Frits
  for ceramic oxides, 309–311
  function, 311
Fox relation, derivation, 349–350
Functional groups, contribution to organic glasses, 352–353
Furfuryl alcohol, 177

G

Gallate-aluminate glasses, properties, 200–201
Galloaluminosilicate glasses, properties, 191
Garnet, radiation effects on, 89
Gases, dissolved in glasses, 59
GAST, as switching glass, 235

SUBJECT INDEX

General Electric optical glasses, properties, 118
General Electric Type 151, 77
Germanate glasses
  acoustic losses in, 168
  properties and compositions, 196–200
Germanium
  in chalcogenide glasses, 249–256
  –selenium glasses, 251–254
  –sulfur glasses, 249–251
  –tellurium glasses, 254–255
Germanosilicate crystals, 199
Glass(es)
  bonding in, 1
  critical cooling rates, 22
  crystal structure in, 405–413
  formation of, *see* Glass formation
  homogeneous and heterogeneous nucleation in, 406–409
  low-expansion type, composition, 179
  material properties, 41–44
  metastable state of, 6
  multicomponent glass coating of, 59
  nucleation through microimmiscibility, 409–413
  reheated, crystallization, 37–41
  structure of, 405–406
  transition temperature of, 5
  unusual methods of production, 49–103
Glass beads for road signs, titanate glass use in, 203
Glass-ceramics, 403–445
  applications, 441–444
  commercial types, 442–443 (*table*)
  crystallization-heat treatment, 441
  crystallization phenomena, 413–422
  description of, 404–405
  finishing, 440
  forming, 439–440
  glass melting, 437–439
  handling and mixing, 436–437
  machinability, 429–433
  mechanical strength, 425–429
  metastable phases and transformation, 416–417
  processing, 436–441
  properties, 422–436
  raw materials for, 436
  surface compression from differential crystallization, 427–428
  surface strengthening of, by ion exchange, 426–427
  transparency, 424–425
  ultralow thermal expansion, 422–423
Glass-film formation, 67–73
Glass formation, 1–47
  ability for, 41–44
  in alkali silicate systems, 30, 122–147
  cation classification in, 3, 7
  critical cooling rate, 26–28
  crystal growth in, 10–12
  crystallization with change of composition, 12–14
  crystal nucleation in, 8–10
    rate, 16–20
  by deposition methods, 50, 60–86
    comparison, 61, 66
  by evaporation, 60–62
  glass-films, 67–73
  by high-pressure melting, 58–59
  inorganic systems for, 105–299
  by ion implantation, 61, 64–65
    comparison with other methods, 66
  kinetic models of, 14–41
  by laser film melting, 58
  by laser spin melting, 58
  by liquid quenching, 50–51
  material properties, 41–44
  by metal evaporation, 59
  in metallic systems, critical cooling rates, 53
  nucleation frequencies, 32–37
  by quenching, 50–58
  by radiation, 89–90
  random array picture of, 4
  random network model for, 3
  by reactive deposition, 66–80
  soda–lime glasses, 147–153
  shock induction of, 50, 90–97
  by solid state transformations, 50, 86–89
  by solution methods, 50, 86–89
  by splat cooling, 51–58
  by sputtering, 61, 63–64
    comparison with other techniques, 66
  structural models for, 2–5
  thermodynamic models of, 5–8
  time–temperature transformation (TTT) curves, 21
  by unconventional melting, 50–59

by vitreous silica, 107–147
Zachariasen–Warren model for, 2–3
"Glassiness," definition of, 15
Glass transition temperature, 339–340
   derivation, 342–348
   of simple substances, 340
   of mixtures, 348–350
Glazes, 301–338
   ceramic, 303–328
   definition and description, 301–302
Gloss glazes
   lead-containing, 317–322
   leadless, 311–317
      high-temperature type, 312
Glow discharge, glass formation by, 120
Glucose glass, thermal expansion coefficient, 356
Gold in silicate glasses, 160
Granodiorite
   composition, 91
   shock effects on, 93
Ground-coat porcelain enamels, 330–332

H

Halide glasses, 226–231
   fluoride glasses, 226–228
   oxide glasses, 228–231
   with silicates, 229
Halides
   multi component glasses from, 87
   tellurite glasses containing, 207
Hall effect, 233
Henry's law, 360–361
Homosil, optical properties, 118, 120
Hydrate glasses, formation and properties, 216–223
Hydrogen bonding in glasses, 1

I

Infrared optical material
   arsenic sulfide glass as, 245
   chalcogenide glass use in, 236
Infrasil, optical properties, 118, 120
Inorganic glass-forming systems, 105–299
Inside Vapor Phase Oxidation (IVPO), process, 81–82, 84
   diagram, 83
   microwave plasma use in, 84

Intimate valence alternate pairs (IVAP), of chalcogenide glasses, 234–235
Iodine, antimonate glasses containing, 202
Ionic bonding in glasses, 1
Ionic salt and solution glasses, 209–226
   aqueous hydrate glasses, 216–223
   dichromate glasses, 216
   formate and acetate glasses, 213–214
   general description, 209–210
   nitrate and nitrite glasses, 210–212
   sulfate glasses, 215–216
   thiocyanate glasses, 215
Ion implantation
   glass formation by, 64–65
      comparison with other methods, 66
Iron, ferromagnetic film formation, 62
Iron phosphate glasses, 197
Iron silicate glasses, 159

J

Jadeite composition, 91
Jena glasses, 180
Johnson–Mehl–Avrami equation, 40

K

K_{gl} values, 38
Kalsilite, 187
   ion exchanging on surface of, 426
Keatite as silica phase, 109
Kinetics of glass formation, 14–20
Kondo effect, 373
"Kurrol," 193

L

Lacy's model for aluminosilicate glasses, 182, 187–188
Langmuir adsorption isotherm, 360–361
Lanthanide oxides, 207, 208
   glass formation from, 58
Lanthanides
   niobate formation, 58
   tantalate formation, 58
   titanate formation, 58
Lanthanite titanate glasses, 204
Lanthanum borate glasses, properties, 170–171
Lanthanum borogermanate glasses, 199
Laser film melting, glass formation by, 58

Laser glasses
  comparison of properties, 228
  fluoride glass use in, 227
Lasers, lithium aluminosilicate glasses as hosts for, 187
Laser spin melting, glass formation by, 58
Lead, in silica glasses, 155–158
Lead bisilicate, as glaze raw material, 310
Lead borate glasses, properties, 168–169
Lead germanate, 199
Lead monosilicate, as glaze raw material, 309, 310
Lead oxide
  as fluxing agent for glazes, 307–308
  poisoning by, frit use to lessen, 311
  raw materials for, 310
Lead poisoning
  from lead glazes, 308
  from pottery, 319–320
Lead titanate glasses, 204
Leucite, melting point, 186
Lime–alumina glasses, properties, 200
Liquid(s)
  cooling rate vs. glass formation, 43–44
  free volume model of, 6–7
Liquid quenching, glass formation by, 50–51
Liquid-state cooling, glass formation by, 1, 2
Litharge, as glaze raw material, 309, 310
Lithium aluminosilicate glasses, properties and compositions, 187
Lithium aluminosilicates, structure, 186
Lithium oxide (lithia)
  as fluxing agent for glazes, 306
  raw material for, 310
Lithium silicate glasses, 125, 127, 135
  halogens in, 231
  insolubility, 145–146
  phase diagram, 128, 129
  phase separation, 131
Litt's free-volume model, 339–340

M

Magnesium
  in alkaline earth glasses, 168
  in silicate glasses, 153

Magnesium oxide (magnesia)
  as fluxing agent for glazes, 307
  raw materials for, 310
Magnetic behavior of glass metals, 55
Magnetic bubbles, in rare–earth—transition–metal alloys, 55
Magnetic properties of arsenic selenide glasses, 246
Matte glazes, description and formulas for, 326
Mechanical properties of amorphous metals, 54–55
Melt extraction of metal alloys, 53
Melting, unconventional, glass formation by, 50–59
Memory effects of arsenic selenide glass, 245
Metal alloys, conditions for glass forming, 42
Metal alloy glasses, structures of, 4
Metallic bonding in glasses, 1
Metallic glasses, 365–402
  chemical stability, 373–376
  dislocation lattice model, 383
  formation and structure of, 54
  at high temperatures, 393–398
  magnetic and electrical properties, 55, 373
  mechanical properties, 376–380
  crystallization effects on, 385–393
  properties and behavior, 55–56, 373–380
  theoretical mechanisms of deformation and fracture, 381–385
  theories of formation, 369–373
  types, 366–367
Metallic ribbons, method for making, 53
Metal–metalloid glasses, structure of, 54
Metals, amorphous, mechanical properties, 54–55
Metaphosphates in glasses, 193
Metglas alloys, 56
Methacrylic acid, glass-transition temperature, 348
α-Methyl pyridinium chloride glass formation by, 223
Mica glass-ceramics, machinability and structure, 429, 431
Microcline composition, 91

Minerals
  in ceramic glazes, 309
  composition, 91
Mirror blank, from fused silica, 77
Mixed alkali silicate glasses, properties, 142–144
"Mobility edge" in crystalline solids, 233
"Mobility gap" in crystalline solids, 233
Modified Chemical Vapor Deposition (MCVD) process, 81
Molecular glasses, *see* Organic glasses
Molybdate glasses, properties and structures, 208
Moon
  glassy materials on surface of, 91
  rocks, 23
    critical cooling rates, 29
    crystallization temperatures, 39
    oxide composition, 23
MOTT–CFo model, 232
Mott–Davis–Street model, 233
Mullite
  crystallization of, 182, 183
  in glass-ceramics, 409, 411
Multicomponent glasses, solution methods for, 87–88

N

Nabal glasses, properties, 170
Nabarro–Herring creep, 395
Nepheline
  as glass-ceramic raw material, 436
  ion exchanging on surface of, 426
  structure, 185–186
Nepheline glass ceramics
  composition, 186
  sodium aluminosilicate glasses as precursors of, 185
Nepheline syenite, as glaze raw material, 310
Niobate glasses, properties, 207–208
Nitrite glasses, 210
  properties and structures, 210–212
Nitrates, multicomponent glasses from, 87
Nitrogen
  in alkali borate glasses, 171
  injection into silicon surface, 65
  in silicate glasses, 160

Nucleation frequencies, determination, 32–37
Nucleation rate, in glass formation, 16–20

O

Oligoclase, composition, 91
Olivine, composition, 91
Opacifiers, for glazes, 324
Opal glasses, formation and properties, 230
Opaque glazes, description and composition, 322–326
Optical fibers, doped silicas for, 121
Optical flint glasses, 180
  composition, 157
Optical glasses
  composition, 180
  shock-produced, 97
Optical properties
  of arsenic selenide glasses, 246
  of phosphate glasses, 195–196
Optical waveguides
  cross sections, 85
  formation, 79–86
    by sputtering, 64
  with K_2O–SiO_2 core, 59
Optosil, optical properties, 118, 120
Organic glasses, 339–363
  admixture effect on, 358–359
  cooling rate effects on, 342
  equilibrium properties, 356–358
  functional group contributions to, 352–353
  glass transition temperature of, 342–348
    of mixtures, 348–350
  heterogeneous mixtures, 350
  molecular structure effects, 351–355
  transport phenomena in, 359–361
Organic plastic crystals, 17
Organic salt glasses, formation and properties, 222–223
Orpiment, structure of, 241
Orthoclase
  composition, 91
  glass, 186–187
Orthosilicates, shock effects on, 93
Outside Vapor Phase Oxidation (OVPO) process, 80, 81
  diagram, 81

Oxide glasses
  chalcogens in, 258–259
  devitrification, 38
  shock-produced, 97
Oxide glass films, production of, 64
Oxides
  in moon rock samples, 23
  role in glaze formulation, 304–309
Oxyfluoride glasses, 228
Oxygen, injection into silicon surface, 65
Oxygen-rich polycrystalline silicon (ORPS), 61

P

Phosphate glasses, 191–196
  alkali-free, 193–194
  chlorine or bromine in, 230
  high-alkali-type, 194
  phosphorus pentoxide in, 193
  properties and applications, 195–196
  silver phosphate in, 195, 196
  structure, 191–195
Phosphorus pentoxide glass, 193
Phosphorus pentoxide systems, glass formation in, 228–229
Photochromic glasses, 196
Photoconductivity in chalcogenide glasses, 257
Photocopiers, selenide glasses as, 236
Photodissolution in chalcogenide glasses, 257
Photography, "dry," chalcogenide glass use in, 236
Photoluminescence
  of arsenic selenide glasses, 246
  theoretical aspects, 234–235
Photovoltaic cells, antireflection coatings on, 62
α-Picolinium ion, in glasses, 210
Plagioclase
  composition, 91
  shock effects on, 92, 93
Plasma, droplet melting in, 56
Plasma torch method for fused silica boules, 77–78
Plutonium in glasses, 159
Polk's model of hole filling, 369
Polyacrylic acid glass(es), 341
  thermal expansion coefficient, 356

Poly(ethylene terephthalate) thermal expansion coefficient, 356
Polymers
  aromatic, structural corrections in, 354
  glass transition temperatures, 347
Polyphosphates in glasses, 193
Polystyrene glass
  equilibrium properties, 357–358
  thermal expansion coefficient, 356
Porcelain, hard-paste type, formula for, 312
Porcelain enamels, 329–336
  adherence to substrates, 328–330
  cover-coat type, 332–336
  ground-coat type, 330–332
Porous glasses, formation, 87
Potassium oxide (potash)
  as fluxing agent for glazes, 306–307
  raw materials for, 310
Potassium silicate glasses, 125, 127, 130
  phase diagram, 129
Potassium sodium silicate glasses, properties and structures, 185–187
Power transformer cores, metallic glass use in, 399
Pressure quenching, glass formation by, 1
Pursil, 118, 120, 453
Pursil Ultra, 118
Pyrex glass, 175, 176, 180
  compositions, 178, 179
Pyridine hydrochloride, glass formation by, 222
Pyridinium ion, 224, 225
Pyroceram, as lithium aluminosilicate glass, 187

Q

Quartz
  phases, 109
  radiation effects on, 90
  shock effects on, 96
  structure, 107
α and β Quartz, structures, 417
β-Quartz, in aluminosilicate glasses, 444
Quartz glasses, shock effects on, 92, 96
Quartz sand, as glaze raw material, 310
Quenching, glass formation by, 50–58

R

Radiation glass formation by, 89–90
Random network model of glass formation, 3
Rapid cooling apparatus diagram of, 51, 52
Rare earth acetates glass formation in, 214
Rare earths
 effect on borosilicate glasses, 177
 in silicate glasses, 159–160
Reactive deposition grass formation by, 66–80
Reactive sputtering, glass film formation by, 68, 120
Refractory oxides, glasses from, 57–58
Rocket casings, organic glasses in, 359

S

Sandstone, shock effects on, 93
Sanidine, composition, 91
Sanitary ware, glazes for, 313
Satin glazes, composition and formulas, 326–328
Schottky (vacancy) defects, lack of, in glass, 145
Sealing glasses, 180
Seger molecular formula for earthenware glazes, 304
Selenide glasses, application, 236
Selenium
 in alkali borate glasses, 259
 in chalcogenide glasses, 237–240
  copolymerization in, 239
  doped selenium, 238–239
  electrical properties, 239–240
  optical properties, 240
  structure, 237–239
Semiconductors, amorphous, 232–233
Shock waves, glass production by, 50, 90–97
Sialon glass, nitrogen diffusion through, 160
Silica(s)
 in ceramic glazes, 305–306
  raw materials for, 310
 doped, 121
 fused, 75–79
 noncrystalline, high-density forms of, 122
 polymorphs of, 108, 109
 radiation effects on, 90
 vitreous, see Vitreous silica
Silica gel, formation of, 86
Silica glass(es)
 cations in, 153–160
 composition, 179
 electromelted, 120
 flame-fused, 120
 halides in, 229
 preparations and characteristics, 120
 radiation effects on, 90
 shock- and radiation-produced, comparison, 93–94
 thermal expansion coefficient, 356
Silica M, 109
Silicates, fused, 75–79
Silica W, formation, 109
Silicogermanate glasses, 199
Silicon
 in chalcogenide glasses, 249–256
 film formation on, 68
 oxidation of, glass formation by, 120
 oxide films from, 65
Silicon dioxide films, 68, 69
Silicon nitride films, 64, 68
Silver
 in chalcogenide glasses, 256–258
 in silicate glasses, 160
Silver borate glasses, properties, 169
Silver iodide–oxide glasses, formation and properties, 231
Silver phosphate glasses, properties and structure, 195, 196
SiO_2 fibers, production from solution, 88–89
Slips for glazes, 309
Small-angle scattering (SAXS) of silica glass, 110
Soda–borate–sulfur system, glass formation in, 258–259
Soda–silica–sulfur system, glass formation in, 258
Soda–lime glasses, 147–153
 additives to, to promote fining, 151–153
 conventional type, 149–150
 phase diagrams, 148
 phase separation, 150–151
Sodium calcium aluminosilicate glasses, structure, 188
Sodium calcium silicate glasses, fluorine in, 230

SUBJECT INDEX

Sodium oxide (soda)
  as fluxing agent for glazes, 306
  raw materials for, 310
Sodium phosphate glass, thermal expansion coefficient, 356
Sodium silicate glasses, 123
  alkali introduction, 130
  phase diagram, 128, 129
  phase separation, 131–135
  properties, 183
  x-ray diffraction, 126, 127
Solid state transformations, glass formation by, 89–97
Solution hydrolysis, glass formation by, 1
Solution methods, for glass production, 86–89
Sonic applications of chalcogenide glasses, 236
Soot
  melt process, 85–86
  particulate glass, 73, 74, 75, 80–82
SPAT, as switching glass, 235
Spectrosils, 77, 118, 120
Spectrosil WF, 77
Splat cooling, glass formation by, 51–58
Spodumene, 444
  as glass-ceramic raw material, 436
  as glaze raw material, 310
  melting point, 186
  structure, 187
β-Spodumene, growth behavior, 418–419, 421
Sputtering, glass formation by, 63–64
STAG, as switching glass, 235
Stishovite, 109, 122
  SiO_6 in, 108
Stokes–Einstein relation, 8
Strontium in silica glasses, 154
Strontium carbonate, as glaze raw material, 310
Strontium oxide (strontia)
  as fluxing agent for glazes, 307
  raw material for, 310
Sulfates, tellurite glasses containing, 207
Sulfur, in soda-lima-silica glasses, 259
Suprasil, 77
  optical properties, 118
Suprasil W, 77
Switching in chalcogenide glasses, 235–236
Synsil, 77, 118

T

Talc, as glaze raw material, 310
Tear glazes, 328
Tellurite glasses
  containing halides and sulfides, 207
  properties and structures, 204–207
Tellurium
  in alkali borate glasses, 259
  in chalcogenide glasses, 240–241
Tellurium
  –germanium glasses, 254–255
Tetraethyl orthosilicate, silica gel from, 86–87
Tetrasil, 77, 118
Thallium
  in chalcogenide glasses, 256–258
  in silicate glasses, 155–158
Thallium glasses, 125
Thermal oxidation, film formation by, 68–69
Thermal spike, as irradiation effect, 89
Thermodynamic models of glass formation, 5–8
Thermometers, secular rise and ice-point depression of, 135
Thetomorphic glasses, 91
Thiocyanate glasses, properties and formation, 215
Thorium dioxide in silicate glasses, 158
Tiles
  Bristol glazes for, 314
  opaque glazes for, composition, 325
Time–temperature–transformation (TTT)
  curves, 28
  description, 21
  uses, 25
Tin, in silicate glasses, 158
Tin borate glasses, 171
Tin germanate glasses, 199
Titanate glasses, properties and structures, 203–204
Titania, 444
Titanium dioxide, in silicate glasses, 158
Titanium isoperoxide, use in titanate glass production, 203
Titanogermanate glasses, 199
Toluene, 1

Topaz
  heat effects on, 428
  radiation effects on, 89
Tourmaline, radiation effects on, 89
Transition elements, in silicate glasses, 159
Transition metals in metallic glasses, 367
Transparency of glass-ceramics, 424–425
Transport phenomena in organic glasses, 359–361
"Triclusters," in structure of aluminosilicate glasses, 182, 187–188
Tridymite, 186, 194
  phases, 109
  radiation effects on, 90
  structure, 107
  viterous silica structure and, 110
Trifluoracetate glasses, 214
1,3,5-Trinaphthylbenzene glass, thermal expansion coefficient, 356
Tungstate glasses, properties and structures, 208

U

U-235 fission fragments, glass formation using, 90
Ultraphosphates in glasses, 193
Ultrasil, optical properties, 118
Uranium in glasses, 159
Unconventional melting, glass production by, 50–59

V

Valence alternate-pair model (VAP) of chalcogenide glasses, 234
Valentinite, 202
Vanadate glasses, properties and compositions, 202–203
van der Waals bonding in glasses, 1
Vapor Axial Deposition (VAD) process, 80–81
  diagram, 82
Vapor condensation, glass formation by, 1
Vapor deposition of coatings, comparison with other methods, 66
Verneuil crystal-growing apparatus, 51
Viscosity, related to glass-forming properties, 42

Vitreosils, 118, 120
Vitreous silica
  coefficient of expansion, 114
  compressibility, 115
  defect centers, 112
  dielectric constant, 116
  dielectric loss, 116
  dielectric relaxation, 116
  diffusion through, 119
  elastic moduli, 115
  electronic properties, 116
  electronic structure, 110–111
  fluorescence, 118
  in glass-forming systems, 107–147
  hardness, 115
  optical properties, 117–118
  Poisson's ratio, 115
  preparation and characterization, 119–120
  refractive indices, 114, 118
  shear modulus, 115
  specific volume, 113–114
  strength, 115
  structural principles, 107–113
  structure and properties, 113–122
  temperature effects, 115
  thermal properties, 116
  two-dimensional schematic, 108
  Young's modulus, 115
Vogel–Fulcher relation, 17, 26
Volterra dislocations in metallic glasses, 381
Vycor glass, 171, 176
  compositions, 172, 177, 179
  sintering, 178
  "thirsty," 177

W

Water, in alkali silicate glasses, 146–147
Willemite, 326
Williams–Landel–Ferry form, in glass transition, 6
Wollastonite, as glaze raw material, 310

X

X-ray photoelectron spectra (XPS) of alkali aluminosilicate glasses, 183

X-ray shielding glasses, 157
m-Xylene, 43

Z

Zachariesen concept of glass structure, 304
Zachariasen–Warren model for glass
  formation, 2–3
Zeldovich expression, 10
Zinc, in silica glasses, 154
Zinc borate in glasses, 168
Zinc chloride, halides based on, glass
  formation by, 223–225
Zinc oxide
  as fluxing agent for glazes, 306, 307
  raw materials for, 310
Zinc polyacrylate glass, thermal expansion
  coefficient, 356
Zircon
  composition, 91
  as glaze raw material, 310
  radiation effects on, 89
  shock effects on, 93, 94, 95
Zircon bodies, glasses for, 317
Zirconia, 444
  as pacifier, 324–326
Zirconium dioxide in silicate glasses,
  158
Zirconium oxide (zirconia)
  as fluxing agent for glazes, 309
  raw materials for, 310